WATER RESOURCES ENGINEERING

S.R. Sahasrabudhe

WATER RESOURCES ENGINEERING

by

S. R. Sahasrabudhe
(Formerly Assistant Professor,
Guru Nanak Engineering College)
Retired Commissioner (Projects)
Union Ministry of Water Resources,
Government of India

S.K. KATARIA & SONS®
Publisher of Engineering Books
4885/109, Prakash Mahal, Dr. Subhash Bhargav Lane,
Opposite Delhi Medical Association, Daryaganj, New Delhi–110002 (INDIA)
Phone: +91-11-43551243; *Mobile:* +91-9871775858, 7678425209
e-mail: katariabook@gmail.com; katariabooks@yahoo.com
Website: www.skkatariaandsons.com
(e-books, subscriptions and bulk buying also available on our website)

Published by:
S.K. Kataria & Sons®
4885/109, Prakash Mahal, Dr. Subhash Bhargav Lane,
Opposite Delhi Medical Association, Daryaganj, New Delhi-110002 (INDIA)
Phone: +91-11-43551243; *Mobile:* +91-9871775858, 7678425209
e-mail: katariabook@gmail.com; katariabooks@yahoo.com
Website: www.skkatariaandsons.com

Third Edition: July, 2011; Fourth Edition: March, 2014, Reprint: 2025

Laser Typesetting by : Plus Computers, Delhi *Printed at :* New A.S. Printing Press, Delhi

Preface to the Fourth Edition

It is really a measure of utility of this book that a new edition has to be brought out so early.

The book has been comprehensively edited to remove shortcomings. Also two new Chapters, one on "Design of Diversion Headworks and other Hydraulic Structures" and second on "Hydropower Stations" have been added to make the book more comprehensive.

S.R. SAHASRABUDHE

Preface to the Third Edition

Water is one of the most vital requirements of human being. The subject of water is so vast that whole life may not be sufficient to study it in entirety.

Students of sciences and especially civil engineering have to grasp the subject of Hydrology and Irrigation in a small period of one or two semisters while preparing for university examinations.

This book has been objectively designed to enable the students to learn the subject of Water Resources Engineering in the limited time at thier disposal while pursuing the degreee course in our universities.

Attempt has been made in this edition of the book to cover basic principles as succinctly as possible. It is hoped that the students of engineering will find this endeavour useful.

S.R. SAHASRABUDHE

Contents

1

Elements of Hydrology

1.1. DEFINITION AND SCOPE OF HYDROLOGY

It is the science that deals with all aspects of the water available on the earth. It includes study of occurrence of water, its properties, its distribution and circulation and also its effects on the living beings and their surroundings. It is not entirely a pure science because it has many practical applications and it utilises knowledge of other sciences greatly.

Broadly, the whole subject matter can be expressed in the form of a mathematical equation. The equation is

$$P = R + L \quad \text{or}$$

$$\text{Precipitation} = \text{Runoff} + \text{Losses}$$

In the above equation precipitation indicates total supply of water from all forms of falling moisture and mainly includes rainfall and snowfall. The runoff represents surplus water that flows over the surface to join some river or sea. The term losses includes that portion of water which goes to atmosphere and underground by the processes like evaporation and percolation respectively. For practical reasons hydrology does not cover all studies of oceans and medical uses of water.

After studying this equation with the background of hydrologic cycle it will be clear that the term losses never implies that this water is lost and can not be used again. It is the water which temporarily disappears from view (*e.g.*, evaporation, seepage, etc.) and given favourable conditions, reappears to perform various duties. Hence, it is necessary to study all the three terms of the equation, namely rainall, runoff, and losses.

The subject of hydrology mainly involves:

- assessment of catchment rainfall;
- assessment of dependable flow;
- design flood computations for safety of hydraulic structures;
- assessment of losses; and
- calculation of useful life and capacity of reseroirs.

1.2. HISTORY OF DEVELOPMENT

The modern science of hydrology may be considered to have begun with measurements of rainfall, evaporation, river discharge by area-velocity method etc. From these measurements which started in 17th century scientists were able to draw correct conclusions on the observed hydrological phenomenon. The advent of hydrology cannot, however, be said to be only since

17th century. In fact concept of hydrological cycle was professed by many philosophers from time immemorial. The chronology of various phases of development of the science of hydrology can be broadly visualised as follows:

- Speculation of concepts — upto 14th century
- Observations — 15th to 16th century
- Measurements — 17th century
- Experimentations — 18th century
- Modernisation — 19th century
- Quantification of empirical formulae — 1900 to 1930
- Rationalisation of hydrologic theory — 1930 to 1950
- Theorization by mathematical analysis — 1950 to date

Although upto the end of 14th century hydrological concepts were only speculated many hydraulic structures were constructed. The great works known in the history are the Abassinian wells, the Persian kanats, the Egyptian and Chinese irrigation systems, the water supply and drainage systems of Indus valley, the Roman aqueducts, Chinese flood control works etc. These were naturally constructed on the basis of the practical knowledge of hydrology though it was not widely enunciated.

In the subsequent two centuries the trends of mere speculation changed to close observation. During this period Leonardo da Vinci recognised the hydrological cycle as is acepted today.

The seventeenth century saw development of techniques for measurements of rainfall, evaporation, river discharge etc., which provided documented proof of the principle of hydrological cycle. Names of Pierre Perrault and Edme Mariotte are noteworthy in this context.

In the eighteenth century number of hydraulic experimental studies in the field of hydrology were performed. As a result various hydraulic principles were discovered. Notable among them are Bernoullis piezometer, the Borda tube, the Pitot tube, Bernoulli's theorem, Chezy's formula etc. These developments vastly contributed towards taking up of quantitative hydrologic studies.

In the nineteenth century the experimental studies were greatly modernised. All these activities laid a firm base of modern science of hydrology. Majority of the contributions were related to groundwater hydrology and surface water measurement. Darcy's law of groundwater flow, Dupit's well formula, Hagen-Poiseuille's equation of capillary flow, Francis weir discharge formula. Ganguillet and Kutter's determination of Chezy's coefficient, Manning's flow formula, development of price current-meter, Dalton's law are some of the notable developments of this century.

Upto the end of the nineteenth century the science of hydrology was largely empirical. It was so because the physical basis for many quantitative hydrologic determination was not well known. The selection of coefficients and parameters to be used in empirical formulae had to depend upon experience and judgement. Thus, ad-hocism in hydrology became more evident. In the first three decades of the twentieth century increasing hydrological investigations were taken up for advancement of the science of hydrology.

During the period from 1930 to 1950 many great hydrologists emerged who gave rational basis to solve hydrological problems in place of empirical solutions. To name a few, Sherman gave unit hydrograph theory, Horton gave method of determining rainfall excess on the basis of infiltration theory, Gumbel proposed the use of extreme value distribution for frequency analysis, Einstein developed the bedload function for use in the theoretical analysis of sedimentation problems.

Since 1950, increasingly theoretical approaches have been adopted in hydrologic problems. Now such problems are readily subjected to mathematical analysis. With the development of computers solution of complicated mathematical hydrologic theories has become a reality.

1.3. APPLICATION IN ENGINEERING

Success of any water resources development project depends on timely and sufficient availability of water. Naturally proper assessment of this natural resource assumes great importance. By assessment we try to know in detail from where the resource comes, where it goes, at what time or when it comes and how much of it is really available. Therefore, hydrological investigations form the first step in any water resources development scheme involving design, construction and operation of hydraulic structures. The history of hydraulic structures which have failed shows that majority of failures is due to insufficient hydological analysis done while the structures were designed and constructed rather than due to structural weakness. The cost of collecting sufficient hydrological data and its analysis constitutes an insignificant part of the total cost of the water resources development project but it ensures successful operation and life of the project and therefore becomes indispensible activity. Although water is one of the most vital natural resources soemetimes it brings destruction by way of storms and floods. An engineer is expected to forecast floods, to assure adequate storage capacity for irrigation, hydropower generation, industrial and domestic water supply, flood control etc. The practical applications of the knowledge of hydrology are the following:

- Peak flow and future conditions of flow, at any point in the drainage valley can be correctly estimated for any basin or area.
- Spillway capacity can be accurately designed by estimating design flood.
- Design of river training work is facilitated.
- Dependable yields from the stream for generation of hydroelectric power can be calculated.
- Water supply to township and sewerage schemes can be properly designed.
- Water resources account of a river basin can be prepared.
- Reservoir capacity can be determined accurately.
- Operation of reservoirs can be done in an efficient manner.

1.4. HYDROLOGIC CYCLE

Entire moving water of the universe whether in atmosphere or on the surface of earth or below the surface is constantly changing its state and passing from one state to the other. As a whole it forms an endless chain of various processes and is known as a hydraulic cycle. Sun's heat energy is responsible for the entire process. In short, the hydrologic cycle represents circulation of water from the oceans to the land via atmosphere and back to the oceans by overland and underground routes. It may be properly considered as constituting of two main divisions, they are:

- atmospheric division, and
- surface division

Each division consists of three main phases, namely: (*a*) Transportation of water in some form, (*b*) Temporary storage, and (*c*) Change of state. The atmospheric division comprises: (*a*) vapour flow, (*b*) vapour storage in the atmosphere, and (*c*) change from vapour to water. It may be rainfall or change may be in some other form for example, snowfall, frost, dew etc.

The surface division comprises: (*a*) surface runoff, infiltration or groundwater flow, (*b*) surface, groundwater, soil moisture storage, and (*c*) change from water to vapour or evaporation due to heat. The hydrological cycle can be best examplified pictorially as shown in Fig. 1.1.

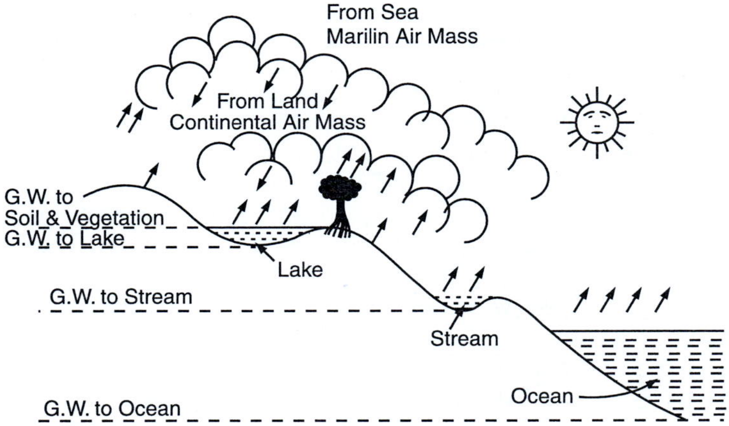

Fig. 1.1. Hydrologic cycle

From the above discussion it is clear that the rain in hot regions and snow in cold regions are the primary sources of water. Subsequently they give rise to other sources of water, for example rivers, streams, tanks, wells etc.

The pictorial diagram given in Fig. 1.1 shows the conceptual hydrologic cycle. The elements of the hydrologic cycle taking the case of a catchment area like that of a river basin can be shown by a flow diagram given in Fig. 1.2.

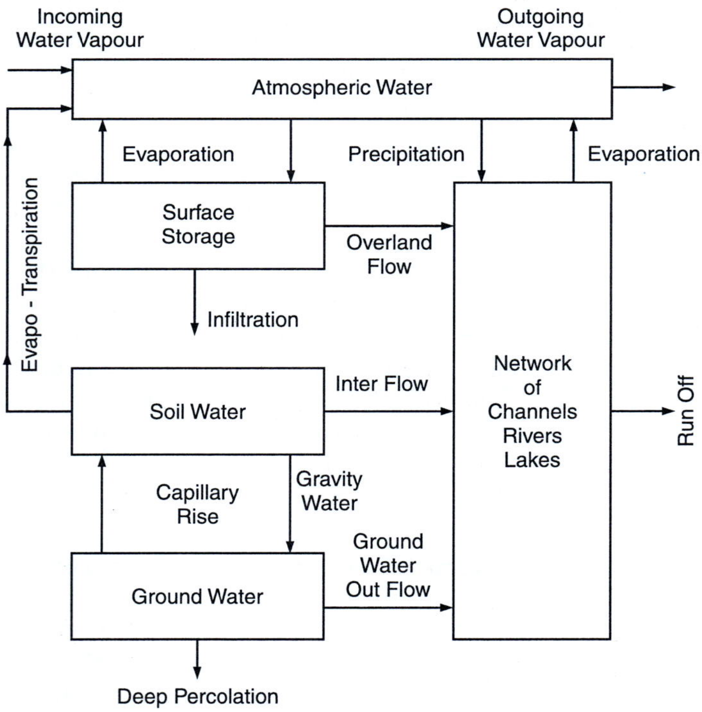

Fig. 1.2. Elements of hydrologic cycle

1.5. SURFACE OR DIRECT SOURCES OF WATER

Surface sources appear at any site naturally. This type includes the following:

1. Rivers: Rivers are formed along more or less defined channels to drain from land all the water received in the form of rainfall and melting of snow from high altitudes. Their development is the work of ages.

2. Streams: Rainwater infiltrates into the soil and subsequently joins the ground water storage. When the natural relief is such that the ground surface at any point falls below the top surface of groundwater reservoir then there exists greater hydrostatic pressure in the soil mass. The groundwater under pressure then finds way through the soil into the depression forming a stream.

Figure 1.3 shows formation of a stream.

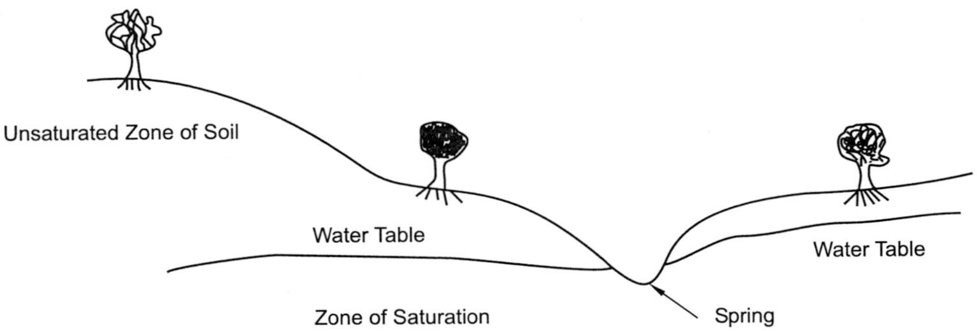

Fig. 1.3. Formation of a stream

Streams may be divided into three types:

- Torrents
- Intermittent streams
- Springs

Torrents are those which carry only the surface runoff and hence water flows through them only during the period of rainfall and subsequent runoff. In Intermittent streams flow of water is only in rainy season when the whether is wet. After rainy season the intermittent streams dry out. It is so because frequent rains in rainy season raise the water-table temporarily. Thus ground water level is above the bottom of the stream temporarily. This causes flow of water from the soil to a ditch so long as watertable is high.

Springs are perennial streams. Water flows through them throughout the year since the water-table is higher than the depressed ground level permanently.

4. Lakes: Rainwater in excess of all sorts of losses runs off the surface of the earth. When this water is caught in very big natural depression a lake is formed. Lakes also derive water from groundwater. This water is also available for use whenever required.

5. Reservoirs: When some obstruction like a dam or a weir is constructed at the narrowest point of a valley to store large quantity of water behind it reservoir or artificial storage is formed. This water can be very conveniently used for various purposes with provision of suitable hydraulic structures.

1.6. INDIRECT OR UNDERGROUND SOURCES OF WATER

When water is not available naturally at the time of necessity, from the knowledge of subsurface, underground water can be made available for use. In this category the following sources can be listed:

1. **Open wells:** If it is ascertained that sufficient water is entrapped in some water bearing stratum below the ground surface, the entrapped water can be made available for use. A hole is sunk into the ground till it reaches such a depth as to hold sufficient water for use. Water should be available at lesser depth for economic justification.
2. **Tube wells:** If there are water bearing rocks or soil layers alternate to impervious layers or water bearing strata of indefinite extent then a metal tube with suitable perforations may be sunk in the ground to derive the water for use.
3. **Artesian wells:** When a permeable stratum is confined between impervious strata at the top and bottom artesian condition exists. The outcrop of the permeable stratum should be at a height enough to produce sufficient hydrostatic pressure on the water at lower points. Then if a bore is drilled at proper position through the upper impermeable strata the water under pressure will rise in the bore. When the pressure is more water even overflows at the surface and can be utilized. Figure 1.4 shows the typical situation where artesian well is adopted.

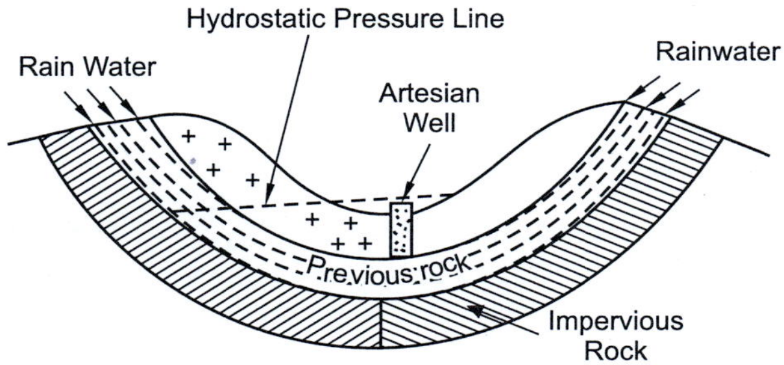

Fig. 1.4. Artesian well

4. **Infiltration gallery:** When water can be obtained within a reasonable distance below ground level, for example, below the river bed, horizontal porous pipes with open joints can be laid under the ground. It is apparent that a very large proportion of the groundwater will be intercepted by galleries than by a vertical well. The intercepted water can be collected at suitable points in vertical collecting wells and can be used. Inverted filter* surrounding the pipes can be provided to prevent clogging of pipes.

   Horizontal galleries laid at a depth more than 8 metres are uneconomical.

   Figure 1.5 shows cross-section and longitudinal section of the infiltration gallery.

* Filtration is the process in which suspended impurities are removed by passing the liquid through a close-grained medium. Texture of the medium and arrangement of particles of the medium is such that it does not permit suspended impurities to pass through them but allows the liquid to pass readily.

The layer of the close-grained medium is called a filter. Now if filter media is formed of various layers of different texture and if the arrangement of layers is such that the permeability of each layer increases in the direction of flow the arrangement is termed as inverted filter.

The function of the first close textured layer is to hold all suspended fine particles or impurities in liquid. The function of the subsegunet layers is to allow the passage of water or liquid with lesser and lesser resistance to increase the rate of flow or infiltration.

Inverted filter provided on all sides of the porous pipes consists of fine grained sand in first layer, coarse sand and selected gravel in subsequent layers. The layer of fine sand though very much permeable to soil-water is capable of arresting all the soil particles coming with the water.

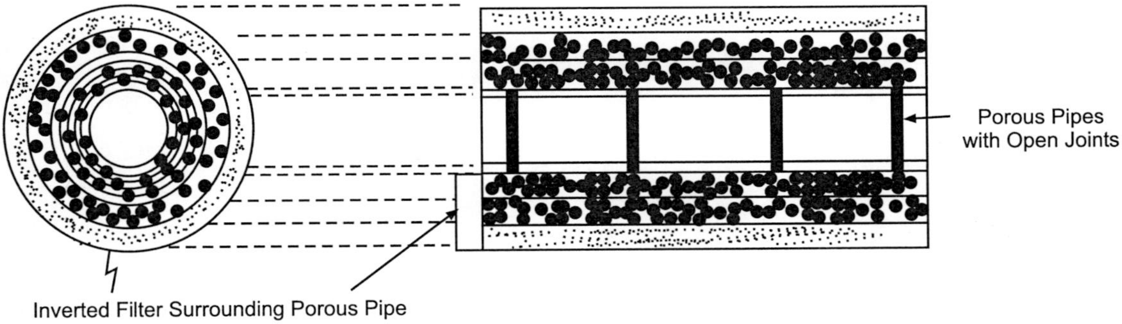

Fig. 1.5. Infiltration gallery

5. **Infiltration wells:** Sometimes water can be made available by sinking infiltration wells in the porous soil, for example, in the dry bed of a river. The infiltration well can be joined to vertical collecting wells or jack wells sunk on the bank of the river by means of horizontal underground porous pipeline. It intercepts water also and is called **Infiltration gallery**. Thus it can be recognized that infiltration wells and galleries supplement each other. Figure 1.6 shows the cross-sectional elevation and plan of an infiltration well. While carrying out sinking operation it is very essential to see that infiltration well does not tilt. Tilt may cause breakage of pipe line which is laid horizontal. It is also essential to see that after sinking of well there is no appreciable settlement. It is also essential to see that after final sinking the well does not settle or sink under its own weight.

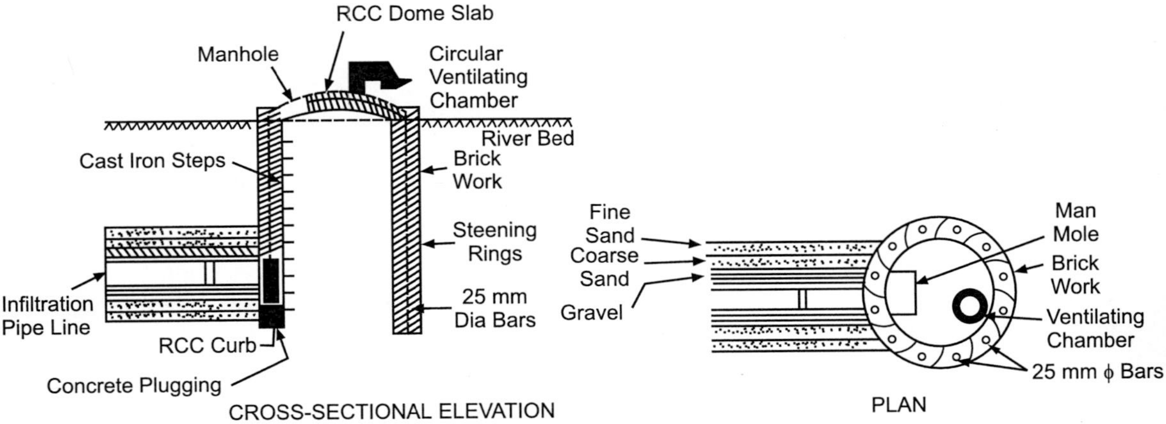

Fig. 1.6. Infiltration well

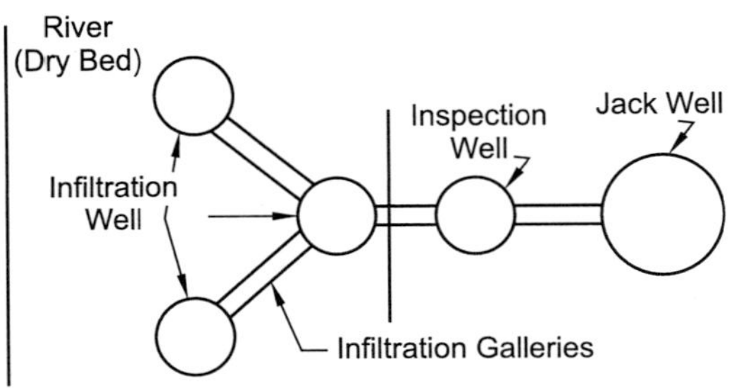

Fig. 1.7. Plan showing arrangement of wells and galleries

1.7. RIVER SYSTEMS OF INDIA

The river systems of India can be classified into four groups as follows:

- Himalayan rivers
- Rivers of Deccan plateau
- Coastal rivers, and
- Rivers of desert.

1.7.1. Himalayan Rivers

They are formed due to melting of snow and glaciers and therefore carry continuous flow throughout the year. During monsoon months the region receives heavy rainfall and the rivers cause floods. The Indus, the Ganga and the Brahamputra are the three main Himalayan river systems. The Indus rises near Mansarover in Tibet and flows through Kashmir and Pakistan before falling in Arabian Sea. Important tributaries of the Indus are the Jhelum, the Chenab, the Ravi, the Beas and the Satluj. The Ganga rises near Gangotri and flowing through Uttarakhand, Uttar Pradesh, Bihar, West Bengal and Bangladesh falls in Bay of Bengal. Important tributaries of the Ganga are the Yamuna, the Ramganga, the Ghaghra, the Gandak, the Kosi, the Sone, the Damodar and the Mahananda. The Yamuna has the Chambal and the Betwa as its two important sub-tributaries. The Brahmaputra rises in Tibet and flowing through Arunachal Pradesh, Assam and Bangladesh joins the Ganga in Bangladesh. Important tributaries of the Brahamputra are the Dibang, the Lohit, the Subansiri, the Manas and the Tista. The Barak rises in Nagaland and flowing through Manipur and Assam flows into Bangladesh to join combined limb of Ganga-Brahmaputra.

1.7.2. Deccan Rivers

Unlike Himalayan rivers the rivers of the Deccan Plateau receive flows from rainfall only. They, therefore, carry huge discharges during monsoon season and the flow reduces drastically in the non-monsoon months. Many rivers become completely dry after rainy season. Important west flowing river systems are the Narmada and the Tapti which fall into Arabian Sea. The important east flowing rivers systems are the Mahanadi, the Godavari, the Krishna and the Cauvery which fall into Bay of Bengal.

1.7.3. Coastal Rivers

There are many small rivers which join Arabian sea and Bay of Bengal directly on west and east coast respectively. There are about 600 small rivers on west coast whereas on east coast their number is less. These rivers drain off enormous flows during monsoon months.

1.7.4. Rivers of Desert

There are few rivers of desert which belong to West Rajasthan. They are separate basins and either drain salt lakes like Sambhar or get lost in sands without falling into the sea. The Luni is the only river of West Rajasthan which drains into Rann of Kutch.

There are few other rivers in eastern most part of the country like Karanphali, Kaldan, Imphal etc., which originate in India but flow down into Burma.

On the basis of the annual run-off from all the rivers in India the total surface water resources of the country has been assessed to be 170.9 million hectare metres (M ham). The contribution made by various rivers systems has been indicated in Table 1.1.

Table 1.1. Water Resources Potential in River basins of India

	River systems	*Catchment area in India (M ha)*	*Annual flow (M ham)*
(*a*)	Himalayan rivers		
	Indus	32.13	7.3
	Ganga	86.14	52.5
	Brahmaputra	19.44	53.7
(*b*)	Deccan rivers		
	Sabarmati	2.17	0.4
	Mahi	3.48	1.1
	Narmada	9.88	4.6
	Tapti	6.51	1.5
	Subernarekha	1.93	1.1
	Brahmani and Baitarni	5.18	2.9
	Mahanadi	14.16	6.7
	Godavari	31.28	11.1
	Krishna	25.89	7.8
	Pennar	5.52	0.6
	Cauvery	8.12	2.1
	Other rivers	32.28	
(*c*)	Coastal rivers	9.65	
(*d*)	Desert rivers	9.94	24.4
(*e*)	Rivers flowing into Burma & Bangladesh	3.62	

1.8. DEFINITIONS OF GENERAL TERMS

(*i*) **Climate:** Climate is the general aggregate of the weather closely related to meteorology and geography. It is a composite of weather conditions over a long period of time truely portrayed. It includes details of variations, extremes, frequencies, sequences of the weather elements which occur from year-to-year, particularly in temperature and precipitation.

(*ii*) **Weather:** Weather is the state of atmosphere at any given time and place. Weather, generally defines status of sunshine, temperature, pressure, winds and moisture of short durations like a day, a week or even a season.

(*iii*) **Meterologically homogeneous area:** A region is said to be meteorologically homogeneous if stations therein have very nearly the same rainfall frequency ditribution.

(*iv*) **Water year:** It is a continuous twelve months period selected for presentation of hydrological data. The period is selected in such a way that all the precipitation that has fallen within this period will also run-off. The year is named after the calendar year to which most months belong. In India water year ranges from June 1 to May 31 of next year. This arrangement helps in statistical treatment of hydrological data. It is also called hydrological year sometimes.

(*v*) **Depression:** Depressions are the cyclonic circulations in which the maximum wind speed near the centre is less than 34 knots over any part of the cyclonic disturbance.

(*vi*) **Storm:** It is a meteorological disturbance accompanied by such phenomena as rain, snow, hail, thunder, lightening and wind.

(*vii*) **Hydrometer:** It is a meter consisting of an ensemble of liquid or solid water particles falling through or suspended in the atmosphere, blown by the wind from the earth's surface or deposited on objects on the ground or in the free air. It includes rain, drizzle, snow, sleet, hail, fog, and blizzard.

(*viii*) **Precipitation:** It is the total supply of water of all forms of falling moisture expressed as depth of liquid water, on a horizontal surface. It includes rainfall, snowfall etc. In hydrologic cycle it represents change of state from vapour to water.

(*ix*) **Probable maximum precipitation:** The probable maximum precipitation is the storm which after maximisation yields highest precipitation amounts for the critical duration and area of the project basin. It represents an estimate of physical upper limit of storm rainfall and is obtained after study of all storms and maximising them for highest precipitation.

(*x*) **Point rainfall:** Rainfall at a specified location, measured at a single rain gauge is called point rainfall.

(*xi*) **Effective rainfall:** It is the total net rainfall which is volumetrically equal to total direct surface runoff. It is also called rainfall excess.

Rainfall excess = Total rainfall – abstractions

Here abstractions include interception, evaporation, depression storage, transpiration and infiltration.

(*xii*) **Isohyet:** A line joining points on a map where the amount of preciptation, recorded in a given period, is the same is called an isohyet.

(*xiii*) **Mass curve of rainfall:** It is a graph showing the accumulated amounts of rainfall progressively against time.

(*xiv*) **Rain shadow area:** It is a region, situated on the lee side of a mountain or mountain range where the rainfall is much less than on the windward side.

(*xv*) **Arid region:** It is an area where precipitation is deficient in quantity or occurs at such times, that agriculture is impractical without irrigation.

(*xvi*) **Unit storm:** This is defined as a rain of such duration that the period of surface runoff is not appreciably less for any rain of shorter duration.

(*xvii*) **Critical position of storm:** The position of the transposed storm centre is so adjusted as to give the maximum average depths of rainfall over the basin. This particular adjustment of the storm with respect to the basin is called the critical position of the storm.

(*xviii*) **Storm transposition:** It is the method of application of a storm from one area to another area having the similar homogeneous meteorological characteristics.

(*xix*) **Snow line:** The line or zone on land that separates areas in which fallen snow disappears in summer from areas in which snow remains throughout the year.

(*xx*) **Runoff:** It is that part of precipitation as well as any other flow contributions, which appear in surface streams of either perennial or intermittent forms.

(*xxi*) **Interception:** It is a portion of the rainfall absorbed or intercepted by leaves, branches, trunks of dry trees and vegetation.

(*xxii*) **Infiltration:** It is a process in which the rain water enters the surface strata of soil and moves downward to join the groundwater. On the way it first fulfills the moisture requirement of unsaturated zone of soil.

(*xxiii*) **Transpiration:** It is that portion of water which after circulation through plant structure goes to atmosphere in the form of vapour from the surfaces of leaves, branches etc.

(*xxiv*) Interflow: It is that part of infiltrated water which reappears in the streams on the surface of the earth. It becomes possible because infiltrated water moves horizontally as well as vertically.

(*xxv*) Evaporation: Owing to heat energy of the sun, water is constantly changing its form to vapour from soil and water surfaces. This vapour goes to atmosphere as evaporation.

1.9. WORLD WATER BALANCE

According to the estimate made by United Nations the total amount of water on the earth is about 1400 million BCM (Billion Cubic Metres). Most of it is, however, contained in the oceans, which cover almost three-fourth of earth's surface. This quantity of water is enough to cover earth's surface with a 3 km deep layer of water. But irony is 97.3% of the water is saline and not fit for human consumption. Only about 2.7% of the total water available on the earth is fresh water. Out of this nearly 75.2% of the total water available on the earth is freshwater. Out of this nearly 75.2% lies frozen in polar regions. Another 22.6% is present below ground as groundwater. What remains is only 2.2% of the freshwater resource which is available in atmosphere, lakes, rivers, moisture, vegetation, soil etc. Following Table 1.2(*a*) and 1.2(*b*) gives breakup of total water of the world:

Table 1.2: (a) Distribution of World's water

S. No.	*Item*	*% of Total*	*Volume (Million BCM)*
I	Total water resource	100%	1385.5
II	Ocean water (Salty-unavailable for human consumption)	97.3%	1348.0
III	Freshwater	2.7%	37.5

From Table 1.2(*b*) it is clear that water available for human consumption is only in rivers, lakes and not too deep groundwater. Thus effectively only about 0.35 % or 4.84 million BCM water from earth's total water resource can be harnessed for beneficial uses. But even this water has uneven spatial distribution over the earth's surface. Here lies the importance of making most efficient and economic use of water resource.

Table 1.2: (b) Source-wise break up of freshwater

S. No.	*Item*	*% of Earths Total resource*	*Volume (Million BCM)*
III	Freshwater [same as in Table 3.1(*a*)]	2.70 %	37.50
III (*a*)	Polar ice and glaciers	2.04 %	28.20
III (*b*)	Groundwater (at depth > 800 m)	0.26 %	3.74
III (*c*)	Groundwater (at 400 – 800 m depth)	0.34 %	4.71
III (*d*)	Lakes and rivers	0.01 %	0.13
III (*e*)	Atmospheric vapour and soil moisture	0.05 %	0.72

1.10. COUNTRY'S WATER BALANCE

Annual precipitation including snowfall is the main source of water of our country. It is estimated to be about 4000 BCM (billion cubic metres). The water resources potential of the country is about 1869 BCM as estimated by Central Water Commission, considering both surface and ground water as one system. Out of this quantity nearly 59% is available in Ganga-Brahmaputra-Meghna system. Due to various constraints of topography, uneven distribution of water resource over space and time, it has been assessed that only about 1123 BCM of total potential of 1869 BCM can be put to beneficial use. Out of 1123 BCM of available potential 690 BCM is share of surface water potential. Balance 433 BCM is total replenishable ground water potential of the country. It is to be remembered that this potential is available year after year as a consequence of operation of hydrologic cycle.

1.11. GENERAL AIR MASS CIRCULATION

Basically the suns heat energy is responsible for maintaining circulation of air mass on the earth's surface. The areas lying on the equator receive more solar radiation than those lying on the higher latitudes. As a result the air on the equatorial belt gets warmer and lighter and it rises in the sky. As the air mass on the equator rises it is replaced by cooler air from higher latitudes. In turn, the air of higher latitudes gets replaced by the warm equatorial air which moves towards pole from above. In practice, however, mechanics of circulation is not so simple and straight forward. Two factors are responsible for modifying the basic thermal circulation. They are :

- earth's rotation; and
- distribution of land and seas on the earth and landforms.

It is known that earth rotates from west to east. A point on the equator moves with a velocity of 1670 km/hour speed. Applying the principle of conservation of angular momentum it can be shown that a small air body at rest relative to earth's surface at equator would attain a theoretical eastward velocity of 2505 km/hour relative to earth's surface if moved northward to 60° latitude. Similarly on the basis of same principle it can be shown that a small air body from north pole if moved south to 60° latitude would attain a theoretical velocity of 835 km/hour in westward direction. It is true that due to friction air masses do not attain these velocities any time. Assuming the earth's surface to be smooth and of uniform composition the ideal circulation of air mass in the northern hemisphere can be visualised as shown in Fig. 1.8.

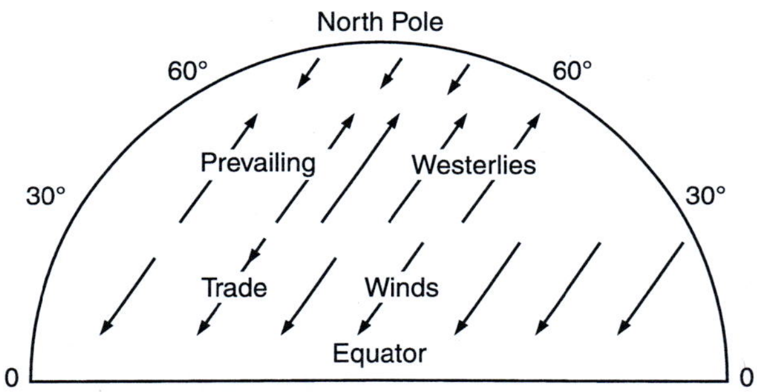

Fig. 1.8. Ideal circulation of air mass

The heated air mass at equator rises and moves northward acquiring eastward component as it moves. At about 30° latitude this mass cools down and descends to lower levels. Here the air mass divides into two components. Once component continues in its original direction. The second component turns southward and starts moving as northwest trade winds.

Towards pole between 90° and 60° latitude the cooled air mass moves towards 60° latitude in south-westwardly direction in lower layers. This air mass at about 60° latitude gets warmed up rises again and moves towards pole as southwesterly current.

From this general air circulation it becomes obvious that belts of low pressure are created at the equator and about 60° latitude whereas high pressure belts are formed at about 30° latitude and the north pole.

This ideal air mass circulation pattern gets further distorted due to effects of distribution of water bodies and landmasses and land forms. The main features which contribute towards distorting the air circulation can be enumerated as follows:

(*i*) Landmass is heated or cooled only at the surface and as a result the air above gets more heated or cooled.

(*ii*) The large water bodies owing to their intermixing property gets heated or cooled to larger depths.

(*iii*) The soil has lower specific heat than water.

(*iv*) The landforms create barriers to free airflow.

These features create differential heating and cooling over the earth. For example, in winter air masses over land surfaces are dense and cold whereas those on the oceans are warm. The situation is just the reverse in the summer season. The general air mass circulation, therefore, gets modified locally depending upon the extent of interplay of various factors mentioned above.

1.12. CYCLONES AND ANTICYCLONES

The phenomenon of general air mass circulation with its likely distortion has been explained earlier. Mostly all weather phenomena are caused as a result of circulation of either large or small air masses. The features of air mass circulation are called cyclones or anticyclones. The features could be either semipermanent, *i.e.*, occurring more or less regularly or migratory.

Cyclones: A cyclone is an area of low pressure in which the winds blow in counterclockwise direction in the Northern Hemisphere. Cyclones are usually accompanied by extensive cloud systems and precipitation. That is why the word cyclone is almost synonymous with the word storm which denotes a period of heavy rainfall over a specific area. In tropical belt cyclones originate at low latitudes in the oceans from thermal convection. It develops into a voilent whirling air mass having horizontal dimensions of 150 to 500 km and wind speed exceeding 120 km/hour. The tropical revolving storms of this type are called **Hurricane** in general but are known by name Cyclone in India and Typhoons in Far East region. The cyclones which are formed in midlatitude belt (30° to 60°) are called **extratropical cyclones**. They form along the boundaries between warm and cold air masses (The boundary is known by a technical name 'Front').

Anticyclones: An anticyclone is an area of high pressure in which the winds blow spirally outword in a clockwise direction in the Northern Hemisphere. Unlike cyclones, in general anticyclones are accompanied by fair weather and the winds are generally moderate.

1.13. MONSOONS

The word Monsoon is derived from an Arabic word which means season. Monsoons are systems of seasonal wind which blow with consistency and regularity from oceans to continents in summer and in the reverse direction in winter. These wind changes are caused due to differential annual heating and cooling of continents and adjoining oceans. These winds are largely confined to tropical belt. They are especially more pronounced over Indian sub-continent betweeen June and September each year. The monsoons associated with other continental masses are not so marked as the Indian monsoons. The Indian monsoons are widely known owing to its excessive rainfall during summer and equally excessive dryness during winter. From October to December, however, a branch of north-east trade winds sweep across the Bay of Bengal and causes precipitation over Southern half of the Indian Peninsula. This is called North-East or Winter Monsoon of India. Due to its vagaries the monsoons create great impact on the Indian Agriculture. It calls for greater understanding of the processes involved in creation of monsoons.

1.14. CLIMATOLOGY OF INDIAN SUMMER MONSOON

The main characteristics features of the Indian Summer Monsoon are the following:

(*i*) Moonsoon winds approach India from a southwesterly direction. The summer monsoon sets in over the extreme south of the Indian Peninsula on the first of June.

(*ii*) The arrival of the monsoons is a gradual process. The period of transition may extend over a week or even more. The depth of fully established monsoon current off the west coast of India is about 6 km.

(*iii*) The monsoon advances over Indian subcontinent along two branches namely, an Arabian Sea branch and Bay of Bengal branch.

(*iv*) The Bay of Bengal branch which advances into West Bengal, Assam and North-East India is deflected by the orientation of the mountains in the Indo-Gangetic plains of north India.

(*v*) The normal duration of the monsoon is about four months. It begins to withdraw from northwest India by mid-September.

(*vi*) Although convective phenomena also play an important role substantial rainfall is caused by the rather fortuitous orientation of the mountain barriers.

(*vii*) The variability of the monsoon rain is highest over north-west India including Rajasthan. These are the areas which receive small amounts of rain during monsoon.

(*viii*) An extension of the seasonal low into the Indo-Gangetic plains is known as the Monsoon trough. The axis of this trough shows periodic movements of the trough to the north or to the south of the Indo-Gangetic plain.

(*ix*) Southerly position of the axis of the trough is favourable for the general activity of the monsoon and wide spread rainfall occurs over most of the country.

(*x*) On the other hand northerly shift of the axis of the trough results in shift of rainfall further northward into the Himalayan foot hills. Also activity of the Arabian Sea branch of the monsoon is decreased while the Bay of Bengal branch remains confied to north-east India. This type of situation is known by term "break" in the monsoon.

1.15. CLIMATOLOGY OF NORTH-EAST OR WINTER MONSOON

The main characteristic features of the north-east monsoon are the following:

(*i*) Like south-west monsoon onset of winter monsoon is also gradual and has a period of transition. Only difference is that winter monsoon's onset is not well defined.

(*ii*) The duration of winter monsoon is from October to December.

(*iii*) The upper winds of the north-east monsoon sweep across the Bay of Bengal in a clockwise direction.

(*iv*) The state of Tamil Nadu which is shielded by Western ghats mountain range from south-west monsoon receives rainfall from north-east monsoon during winter.

1.16. CLASSIFICATION OF CLOUDS AND MONSOON CLOUDS

The first classification of clouds is due to Luke Howard who, in 1803, introduced the Latin names *Cirrus*, *Cumulus*, *Nimbus* and *Stratus* for different forms of clouds. The *International Cloud Atlas* published by the World Meteorological Organisation recognises ten main types of clouds. Their main features are:

1.16.1. High Clouds

Cloud base at 6 km or higher:

- **Cirrus (meaning curl):** It is a detached fibrous cloud in the form of white feathers of narrow bands. All cirrus type clouds are composed of minute ice crystals and do not cause precipitation.

- **Cirrocumulus:** They are thin white layers of high clouds, without shading.
- **Cirrostratus:** They are transparent white clouds through which halos are often seen.

1.16.2. Medium Clouds

Cloud base at or above 2 km.

- **Altocumulus:** It is a white or grey layer cloud sometimes seen in the form of rolls or round globules.
- **Altostratus:** It is a greyish cloud sheet. Halos cannot be seen through Altostratus clouds.

1.16.3. Low Clouds

Clouds extending from the surface to 2 km.

- **Stratus:** It is generally a grey cloud layer of uniform base. It resembles fog but does not rest on ground.
- **Stratocumulus:** It is in the form of grey or white patches. They often appear as rolls or rounded masses of clouds.

  Another three species which are frequently observed during the monsoon months are the following:
- **Cumulus:** They are detached clouds with sharp outlines. Rising towers or domes are often seen within a Cumulus cloud.
- **Cumulonimbus:** They are heavy and dense shower clouds with tops spread out in the form of an anvil.
- **Nimbostratus:** They are grey or dark cloud layers from which we observe continuous rain.

1.17. MONSOON WEATHER SYSTEMS

Over 70 per cent of annual rainfall over India is recorded during southwest monsoon from June to September. The heaviest rainfall is recorded along west-coast, in which orographic features play important role. It may be mentioned that it is however, not the sole factor of importance for monsoon rain. Rainfall also comes in association with thunder storms, for example, in April and May. This type of rainfall is generally of transient nature. Often the intensity of precipitation is very high. On the contrary monsoon rain is continuous rain spread over days and its intensity is not as high as that of convective rain. During monsoon there are occasions of "cloud bursts". The cloud bursts are the result of violent convective activity over small areas of limited extent, of course such occasions are comparatively rare.

A significant portion of monsoon rainfall over the greater portion of India is closely associated with movement of cyclonic depressions from the Bay of Bengal. When a cyclone or a depression representing a region of low pressure forms in the Bay of Bengal the pressure begins to fall over a wide region of several hundred kilometres. The wind, in response to the pressure gradient adopts anticlockwise cyclonic circulation. As the depression moves from the head of the Bay of Bengal towards the coast, a belt of heavy rainfall extends to the southern and south-eastern parts of West Bengal and lower Assam. With further movement of the storm westward the rain belt extends to Orisa and Bihar. As the storm proceeds and enters Madhya Pradesh the presence of this depression strengthens the Arabian Sea branch of the monsoon. This causes another spell of moderate to heavy rain over Madhya Pradesh and southern parts of Uttar Pradesh. The depression then may carry rainfall further to Rajasthan and Gujarat. Sometimes the depression may recurve towards the north and breakup in the sub-himalayan region of Punjab. In such situation the Arabian Sea branch of the Monsoon feeds extra moisture in the storm area and heavy rainfall is recorded in Punjab.

Generally after the passage of a depression the monsoon weakens and the rains slacken. After a gap of few days, however, the monsoon tends to revive and another depression is formed at the head of the Bay of Bengal. The similar pattern of rainfall associated with depression as explained above follows this time also. Thus, with a certain degree of periodicity monsoon rainfall occurs over northern and central parts of India. On an average about 8 cyclonic depressions pass from the Bay of Bengal over the Indian land mass between June and September.

The monsoons are also frequently accompanied by violent storms of hurricane type. Sometimes the atmospheric pressure falls much below than what is observed in depression. The region of low pressure is more localised and air acquires much greater cyclonic spin. The depression is then said to concentrate into a tropical storm or a cyclone.

QUESTIONS FOR REVISION

1. What is hydrology? Explain the scope of hydrology and its application in water resources development programmes.
2. Draw pictorial diagram of global hydrologic cycle. Mention various sub-systems of the cycle and name four water storage places and modes of water transport each.
3. Write short notes on the following:
   (*i*) Cyclones and Anticyclcones (*ii*) Water year
   (*iii*) Rainfall excess (*iv*) Infiltration gallery
4. (*a*) What are monsoons ?
   (*b*) Give climatological features of south-west monsoon.
   (*c*) Why monsoons are important for India?
5. Explain world's water balance, showing how it is important to put available water resource to most economic use.

❑❑❑

2

Precipitation and its Measurement

2.1. THEORY OF RAINFALL

Precipitation is the natural process of conversion of atmospheric vapour into water. The water so formed then falls to the earth in the form of a rainfall. In terms of hydrology rainfall constitutes the third phase of atmospheric division of the hydrologic cycle, "the change of state". The term precipitation is also used for rainfall. Precipitation is however, a general term and includes all forms of falling moisture viz., rainfall, snowfall, sleet, hail etc.

During summer season evaporation loss is highly accelerated from all types of free water surfaces. The water lost in evaporation finds room in an air mass. It adds to the atmospheric vapour storage. Although evaporation loss is excessive in hot season the capacity of the air mass is also more. This large quantity of vapour makes the air mass moist. The change of state from atmospheric vapour to water takes place when the capacity of the air mass to hold the vapour particles exceeds. Following two main reasons are responsible for this change of state:

- Hot air mass has large capacity to hold the vapour particles in suspension. When by some means this moist and warm air mass cools down its capacity to hold vapour particles is reduced. Finally vapour precipitates in the form of rainfall.
- Some times variation in pressure brings about the change of state from vapour to rainfall.

It is interesting to know the actual mechanism of precipitation. By a process known as nucleation, ice or water crystals are formed upon the floating particles, in the air mass (*e.g.*, dust particles, salt particles, etc.). The small crystals then grow in the size by combining with other crystals. A stage comes when they fall down on the earth as snow or as rain water.

2.2. FORMS OF PRECIPITATION

The precipitation takes place in many different forms in the regions located in the middle latitudes. Typical characteristics of various forms of precipitation are explained below.

(*i*) **Rain:** It consists of water drops mostly larger than 0.5 mm in diameter.

(*ii*) **Drizzle:** They are tiny water droplets of size between 0.1 to 0.5 mm which fall with such slow settling rates that they occasionally appear to float.

(*iii*) **Snow:** It is that type of precipitation which results from sublimation, *i.e.*, water vapour directly changes into ice. It falls as white or translucent ice crystals often agglomerated into snow flakes. The specific gravity of snow is often taken to be 0.1.

(*iv*) **Hail:** It is the precipitation in form of lumps of ice. The hail stones are produced in convective clouds mostly cumulonimbus. Their shape may be conical, spheroidal or irregular. The size of hail stones may be anything more than 5 mm. The specific gravity of hail stone is about 0.8.

(*v*) **Snow pallets:** Some times they are called soft hail also. Snow pallets are more crisp and are of size 2 to 5 mm. Due to their, crispness upon hitting the hard ground they often break up.

(*vi*) **Sleet:** When the rain drops fall through the layer of sub-freezing air near the earth's surface the rain drops get frozen to ice stage. It is called sleet or grains of ice.

2.3. TYPES OF RAINFALL

Different types of rainfall can be recognised according to the process by which warm and moist air mass gets lifted and subsequently cooled. Broadly speaking there are three types of rainfall.

2.3.1. Cyclonic precipitation or rainfall

This type can be sub-divided into (*a*) Frontal and (*b*) Non-frontal precipitation. This type of rainfall occurs from lifting of air which converges into a low pressure area or cyclone. This type of rainfall generally occurs in plain regions.

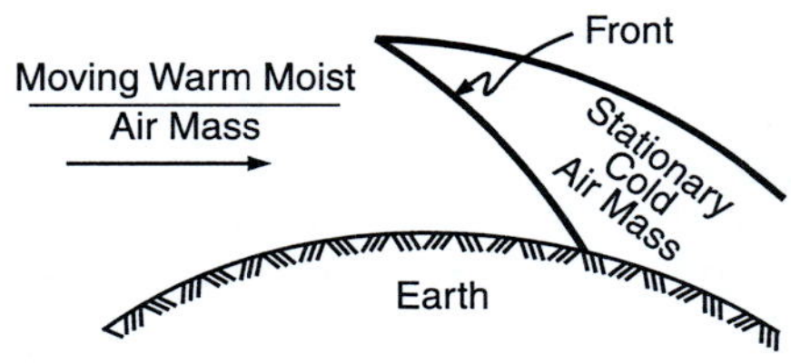

Fig. 2.1. Frontal precipitation

(*a*) **Front type of precipitation:** Front is a boundary joining warm moist air mass and cool air mass.

When a moving warm moist air mass is obstructed by a stationary cold air mass, the warm air mass rises up as it is lighter than the cold air mass. Sometimes cold moving air mass meets stationary warm air mass with similar results.

The lifted air mass cools down at high altitudes and precipitation occurs. This process continues till the whole warm air mass passes over the cold air mass. A showery type precipitation occurs in case of cold front whereas in case of warm front precipitation continuous rainfall occurs till the warm front passes over the cold air mass.

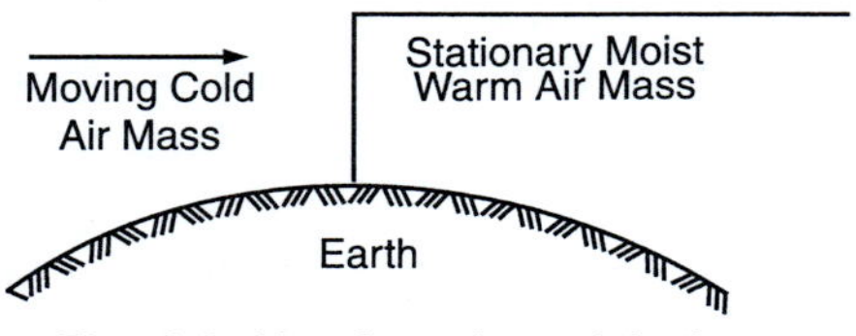

Fig. 2.2. Non-frontal precipitation

(*b*) **Non-frontal precipitation:** This type of rainfall is not related to fronts.

When the moving cold air mass meets the warm moist air mass moist and warm air mass gets lifted up being lighter than the cold air mass. When the warm air mass cools down at high altitude precipitation occurs.

2.3.2 Convective Precipitation

Due to some local effects air gets heated up and stores more vapour particles. Then it rises up in the atmosphere as it is lighter than the cold air surrounding that area. At high altitudes it gets cooled and precipitation occurs. The intensity of this type of precipitation may range from light showers to cloud bursts.

2.3.3. Orographic Precipitation

When a moving warm moist air mass is obstructed by some type of barrier like mountains, the warm moist air mass, finds its way upward and rises to sufficiently high altitudes automatically. There it gets cooled and finally precipitation occurs.

Seaward side mountain slope gets ample rainfall but landward side mountain slope and some portion of the plain receives very little rainfall. The reason for this peculiar phenomenon can be explained as follows. When the elevated warm air mass gets cooled change of state from vapour to water takes place and rainfall occurs on the area below.

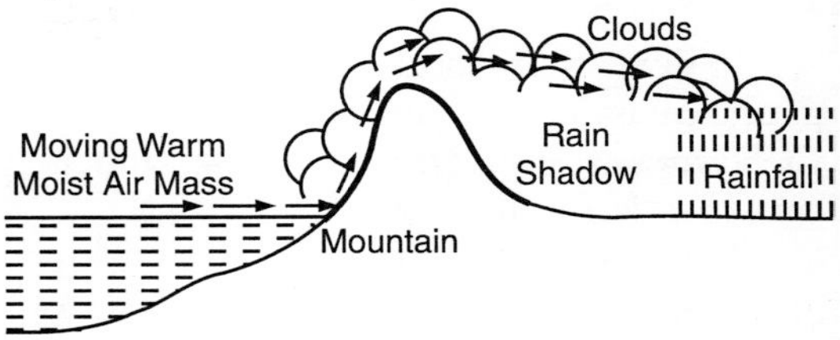

Fig. 2.3. Orographic precipitation

Obviously by the time clouds travel past the barrier they get relieved of the moisture and become weak. It takes some time before the clouds regain sufficient moisture to cause precipitation. During this interval the clouds leave some area dry. The area which does not receive rainfall in this process is called the area of rain-shadow. Figure 2.3 gives lucid illustration of the phenomenon. This area is called the area of rain shadow because one can see the clouds readily passing overhead but the land below does not receive rainfall but only gets shadow of it.

2.4. ARTIFICIAL RAIN-MAKING OR CLOUD SEEDING

Artificial stimulation of rainfall has now become a reality. It can play a very useful role in needy areas. The objective of the process is to identify right type of rain-bearing clouds and to seed them extensively at the right time with a view to intercepting them before they pass over the tract on which rainfall is required to be made. The process can not assure rain from any type of clouds on a particular area. The process consists in seeding the clouds with minute particles of some substance which can bring down the temperature of rain-bearing clouds and also provide innumerable nuclei for forming rain drops. There are two methods of seeding the clouds. In the Bergeron or ice crystal process dry ice-crystals are sprayed over the clouds to lower their temperature to (–) 15°C. The sprayed crystals form nuclei and is a result the rainfall process begins.

In the silver iodide method a 'smoke' of silver iodide is dispersed on supercool clouds. It is done by burning a solution of silver iodide on acetone in which small quantity of sodium iodide has already been dissolved. The seeding of the clouds with silver iodide smoke can be done by releasing it from the burners stationed on the ground or also by carrying the burners in an aircraft. From the point of view of both the quantity of seeding material required and the method of dispersing, the silver iodide method is more attractive. The method of seeding the clouds is still under experimental stage in our country.

2.5. MEASUREMENT OF RAINFALL

The yearly rainfall, is expressed in centimeters or millimetres of depth over a particular over which precipitation takes place.

Definition of one centimetre of rainfall: Suppose the water precipitated on a certain plain area in the form of a rainfall is not lost by any means and if there is no runoff and evaporation what so ever then all the water will go on accumulating on the surface of the area in the form of a sheet of water. When the sheet of this deposited water is one centimetre thick, one centimetre of rainfall is said to occur.

Rainfall can be measured by a **rain-gauge**. A rain-gauge may be of two types:

- Non-recording type, and
- Recording type or automatic.

2.5.1. Non-Recording Type Rain-Gauge

It gives only total rainfall occurred during particular time period. Recording type rain-gauge gives hourly rainfall. Under non-recording type rain-gauges, one most commonly used in **Symon's rain-gauge**. This type is mentioned below. It is the simplest in principle, construction and working.

Principle: From the definition of unit rainfall it is clear that the definition is independent of extent of area. So far as only measurement of rainfall is concerned area under consideration may be large or small. Now taking smallest possible area, if the water, which comes down as rainfall, is collected before the losses take place or water runs off then the depth of this water over the small area can be quite accurately determined to give the amount of rainfall occurred in proper units (centimetres). The small area should be selected in such a way that its meteorological characteristics are similar to that particular large area which it represents.

Construction: It consists of a funnel and a receiver mainly. The receiver is a cylindrical (zinc) metal bottle. The diameter of the bottle and the topmost diameter of funnel is 127 mm. The funnel is fitted in the neck of the bottle.

Both are then placed in a metal casing with suitable packings. The base of the metal casing is enlarged to 210 mm. The capacity of the bottle is such as to measure extremes of rainfall likely to occur in 24 hours. Zinc receivers hold 175 mm to 1000 mm according to size. Gauge is provided with one measuring graduated jar which measures the water in mm. The smallest division on the jar is 0.2 mm. The rainfall should be estimated to the nearest 0.1 mm. See Fig. 2.4.

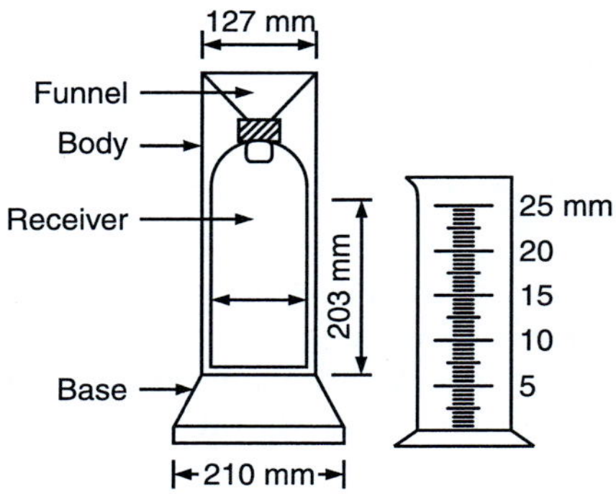

Fig. 2.4. Symons non-recording type rain-gauge

At the site where rainfall is to be measured concrete block us constructed. The base of the gauge is permanently fixed in the block in such a way that the top of the casing is about 30 cm above natural surface level. While fixing the base of the gauge precaution is taken to level it perfectly. The rain-gauging station is protected by barbed wire fencing with a gate. The size of the concrete foundation block should be 60 cm × 60 cm. The necessity to keep the rim of the funnel above the natural surface by 30 cm is two fold:

(*i*) It prevents splashing of water into the funnel almost to a negligible amount.

(*ii*) If the height is kept more than 30 cm the amount of rain water collected decreases owing to wind eddies set up by the gauge itself.

Working: The gauge is adjusted every day for measurement of rainfall. When rainfall occurs the rainwater covering area of the funnel passes to the receiver before any sort of loss takes place. After every 24 hours the rainfall is measured. Usually the measurement is taken at 0830 hr. I.S.T. The received water is poured carefully in the measuring jar to measure daily rainfall. If it is raining at the time of observation, it is necessary to do the measurements very quickly. If found essential spare receiver may be placed immediately in the body after previous receiver is taken out. The total amount of rainfall measured during the previous 24 hours should invariably be entered against the date of measurement irrespective of the fact whether the rainfall was received on the date of measurement or on the previous date after yesterday's measurement.

Totaliser: Sometimes unavoidably rain-gauge is to be installed at such a location which is not easily accessible in unfavourable climate. Then it is not possible to measure the rainfall every day at 0830 hr I.S.T.

In such cases a different type of non-recording gauge is used. It is called **totaliser**. It is in the form of a can. To accommodate 1220 mm of rain upper and lower diameter of the can is kept 203 mm and 610 mm respectively. Since, the observer goes at a longer interval it is essential to provide some arrangement to minimise evaporation loss. Generally a wind screen is mounted on the can to stop evaporation loss. Some times a thin layer of oil is also kept floating at the water surface in the can to reduce evaporation loss.

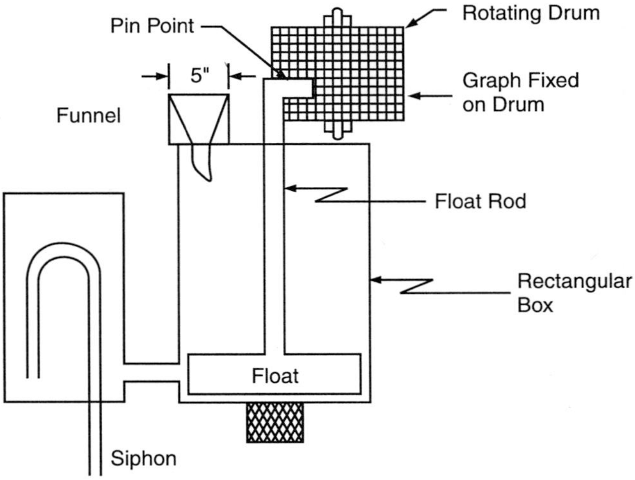

Fig. 2.5. Recording type rain-gauge

2.5.2. Recording Type Rain-Gauge

The recording gauge consists of a funnel 127 mm in diameter fixed on one side of a rectangular box. It is called receiver also. In the rectangular box a float is adjusted. The float is connected by means of a float rod to a pin point (or a recording pen). The pin point touches a graph paper mounted on a rotating drum. The drum is mounted on the top of the receiver on the other side. A clock work arrangement revolves the drum once in 24 hours.

At the bottom the box is connected to a siphon. The siphon comes into action and releases the water as soon as box is filled to a certain level. Figure 2.5 shows complete arrangement. it is called natural siphon type recording rain-gauge.

As the rainfall starts rain water passes through the funnel into the box. As the water level in the box rises the float is also raised. In turn the pin point moves on the graph to plot a mass curve of rainfall.

When the box is filled to such an extent that the float touches the top, the siphon starts working and the rainwater collected in the box is drained out.

Mass curve principle of integrating rain-gauge. The recording type rain-gauge is also called integrating rain-gauge. The reason is that the curve obtained on the graph is a cumulative curve in respect of rainfall. On y-axis we get accumulated or integrated rainfall and on x-axis we have equal time increment. This type of curve in which one ordinate gives accumulated values is called a mass curve. On the graph mounted on the rotating drum we get the mass curve of rainfall (Fig. 2.6).

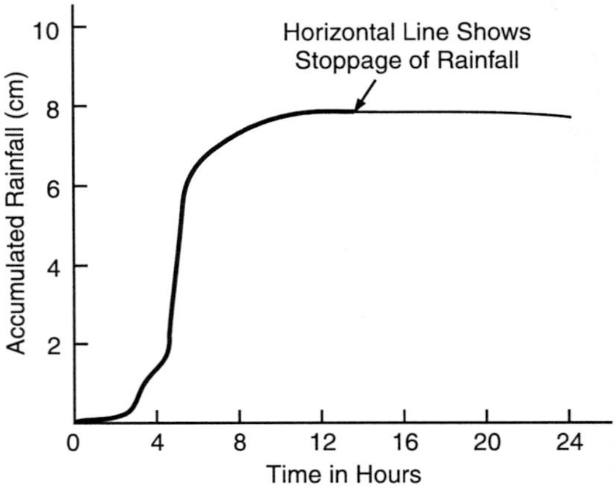

Fig. 2.6. Mass curve of rainfall

Tipping bucket type rain-gauge for remote recording: To facilitate remote recording of rainfall a new type of rain-gauge is used. It is called tipping bucket type rain-gauge. In this type a pair of tipping buckets is placed below a funnel. The bucket gets filled up by 0.25 mm of rainfall and immediately it tips and empties the water into a chamber below. At that very instant other bucket comes below the funnel to receive rainwater. The tipping of the bucket actuates an electrical circuit which moves a pointer to register the rainfall on a graph. The water collected in the chamber below could also be measured by a measuring jar.

2.6. LOCATION OF SITE FOR RAIN-GAUGING STATIONS

While selecting the site for rain-gauging station following points are taken into consideration:

1. The spot at which rain-gauge is to be installed should be truely representative of the area, of which it is supposed to give depth of rainfall.
2. The rain-gauging station should be easily accessible to the observer at all times.
3. The gauge should be erected on level ground, not upon a slope or a terrace and never on a wall or a roof.
4. All other conditions satisfied, a position sheltered from the wind is preferable to an exposed one. In mountains and near sea coasts it is very essential to ensure that the gauge is not undully exposed to the swept of wind.
5. The gauge should be properly secured by a barbed wire fencing and locking arrangement.
6. The rain-gauging station should not be too close to the buildings or trees etc. The nearness of such objects affects entry of rainfall in the funnel. Its distance from other objects should not be less than twice the height of the object above the rim of the gauge.

2.7. NETWORK OF RAIN-GAUGES

The network of observation stations has to be properly designed to cover all the representative portions of the basin area with due care to the topography, orography so as to get equitable coverage of spatial variations in the basin. The broad guide lines set up for the required network of rain-gauge stations as per World Meteorological Organisation (WMO) is as follows:

For plain areas of the basin:

1 rain-gauge for every 500 km^2.

For hilly areas of the basin:

1 rain-gauge for every 150 km^2.

Installation of self-recording gauge also needs due importance in a network of rain-gauge stations of the basin. Generally not less than 10% of the stations should have self-recording rain-gauges for proper evaluation of storm. The number of rain gauging stations in any basin will obviously depend upon extent of area in addition to the factors mentioned above. The number of rain-gauging stations for various sizes of the basin are given below to give broad indication (Table 2.1).

Table 2.1.

Area of basin (km^2)	*Number of rain gauges*
Less than 125	1
125 to 250	2
250 to 500	3
500 to 1000	4

2.8. USES OF RAINFALL RECORDS

At each rain gauging station the rainfall is measured after 24 hours. Usually the measurement is taken at 0830 hours IST obviously total rainfall occurred in the past 24 hours is entered against the date on which measurement is done.

The rainfall records are maintained on daily, monthly, seasonally or yearly basis for any basin. The rainfall varies from year-to-year. The average of the series of yearly records gives mean rainfall value. The long term mean is called the normal rainfall. It is not possible to determine normal rainfall from the rainfall records covering short period. The question now arises is how long should be the rainfall series for obtaining useful results? After the extensive studies Alexander Binnie concluded that the average percentage of deviation from the true mean for 5 years records was ± 15 per cent. For 30 years records it was found to be ± 2 per cent and remained so for longer period also. Thus for obtaining satisfactory results the length of record should be at least for 30 past years.

Following are the main uses of the rainfall records:

1. The trend of rainfall can be studied from rainfall records. Knowing the trend of rainfall future predictions can be done.
2. Runoff over the basin can be calculated.
3. Maximum flood due to any storm can be calculated, and predicted.
4. Rainfall records help in estimating irrigation requirements.

2.9. VARIATION OF RAINFALL

Factors responsible for inequitable distribution of rainfall over large area are the following:

1. **Nearness to sea:** From the sea very large quantity of water goes to the atmosphere in the form of vapour. Naturally when excessively moisture laden clouds pass over the sea coast, clouds drop off some of their load. As a result coastal area receives more rainfall.
2. **Presence of mountains:** It is already mentioned that windward side slope of the side towards which clouds travel gets excessive rains whereas on the other or lee-ward side slope there is area of rain-shadow. Mountainous region receives more rainfall than plain areas.
3. **Direction of wind:** Clouds are driven by wind. It is clear that the area over which wind brings clouds will get rainfall.
4. **Development of forest:** The forests also behave to some extent as a barrier and intercept the clouds to derive rainfall. The area with thick forest gets more rainfall.
5. **Height of a place above sea level or altitude:** The places of high altitude receive more precipitation. At high altitudes temperature of atmosphere is low and when clouds reach that area they get cooled and precipitation occurs.

2.10. AVERAGE RAINFALL IN INDIA

On Indian sub-continent the rainfall is not equitably distributed. The reasons for this inequitable distribution are now obvious. The yearly rainfall actually varies from 125 mm in pockets of western Rajasthan to about 10,000 mm in Cherapunji in Assam. Average manual rainfall in Indian States is given in Table 2.2. India receives an average 890 mm rain in the monsoon season.

2.11. CONSISTENCY OF RAINFALL DATA

It is already mentioned that as the area of the basin goes on increasing the number of rain-gauging stations is also increased. Each station is located in such a way that it represents a particular part of the total area. The total area is selected in such a way that the storm and basin characteristics are same and as a result rainfall data is consistent at all the stations. Sometimes because of certain reasons the rainfall data may not be consistent. The reasons for the lack of consistency may be any one or more of the following:

- Shift in a rain recording gauge station.
- Change in the method of exposure.
- Change in the surroundings.
- Instrumental error.
- Personal error.
- Accidental error.

Table 2.2.

State	*Rainfall in mm*
Rayalaseema region (A.P.)	700
Rest of Andhra Pradesh	1000
Assam (including Manipur, Tripura, Meghalaya, Nagaland and Arunachal Pradesh)	2500 – 3000
Bihar (including Jharkhand)	1000 – 1500
Gujarat region	1000
Kutch and Saurashtra (Guj.)	500 and below
Kerala	3000
Madhya Pradesh (including Chhattisgarh)	1000 – 2000
Western Ghats (Mah & Kar.)	2500 – 3800
Rest Maharashtra	700 – 1100
Rest Karnataka	1000
Orissa	1500
Punjab (including Haryana, Himachal Pradesh and Jammu and Kashmir)	600 – 1000
Western Rajasthan	150 – 300
Eastern Rajasthan	500 – 700
Tamil Nadu	1000
Uttar Pradesh (including Uttarakhand)	900 – 1300
Gangetic West Bengal	1500
Sub-Himalayan West Bengal	3000

To avoid the lack of consistency in the rainfall data above mentioned causes should be avoided so far as possible. It has been noted that if each area contains at least one self-recording gauge station then it can be used as a standard for the corrections to be done in the rainfall values of other stations. Sometimes rainfall data can be made consistent by adopting double mass curve technique. In short the procedure is as follows:

Rainfall data of at least 6 rain gauging stations is taken. The stations should be meterologically similar. Then double mass curve is plotted. It is a graph in which abscissa represents mean cumulative precipitation of all the stations under consideration. Ordinate represents cumulative precipitation of a station whose consistency is to be checked. If the curve comes out to be straight line then the rainfall data of the station under consideration is consistent. This is based on the principle that the mean accumulated precipitation of a number of stations which are meteorologically similar is not materially affected. If the curve is not a straight line data is inconsistent. Knowing the readings which are fictitious necessary corrections can be made in the data.

Once it is ascertained that the rainfall data is consistent then general tendency or trend of the rainfall values can be satisfactorily studied to forecast future rainfall that may occur. This can also be done by plotting what is known as rainfall frequency curve. By frequency we mean the average time interval that elapses between successive occurrences of rainfall of same depth or value.

2.12. AVERAGE OR MEAN DEPTH OF PRECIPITATION

In a particular basin if the rain gauging stations installed are more than one then while finding out the quantity of rainfall, question may arise as to which value of rainfall should be taken. From the existing values no one can be chosen for the purpose because each rain gauging station has its own domain to which it correctly represents. Therefore, it is essential to find one such value of rainfall which represents the whole area. When there is only one rain-gauging station then the value of rainfall recorded by the gauge will obviously by the average value. There are three main methods of calculating average depth of precipitation upon the area of the basin.

- Arithmetic mean
- Theissen polygon method
- Iso-hyetal method

2.12.1. Arithmetic Mean

When the area of the basin is less than 500 km^2 this method implies summing up of all the rainfall values from all the raingauging stations and then dividing it by the number of stations in that basin. The method becomes very clear by the use of a tabular forra.

Table 2.3.

Rain gauging station	*Rainfall in cm*
A	5.6
B	4.9
C	5.2
D	5.5
Σ 4	Σ 21.2

$$\text{Now average depth} = \frac{\Sigma \text{ Rainfall values}}{\text{Number of stations}} = \frac{21.2}{4} = 5.3 \text{ cm.}$$

To explain, there are in all four rain-gauging stations A, B, C, D in the basin, whose rainfall values are given in the table 2.3. Sum of the rainfall values comes out to be 21.2 cm. It is divided by number of stations to give average depth of precipitation which comes out to be 5.3 cm.

This method gives accurate results if the stations are uniformly distributed over the area. There should not be much variation in the rainfall values of the stations under consideration.

Drawback of this method is the stations just outside of the basin are not considered although these stations might have some influence on the basin under consideration.

2.12.2. Theissen Polygon Method

This method is very accurate. It is used for basins having areas between 500 – 5000 km^2.

This method can be best understood with the help of Fig. 2.7.

The firm closing line shows a basin having an area between 500 to 5000 sq km. Let there be rain-gauging stations A, B, C, D, E and F. It is assumed that each stations has its own domain in the total area. While finding out the average value of rainfall it is very essential to divide the total basin area in such a way that each station enclosed in a particular area represents that area in true sense.

The domain of each rain recording gauge station can be marked as now mentioned here. Join all the stations to each of the adjacent stations by dotted line so as to form a system of triangles. Rain-gauging stations form vertices of the triangles. Then draw the perpendicular bisector of each of the sides of all the triangles. In Fig. 2.7 triangles are shown by dotted lines and perpendicular bisectors by firm lines. As a result the whole basin area is divided into number of polygons. Remarkable thing is one polygon enclose only one rain-gauging station. Each polygon is the domain of the rain-gauging station which is enclosed in it. The justification for it can now be given. Each firm line is perpendicular bisector of the line joining two stations. So any point on this line will be equidistant form both the stations. If we go slightly this or that side of the bisector our position will distinctly fall in the domain of that station to which our position is now nearer.

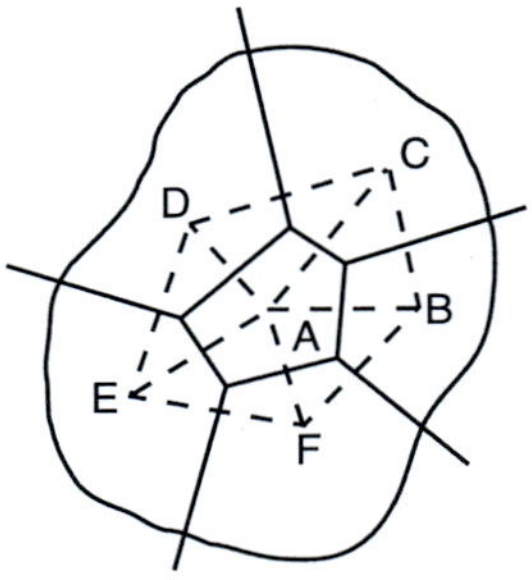

Fig. 2.7. Theissen polygon method

Naturally perpendicular bisector will mark the boundary of the domain. As all the sides of the polygons for all the stations are perpendicular bisectors the new polygons system drawn by firm lines in Fig. 2.7 represents the domain of various stations.

Thus the domain of each station can be plotted. Then the area of each domain can be found by use of a graph paper or a planimeter.

The values can now be tabulated as shown below:

Table 2.4.

Station	*Rainfall in cm*	*Area of Polygon in sq km*	*Weighted depth of rainfall*
1	*2*	*3*	*4 = (2) × (3)*
A	5.6	a	$5.6 \times a$
B	4.9	b	$4.9 \times b$
C	5.2	c	$5.2 \times c$
D	5.4	d	$5.4 \times d$
E	5.5	e	$5.5 \times e$
F	5.2	f	$5.2 \times f$
		Σ	Σ

To explain the procedure, column: (*i*) shows the various rain-gauging stations, column (*ii*) the amount of rainfall at each station, column, (*iii*) gives area of each polygonal domain of the stations and column, (*iv*) gives weighted depth of rainfall which is obtained by multiplying values in columns 2 and 3.

$$\text{Now, mean depth of rainfall} = \frac{\Sigma \text{ column No. 4}}{\Sigma \text{ column No. 3}}$$

$$\Sigma \text{ Column number 3} = \text{Total area of basin} = a + b + c + d + e + f$$

$$\therefore \text{ Mean depth of rainfall} = \frac{5.6a + 4.9b + 5.2c + 5.4d \times 5.5e + 5.2f}{a + b + c + d + e + f}$$

2.12.3. Iso-Hyetal Method

As contours are lines joining points of equal height, **iso-hyetes** are the lines joining the points of equal depth of precipitation. The properties of iso-hyetes are similar to that of contours. For example:

- two different iso-hyetes do not cross each other;
- iso-hyet of higher value shows the places which receive more rainfall;
- each iso-hyet must close on itself or must go out of the area under consideration.

Iso-hyetal method is used for basins having area more than 5000 km^2.

For a given basin iso-hyetes are drawn by joining the points of equal depth of precipitation as shown in Fig. 2.8. The points of equal depth of precipitation can be computed by the method of estimation from the rainfall values of rain-gauging stations.

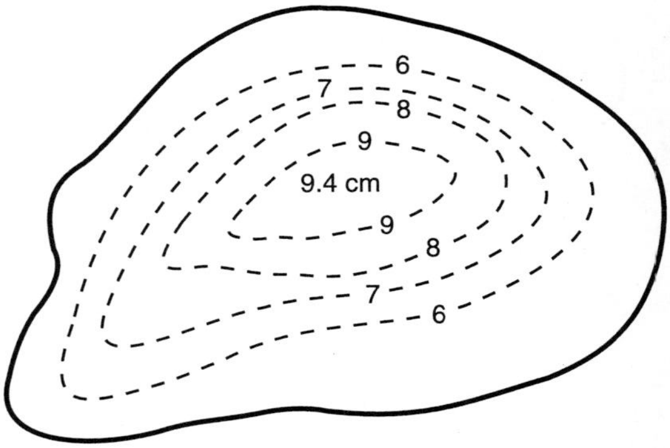

Fig. 2.8. Iso-hyetal diagram for a basin

In Fig. 2.8 dotted lines show iso-hyetes and the outer most firm line is basin boundary. The interval of iso-hyetes is 1 cm. The highest spot rainfall value in area is 9.4 cm. Now the areas between two successive iso-hytes can be found by using a graph paper or a planimeter.

The rest of the procedure of finding out average or mean depth of rainfall is done by tabulating the values as shown in Table 2.5.

Table 2.5.

Iso-hyetal interval	*Mean*	*Area between two successive iso-hyetes*	*Mean × area*
(1)	*(2)*	*(3)*	*(4) = (2) × (3)*
9.4 – 9	9.2	a	$9.2 \times a$
9 – 8	8.5	b	$8.5 \times b$
8 – 7	7.5	c	$7.5 \times c$
7 – 6	6.5	d	$6.5 \times d$
6 – boundary	5.5	e	$5.5 \times e$
		Σ	Σ

To explain, column (1) shows the iso-hyetal interval of successive iso-hyetes, column (2) gives the average of the two extreme values of interval, column (3) gives the area enclosed between two successive iso-hyetes and column (4) shows mean of interval multiplied by the area of interval.

Now,

$$\text{Mean depth of rainfall} = \frac{\Sigma \text{ column No. 4}}{\Sigma \text{ column No. 3}}$$

$$\Sigma \text{ column number } 3 = \text{Total area of basin} = a + b + c + d + e$$

$$\therefore \text{ Mean depth of rainfall} = \frac{9.2a + 8.5b + 7.5c + 6.5d \times 5.5e}{a + b + c + d + e}$$

PROBLEM 2.1. A drainage basin has the catchment area of 626 km^2. There are in all 11 rain-gauging stations of which 6 are within the catchment and 5 are in the vicinity but outside the catchment. The point rainfall observed during a particular storm at various stations has been shown in Fig. 2.9.(*a*).

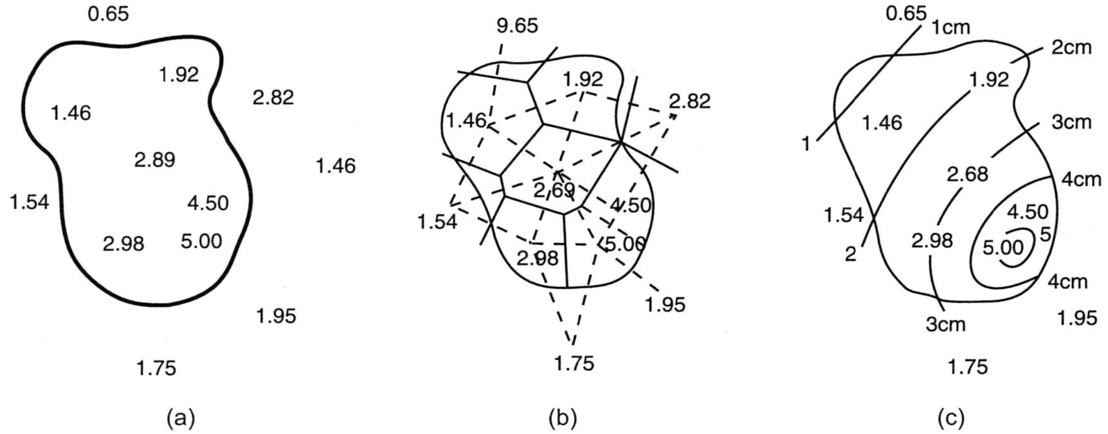

Fig. 2.9. Problem

Figure 2.9. (*b*). It is given that the areas of the polygons and the rainfall station with its value is as follows:

Rainfall (cm)	0.65	1.46	1.92	2.69	1.54	2.98	5.00	4.50
Area of polygon (km^2)	7	120	109	120	20	92	82	76

Refer Fig. 2.9 (*c*). It is given that the areas enclosed within isohyets are as follows:

Isohyet	5	4	3	2	1
Area enclosed (km^2)	13	90	206	402	626

Calculate average depth of precipitation over the catchment by arithmetic mean, Thiessen polygon and Isohyetal methods and compare the results.

Solution: Step 1: Arithmetic average method.

Referring Fig. 9(*a*). There are 6 stations within the catchment.

$$\therefore \text{ Average depth of ppt.} = \frac{1.46 + 1.92 + 2.69 + 4.50 + 2.98 + 5.00}{6} = 3.09 \text{ cm}$$

Step 2: Thiessen polygon method,

Referring to Fig. 9(*b*) the precipitation and areas can be tabulated as in Table 2.3.

$\therefore$ Average depth of ppt. = 2.84 cm

Step 3: Isohyetal method

Referring to Fig. 9(*c*) the iso-hyetal interval and area enclosed can be tabulated as follows to work out the average depth as on in Table 2.4.

$$\text{Average depth of precipitation} = \frac{\text{Volume of ppt.}}{\text{Area of catchment}} = \frac{1622}{626} = 2.59 \text{ cm}$$

Answer: Average depth of precipitation by the three methods come out to be :

Arithmetic mean method = 3.09 cm

Thiessen polygon method = 2.84 cm

Isohyetal method = 2.59 cm

Table 2.6.

Observed ppt. (cm)	*Polygon Area*	*% of Total area (km^2)*	*Weighted ppt (cm) col. 1 × col. 3*
1	*2*	*3*	*4*
0.65	7	1	0.01
1.46	120	19	0.28
1.92	109	18	0.35
2.69	120	19	0.51
1.54	20	3	0.05
2.98	92	15	0.45
5.00	82	13	0.65
4.50	76	12	0.54
Total	626	100	2.84

Table 2.7.

Isohyet interval (cm)	*Enclosed area (km^2)*	*Net area (km^2)*	*Average ppt. (cm)*	*Volume of ppt. col. 3 × col. 4*
1	*2*	*3*	*4*	*5*
5	13	13	5.0	65
4	90	77	4.5	346
3	206	116	3.5	406
2	402	196	2.5	490
1	595	193	1.5	290
		31	0.8	25
Total	626	626		1622

2.13. HYETOGRAPH OF RAINFALL

The hyetograph shows the average rainfall rates, over the specified drainage catchment, during successive units of time during a particular storm (Fig. 2.10).

To prepare the hyetograph from given storm, the rainfall quantities during successive units of time are measured from the mass curves of stations in and near the drainage basin. A unit time of 1 to 6 hours as convenient may be selected. The average rainfall depths over the basin for successive units of time are computed from the tabulated data by the Theissen polygon method or isohyetal method. The hyetograph is then drawn by plotting the average rainfall depth per unit of time as shown in Fig. 2.10.

The hyetograph is very convenient in relating the rainfall over the basin with the resulting flood hydrograph. It is usually plotted on the same sheet where hydrograph is plotted. Only thing is it plotted upside down whereas hydrograph is plotted errect (Fig. 2.11).

The hyetograph of a storm when plotted by the side of the flood hydrograph gives the time lag between rainfall and peak flow. It also gives important information about effective duration of storm which produces run-off. The use of hyetograph is commonly made in flood estimation by unit hydrograph method.

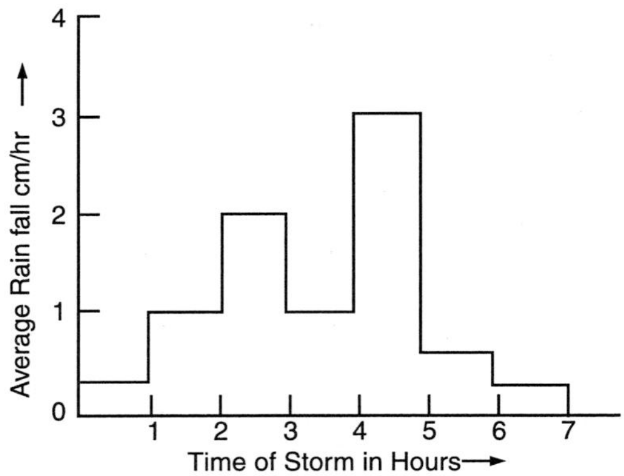

Fig. 2.10. Hyetograph of rainfall

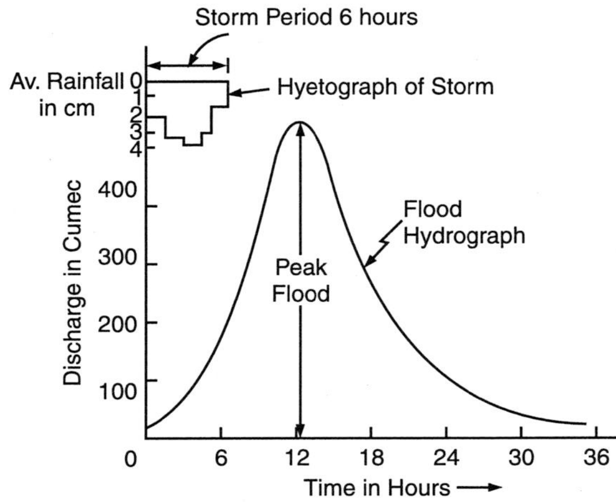

Fig. 2.11. Rainfall-peak flow relation

2.14. ESTIMATION OF MISSING RAINFALL DATA

It is a matter of common experience that due to failure of the instruments or absence of a gauge reader the rainfall record at quite a number of stations is incomplete. In such cases it is necessary for a Hydrologist to estimate the missing rainfall data to accepted accuracy. There are two methods by which estimation can be done. They are:

(*i*) Simple arithmetic average method; and

(*ii*) Method of weightage or normal ratio method.

(*i*) Simple arithmetic average method: In this method simultaneous rainfall records of three closeby stations are made use of. The stations should however be evenly spaced around the station with missing records. A simple arithmetic average of the rainfall of the three selected stations gives the estimate of the missing value. This method can be used to calculate monthly as well as annual missing rainfall values. This method should be used only when normal annual precipitation at each of the selected stations is within 10% of that station for which records are missing.

(*ii*) Method of weightage or normal ratio method: When the normal annual rainfall of any of the selected stations is more than 10% of that station with missing records simple average method cannot be used, then the method to be adopted consists of weighting the rainfall value by the ratios of the normal annual rainfall values. To make the procedure clear–

Let, x be a station with missing records

A, B, and C are three selected stations from homogeneous area

P is rainfall value of certain period under consideration

N is normal annual rainfall.

Then, $$P_x = \frac{1}{3}\left(\frac{N_x}{N_A}\cdot p_A + \frac{N_x}{N_B}\cdot P_B + \frac{N_x}{N_C}\cdot P_C\right)$$

or $$\frac{P_x}{N_x} = \frac{1}{3}\left(\frac{P_A}{N_A} + \frac{P_B}{N_B} + \frac{P_C}{N_C}\right)$$

This method is quite useful in regions where large orographic variation occurs in precipitation.

PROBLEM 2.2. Calculate for station *K* the missing rainfall data for the month of July 83 when the rainfall data for other three stations J, P and N are available as follows:

Station	*Normal annual rainfall (cm)*	*July 83 rainfall (cm)*
K	26.2	?
J	35.0	14.9
P	28.7	4.9
N	30.2	13.1

Solution: Using the formula

$$\frac{\text{July 83 ppt. of } K}{\text{Normal annual ppt. of } K} = \frac{1}{3}\left(\frac{P_J}{N_J} + \frac{P_P}{N_P} + \frac{P_N}{N_N}\right) = \frac{1}{3}\left(\frac{14.9}{35.0} + \frac{4.9}{28.7} + \frac{13.1}{30.2}\right)$$

$$\text{or July 83 ppt. of } K = \frac{26.2}{3}\left(\frac{14.9}{35.0} + \frac{4.9}{28.7} + \frac{13.1}{30.2}\right) = 9.2 \text{ cm}$$

2.15. EXTENSION OF RAINFALL SERIES

Extension of rainfall series is done by what is called a Station-year method. This is a method of extending the length of record for a frequency curve at a station. It is based on the assumption that the records for the same or different periods at a number of stations may be considered as a composite record for a single station for a period equal to the total number of years involved. For example, if 50 years record was available for 100 stations it might be assumed that this was equivalent of 5000 years at a single station.

Limitations:

1. This method cannot be applied very rigorously unless it is expected, that over very long periods of time say thousands of years, the stations will have very nearly the same frequency distribution. In that case the stations are said to be meteorogically homogeneous. This requirement of homogeneity, practically excludes use of the method in mountainous areas.
2. If the maximum amounts at different stations tend to occur in the same storms or even in the same group of storms or in any other non-random way the method is not strictly applicable.

In spite of the limitations it is probably as reliable a method as can be developed for estimating the amounts of rainfall for a station for return periods that are several times the length of record at any of the stations. However, it has to be used cleverly knowing pitfalls involved in statistical analysis.

In areas where maximum amounts are associated with geographically separated intense local storms, the method may be considered more applicable.

2.16. DEPTH-AREA-DURATION CURVES

Once the sufficient rainfall records for the region are collected the basic or raw data can be analysed and processed to produce useful information in the form of curves or statistical values for use in the planning of water resources development projects.

Many hydrologic problems require an analysis of time as well as areal distribution of storm rainfall. Depth-Area-Duration (DAD) analysis of a storm is done to determine the maximum amounts of rainfall within various durations over areas of various sizes. The preparation of DAD curves is done in following steps:

1. Examine the rainfall records of the region in which catchment area under consideration is located. Also consider records of meteorologically similar regions. From it prepare a list of most severe storms with their dates of occurrence and duration.
2. For the listed severe storms prepare isohyetal maps and determine the rainfall values over the area of each isohyet (rainfall contour).
3. Draw on a graph curves connecting area and rainfall values for different durations say 1 day rainfall, 2 day rainfall, 3 days rainfall (Refer Fig. 2.12).

The curves shown in Fig. 2.12 are called **Depth-Area-Duration** curves.

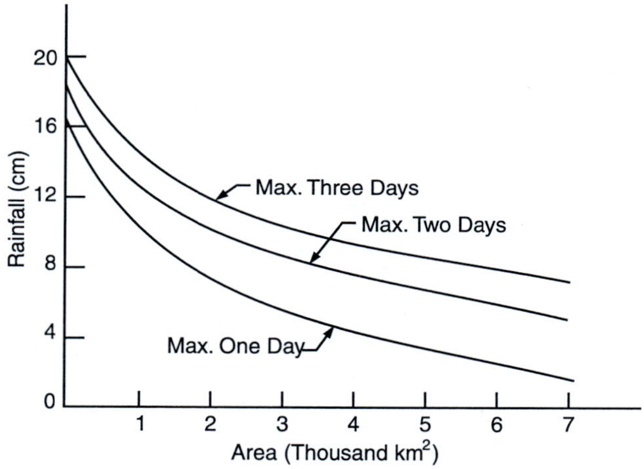

Fig. 2.12. Depth-area-duration curves

Use of DAD curves: Although most severe storm in the listed storms may not have occurred right over the catchment under consideration there is possibility of such occurrence. So from DAD curves 1 day, 2 day, 3 day rainfall depths for the catchment area of the proposed project are read. These give the rainfall depths when the storms is centered over the catchment.

20.17. PROBABLE MAXIMUM PRECIPITATION (PMP)

Hydraulic structures like dams are designed to withstand calamitous precipitation resulting from unusually worst weather condition which is capable of producing severest storm over the basin area. Statistically probable maximum precipitation (PMP) occurs very rarely but has to be taken into consideration while designing important hydraulic structures. Estimation of PMP require thorough knowledge of hydrometeorology and sufficient record of storm rainfall. In addition other meteorological parameters like humidity, temperature, wind velocity etc., have to be available for study. Then PMP is estimated by maximising these parameters over the given area.

Many times such data is not available for study. In such situations the estimate of PMP is done for a neighbouring or similar basin area where required data may be available. Then it is transposed on the basin area under study with suitable care.

Statistically PMP may be estimated using following equation:

$$PMP = \bar{P} + K\sigma$$

where $\bar{P}$ is mean of annual maximum rainfall series;

K is frequency factor (normally equals 15); and

σ is standard deviation of the rainfall series.

India Meteorological Department (IMD) has prepared suitable charts for the whole country comprising isohyetal lines showing maximum storm depths, absolute for any frequency, that may result from severest storm of given duration. From these charts maximum rainfall for given duration can be obtained. These charts serve the purpose of guiding curves. However, services of expert hydro-meteorologist are necessary to derive PMP after analysing available data.

20.18. INTENSITY DURATION FREQUENCY CURVES

In hydrology, frequency analysis of station rainfall data is done for use in design of bridges and culverts on highways, design of storm drains etc. With the advancement of science of hydrology rainfall frequency analysis is done using Gumble's extreme-value distribution and annual series data. Now the frequency analysis concept is applied on a seasonal basis and for areal frequency. The rainfall records of deficient length have to be extended by station year method. The results of frequency analysis are plotted on the log-log paper. The typical intensity-duration frequency curves are given in Fig. 2.13.

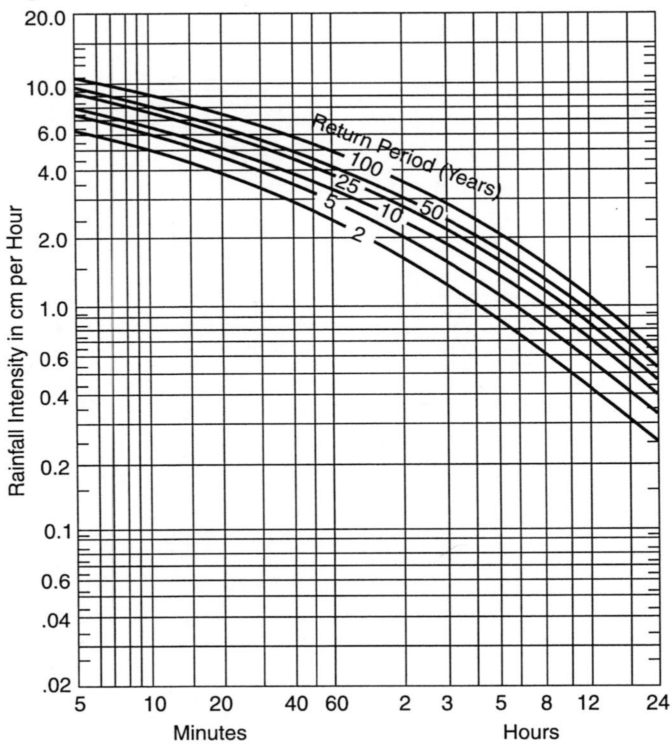

Fig. 2.13. Typical intensity duration frequency curves

2.19. MEASUREMENT OF SNOW

Since snowfall is also one of the forms of precipitation it is also measured at the time of fall and water equivalent of the snow is included in the total precipitation amounts of a particular station to compile seasonal and yearly records. Measurement of water equivalent of snow at the time of fall can be done in two ways.

1. **Measurement of falling snow by catching it in the way:** It is done by means of a non-recording rain-gauge type snow gauge of bigger diameter (say 203 mm). For snow measurements, however, the funnel and bottle receiver is ordinarily removed and only a sharp edged cylindrical metal receiver of 203 mm diameter and 60 cm height is used to collect the falling snow. The receiver is installed over a tower within a Nipher pattern wind-shield of 75 cm diameter. The height of the tower is so adjusted that the top of the receiver is about 1.8 m above ground level. A metal funnel with measuring glass is supplied with the snow gauge (Fig. 2.14). The wind-shield consists of a cone, with 75 cm diameter at top joined to hollow vertical cylinder to house the receiver.

   The collected snow is then brought in a warm room. The sample collected in the receiver is melted by adding measured quantity of hot water. The water from melted snow is measured in the usual manner alongwith hot water. From it water equivalent of snowfall can be calculated.

2. **Measurement of snow deposited on ground:** This can be done by use of snow boards or snow stakes.

Snow boards: They are nothing but square boards of 40 cm × 40 cm size used to collect the falling snow. The snow boards can be placed on the previous accumulations of snow. Thus only freshly fallen snow is identified and collected. The snow samples can be cut from the board. This method can only used in places where strong winds do not drift fallen snow and accumulated snow truely indicates snowfall at that site. In practice, however, it is difficult to select such a location and therefore this method is not considered reliable. To obtain representive data it is necessary to install several such boards.

Snow stakes: They are nothing but wooden gauges square in section of say 4 to 6 cm square section. They are erected on ground supported by angle-iron supports. They are marked in tenths of metre with convenient subdivision to indicate the vertical depth of freshly fallen snow on the ground. Thus, depth of unmelted fresh snow can be directly measured on ground. Assuming average density of fresh snow to be 0.1 water equivalent of snowfall can be calculated.

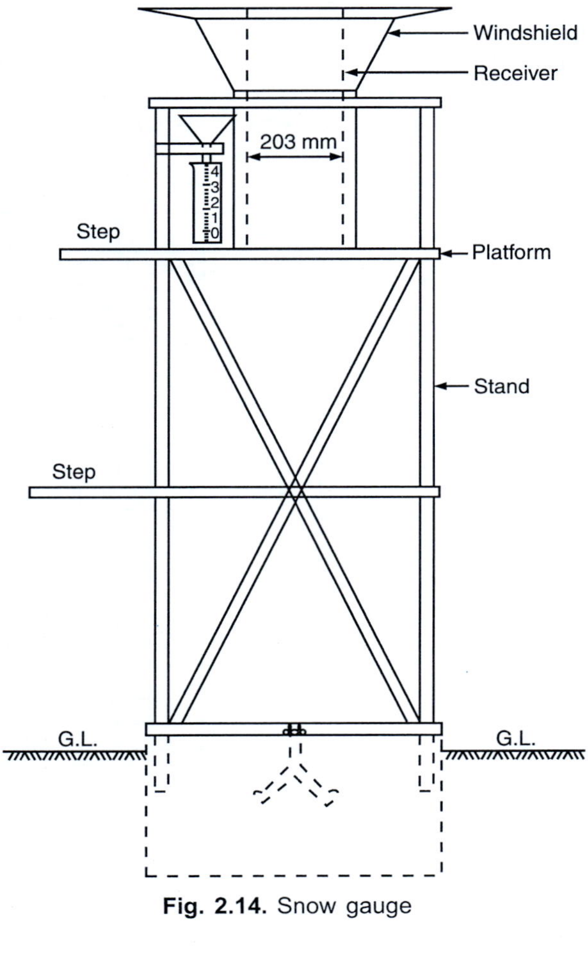

Fig. 2.14. Snow gauge

These two methods, however, suffer from the following two disadvantages:

(*a*) Density of snow varies with depth and texture of snow.

(*b*) In strong winds the results are unreliable because the fallen snow is likely to be flown away.

Considering the above disadvantages first method of precipitation gauge appears more accurate.

2.20. SNOW SURVEYS

Unlike rainfall heavy snowfalls of hydrologic importance occur on the higher altitudes in sparsely inhabited places. Also density of snow which is a ratio of volume of melted water from sample of snow to initial volume of snow sample varies from 0.004 for fresh snow at high altitudes to 0.91 for compacted snow of glaciers. In the regions of heavy snowfall, by the spring time the snow gets compacted and its density varies from 0.4 to 0.6. Further accumulation of snow is of greater interest in hydrologic sense than pure precipitation because this information helps water users in number of ways. For example:

- Irrigation engineer is benefitted in estimating total quantity of water that may be available to him;
- Designer or flood control engineer can estimate component of snow-melt in peak rates of run-off and its distribution.
- Hydroelectric power system engineer can get correct assessment of minimum flows that may be available.

Snow surveys are essential to understand following characteristics of snow pack at different places at desired time:

- depth of snow,
- density of snow,
- water equivalent of snow.

Layout of snow course network: The snow surveys are conducted on the network of selected sampling points. The chains of sampling points are called snow courses. The network could correspond to any geometric pattern as a result of sampling points. The network may be rectangular, radial or triangular in shape. Before initiating snow surveys reconnaissance is done during winter to decide the coverage of area by the snow courses. The sampling points are however, selected during fair weather when the area is clear. The interval between sampling points may vary from 15 to 30 m. One network may cover an area of 250 to 1000 km^3. Each selected sampling point is properly levelled and cleared. At each sampling point markers or reference points are errected. The height of the markers should be such that they remain visible even after maximum amount of snow falls. The marker is something like a name plate fixed on some high object like a tree or a post specially erected for the purpose. Once the snow courses are laid it is possible to collect snow pack data year after year at the same locations. At least 10 years record is considered essential to do useful analysis.

Snow sampling procedure: Snow survey is performed by taking samples of snow pack with a suitable core cutting equipment. Water equivalent of the sampled snow is determined from its weight and not melting the snow because it is tedious to melt snow in the field. The snow sampling equipment consists of the following accessories:

- A telescopic sampling tube 38 to 90 mm in diameter with toothed cutting edge. It has holes for inspecting the core length cut from snow pack.
- A handle to rotate the sampling tube.
- Weighing scale with sample carrier.

- Sounding rod.
- Miscellaneous tools, cleaning equipment and hard wax or parafin to avoid plugging of the tube when the temperatures are much below the freezing point.

Before snow core is cut from the sampling point, with the help of sounding rod depth of snow is measured in the vicinity of the sampling point to ascertain snow depth, layers and compactness of snow pack. Then the core is cut by turning the sampling tube in the clockwise direction till it reaches the ground level. The length of the sampler can be increased by screwing the pieces together. Once the core is cut the sampler is weighted alongwith the snow core. By deducting the weight of the sampler weight of snow core and its water equivalent is determined. The recording of the data is done simultaneously.

It may be seen that snow survey is a strainuous process and requires experienced and healthy surveyors who can withstand hazardous weather conditions. The surveyor is equipped with special kit to negotiate snow covered areas. The depths of snow at inaccessible points can be measured by using field glasses or telescopes. The gauges are graduated by cross bars at every 50 cm height. In developed countries now help of radio isotopic snow gauges, over-snow vehicles, helicopters, aerial photography etc. is progressively taken to perform snow surveys.

Uses of snow survey data: The snow survey records are analysed to gain useful information required for operation of existing projects and for designing future projects. Some of the important aspects evolved from the analysis are:

- Snow lines determination;
- Seasonal water yield forecasts;
- Assessment of rate of run-off for river-regulation;
- Floods due to snow melt etc.

QUESTIONS FOR REVISION

1. (*a*) What is precipitation, what causes precipitation?
   (*b*) What are different forms of precipitation?
   (*c*) Mention different types of precipitation.
2. Name various methods adopted for finding out mean depth of precipitation. How will you categorise them from the point of view of accuracy and speed in calculation?
3. (*a*) Draw a sketch of recording type rain-gauge and explain its working.
   (*b*) Draw a sketch of a snow gauge and explain its working.
4. (*a*) Discuss the procedure of estimating the missing rainfall data which became in-operative for a short duration.
   (*b*) Explain the principle of extending rainfall records of a rainfall station.
5. Write short notes on the following:
   (*i*) Depth-Area-Duration curves
   (*ii*) Cloud seeding
   (*iii*) Double mass curve technique
   (*iv*) Normal yearly rainfall
   (*v*) Hyetograph of rainfall
6. (*a*) What is meant by snow surveys? Why it is necessary to perform snow surveys?
   (*b*) Explain briefly the snow survey procedure.

3

Interception, Infiltration and Evaporation

3.1. TYPES OF LOSSES

It is now clear that all the water available on the earth is constantly kept in circulation. In the process of circulation the water comes down on the earth as rainfall or snow and finally goes back to the atmosphere in the form of vapour. Also, some water goes deep into the soil and joins groundwater reservoir. But this water is many times available for use. Evaporation of water takes place not only from water surfaces but also from vegetation and soil mass. So we can say that all the losses, that is the water lost from the surface of the earth, are evaporative in final analysis. However, considering immediate causes losses may be recognized in various ways and are mentioned below.

(i) **Infiltration:** It is a process in which rainwater enters the surface strata of the soil and moves downward to join the groundwater reservoir. The water first fulfills the soil moisture deficiency and then the remaining portion goes down. If it encounters a relatively impervious substratum only a port of the absorbed water continues to flow downwards, the remaining starts moving horizontally or rather laterally in the soil itself towards the stream channel. The infiltration affects run-off so its knowledge is necessary for a water resources engineer.

(ii) **Seepage:** The loss of water due to seepage from the water bodies like canals, reservoir and weir ponds is of great concern to water resources engineers. It is so not only because the water is lost but also due to the fact that when the water seeps through canal embankments, body of the dam and weir and the foundations of hydraulic structures it tries to undermine the media through which it flows. Also, seeping water exerts upward pressure thereby endangering the stability of the hydraulic structure.

(iii) **Watershed leakage:** The geological structure of the subsurface below a particular catchment area may be such that part of rain water may find its way through fissures, cracks and outcrops of water bearing strata to some outlet. This outlet may be in some other basin or may be in the sea. This type of loss is called watershed leakage.

Actual amount of loss in any drainage basin depends upon the following factors:

1. Nature of precipitation.
2. Type and development of vegetal cover.
3. The area covered by buildings, pavements and other permanent structures.
4. The climatological factors such as temperature of the area, humidity and wind velocity.

This type of loss assumes serious proportion when it occurs in storage reservoir.

(*iv*) **Interception:** Before the rain starts, due to the heat of the sun the surfaces of leaves, branches and trunks of trees and the stems of vegetation are kept dry and made capable of soaking some more moisture. When the rain starts part of rain water is retained back as the interception storage by the vegetation.

(*v*) **Transpiration:** It is the process in which the water circulates through the plant structure and afterward returns to the atmosphere in the form of vapour from the surfaces of leaves, trunks and branches of trees. The amount of transpiration depends upon temperature, humidity, wind, light and soil moisture.

(*vi*) **Soil evaporation:** The soil moisture is also lost by direct evaporation from the soil mass. This phenomenon is different from the evaporation taking place from water surfaces. It is so because the moisture is available in reducing proportion in the soil pores from the ground water table to the surface of earth. Also the water is held in the soil pores by force of attraction. The soil evaporation increases with the length of dry period and temperature changes. It is also affected by the vegetation grown on the soil.

(*vii*) **Evaporation from water surface:** Owing to the heat energy of sun, water is lost in evaporation from the free surfaces of lakes, ponds, canals, rivers, reservoirs, etc. The loss of water by evaporation is a very serious problem in semiarid and arid regions. The evaporation loss is generally taken as direct function of the exposed area. The rate of evaporation depends upon temperature, wind velocity, relative humidity, sunshine and altitude of the place. Of all the losses mentioned above interception, evaporation including evapotranspiration and infiltration which are of hydrologic importance are dealt with in this chapter in subsequent paragraphs.

3.2. INTERCEPTION

Interception deals with the amount of water that is caught and stored on the leaves and stems of vegetation. The term vegetation includes forest tree cover, crops and low level vegetation like underbush, grass etc. Part of the precipitation which falls either as rainfall or snowfall is caught by vegetative cover before it reaches ground. This is a matter of common experience when the people take shelter below a tree during rainfall. The rain drops or snow flakes are retained by the leaves as droplets or thin layers on their surface or in the depression of the leaves. After this storage capacity is exhausted and the rain or snow starts dripping down some water also flows down along the stems of the tree or the plant.

The rate of interception is high at the beginning of the rain in case of a dense vegetal canopy during summer. As time progresses and if storm continues the rain starts falling through the tree canopy. The water caught by the vegetation gets disposed off in three ways namely:

- through fall;
- flow along the stem; and
- evaporation.

Obviously after the storage capacity is exhausted the rate of interception reduces considerably and is equal to that amount which gets evaporated from the vegetation.

The amount of precipitation intercepted can be measured by placing several rain-gauges below the vegetal canopy on the ground. Average precipitation that reaches this gauge can be compared with the precipitation measured from a rain-gauge placed in an open area. The difference between the two gauge readings gives the precipitation intercepted by the vegetation. It may, however, be remembered that all the intercepted water is not lost in evaporation. Depending upon site conditions some water falls down either as a through fall or a stem flow.

It is estimated that a dense forest cover intercepts about 10 to 25% of the annual precipitation.

3.2.1. Factors Affecting Interception

The amount of water intercepted is greatly variable and depends on many things. Since the interception affects the distribution of rainfall or snowfall and subsequent run-off, it is necessary to understand the factors which affect interception. Following are some of the important factors:

(i) **Type of Vegetation:** Interception varies with the species, its age and density of stands. About 10 to 20% of precipitation occurring in the growing season is intercepted. It is lost substantially by way of evaporation from leaves. In dense tall vegetation interception is quite substantial. So interception by low-lying vegetation is usually negligible for hydrologic studies. But if such vegetation cover exists below the forest canopy it affects runoff substantially. Dense grasses and herbs approaching full growth intercept as much precipitation as forest cover. However, since their season is short total amount intercepted is considerably less than the forest cover.

(ii) **Wind Velocity:** If the wind accompanies the precipitation the leaves become incapable of holding much water as compared with the still air condition. On the other hand due to blowing of wind the evaporation rate also increases. It is thus seen that during short storm interception decreases but during long storm interception is augmented due to increased evaporation.

(iii) **Duration of Storm:** Interception storage is filled up in the first part of a rain storm. Therefore, if yearly precipitation is made up of several small duration storms separated by dry spells evaporation will be high and interception will be consequently more. However, if storms of long duration occur and if weather remains cloudy, relatively interception loss will be less.

(iv) **Intensity of Storm:** When precipitation occurs in still air conditions with low intensity interception will be more. On the contrary if rain drops come with great speed their impact dislodges intercepted drops and leaves can not hold much water.

(v) **Season of the Year:** During summer or dry season the interception rate is quite high because of high evaporation. Summer interception is 2 to 3 times more than the winter season interception.

(vi) **Climate of the Area:** In arid and semiarid regions due to prevailing dry conditions the interception loss is more than that occurring in humid regions.

3.3. INFILTRATION

It is a process in which the water enters into the soil surface. It is different from the term percolation because percolation means movement of water in the soil mass. From the definition it follows that the process of infiltration will stop unless percolation removes infiltrated water. Thus although the two phenomenon viz. infiltration and percolation are different they are closely related.

During the rain infiltration loss occurs quickly almost exclusively from the water that has reached the ground surface. The water infiltrating into the soil moves downward through larger soil pores under the force of gravity. The smaller surface pores take in water by capillarity. The downward moving water is also sucked in by capillary pores. The gravitational water moves towards the ground water following the path of least resistance. When the capillary pores at the surface are filled and intake capacity reduced infiltration rate decreases. As a trend the rate of infiltration is high in the beginning. It decreases rapidly in the initial stages and then slowly till it approaches a nearly constant rate in about 30 to 90 minutes depending upon the type of soil. The typical infiltration curves for soils with different land uses viz. bare soil, soil with grass cover and soil covered with grain crop are shown in Fig. 3.1.

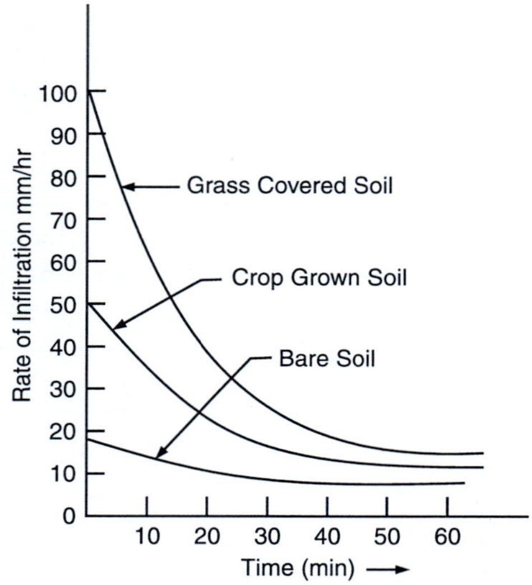

Fig. 3.1. Infiltration curves

This trend of infiltration process is observed not only because of filling up surface capillary pores with water but also due to changes in the soil such as dispersion of aggregates, puddling of the surface layer, impact of the rain drops, swelling of colloids, closing of soil cracks, type of vegetal cover etc.

3.3.1. Factors Affecting Infiltration

The process of infiltration is affected by many different factors. Important factors are the following:

(*i*) **Soil texture and structure:** It is already made clear that the water can not continue to enter soil more rapidly than it is transmitted downward. The conditions at the surface, therefore, can not increase infiltration unless the transmission capacity of the soil profile is adequate. The continuity of non-capillary or large pores provide easy paths for percolating water. If the subsoil formation has coarse texture the water may infiltrate into the soil so quickly that no water will be left for runoff even if rainfall is quite heavy. On the contrary clayey soils after soaking some water in the initial stages of the rainfall may swell considerably. It makes the soil almost watertight and infiltration may get reduced to practically negligible extent.

(*ii*) **Conditions at soil surface:** Even if the subsoil has excellent under drainage but at the surface soil pores are sealed due to turbid water or by in wash of fine soil particles it may prevent entry of water into the soil and infiltration rate will be low.

(*iii*) **Soil-moisture content:** When the soil is fairly dry the rate of infiltration into the soil is quite high. The infiltration rate diminishes as the soil-moisture storage capacity is exhausted. After this infiltration rate equals transmission rate. The rate of infiltration in early phases of a rainfall will be less if the soil pores are still filled from previous rain storm.

(*iv*) **Type of vegetative cover:** Vegetative cover affects surface entry of water significantly. The vegetation or mulches protect the soil surface from impact of rain drops. The lengthy and extensive root system penetrate the soil and increases its porosity. Organic matter from crops promotes a crumbly by structure and improves soil permeability. Forest canopy protects soil surface whereas row crops provide less protection to soil.

(*v*) **Soil temperature:** If saturated soil mass gets frozen due to severe low temperature it becomes nearly impermeable. It affects the infiltration.

(*vi*) **Human activities on soil surface:** If the soil surface gets compacted due to construction of roads, operation of tractors and other farm implements and machinery the porosity of the soil is decreased. As a result bigger pores are almost eliminated making soil impermeable. It reduces the infiltration rates appreciably.

3.4. INFILTRATION INDICES

Various infiltration indices give rates of infiltration in different ways to help assessment of the water lost by way of infiltration. The important among them are the following:

(*i*) **Infiltration capacity:** It is the *maximum rate* at which water can enter through the soil surface at a particular point of time under the given set of conditions. By now it is clear that actual rate of infiltration will be less than infiltration capacity unless the net rainfall rate that reaches the ground after fulfilling retention (*i.e.*, interception + depression storage) is equal or more than infiltration capacity. The infiltration capacity goes on reducing as the soil profile becomes saturated. Like infiltration the infiltration capacity also depends on soil type, moisture content, organic matter present in the soil, vegetal cover and season.

Horton gave the following mathematical expression to find out the value of infiltration capacity at any time:

$$f_p = f_c + (f_o - f_c)\, e^{-Kt}$$

where f_p is infiltration capacity.

f_o is infiltration rate at the beginning of storm.

f_c is constant infiltration rate which is achieved after the soil profile becomes saturated.

e is base of natural logarithms (Napierian base).

t is time from beginning of rainfall and K is a constant.

It may be remembered that this equation can be applied only when rate of net rainfall reaching the surface is more than infiltration capacity throughout the storm rainfall.

(*ii*) **ϕ index:** The ϕ index is that portion of average rate of rainfall during any storm which gets lost by the processes of interception, depression storage and infiltration taken together. It can, therefore, be defined as that rate of average rainfall during any storm beyond which the volume of remaining rainfall equals the volume of direct surface runoff. The index can be calculated from a hyetograph (time versus intensity of rainfall graph) of the storm in such a way

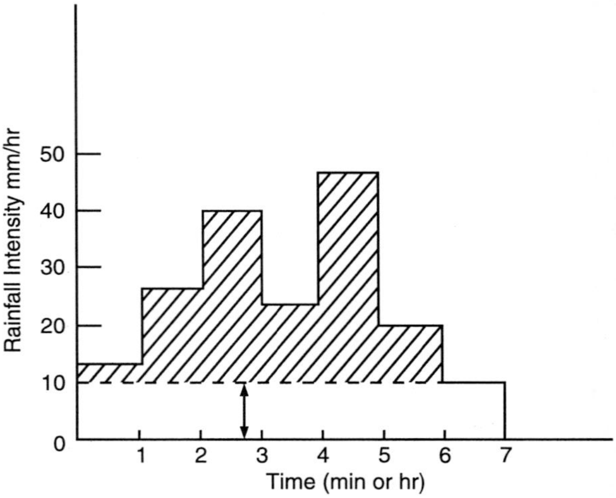

Fig. 3.2. ϕ index from hyetograph

that the rainfall volume in excess of this rate will equal the volume of the storm runoff Fig. 3.2.

If the rainfall intensity throughout the storm remains equal to or more than ϕ index then the ϕ index represents basin recharge because ϕ index represents sum total of infiltration, interception and depression storage.

(iii) **W index:** This index gives the average rate of infiltration for that time period of the storm rainfall during which rainfall intensity is greater than W. Thus it can be said to be refinement over ϕ index which apart from infiltration also includes interception and depression storage. The W index can be obtained from the following equation:

$$W = \frac{P - Q - S}{t}$$

where W is average rate of infiltration

P is total storm rainfall corresponding to t

Q is total storm run-off.

t is time during which rainfall intensity is more than W and

S is effective surface retention.

$$\therefore \quad W = \phi - \text{average rate of retention}$$

where retention includes interception and depression storage.

For all practical purposes ϕ index can be taken to represent average rate of infiltration.

Since ϕ and W indices assume average rate of infiltration which in fact is less than initial infiltration rate and more than ultimate infiltration rate their utility is limited to major flood producing storms. Such storms generally occur on wet soil and storms are of such intensity and duration that the infiltration rate could be very nearly taken to be constant for whole storm or majority period of storm. Obviously for short isolated storms ϕ and W indices are not useful.

PROBLEM 3.1. A drainage basin has a catchment area of 0.5 km². A five hour storm has occurred on the basin with the following rainfall intensities:

Hour	1	2	3	4	5
Rainfall (mm)	5	11	23	18	3

The volume of direct surface run-off observed as a result of this storm at the outlet of the basin was 0.232 cumec-day.

Calculate ϕ index for the basin.

Solution: Step 1: Using the data given rainfall hyetograph can be drawn as shown in Fig. 3.3.

The line XX shows average rate of infiltration and surface retention together, i.e., ϕ index.

We have to find out value of ϕ.

Step 2: From the given hyetograph total rainfall in 5 hour storm is 60 mm.

$$\therefore \text{ Average rate of rainfall} = \frac{60}{5} = 12 \text{ mm/hour}$$

$$\text{It gives volume of rainfall} = \frac{60}{1000} \times \frac{5}{10} \times 10^6 = 30{,}000 \text{ m}^3$$

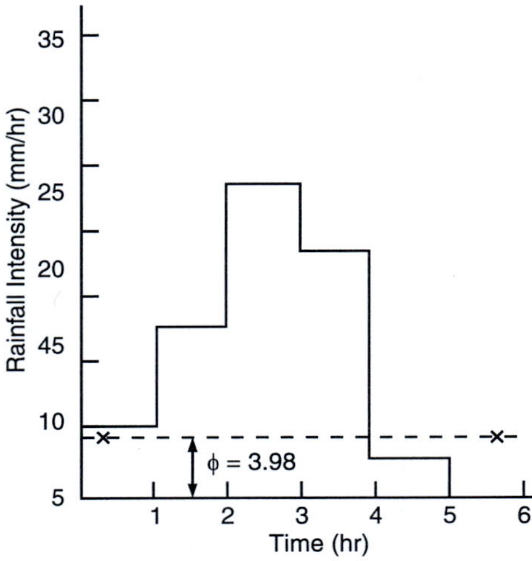

Fig. 3.3. Problem 1 (ϕ index)

Step 3: Total volume of direct surface runoff = 0.232 × 60 × 60 × 24 = 20,045 m^3.

Step 4: From definition of ϕ index it is that portion of the average rate of rainfall above which volume of remaining rainfall equals volume of run-off.

∴ Volume of water lost = (Total rainfall volume) – (Total run-off volume)

i.e.

(infiltration + retention)

$$= 30{,}000 - 20{,}045 = 9955 \text{ m}^3$$

As a depth of water over the basin $= \dfrac{9955}{0.5 \times 10^6} = 0.00398 \text{ m} = 3.98 \text{ mm.}$

∴ ϕ index = 3.98 mm.

3.5. MEASUREMENT OF INFILTRATION

Infiltration can be measured by two methods namely:

1. **Indirect methods:** They involve artificial application of water over a sample area. The mechanism used for the purpose is called infiltrometer. There are two types of infiltrometer viz., flooding type and rain simulators.
2. **Direct method:** It consists of analysis of run-off hydrograph resulting from a natural rainfall over a basin under consideration.

(*a*) Flooding type infiltrometer: It consists of about 25 cm diameter 50 to 65 cm long cylinder. The cylinder is sunk into the ground to a depth of 40 to 50 cm. The water is then applied through graduated burettes to maintain a constant head of water. The readings on burette at fixed interval of time gives the rate and amount of water infiltrated into the soil. To eliminate the effect of surrounding dry soil on the infiltrometer sometimes two concentric rings, one of same size and another bigger diameter say 35 cm are sunk into the ground. These rings are, however, sunk upto a minimum depth just necessary to avoid leakage from the rings. The space between both the rings is filled upto the same level and maintained at constant level by two different burettes. The burette reading feeding the inner ring gives rate and amount of infiltration. This method is now superseded by rain simulator.

(*b*) Rain simulator: In this method special sprinklers are mounted on both the sides of a 2 m × 4 m experimental plot. The nozzle of these sprinklers direct the spray of water in an inclined fashion to cover the plot fully and to reach a height of about 2 m above the ground. This arrangement ensures application of water in the form of likely rainfall. The intensities of simulated rainfall can be changed by closing and opening the nozzles. The infiltrometer starts working with what is called a rainfall calibration run. For this run a plastic or a metal sheet is placed over the plot so that all the water reaching the ground can be measured without loss of water. This gives the average rate of rainfall. After this the test run begins. This run is allowed to continue till the run-off becomes constant. The difference between simulated rainfall rate and measured run-off rate gives the value of fc (fc is constant infiltration rate which has established itself after the soil is saturated). To eliminate border effect about 0.5 m wide strip all-round the plot is also sprayed with water separately. This method suffers from the following drawbacks:

(*i*) It is difficult to simulate size of rain drops.

(*ii*) The velocity of fall achieved by the water drops do not represent correct rainfall conditions.

(*iii*) The experiment value of infiltration rate tends to be higher than that achieved under natural conditions.

(*iv*) The infiltrometer values can be used to compute run-off from a small watershed only because of the limited area on which the infiltration rate has been computed.

(c) Measurement of Infiltration by hydrograph analysis: The theoretical analysis of run-off hydograph has the advantage that it takes into consideration rainfall pattern, length of overland flow, slope of basin, soil type, vegetative cover, depression storage, surface detention as they tend to occur in actual. However, on a large watershed distribution of rainfall is generally not known in detail to warrant laborious theoretical methods of hydrograph analysis. For practical application it is convenient to separate showers of each storm rainfall into a series of blocks and consider resulting run-off hydrograph independently by transposing recession curves or calculating average infiltration rate. The analysis can be done is the following steps (Refer Fig. 3.4):

- Draw rainfall hyetograph and hydrograph of runoff on the same plot for the storm in the basin.
- Separate each shower of the storm.
- Separate each runoff hydrograph from succeeding one by transposing recession curves.
- Deduct base flow from total flow.
- Obtain storm runoff for each rise.
- Select durations of precipitation excess (T_e) by inspection of hyetograph and hydrograph.
- Plot mass curves of rainfall and obtain cumulative rainfall values (P_{w1}, P_{w2}, P_{w3} etc.).
- Plot mass curves of direct storm runoff and obtain cumulative runoff values (Q_{s1}, Q_{s2}, Q_{s3} etc.).
- Calculate difference of cumulative rainfall and cumulative runoff ($P_w - Q_s$) which gives total infiltration F.
- Divide total infiltration by duration of rainfall excess (T_e) to get average infiltration rate for that shower or block of the storm.

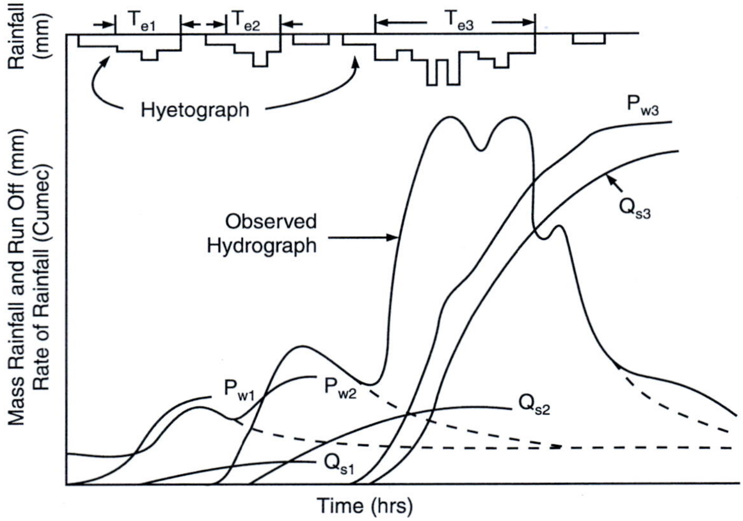

Fig. 3.4. Block method of hydrograph analysis

Suitability of Method of Infiltration Measurement

The various methods of measurement of infiltration direct and indirect mentioned earlier cannot be used on all sizes of watersheds with sufficient accuracy to assess resulting run-off. The commonly adopted methods on various sizes of watersheds are the following:

(*i*) **Small watersheds:** On small watersheds infiltrometer values and storm hydrograph analysis method gives satisfactory results.

(*ii*) **Large watersheds:** In case of large watersheds it is considered convenient to develop standard infiltration-rate curves by studying number of storms on a typical representative watershed considering different land uses.

Another method of practical importance in both cases is adoption of ϕ index which gives average rate throughout the storm. It is well suited for estimating peak run-off from a major storm on wet soils.

3.6. EVAPORATION

Definition: Evaporation is the process during which water is changed from liquid or solid state into vapour through the transfer of heat energy. The process of evaporation of water is one of the basic components of the hydrologic cycle and consists that phase in which precipitation reaching earth's surface is returned to the atmosphere in the form of vapour. There are three types of evaporation processes, namely:

- Evaporation from free water surfaces (for example, reservoirs, streams and ponds and lakes);
- Evaporation from land surfaces; and
- Evaporation from vegetation cover (viz., transpiration).

Evaporation is a process of diffusion in which vapour is transferred from the natural surfaces on the earth to the atmosphere. There are two essential requirements for evaporation to take place. They are:

- **Availability of source of heat energy** to vaporize the water. For vaporization, irrespective of the surface from where evaporation is taking place, an exchange of 590 calories per gram of water evaporated at 20°C is required. The source of heat energy could either be from solar radiation or air blowing over the surface or from within the underlying surface.
- **Existence of vapour concentration** gradient between the evaporating surface and the surrounding air. Evaporation can take place only if vapour concentration at evaporating surface is greater than that present in the overlying air.

3.7. ESTIMATION OF EVAPORATION FROM FREE WATER SURFACE

The change of state from water to vapour occurs when some molecules in the water body attain sufficient kinetic energy to reach the air above. This motion of molecules (aqueous vapour) through the water surface produces a pressure and is called vapour pressure.

Some of the molecules escaping from the water body fall back in the water as the aqueous vapour get condensed. Thus, evaporation from and condensation into the water surface are continuous processes. When the number of molecules that leave water body as vapour is equal to the number that falls back after condensation a saturation condition is said to reach. It indicates a state of equilibrium between the pressure exerted by the escaping molecules and the pressure of the surrounding atmosphere. It is thus clear that evaporation will be more than condensation if the space above the water surface is not saturated.

In short, the evaporation is a function of the difference between vapour pressure of the body of water and vapour pressure of the air above. Dalton (in 1802) showed that under given conditions:

$$E \propto (e_s - e_d)$$

or

$$E = (e_s - e_d)\, \psi$$

where E is evaporation

e_s is saturated vapour pressure at the temperature of evaporating surface (mm Hg)

e_d is saturation vapour pressure at the dew point temperature (mm Hg).

and ψ is a wind factor.

Several empirical equations for estimating evaporation have been developed on the basis of Dalton's law. Some of them are mentioned below. (It may be noted that these equations are in F.P.S. units).

(*i*) Meyer's formula (Developed in 1915)

$$E = c\,(e_s - e_d)\,\psi$$

where E is rate of evaporation in inches per 30 days month

c is a constant = 11 for large deep water bodies, and

= 15 for small shallow water bodies

e_s is maximum vapour pressure in inches of Hg.

(*i*) corresponding to monthly mean air temperature for small and shallow water bodies, and

(*ii*) corresponding to water temperature for large and deep water bodies.

e_d is actual vapour pressure in air in inches of Hg.

(*i*) based on monthly mean air temperature and relative humidity for small and shallow water bodies, and

(*ii*) based on information about 30 ft above water surface for large and deep water bodies.

ψ is a wind factor = $(1 + 0.1\,w)$

w is monthly mean wind velocity in *mph* at about 30 ft above water surface.

(*ii*) Rohwer formula (Developed in 1931)

$$E = 0.771\,(1.465 - 0.0186\,B)\,\psi(e_s - e_d)$$

He considered effect of atmospheric pressure and introduced a factor $(1.465 - 0.0186\,B)$

In the above equation

$$\Psi = 0.44 + 0.118\,w$$

In this equation

E is rate of evaporation in inches per day.

B is mean barometric reading in inches of mercury (Hg) at 32°F.

e_s is maximum vapour pressure in inches of Hg.

e_d is actual vapour pressure in air based on monthly mean air temp. and relative humidity in inches of Hg.

w is monthly mean wind velocity in mph.

(*iii*) Christiansen formula (It is metric units)

$$E_p = 0.473\,R \cdot C_t \cdot C_w \cdot C_s \cdot C_e \cdot C_m$$

where E_p is evaporation loss in mm

R is extra-terrestrial radiation in mm (the value of R varies with latitude and also month by month).

C_m is a coefficient to represent evaporation as mean for the month.

C_t, C_w, C_h, C_s and C_e are coefficients for temperature, wind velocity, relative humidity, percent possible sunshine and elevation all expressed in same units as E_p. For calculating the values of various coefficients Christiansen gave separate expressions. The expressions are complicated one and not within the scope of study.

Limitations of Empirical equations: The above equations suffer from following limitations:

(*i*) Application of these equations is difficult because it may not be possible to obtain the information needed for their solution at desired locations.

(*ii*) Most of the quantities used are average values based on monthly averages whereas in practice evaporation depends on the actual situation at different time.

3.8. FACTORS AFFECTING EVAPORATION FROM FREE WATER SURFACE

It is already mentioned that evaporation takes place because of difference in vapour pressures of the water body and the air above it. Obviously the factors which affect vapour pressures will also affect process of evaporation. The important factors are the following:

3.8.1. Meteorological Factors

(*i*) **Temperature:** As temperature rises the capacity of the air mass to hold vapour molecules increases. Also vapour pressure of the water body increases. If the temperature rise of air and water is equal no increase in the rate of evaporation can be expected. But due to differential rate of heating as temperature increases the rate of evaporation also increases.

(*ii*) **Wind:** Wind plays dual role in affecting the process of evaporation. Firstly incoming fresh air removes the molecules of water vapour and makes space for other vapour molecules. Obviously faster the rate of incoming fresh air faster the removal of vapour molecules. But once the wind velocity is sufficient to remove all rising vapour molecules extra increase in wind velocity has no effect. Secondly, incoming fresh air if hot provides extra heat energy to accelerate process of evaporation. On the contrary if incoming fresh air is cool it reduces evaporation rate.

(*iii*) **Atmospheric pressure:** At higher altitudes there is decrease in atmospheric pressure. Such situation increases the rate of escaping water molecules from free surface as the air above has lower molecules to prevent entry of other molecules.

3.8.2. Physical Factors

(*i*) **Nature of evaporating surface:** Every surface that receives rainfall is a potential evaporation surface. The evaporation from any surface will be limited to that quantity of water required to saturate the surface.

For example, rate of evaporation from saturated soil surface is approximately the same as that from the adjacent free water surface at the same temperature. But as the soil begins to dry, evaporation decreases and eventually it almost stops since there is no possibility of water reaching on its own upto the surface from great depth.

Also evaporation from snow and ice surfaces can occur only when the vapour pressure of the air is less than that of the snow surface. In other words such evaporation to take place the dew point has to be lower than the temperature of snow or ice.

(*ii*) **Shape of evaporating surface:** This is an important consideration when evaporation takes place through small openings, for example, diffusion through stomata in plants.

It is seen that maximum evaporation takes place from a convex surface followed by flat surface and then concave surface.

Also evaporation through small restricted openings is seen to be proportional to their diameter or perimeter (linear dimension) rather than to their areas.

(*iii*) **Water quality:** When soluble solids are present in the water in solution its vapour pressure at a particular temperature is lower than that of pure water at the same temperature. The vapour pressure of sea water which has 35,000 ppm dissolved salts is about 2 per cent less than that of pure water at the same temperature. It is seen that the evaporation rate decreases with increase in specific gravity of the solutions.

3.9. MEASUREMENT OF EVAPORATION

The evaporation is popularly measured by the following instruments:

- Piche evaporimeter,
- Sunken evaporation pan,
- U.S. Weather Bureau class A pan evaporimeter.

3.9.1. Piche Evaporimeter

It consists of a 32 cm long graduated tube of 1 cm inside diameter. It is closed at one end. The other end of the tube is open and has flat edge. There is a disk and clamp arrangement by which this end can be covered by a filter paper as shown in Fig. 3.5.

The tube is filled up with distilled water. The open end is then covered by filter paper supported by the disk and the clamp.

The evaporimeter is then hung in inverted position. The water in the tube is soaked by the filter paper which moves rapidly outward through the paper and ultimately gets evaporated. The rate of loss of water from the tube gives the evaporation rate. This instrument is very sensitive to winds.

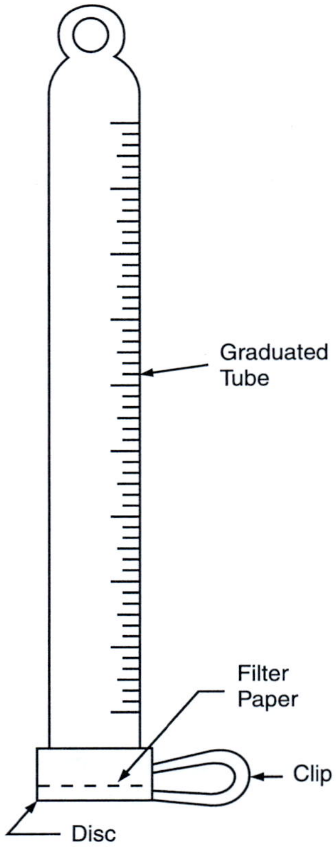

Fig. 3.5. Piche evaporimeter or atmometer

3.9.2. Sunken Evaporation Pan

It is a galvanized iron pan about 183 cm in diameter and 61 cm deep. It is buried in the ground keeping 10 cm height above the ground. The water is filled in the pan the level of which is generally kept 1 cm above ground level. The rate of water loss from the pan gives rate of evaporation. Because of its size this evaporimeter gives good index of lake evaporation. The pan has a coefficient of about 0.95 by which the observed evaporation is multiplied to compute lake evaporation.

Sometimes a floating pan may also be used to simulate natural conditions. The coefficient of floating pan is seen to be about 0.8.

3.9.3. U.S. Weather Bureau Class A Land Pan Evaporimeter

It is a 122 cm diameter (4 ft.) 25.4 cm (10 inches) deep unpainted galvanized iron pan. It is installed on a wooden frame in such a manner that the bottom of the pan is about 15 cm above the ground level. The water is filled to a depth of about 18 cm in the pan. The pan is protected on top by a wire mesh against use by birds. The pan is connected to a stilling well fitted with a hook gauge to measure the water level accurately. The accuracy of the hook gauge is upto 0.025 mm. The water level is measured daily by the hook gauge. The difference of water level gives evaporation in cm. Like rainfall evaporation loss is also expressed as depth of water over the area under consideration either on daily basis or monthly basis or annually.

It is already mentioned that the rate of evaporation from small areas is normally greater than that from large areas. Therefore a pan coefficient of 0.7 is used to convert pan evaporation figures.

It may be mentioned here that although land pan coefficient is lower than sunken pan coefficient land pan evaporimeter is commonly used due to following advantages:

- The land pan can be installed in a convenient place to provide easy access for observation and inspection.
- It is more stable in construction and easy to repair.
- Due to ease of installation more data can be readily collected.
- The cost of the installation of land pan is quite low.
- It provides a stable pan coefficient.
- It is not disturbed by drifting snow, dirt or debris etc.

Many times because of variations in size and shape of evaporation pans, their exposure, quantum of algae present in water, method used in measuring loss of water from pans and insufficient protection against use of water by birds and animals, available pan evaporation data may not be reliable to estimate evapotranspiration. In such cases pan evaporation can be estimated by making use of reliable formula such as Christiansen formula.

3.10. EVAPORATION FROM SOIL SURFACES

The mechanics of evaporation from soil surfaces is similar in principle to one observed for evaporation from the water surfaces. In addition, the escaping molecules of water vapour from the soil have to overcome the resistance due to the attraction of the soil particles for the water. Same factors which affect evaporation from free water surface also affect the evaporation from the land surface, however, the difference that exists is due to degree of wetness of the land surface. The rate of evaporation from saturated soils is almost identical to the rate of evaporation from free water surface. When the moisture content of the surface soil becomes less, the loss of moisture by evaporation decreases and when it becomes fairly low the evaporation practically ceases. It is seen that evaporation from the soil surface will continue as long as shallow surface soil layer say 10 cm for clayey and 20 cm for sandy soils remains moist.

The evaporation from soil surface can be measured by lysimeter. It is described subsequently.

3.11. METHODS OF REDUCING EVAPORATION

It is already mentioned that evaporation does not depend on one single factor but occurs as a result of combined effect of various factors. The typical daily evaporation rates and total annual evaporation at some stations have been given in Table 3.1. These stations have been selected to indicate evaporation rates at high altitudes, latitudes, sea coasts and in arid area, less rainfall areas, high rainfall areas and in the middle of alluvial river plains. The total evaporation values indicate the loss of water that has occurred from the pan evaporimeters over one year. Obviously at all measurement stations a situation was so created that free water surface is available throughout the year without limitation. Such a situation normally does not exist in practice due to insufficient and limited rainfall. Nevertheless it indicates the potential loss that may occur due to evaporation if favourable conditions are available. When artificial lakes and reservoirs are constructed a situation is created when a free water surface is available throughout the year for the evaporation loss to take place. Water lost in evaporation from the free water surfaces like lakes and reservoirs as also land surfaces is a direct loss of stored water of economic value. The loss of water by evaporation is a very serious problem in arid and semi-arid regions where either water is scarce or it is made available at an appreciable cost and effort. It is possible to reduce evaporation from the free water surfaces and land surfaces by adopting suitable methods. Some methods of practical importance are mentioned below:

Table 3.1. Typical evaporation rates (mm/day)

Month	*Station*											
	Kodai-kanal	*Sri-nagar*	*Mum-bai*	*Kol-kata*	*Che-nnai*	*Jodh-pur*	*Amrit-sar*	*Pune*	*Shil-long*	*Delhi*	*Patna*	*Gauhati*
Jan.	3.6	0.9	2.7	1.5	4.0	6.7	1.2	4.1	1.6	4.2	3.2	2.1
Feb.	3.5	0.9	3.6	2.0	4.9	8.2	2.1	6.0	2.3	5.6	4.3	3.3
Mar.	3.7	2.2	4.4	4.0	5.5	10.3	3.5	8.2	3.4	10.0	7.5	4.9
Apr.	2.8	2.3	4.6	4.3	6.8	13.5	4.3	9.3	3.9	11.6	7.6	5.7
May	2.3	3.1	5.3	4.5	7.2	15.9	5.1	9.5	3.4	9.8	9.1	5.0
Jun.	2.3	4.5	4.5	3.7	7.9	13.6	6.0	6.3	2.2	9.6	6.5	4.3
Jul.	2.1	3.8	3.0	3.1	7.3	5.5	3.5	3.8	2.1	4.6	4.2	4.4
Aug.	2.2	3.4	3.5	3.2	5.5	6.7	3.9	3.7	2.2	6.1	5.5	4.2
Sep.	2.0	2.0	3.4	2.6	6.2	6.7	3.1	4.6	2.1	6.0	4.0	3.8
Oct.	1.3	1.8	4.1	2.5	3.2	7.8	2.7	5.0	1.9	6.9	3.6	3.3
Nov.	1.4	0.8	3.3	2.1	3.6	6.5	1.7	3.6	1.7	5.8	2.7	2.7
Dec.	2.7	0.7	2.9	1.5	3.9	4.1	1.1	3.4	1.5	3.3	2.0	2.0
Average Annual	2.5	2.2	3.8	2.9	5.5	8.8	3.2	5.6	2.4	7.0	5.0	3.8
Total Annual (cm)	91	74	139	106	201	322	117	204	88	255	183	139

3.11.1. By Keeping Free Water Surface Area Minimum

Reservoirs and lakes are the water bodies with large surface area. It is possible to reduce the surface area by adopting following measures:

(*a*) The reservoir site may be selected in such a way that area to storage ratio is minimum.

(*b*) To avoid water loss water could be stored below ground.

(*c*) When there are number of reservoirs on one river the water may be stored in one large reservoir when possible rather than in several small reservoirs.

3.11.2. By Spreading Surface Films on the Reservoirs and Lakes

Though still under experiment stage this method when used will prove to be most economical and of practical utility. The method is based on the principle that a thin layer of oil film when spread on the water surface is capable of reducing evaporation to significant extent.

Experiments have shown that a compound called *hexadeconal* or *cetyl alcohol* derived from tallow, sperm oil or coconut oil is capable of forming monomolecular film on the water surface. This material is white, waxy, crystalline, solid and is generally available in flakes or powder form. It is relatively tasteless and odorless. It is pervious to oxygen and carbon dioxide. It is also nontoxic to living being. It is a polar compound in which one end of the molecule is hydrophilic and therefore, shows great affinity for water whereas the other end is hydrophobic and repels water.

Because of this property it forms a monomolecular film on the water surface. The film so formed is only about 15×10^{-4} mm thick but it is tight enough to prevent water molecule from escaping the water body. It is flexible and moves with the motion of water surface without breaking. The film is penetrable by rain drops but the film again closes after allowing rain drops to enter.

It is estimated that such a film may reduce evaporation by one-third. Theoretically only 25 gm of hexadeconal is sufficient to spread a mono-molecular film on one hectare of water surface. In practice, however, due to working problems as much as 100 times of theoretically required quantity may be necessary. Even then it is quite economical and promising. Combination of hexadeconal and octadeconal was also used for better results.

3.11.3. By Providing Mulch on the Land Surfaces

To reduce evaporation from land surface it is a common practice to spread mulch on the land surface. The mulch is nothing but some waste material used to cover the land. By mulch one generally understands a mixture of straw, leaves etc. It restricts air movement, maintains high water vapour pressure near the soil surface and shields the soil from the solar energy and as a result of all these factors reduces evaporation. Considering that the mulch is nothing but a protective blanket sometimes dust mulch of loose dry earth, pebble mulch etc., can be used to suppress evaporation.

3.12. TRANSPIRATION OR EVAPORATION FROM VEGETATIVE COVER

Transpiration is the process by which water vapour leaves the living plant body and enters the atmosphere. It involves continuous movement of water from the soil into the roots, through the stem and out through the leaves to the atmosphere. The process includes cuticular transpiration or direct evaporation into the atmosphere from moist membranes through the cuticle and stomatal transpiration from the leaves.

Transpiration is basically an evaporation process. However, unlike evaporation from a water surface, transpiration is modified by plant structure and stomatal behaviour operating in conjunction with the physical principles governing evaporation.

3.12.1. Factors Affecting Transpiration

There are three main factors which influence transpiration. They are:

- climate factors;
- soil factors; and
- plant factors.

The important climatic factors are light intensity, atmospheric vapour pressure, temperature and wind. The soil factors are those governing the water supply to the roots and the plant factors include the extent and efficiency of root systems in moisture absorption, leaf area, leaf arrangement and structure and stomatal behaviour.

The rate of transpiration depends on the supply of energy to vapourize water, the water vapour pressure or concentration gradient in the atmosphere which constitutes the driving force and the resistance to diffusion in the vapour pathway.

Unlike evaporation from a free water surface vegetation virtually ceases to transpire at night due to stomatal closure (A water surface continues to evaporate, though at a slower rate). Thus, over a period of say a month, evaporation from a free water surface may exceed the transpiration from a crop. Over a few hours during the day, however, the crop may evaporate more than a free water surface.

3.12.2. Measurement of Transpiration

In the field transpiration can be measured by an instrument called phytometer. It is a vessel of big size filled with soil in which plants are transplanted. The soil surface is sealed to prevent direct evaporation. Thus, the water is allowed to escape only by way of transpiration. The amount of water transpired is determined by weighing the vessel after decided interval of time.

3.13. EVAPOTRANSPIRATION OR CONSUMPTIVE USE

In deciding water requirement of crop a common term used is called evapotranspiration. From the name it is clear that it includes both evaporation and transpiration. The term evaporation relates to loss of water taking place from land surface surrounding the plant whereas transpiration relates to loss of water through the plant structure. This term has been evolved because it is difficult to separate the two losses in a cropped field. Generally evapotranspiration is indicated by a symbol (ET).

3.13.1. Difference between Evapotranspiration and Consumptive Use

The term consumptive use is used to designate the losses due to evapotranspiration and the water that is used by the plant for its metabolic activities. Since the water used in the actual metabolic processes is insignificant (about 1% of ET or less) the term consumptive use is generally taken equivalent of ET. It thus includes all water consumed by plants plus the water evaporated from bare land and water surfaces in the area occupied by the crop.

Thus theoretically

$$CU = (ET) + \text{Water lost in metabolic process} = (ET) + 0.01\,(ET)$$

or

$$CU \approx ET$$

3.13.2. Factors Affecting Evapotranspiration

(*i*) **Extent of crop cover:** The relative amount of direct evaporation from the land and water surfaces and transpiration depends usually on the extent of ground cover. For most crops completely covering the soil surface only a very small amount of water is lost from the ground surface as evaporation.

(*ii*) **Climatic factors:** Under field conditions incoming solar radiation supplies the energy for the process of evapotranspiration. Wind is important in removing water vapour from the cropped area. In general, the factors influencing evaporation and transpiration also govern the evapotranspiration process.

(*iii*) **Stage of crop growth:** Other factors being equal, the stage of a crop's growth influences its consumptive use rate substantially. This is particularly true for long duration crops which generally have three rather distinct stages of growth. They are:

(*a*) emergence to development of complete vegetative cover during which time the consumptive use rate increases rapidly from a low value and approaches its maximum;

(*b*) the period of maximum vegetative cover during which the consumptive use rate may be maximum and practically constant if adequate soil moisture is available; and

(*c*) crop maturation stage, when for most crops the consumptive use rate begins to decrease.

(*iv*) **Method of cultivation:** The rate of evapotranspiration also depends on how the crops are planted. The interception of solar energy by row crops is a function of their orientation with respect to the sun. Also if the surrounding fields are bare, more water will be lost than if they were cropped and irrigated.

3.13.3. Potential Evapotranspiration (PET)

The PET is the evapotranspiration that occurs when the ground is completely covered by actively growing vegetation and there is no limitation of the soil moisture, i.e., there is ample water available for meeting evaporation requirements. It may be considered to be the upper limit of evapotranspiration for a crop in a given climate.

As the name indicates it gives the value of continuously occuring unhindered evapotranspiration. It can happen only if the conditions are conducive and ample water is available so that evapotranspiration is not restricted for want of soil water and a large area of land is fully covered with actively growing green vegetation.

Such a situation occurs in practice only in fully grown wet paddy fields or for a short period immediately after rain for other crops. So PET is more or less a theoretical value of evapotranspiration and requires conversion to actual field conditions before adoption.

3.14. MEASUREMENT OF EVAPOTRANSPIRATION

The evapotranspiration can be either measured by direct measurement or estimated from climatological data. The most elaborate and commonly adopted indirect method of estimation is *modified Penman formula.*

The common methods of direct measurement are the following:

- Lysimeter;
- Field experiment;
- Soil moisture depletion study; and
- Water balance method.

3.14.1. Lysimeter

It involves the growing of crops in large containers (lysimeters) installed in cropped fields and measuring their water loss and gains. The soil and crop conditions in these lysimeters should be as close to the surrounding field conditions as possible. The measurements involve weighing of the lysimeter. It may be done with scales or by floating the lysimeters in water or a suitable heavy liquid in which case the change in liquid displacement is computed against water loss from the lysimeter. The technique is useful for measurement of field water losses provided the lysimeters are of appropriate size and necessary precautions are taken in installation. For short crops, the lysimeter should be at least two cubic metre in the volume (2 m deep) and for tall crops like sugarcane it should be four cubic metres or more in volume (3 m deep). The lysimeters must be surrounded by a fairly large guard area of the same crop with similar moisture conditions in order to minimize the cloths line effect, if the data recorded are to bear close relation to the crop water losses in the surrounding field. Water is applied in measured amounts to the lysimeter, as irrigation is applied to the surrounding cropped area. The overflow and deep percolation, if any, are measured. The water received either through irrigation or precipitation, excluding overflow, deep percolation and balance soil moisture, constitutes the water used by the crop in the process of evapotranspiration.

This method often encounters a number of problems. The major limitations are the reproduction of physical conditions such as temperature, water table, soil texture and density etc., within the lysimeter comparable to those outside in the field.

3.14.2. Field Experiment

Measurements of water supplied to the field and changes in soil moisture contents of field plots are sometimes more dependable for computing seasonal water requirement of crops than measurements with small tanks or lysimeters not free from limitations. The seasonal quantities of irrigation water, the effective rainfall received during the season and the contribution of moisture from the soil may be expressed by the following relationship:

$$WR = IR + ER\ (M_s - M_e)$$

In which, WR = seasonal water requirement, mm

IR = total irrigation water applied, mm

ER = seasonal effective rainfall, mm

M_s = moisture percentage at the beginning of the season

M_e = moisture percentage at the end of the season.

The method requires that the amount of water applied to a field is measured accurately. This method, though satisfactory for computing seasonal water requirement does not provide information on intermediate soil moisture conditions, short term use, profile use, deep percolation losses and peak use rate of the crop.

3.14.3. Soil Moisture Depletion Study

The soil moisture depletion method is usually employed to determine the consumptive use of irrigative field crops grown on the fairly uniform soil when the depth to the groundwater is such that it will not influence the soil moisture fluctuation within the root zone. These studies involved measurement of soil moisture from various depths at a number of times throughout the growth period.

Large number of measurements are taken to calculate average consumptive use (CU).

Seasonal consumptive use is calculated by summing of consumptive use values of each sampling interval.

3.14.4. Water Balance Method

The water balance method, is suitable for large areas (watersheds) over long periods. It may be represented by the following hydrological equation:

Precipitation = Evapotranspiration + surface runoff + sub-surface drainage + change in soil water contents

This method necessitates adequate measurement of all factors, except evapotranspiration. This method is also called inflow-outflow method.

3.15. ESTIMATION OF CONSUMPTIVE USE UNDER INDIAN CONDITIONS

Owing to difficulty in accurate field measurements evapotranspiration or consumptive use is generally predicted on the basis of climatological data. Most of the empirical formulae like Blaney-Criddle, Thornthwaite or Christiansen try to relate consumptive use with one or more climatic factor such as temperature, sunshine, wind, humidity, etc. In practice, however, such formulae have to be used under different agro-climatic conditions which are quite different from those for which empirical relations were originally developed. It is therefore clear that in Indian conditions which are so variable from place-to-place purely empirical formula can not give satisfactory results for calculating consumptive use. So it is essential to adopt more theoretical approach which can satisfy set of conditions that may exist at any place. Based on intensive studies of the climatic and measured grass evapotranspiration data from various research stations of the world, Doorenbos and Pruitt proposed a *modified Penman* method in 1975. It gives fairly accurate estimate of the evapotranspiration for the reference crop. By using suitable crop factor evapotranspiration can be calculated satisfactorily. The method is described subsequently in this book.

QUESTIONS FOR REVISION

1. Define interception. On what factors does interception depend ?
2. (*a*) Define potential evapotranspiration ?
   (*b*) Name the instrument which can be used to measure it.
   (*c*) How is it estimated ?
3. (*a*) What do you understand by infiltration ?
   (*b*) How will you measure infiltration capacity by rain simulators ?
4. (*a*) Distinguish between evapotranspiration and potential evapotranspiration.
   (*b*) Distinguish between evapotranspiration and consumptive use.
5. (*a*) Why it is important to control evaporation from free water surfaces ?
   (*b*) Describe various methods of reducing evaporation loss.
6. Write short notes on the following:
   (*i*) ϕ index and W index.
   (*ii*) Pan evaporimeter.
   (*iii*) Infiltration capacity and infiltration rate.
   (*iv*) Piche evaporimeter.

4

Surface Runoff

4.1. CATCHMENT AREA

Whenever rainfall occurs, part of the rainfall is absorbed by the surface of the earth. It is then transmitted downwards to the groundwater storage. This water is said to have lost in infiltration. Some portion of total rainfall is intercepted by the vegetation and depressions. Also, some water gets lost in evaporation. Remaining water passes over the ground surface and subsequently joins torrents, drainages or rivers. This portion is called over land flow or surface runoff.

Natural relief is such that a particular system of natural drainage including torrents, rivers etc., always collect and carry the draining water, underground or overground, from the particular area which surrounds the drainage system. As such **catchment area** can be defined as that area which always contributes the surplus water present over it to the natural drinage grid present in the area. It is the area which is responsible for maintaining flow in the natural drainage like rivers.

The volume of surplus water discharged by a drain, draining the area is called **runoff** from that catchment area. Runoff is generally expressed as a rate of flow (cumec) or volume (cubic metres). Runoff can also be expressed in cm depth of water over the catchment area like rainfall. The runoff can be measured daily, monthly or yearly depending upon the problem in hand.

4.2. COMPONENTS OF RUNOFF

The river flow that appears in the river is not made up of overland flow (surface runoff) only. It is actually total runoff contributed to the river in different ways by the drainage basin. Depending upon the source from where the portion of total flow gets contributed the components of total runoff can be recognised as shown below schematically.

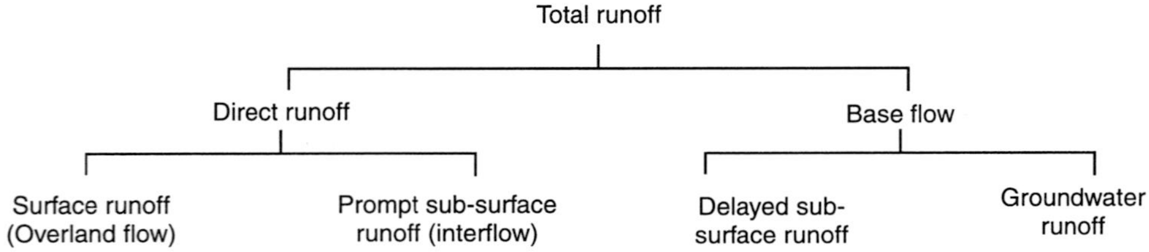

It is clear from above that total runoff consists of two parts namely:

(*i*) Direct runoff caused as an immediate result of precipitation; and

(*ii*) Base flow which is responsible for maintaining a sustained flow throughout the year and is mainly derived from groundwater storage.

Direct runoff is made up of overland flow or surface runoff and that part of infiltration which flows laterally through the unsaturated zone of soil mass and joins the stream flow promptly.

On the other hand **base flow** is made up of groundwater contribution and that portion of infiltration which moves laterally but joins the stream flow quite late after the precipitation has stopped.

Now it may be appreciated that only part of the rainfall is available for producing overland flow or surface runoff. This portion of total rainfall which produces surface runoff is called **rainfall excess**.

$\therefore$ Total rainfall = Rainfall excess + losses

The term losses includes interception, infiltration, evaporation, depression storage etc.

Similarly that portion of rainfall which produces direct runoff is called **effective rainfall.**

$\therefore$ Effective rainfall = Rainfall excess + x

where x is that portion of rainfall which appears in the stream as promt subsurface runoff.

Obviously when promt sub-surface runoff is considered together with delayed subsurface runoff.

Direct runoff = Surface runoff

and Effective rainfall = Rainfall excess

4.3. FACTORS AFFECTING RUNOFF

Runoff is affected mainly by climatic and physiographic factors. The important factors are mentioned below.

1. **Rainfall pattern:** If the rainfall is very heavy the consequent runoff will also be more. If the rainfall is just showery type with low intensity there may not be runoff at all as the rainwater is completely lost in infiltration, evaporation, etc. If the duration of rainfall is more the runoff will also be prolonged.
2. **Character of catchment surface:** If a surface is rocky then the surface absorption will be practically nil and the runoff will be more. If the surface is compact clayey type runoff will be more, but if the surface is sandy then the absorption losses will be more and runoff will be less.
3. **Topography:** If the surface slope is steep runoff will be more as water will pass over the surface rapidly before losses take place. If there are local depressions water will be held up in depressions forming lakes, ponds etc., in the catchment.

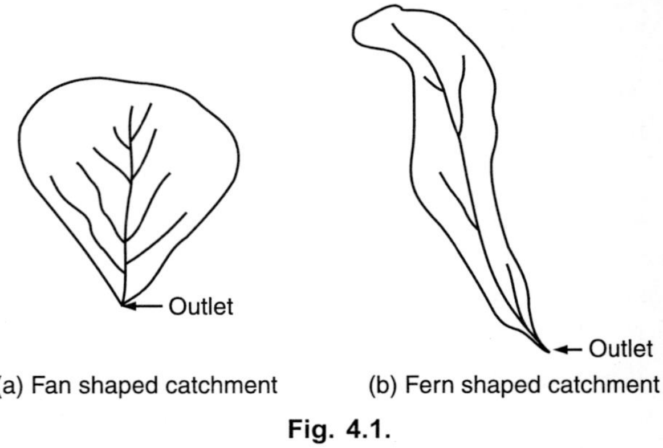

(a) Fan shaped catchment (b) Fern shaped catchment

Fig. 4.1.

4. **Shape and size of the catchment:** If the catchment area is large runoff will be more. If the catchment area is fan shaped runoff at outlet will be more as all the water contributes to the stream practically at the same time.

If the catchment is fern shaped the runoff will be less.

5. **Vegetal cover:** If there is some sort of vegetal cover over the catchment then evaporation loss will be reduced as sun rays cannot reach the ground surface.
6. **Geology of the area:** If there are fissures, cracks, fault zones present in the catchment then rainwater finds its way out through these openings. The water lost may find its way to some other catchment or to groundwater or in the sea.
7. **Weather conditions:** Temperature of the region also affects the runoff to a great extent. If temperature is more it renders surface dry and when rain occurs more water is absorbed by the ground surface. Evaporation rate will also be more if temperature is high.

4.4. MEASUREMENT OF RUNOFF

The quantity of water following down the river can be measured by actual measurements. The methods of discharge measurement have been dealt with subsequently. The records of runoff, i.e., river flow, can be kept on daily, monthly, seasonal or annual basis. Total runoff of the river is also called the yield.

4.5. RAINFALL-RUNOFF RELATIONSHIP

The rainfall as well as runoff are two components of a hydrologic cycle. In fact the runoff occurs mostly as a result of rainfall. As such it is possible to develop a relation between rainfall and runoff. The rainfall-runoff relationship is used to:

- estimate runoff volumes; and
- extension of runoff records or series.

4.5.1. Estimation of Runoff

In the absence of actually observed data it is generally necessary to estimate seasonal or annual runoff volumes. In such cases it is neither necessary nor possible to distinguish between base flow or direct runoff. Rainfall is supposed to be the main or only independent parameter on which runoff depends. A simple plot of annual rainfall versus annual runoff gives a good correlation on a specific area. Such value gives virgin runoff and does not reflect effect of artificial storages intercepting the overland flow if they have come up in between the available rainfall series. If runoff is to be assessed over a shorter period and when various factors affecting the runoff make it difficult to arrive at a satisfactory correlation between rainfall and runoff the relationship can be improved by taking into consideration other influencing parameters like initial soil moisture condition, temperature, storm duration, time of year, storm frequency etc.

Following are some methods derived on the above principle.

4.5.1.1 Empirical Formulae

In this method an attempt is made to derive a direct relationship between the rainfall and subsequent runoff. For this purpose some constants are established which give fairly accurate result for a specific region. Some well known formulae are given below:

Khosla's Formula

$$R = P - 4.811\,T$$

where R = annual runoff in mm

P = annual rainfall in mm

and T = mean temperature in °C

Inglis has given formulae for hilly and plain areas of Maharashtra:

For Ghat region, $R = 0.85\,P - 304.8$

For plain region, $R = \frac{(P-177.8)\times P}{2540}$

where R = runoff in mm

and P = rainfall in mm

Lacey's Formula

$$R = \frac{P}{1+\frac{3048F}{P.S}}$$

where R = monsoon runoff in mm

P = monsoon rainfall in mm

S = catchment area factor

and F = monsoon duration factor

Values of S for various types of Catchment are given in Table 4.1.

Table 4.1.

Type of catchment	*Value of S*
(*i*) Flat, cultivated and black cotton soils	0.25
(*ii*) Flat, partly cultivated, various soils	0.60
(*iii*) Average catchment	1.00
(*iv*) Hills and places with little cultivation	1.70
(*v*) Very hilly and steep, with hardly any cultivation	3.45

Values of F for various durations of monsoon are given in Table 4.2.

Table 4.2.

Class of monsoon	*Value of F*
(*i*) Very short	0.50
(*ii*) Standard length	1.00
(*iii*) Very long	1.50

4.5.1.2 Runoff curves and tables

It is now clear that each region has its own catchment and rainfall characteristics. Naturally formulae given above and coefficients derived therein can not be applied universally. However, for the same region the characteristics mostly remain unchanged. Based on this fact the runoff coefficients are derived once for all. Then a graph is plotted in which one axis represents rainfall and the other runoff.

The curves obtained are called runoff curves. Alternatively a table can be prepared to give the runoff for a certain value of rainfall for a particular region. Many tables and curves are given by research workers for various states, for example, Strange's table for Maharashtra, Sir Alexander Binnie's table for southern part of Madhya Pradesh, Barlow's table for Uttar Pradesh, etc.

Strange's table: The table given by W.L. Strange is given Table 4.3 which gives surface runoff as a percentage of rainfall for various types of catchments.

Barlow's table: T.G. Barlow produced similar tables which are applicable for the areas of Uttar Pradesh. He gave runoff as per cent of average monsoon rainfall. In order to obtain runoff from various storms of different intensities he further suggested coefficients. The average runoff percentages are to be multiplied to obtain runoff for different rain storms. The Barlow table is given below as Table 4.4. The limitation of this table is that it holds mostly for catchments having areas below 140 sq. km.

Table 4.3: Strange's table

Total monsoon rainfall in mm	*Type of catchment*		
	Good	*Average*	*Bad*
	% of runoff to rainfall		
250	4.3	3.2	2.1
500	15.0	11.25	7.5
750	26.3	19.7	13.1
1000	37.5	28.1	18.7
1250	48.8	36.6	24.4
1500	60.0	45.0	30.3

Table 4.4: Barlow's table

Class	*Type of catchment*	*Runoff as % of average monsoon rainfall*	*Coefficients for type of rainfall*		
			Light rainfall	*Average or varying rain*	*Continuous heavy rainfall*
A	Flat, cultivated and Black Cotton soils	10	0.70	1.0	1.50
B	Flat, partly cultivated soils	15	0.80	1.0	1.50
C	Average	20	0.80	1.0	1.60
D	Hills and plains with little cultivation	35	0.80	1.0	1.70
E	Very hilly and steep areas	45	0.80	1.0	1.80

4.5.1.3 Runoff Estimation Using ϕ and W Index

In the previous chapter while defining ϕ index it is already mentioned that if from the rainfall hyetograph ordinates ϕ index is deducted we get direct runoff. Applying this principle volume of runoff can also be calculated. When conditions are wet so that interception and surface retention is negligible and infiltration capacity is low $\phi = W$. This minimum value of W index is used in calculating peak flow.

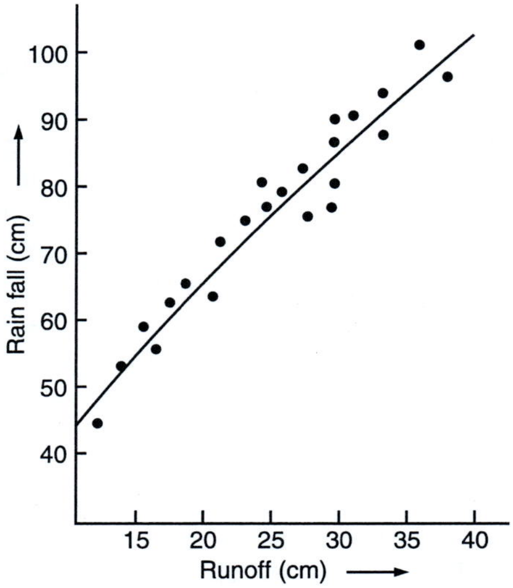

Fig. 4.2. Rainfall-runoff correlation

4.5.2. Extension of Runoff Series

Many times the runoff data at a particular river discharge measurement site is available for a few years, say 5 to 10 years. But the rainfall records at the nearest station may be available for many years. In such situations the runoff data can be extended for other years for which only rainfall data is available. A graph is drawn with monsoon or annual runoff as abscissa and monsoon or annual rainfall as ordinate for the years for which runoff data is available. The points plotted do not provide a perfect curve (Fig. 4.2).

A best fitting curve can be plotted, if necessary, making use of statistical means. Making use of this curve runoff for other years can be read off against their rainfall values.

4.6. RUNOFF HYDROGRAPH ANALYSIS

Hydrograph of a stream is a graphical representation of the discharge variations arranged corresponding to the time. That is, it is a graph of discharge versus time. Figure 4.3 shows the discharge hydrograph of a stream. Hydrograph can also be plotted to indicate stage (river level) variation against time. Then it is called a **stage hydrograph**. Automatic water level recorder produces a stage hydrograph on the strip chart.

When a stream under consideration is a natural drainage which is perennial it will flow for all the twelve months. Suppose there is no surface runoff in rainy season. Then the graph will be almost a straight line parallel to the time axis. It shows that the flow in the stream is due to the sub-surface flow only which is practically the same provided water-table remains static for all the time. But in actual practice discharge of the stream is more due to rainfall and subsequent runoff over the catchment for a certain period. In Fig. 4.3 we can correctly point-out the addition of surface runoff to the stream. Also the groundwater table fluctuates depending upon the water infiltrating into the ground.

In fact, runoff hyrograph of such a stream can be visualised as a runoff cycle the variation of which depends on the nature of supply. With reference to Fig. 4.3 annual river hydrograph can be considered in five distinct phases. They are as follows:

1. First phase indicates minimum flow or dry condition. It represents a rainless period just prior to the beginning of rainfall after an extended dry season. The ground water obviously will be quite low. The river hydrograph is almost a horizontal line.

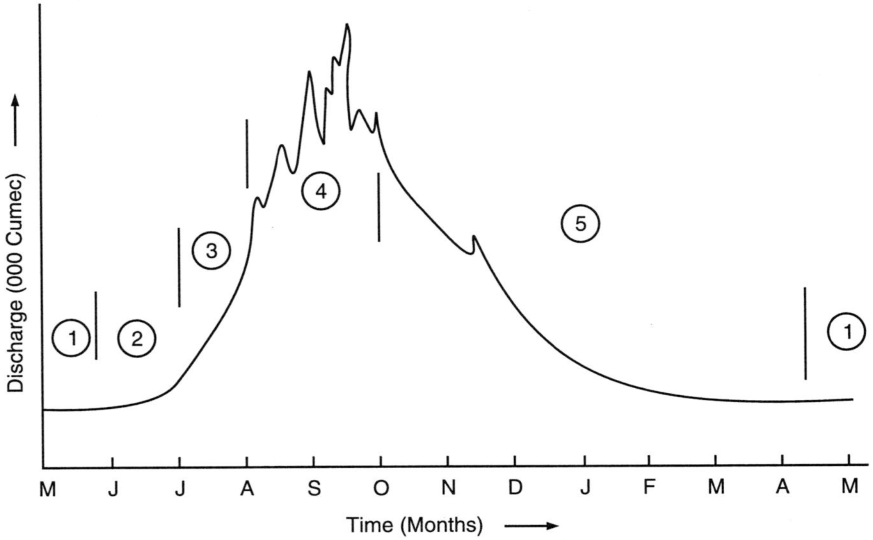

Fig. 4.3. Hydrograph of a stream

2. Second phase indicates initial period of rainfall. Since the soil condition is dry the overland flow reaches the river stream only after satisfying the needs of various abstractions like interception, evaporation, infiltration, depression storage etc. The hydrograph, therefore, shows mild rising trend following the rainfall.

3. Third phase represents quick rising trend of the hydrograph due to setting in of rainy days. By this time the losses also progressively decrease as saturation of soil takes place. Surface runoff becomes more prominent in comparison with abstractions especially infiltration.
4. Fourth phase represents sudden rise in river flows with falling rain. This stage indicates that the surface runoff is more or less equal to rainfall. Obviously water losses are practically nil and whatever abstraction takes place by way of infiltration also reaches river as prompt subsurface runoff. In this phase following every intense rain hydrograph shows sharp rise and fall.
5. Fifth phase starts after rainy season is over. The river flow starts receding. This condition continues for such a time till the abstracted water continues to reach the stream. If phases 2 and 3 are taken as rising limb, the fifth phase represents falling or receding limb.

4.7. COMPONENTS OF SIMPLE HYDROGRAPH

What has been described above is an annual river hydrograph. In hydrologic analysis, however, many times hydrographs resulting from isolated periods of rainfall are to be studied and analysed to calculate volume of runoff. Therefore it is essential to understand various components of total runoff and parts of the hydrograph. As already mentioned the runoff hydrograph is the sum total of three components viz.

- surface runoff (overland flow);
- subsurface runoff (inter flow); and
- groundwater runoff or base flow.

Any typical simple hydrograph with one peak can be divided into five parts (Fig. 4.4.) as follows:

- Approach segment (1 – 2)
- Rising limb (2 – 3)
- Crest segment (3 – 4) with peak at P
- Falling limb (4 – 5)
- Departure segment (5 – 6)

Such a hydrograph results from an isolated and concentrated storm rainfall. The approach segment (1-2) indicates a condition when the stream is fed by groundwater runoff or base flow. As rainfall starts the increasing contribution to the stream occurs by way of surface runoff and prompt sub-surface runoff as shown by the total rising limb (2-P). The rising limb is the function of the time and the rainfall pattern feeding the basin. The crest segment extends from the point of inflection (3) on the rising side of the hydrograph to the another point of inflection (4) on the recession or falling side (P-5) of the hydrograph. The crest point P indicates the peak of hydrograph and represents the highest instantaneous flow in the stream or the **peak flow**. The point of inflection on the falling side of the hydrograph is commonly assumed to indicate the time at which the overland flow or surface runoff stops. From this point onward the falling or recession limb (4-5) indicates withdrawal of water from the **valley storage** or **channel storage** as the case may be.

Valley storage or channel storage is that part of the overland flow which does not get dissipated instantaneously. This volume of water temporarily occupies the space in the river valley or within the banks of the river channel till such time that the passage becomes free enough for this flood to get depleted.

The lower portion of the falling or recession side (5-6) shows decreasing rate of ground water runoff. It can be seen from Fig. 4.4 that the peak flow occurs some time after the rainfall has stopped. The time difference between the centeroid of rainfall excess and the peak of the hydrograph is called the **lag time**. In some situations when after a concentrated heavy rainfall rain further continues with low intensities for extended time, it is possible that the peak flow may occur even before the rainfall has completely ceased. The shape of the falling or recession limb (P-5) is generally not dependent on the rainfall characteristics but mostly depends on the basin characteristics.

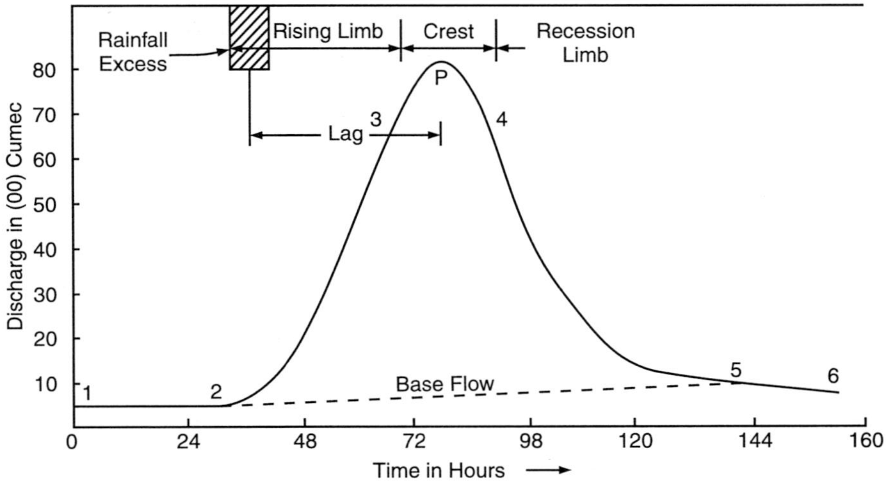

Fig. 4.4. Components of a simple hydrograph

4.8. SEPARATION OF RUNOFF COMPONENTS

Owing, to the complexities of runoff processes it is difficult to precisely separate out various components of total runoff. However, taking advantage of the fact that the runoff processes mostly follow exponential law of flow, various runoff components can be separated rather approximately but with practical utility. The hydrograph analysis given earlier provides sound basis for such separation. When a typical single-peaked hydrograph is plotted on a semi-log paper with discharge values on the logarithmic scale on the ordinate, various parts of hydrograph become more pronounced and are indicated by straight lines. By extending the departure segment line (5-6)

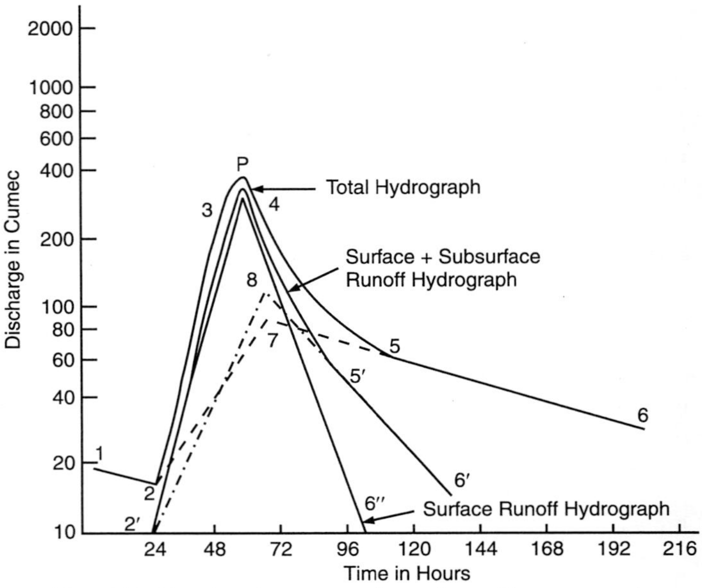

Fig. 4.5. Step-by-step separation of runoff components

backward upto a point (7) lying vertically below the second inflection point (4) and joining this point to the point (2) the groundwater runoff component can be separated out from the total runoff (Fig. 4.5). The balance runoff hydrograph (2′ – 5′ – 6') representing total of surface runoff and prompt subsurface runoff can now be plotted. Using similar process subsurface runoff component can also be separated by extending 5′ – 6′ backward upto point (8) lying vertically below point (4). Now curve (2′ – P – 6″) gives surface runoff hydrograph. Thus step-by-step procedure can be adopted to separate various components of total runoff.

4.9. SEPARATION OF BASE FLOW

The above mentioned step-by-step separation procedure is generally not necessary because prompt subsurface runoff is also contributing to the direct runoff. In practice base flow separation is done in an arbitrary manner. In most cases the base flow constitutes a small percentage of critical peak flow and error that may be involved in arbitrary determination is not significant. It is already mentioned that

$$\text{Base flow} = \text{G.W. runoff} + \text{Delayed subsurface runoff}$$

A simple way to separate base flow from direct runoff is to draw a straight line from point (2), with reference to Figs. 4.4 and 4.6, to an arbitrary point (5) on the lower portion of the recession or falling limb of the hydrograph. The separation line is shown by dotted line (*i*) on Fig. 4.6.

While drawing the line care should be taken to assure that neither the separation line is too long nor it rises too high. Within these constraints a reasonable line can be fitted arbitrarily to separate the bas flow. As a rough guide in selecting the point (5) on the lower portion of the recession limb at which direct runoff stops and beyond which only ground water runoff exists following equation may be used:

$$N = 0.8\, A^{0.2}$$

where N is number of days after the peak of the hydrograph has occurred

and A is drainage area in km^2.

This method of locating the point of termination of direct runoff is adhoc and should only be adopted as guidance in selecting the point by visual inspection of the hydrograph.

For separation of base flow other two different concepts may also be advanced. They are the following:

(*i*) As the river water level starts rising fast due to increased overland flow the groundwater runoff into the stream decreases. At this stage due to water level difference the water flows from the stream into the ground and creates what is called **bank storage**. After some time when the water level in the river starts falling below the groundwater the groundwater runoff starts increasing and bank storage gets depleted. The base flow separation line on this concept is shown by dotted line (*ii*) in Fig. 4.6. To obtain the separation line approach segment is extended forward upto a point vertically below point *P*. The drawback of this concept is that although the reasoning appears justified it is neither possible to locate

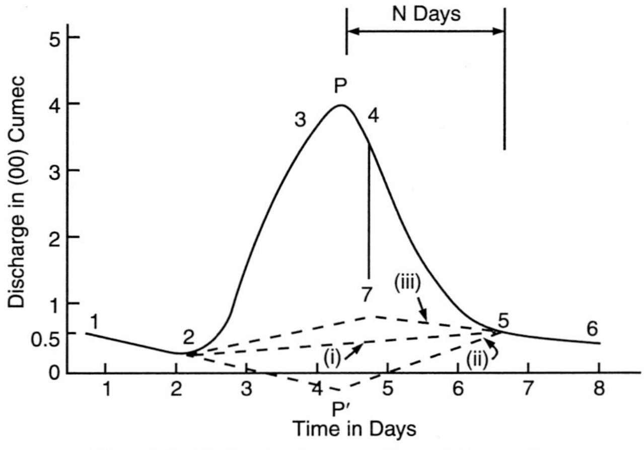

Fig. 4.6. Method of separation of base flow

the point from where again ground water runoff starts nor slope of recession or rise of base flow contribution can to correctly given. This method is, therefore, equally arbitrary.

(*ii*) According to this concept it is more justified to extend departure segment of the recession limb backward upto a point vertically below the inflection point on the recession limb to separate the base flow. The base flow separation line is shown by dotted line (*iii*) in Fig. 4.6. This type of separation is valid where terrain is extremely porous and groundwater is plentifully available and reaches the stream quickly. Since such terrains (*e.g.*, limestone terrains) are not always met with this concept is not generally applicable. Considering all, first method of drawing straight line of separation from point 2 to 5 appears quite practical.

4.10. SEPARATION OF COMPLEX HYDROGRAPH

Analysis of simple single peak hydrograph has been described in the earlier paragraphs. It may be noted that generation of such a hydrograph requires isolated intense storm of such nature that it is not followed by another rain storm till the direct runoff has left the drainage basin. In practice, however, many times an intense storm is followed by another storm after a short interval. Such rainfall pattern generates a multi peak or complex hydrograph as shown in Fig. 4.7.

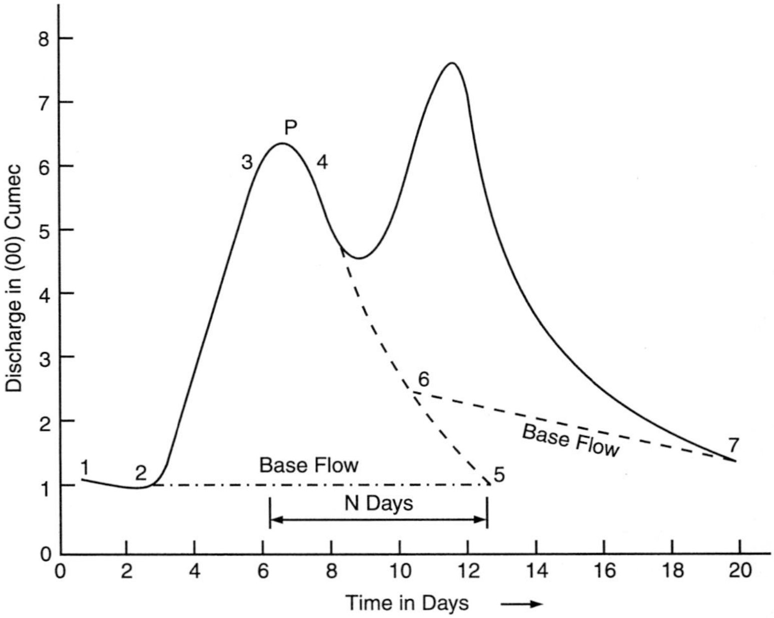

Fig. 4.7. Complex or multipeak hydrograph

In such cases to calculate the direct runoff caused by each rain burst it is necessary to separate two peaks of hydrograph and outline two simple single peak hydrographs. This can be achieved by projecting the short recession limb (P-4) of the first hydrograph as shown in Fig. 4.7. This projection is taken upto the point (5) the location of which is given by distance N (= 0.8 $A^{0.2}$) days on the abscissa from point P. The base flow is separated in usual arbitrary manner. The base flow of second hydrograph is separated by extending the departure segment backward as shown by dotted line 6 – 7 in Fig. 4.7.

4.11. UNIT HYDROGRAPH THEORY

The problem of flood and its estimation constitutes important aspect of hydrological study. The peak flow gives an instantaneous-value of discharge. It alone does not provide sufficient information about the runoff or the volume of water discharged during the flood. This information can be obtained from the direct runoff component of the flood hydrograph. Introduction of unit hydrograph theory in 1932 by Sherman made it possible to predict a runoff hydrograph corresponding to an observed or hypothetical storm. The basic concept of unit hydrograph theory is that for any drainage basin whose topographical features (*e.g.*, shape, size, slope) are not undergoing change, the shape of hydrographs of runoff from storms of similar rainfall characteristics would be considerably similar.

Definition unit hydrograph: It is a typical hydrograph of direct runoff which gets generated from one centimetre of effective rainfall falling at a uniform rate over the entire drainage basin uniformly during a specific duration.

It is already mentioned, that effective rainfall is that portion of rainfall which fully contributes towards direct runoff. Therefore, unit hydrograph can also be defined as the hydrograph of a drainage basin which gives one centimetre of direct runoff from a rain storm of specific duration[1].

4.11.1. Assumptions in Unit Hydrograph Theory

The theory of unit hydrograph is based on certain assumptions. The main assumptions are following:

(*i*) The effective rainfall is uniformly distributed over the entire drainage basin.

(*ii*) The effective rainfall occurs uniformly within its specifier duration.

This requirement calls for selection of storms of so small a duration which would generally produce an intense and nearly uniform effective rainfall and would produce a well defined single peak of hydrograph of short time base. Such a storm can be termed as "unit storm".

(*iii*) The effective rainfalls of equal (unit) duration will produce hydrographs of direct runoff having same or constant time base.

(*iv*) The ordinates of the direct runoff hydrographs having same time base (*i.e.*, hydrographs due to effective rainfalls of different intensity but equal duration) are directly proportional to the total amount of direct runoff given by each hydrograph.

This important assumption is called principle of linearity or proportionality or superposition.

(*v*) The hydrograph of runoff from a given drainage basin resulting, from a given pattern of rainfall reflects all the combined physical characteristics of the basin.

In other words the hydrograph of direct runoff resulting from a given pattern of effective rainfall will remain invariable irrespective of its time of occurrence.

This assumption is called principle of time invariance.

4.11.2. Limitations of Unit Hydrograph Theory

(*i*) In theory, the principle of unit hydrograph is applicable to a drainage basin of any size. In practice, however, uniformly distributed effective rainfall rarely occurs on large areas. Also on large areas effective rainfall is very rarely uniform at all locations, within its specified duration. Obviously bigger the area of the drainage basin lesser will be the chances of fulfilling the assumptions enunciated above. The limiting size

1. Specific duration or unit time as originally called is linked with an intense storm excess. It varies from one day for large catchments (over 5000 km^2) to a few hours for small ones (say 3 hours for a catchment area of 750 km^2). Actually unit time or specific duration should be less than time of concentration[2] or peaking time. As a thumb rule unit time may be $\frac{1}{3}$ to $\frac{1}{4}$ of the peaking time.
2. Time of concentration or peaking time is the time in hours taken by the rainwater that falls at the farthest point to reach, the outlet of a catchment.

of the drainage basin is considered to be 5000 km^2. Beyond it the reliability of the unit hydrograph method diminishes.

When the area of the drainage basin exceeds a few thousand km^2. The catchment has to be divided into sub-basins and the unit hydrographs developed for each sub-basin. The flood discharge at the basin outlet can then be estimated by combining the sub-basin floods adopting flood routing procedure.

(*ii*) The unit hydrograph method can not be applied when appreciable portion of storm precipitation falls as snow because snow-melt runoff is governed mainly by temperature changes.

(*iii*) Also when snow covered area in the drainage basin is significant the unit hydrograph method becomes inapplicable. The reason is that the storm rainfall gets mixed up with the snow pack and may produce delayed runoff differently under different conditions of snow pack.

(*iv*) The physical basin characteristics change with seasons, man-made structures in the basin, conditions of flow etc. Obviously the principle of time invariance is really valid only when the time and condition of the drainage basin are specified.

(*v*) It is commonly seen that no two rain storms have same pattern in space and time. But it is not practicable to derive separate unit hydrograph for each possible time-intensity pattern. Therefore, in addition to limiting drainage basin area upto 5000 km^2 if storms of shorter duration say $\frac{1}{3}$ to $\frac{1}{4}$ of peaking time are selected it is seen that the runoff patterns do not vary drastically.

(*vi*) The principle of linearity is also not completely valid. This is so because due to variability in proportion of surface, subsurface and groundwater runoff components during smaller and larger storms of same duration, the maximum ordinate (peak) of the unit hydrograph derived from smaller storm is smaller than the one derived from larger storm. Obviously the character and duration of recession limb which is a function of the peak flow will also be different. When appreciable non-linearity is seen to exist it is necessary to use derived unit hydrographs only for reconstructing events of similar magnitude.

(*vii*) The unit hydrograph can be used theoretically to construct a flood hydrograph resulting from a storm having same unit duration. Obviously it necessitates construction of several unit hydrographs to cover different durations of storms. In practice however it is seen that a tolerance of ± 25% in unit hydrograph duration is acceptable. Thus a 2 hour unit hydrograph can be applied to storms of 1.5 to 2.5 hours duration.

4.11.3. Advantages of Unit Hydrograph Theory

The limitation to the theory of unit hydrograph can be overcome to a large extent by remaining within the various ranges and restrictions indicated above. The unit hydrograph theory has several advantages to its credit which can be summarised as below :

- Flood hydrograph can be calculated with the help of very short record of data.
- In addition to peak flow unit hydrograph also gives total volume of runoff and its time distribution.
- The unit hydrograph procedure can be computerised easily to facilitate calculations.
- It is very useful in checking the reliability of flows obtained by using statistical methods.

4.12. CONSTRUCTION OF UNIT HYDROGRAPH

In order to construct unit hydrograph for any drainage basin it is necessary to collect concurrent basic rainfall and discharge data. The rainfall data should be obtained from well distributed rain gauging stations equipped with self recording raingauges which give areal and time distribution of rainfall. Also continuous stage hydrograph or discharge hydrograph data from water level recorder which gives complete shape of hydrograph at the outlet of the drainage basin is required.

The collected rainfall data is analysed to obtain hyetograph of effective rainfall of specific or unit duration. Similarly corresponding direct runoff hydrograph is obtained from the total discharge hydrograph. These two things form the two basic requisites in derivation of unit hydrograph. The construction of unit hydrograph can be accomplished in the following steps:

1. From the past rainfall records select an isolated intense storm-rainfall of specific or unit duration,
2. For this storm using isohyetal or Thiessen polygon method calculate average depth of precipitation over the drainage basin.
3. Using (SRRG) self recording rain gauge data of all the available stations plot mass curves of rainfall for this storm and obtain, average mass curve of rainfall.
4. From the average mass curve of rainfall construct hyetograph. To construct hyetograph incremental rainfall quantities during successive units of time are obtained from the mass curve. The average depths of rainfall per unit of time are then plotted on ordinate against time as abscissa.
5. Using stage hydrograph and stage-discharge relationship obtain a complete discharge hydrograph (covering some time before as well as after the storm) at the drainage outlet for the selected storm.
6. If the recession limb is not smooth and contains bumps, make recession limb smooth or normal by removing the bumps.
7. Separate the base flow from total storm hydrograph using suitable empirical method described earlier in para 4.9.
8. Subtracting the base flow components plot and obtain ordinates of direct runoff hydrograph.
9. Either by planimeter or by mathematical calculations find out the volume of the direct runoff.
10. Convert this volume into cm of runoff over the catchment area.

    For example:

    Volume of direct runoff = Area within the hydrograph = Σ Ordinates $\times t = \Sigma 0 \times t$

    where Σ Ordinates = Sum of the ordinates of direct runoff hydrograph at equal time interval

    t = Time interval between successive ordinates

    To get volume of runoff in m^3, ordinates have to be converted in cumec and time interval in seconds.

    This volume can be converted into cm of runoff by dividing the same by the area of the drainage basin $m^2 \times 100$.

    Note: It is already made clear earlier that:

    Depth of direct runoff = Effective rainfall

11. The duration of effective rainfall of the storm of specific or unit duration is determined by drawing a horizontal line on the hyetograph in such a way that the area of the hyetograph above the horizontal line is equal to the volume of direct runoff. Obviously the area below the horizontal line gives the abstractions. This is an arbitrary method. For elaborate analysis infiltration indexes and curves will have to be used.
12. Measure the ordinates of direct runoff hydrograph.
13. Divide these ordinates of direct runoff hydrograph by the obtained depth of runoff in cm to get ordinate of unit hydrograph.

    Mathematically,

$$\text{Ordinate of direct runoff hydrograph} = \frac{\text{Ordinate of direct runoff hydrograph}}{\text{Depth of direct runoff in cm}}$$

14. Plot these ordinates against uniform and the same time interval as the one used in direct runoff hydrograph to get unit hydrograph.

    Derivation of unit hydrograph can be done very conveniently by making step-by-step calculations in a tabular form. Following example makes the procedure clear.

PROBLEM ON UNIT HYDROGRAPH

PROBLEM 4.1. On a drainage basin of 400 km^2 size 8 hours storm was selected for construction of unit hydrograph. The rainfall increments observed from the average mass curve of the storm rainfall are as given in Table 'A' below.

Table A. Problem 1

Date	*July 9*			
Time (hours)	800 to 1000	1000 to 1200	1200 to 1400	1400 to 1600
Rainfall increment (cm)	1.07	3.90	3.74	1.09

Assume average loss rate of 1 cm/hr for calculating effective rainfall.

The ordinates of the observed hydrograph and base flow are as given in Table 'B'.

Construct unit hydrograph for the basin from the above data and determine specific or unit duration of the unit hydrograph.

Table B. Problem 1

Date	*Time (hrs)*	*Hydrograph ordinate (cumec)*	*Base flow ordinate (cumec)*
July 9	0800	4	4.0
	1200	93	3.5
	1600	237	3.0
	2000	301	2.0
	2400	234	3.0

Date	*Time (hrs)*	*Hydrograph ordinate (cumec)*	*Base flow ordinate (cumec)*
July 10	**0400**	**97.5**	**3.5**
	0800	**41**	**4.0**
	1200	**22**	**4.0**
	1600	**11**	**4.0**
	2000	**6**	**4.0**
	2400	**4**	**4.0**

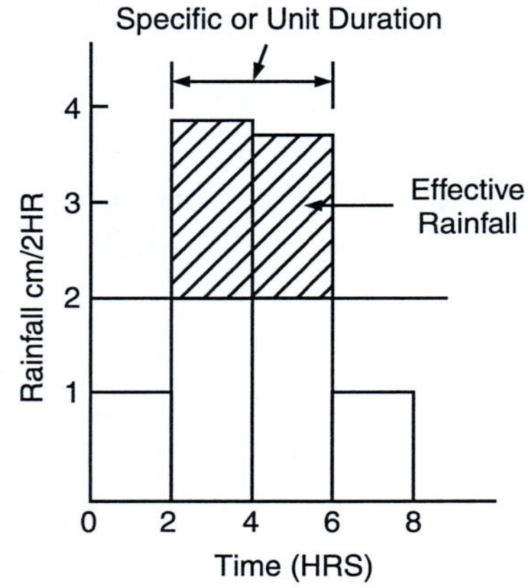

Fig. 4.8. Problem 1 Hyetograph

Solution: Step 1. Using rainfall data hyetograph can be constructed as shown in Fig. 4.8. The horizontal line (ϕ line) can be drawn at 2 cm ordinate. The area above the line gives effective rainfall. It may be also seen that the duration of effective rainfall is 4 hours which is equal to specific or unit duration of the hydrograph.

Step 2. The amount of effective rainfall can be calculated as shown in Table 4.5. The total effective rainfall works out to be 3.64 cm.

Table 4.5.

Date	*Time (hrs)*	*Rainfall increment (cm)*	*Average loss (cm/2 hrs)*	*Effective rainfall (cm)*
July 9	0800 to 1000	1.07	2	—
	1000 to 1200	3.90	2	1.90
	1200 to 1400	3.74	2	1.74
	1400 to 1600	1.09	2	—
Total				3.64

Step 3. The ordinates of the direct runoff hydrograph and 4 hour unit hydrograph can be calculated as shown in Table 4.6. It may be noted that

Direct runoff ordinate = Total runoff ordinate – Base flow ordinate, and

$$\text{Unit hydrograph ordinate} = \frac{\text{Direct runoff ordinate}}{\text{Direct runoff expressed in cm}}$$

Step 4. Volume of direct runoff = $\Sigma\ 0 \times t = 1011 \times 4 \times 60 \times 60 = 4044 \times 3600\ \text{m}^3$

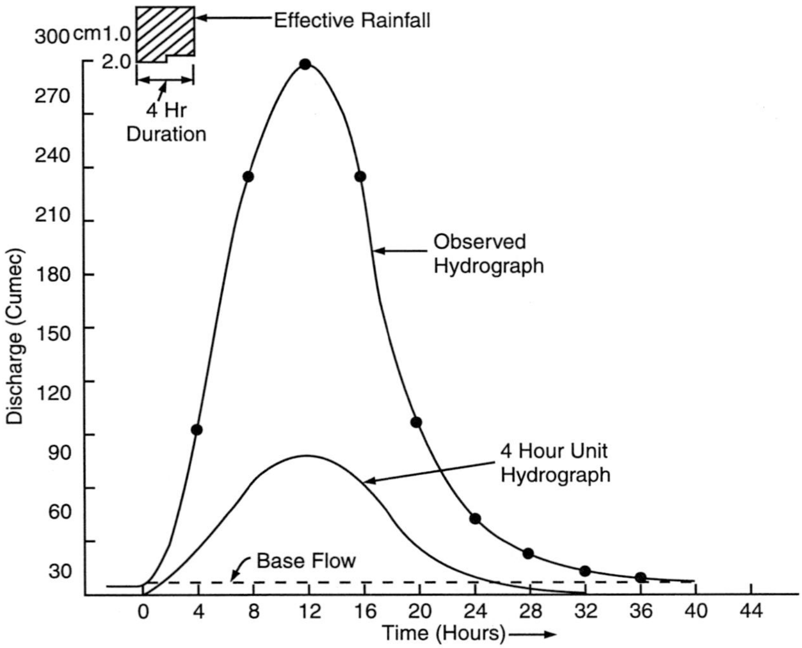

Fig. 4.9. Problem 1 Unit hydrograph construction

Step 5. Depth of direct runoff in cm $= \dfrac{4044 \times 3600}{\text{Area of basin in m}^2} \times 100$

$$= \frac{4044 \times 3600 \times 100}{400 \times 1000 \times 1000} = 3.64 \text{ cm}$$

Table 4.6. Solution Problem 1

Date	*Time (hr)*	*Hours after start*	*Hydrograph ordinates (cumec)*	*Base flow (cumec)*	*Direct runoff ordinates (cumec)*	*Unit Hydrograph ordinates (cumec)*
(1)	*(2)*	*(3)*	*(4)*	*(5)*	*(6) = (4 – 5)*	*(7)* $= \dfrac{(6)}{3.64}$
July 9	0800	0	4	4.0	0	0
	1200	4	93	3.5	89.5	24.6
	1600	8	237	3.0	234	64.3
	2000	12	301	2.0	299	82.1
July 10	2400	16	234	3.0	231	63.5
	0400	20	97.5	4.0	39.5	25.7
	0800	24	41	4.0	37	10.2
	1200	28	22	4.0	18	4.9
	1600	32	11	4.0	7	1.9
	2000	36	6	4.0	2	0.6
	2400	40	4	4.0	0	0
Total					1011	277.8

From step 2 and step 5

Depth of direct runoff =Depth of effective rainfall = 3.64 cm

Hence, stepwise calculations are correct.

Step 6. Volume of runoff from unit hydrograph = $\Sigma\ 0 \times t = 277.8 \times 4 \times 60 \times 60\ \text{m}^3$

$$\text{Depth of runoff from unit hydrograph in cm} = \frac{277.8 \times 4 \times 60 \times 60}{400 \times 1000 \times 1000} \times 100 = 1 \text{ cm}$$

(This is as per definition of unit hydrograph)

Step 7. Now using the values in columns (3) and (7) ordinates of unit hydrograph can be plotted against time as abscissa to construct unit hydrograph as shown in Fig. 4.9.

4.13. ESTIMATION OF RUNOFF AND PEAK FLOW BY UNIT HYDROGRAPH

It can now be recognised that a unit hydrograph is a kind of unit measure of runoff occurring as a result of rainfall of specific or unit duration. A unit hydrograph gives 1 cm of direct runoff obviously resulting from 1 cm of effective rainfall. Accordingly if the effective rainfall in the same duration is double, that is 2 cm, then the ordinates of the direct runoff hydrograph will be double in size the total runoff period remaining the same.

The procedure of obtaining hydrograph from the unit hydrograph is just the reverse of the procedure adopted for constructing unit hydrograph. To obtain direct runoff hydrograph resulting from the storm of same unit duration multiply the ordinates of the unit hydrograph by the runoff value or the effective rainfall expressed in cm.

Similarly peak flow resulting from the storm of same unit duration can be obtained by multiplying maximum ordinate of the unit hydrograph by the runoff value expressed in cm or effective rainfall expressed in cm. Figure 4.10 is self explanatory and shows how a runoff hydrograph is construction from a unit hydrograph.

In may be remembered that always rain storms may not occur with same unit duration. In such cases the unit hydrograph theory permits superimposition. If a storm of longer duration occurs it is possible to divide it into two or more storms of shorter duration for which unit hydrograph is available. Then say if the rain burst producing 1 cm effective rainfall is succeeded by another rain producing 2 cm effective rainfall in equal duration, the total hydrograph can be obtained by drawing the two hydrographs with proper lag time but overlapping each other as if they are two independent hydrographs. The summation the overlapped portion gives the ordinates of total hydrograph.

Generally the calculations are done in a tabular form. Following example makes the procedure clear.

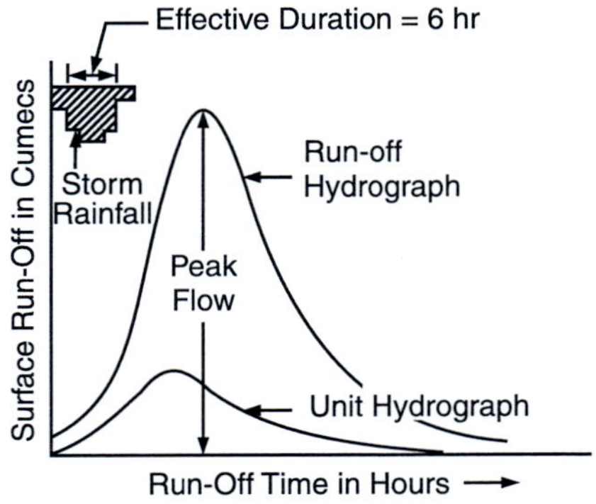

Fig. 4.10. Peck flow by unit hydrograph method

PROBLEM 4.2. The ordinates of a six hour unit hydrograph for a catchment area of 1000 km² are given below.

Time in hours	6	12	18	24	30	36	42
Ordinates in cumec	80	200	146	80	50	15	0

The storm lasted for 18 hours. The effective rainfall increments (rainfall excess) for six hour periods are 0.6 cm, 2 cm and 1 cm respectively. Compute the runoff hydrograph. Also find out the peak flow and the volume of runoff.

A base flow of 100 cumec may be assumed.

Solution: The computation can be done in a tabular form as shown in Table 4.7.

Column-9 of the Table 4.7 gives the ordinates of the runoff hydrograph. The peak flow is 664 cumec being the maximum ordinate. Following graph illustrates the construction of runoff hydrograph. It is obtained by plotting the base flow and the three hydrographs of 6 hour unit time with a lag of 6 hours. The ordinates of the three hydrographs are taken from the columns 4, 5 and 6.

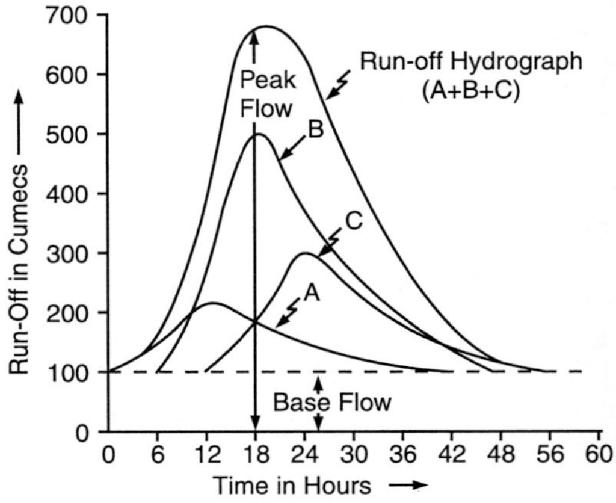

Fig. 4.11. Problem 2

Total volume of runoff = Area of the hydrograph

$$= \Sigma \frac{\text{Sum of two successive ordinates}}{2} \times \text{Unit interval}$$

Table 4.7: Solution Problem 2

(Unless stated all figures in cumec)

Time of start of effective rainfall (hr)	*6 hour unit hydrograph ordinate*	*Effective rainfall increments for 6 hours periods in cm*	*Direct runoff due of effective rainfall increments of*			*Total direct runoff*	*Base flow*	*Run-off hydrograph ordinates*
			0.6 cm	*2 cm*	*1 cm*			
1	*2*	*3*	*4*	*5*	*6*	*7*	*8*	*9*
0	0	0	0	—	—	—	100	100
6	80	0.6	48	—	—	48	100	148
12	200	2	120	160	—	280	100	380
18	140	1	84	400	80	565	100	664
24	80		48	280	200	528	100	628
30	50		30	160	140	330	100	430
36	15		9	100	80	189	100	289
42	0		0	30	50	80	100	180
48				0	15	15	100	115
54					0	0	100	100
		Total				2034	1000	3034

$= 17604$ cumec-hours $= 17604 \times 3600\ m^3 = 63.37$ million m^3

Alternatively from column (7)

Direct runoff volume $= \Sigma\ 0 \times t = 2034 \times 6 \times 60 \times 60$ ...(*i*)

$= 43.93$ million m^3

Base flow volume $= 100 \times 54 \times 60 \times 60$ ...(*ii*)

$= 19.44$ million m^3

Total runoff volume $= 63.37$ million m^3

4.14. CONSTRUCTION OF UNIT HYDROGRAPH FOR OTHER DURATIONS

When a unit hydrograph of a desired unit duration can not be developed by analysis of available data it may be derived by conversion of available unit hydrograph by using **principle of superposition**. It is, however, necessary that the other duration is an integral multiple of available unit duration. The method followed in this case is quite straightforward. For example, it is required to derive a 9 hour unit hydrograph from a 3 hour unit hydrograph. To achieve this the 3 hour unit hydrograph is plotted thrice with a lag of 3 hour between each of them. The three over lapping unit hydrograph ordinates are then summed up to yield a hydrograph of 3 cm runoff in 9 hour effective rainfall duration. The ordinates of the 9 hour summation hydrograph are divided by 3 to get a 9 hour unit hydrograph which represents 1 cm effective rainfall in 9 hour duration.

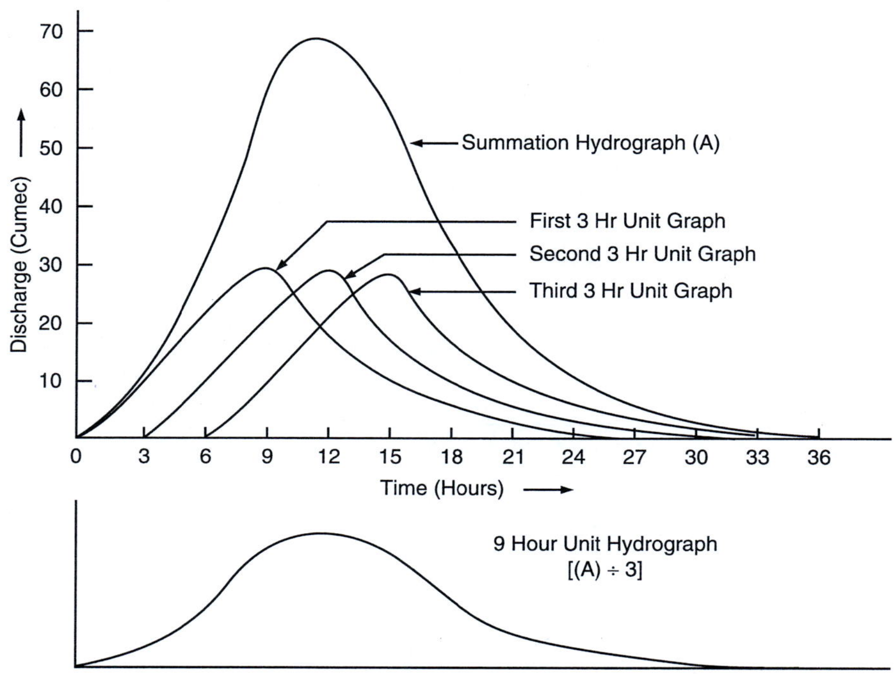

Fig. 4.12. Problem 3

It may be seen from Fig. 4.12 of problem 3 that the runoff period (*i.e.* time base) of 9 hour unit hydrograph is longer than 3 hour unit hydrograph. This is obvious because the effective rainfall intensity for 3 hour unit hydrograph was greater (*i.e.*, $\frac{1}{3}$ cm per hour) whereas for 9 hour unit hydrograph it is lower at the rate of 1/9 cm per hour. Also the peak of 9 hour unit hydrograph will be lower than the 3 hour unit hydrograph. Following example makes the procedure quite clear.

PROBLEM 4.3. Given below are the ordinates of a 3 hour unit hydrograph. Construct 9 hour unit hydrograph from it.

Time(hr)	0	3	6	9	12	15	18	21	24	27	30
Discharge (cumec)	0	10	22	30	16	10	6	3	1	0.5	0

Solution: The 9 hour unit hydrograph can he obtained by performing calculations in a tabular form as shown in Table 4.8. Also refer Fig. 4.12. Column (6) gives 9 hour Unit Hydrograph.

Table 4.8: Solution Problem 3

(Hydrograph ordinates in cumec)

Time (hr)	*3 hr unit hydro. ordinates*	*3 hr unit hydro. with 3 hr offset*	*3 hr unit hydro. with 6 hr offset*	*Summation hydrograph*	*9 hr unit hydrograph ordinates*
(1)	*(2)*	*(3)*	*(4)*	*(5)=(2)+(3)+(4)*	*(6) = (5) ÷ 3*
0	0	—	—	0	0
3	10	0	—	10	3.33
6	22	10	0	32	10.67
9	30	22	10	62	20.67
12	16	30	22	68	22.67
15	10	16	30	56	18.67
18	6	10	16	32	10.67
21	3	6	10	19	6.33
24	1	3	6	10	3.33
27	0.5	1	3	4.5	1.50
30	0	0.5	1	1.5	0.50
33	—	0	0.5	0.5	0.16
36	—	—	0	0	0
Total	98.5	98.5	98.5	295.5	98.50

4.15. 'S' HYDROGRAPH

The method of simple superposition explained above is applicable only when other duration is integral multiple of the available duration of the unit hydrograph. However, many a times it becomes necessary to generate a unit hydrograph of other duration which is not integral multiple of the available duration. In such cases it becomes necessary to evolve a general method based on the principle of superposition. Such a method has been evolved by hydrologists which is called '*S*' hydrograph method.

A '*S*' hydrograph is nothing but a hydrograph generated by a continuous effective rainfall occurring at an uniform rate for an indefinite period. It is called '*S*' hydrograph because the shape of the hydrograph comes out like alphabet '*S*' though slightly deformed. Figure 4.13 shows a typical '*S*' hydrograph.

From the knowledge of rainfall-runoff process it can be visualised that since the rainfall is occurring at an uniform rate indefinitely a stage will reach from when the direct runoff rate will equal effective rainfall rate. Such a point is called a **point of equilibrium**. From the time of equilibrium onwards the ordinate of hydrograph equals the rate of effective rainfall and the hydrograph curve traces a horizontal line parallel to *X*-axis.

Following the Unit Hydrograph theory when effective rainfall occurs at uniform rate continuously for indefinite period it (horizontal line) can be divided into several equal durations of convenient magnitude so that effective rainfall in that duration totals up to 1 cm. In other words it gives a particular unit hydrograph duration. Now if the same unit hydrograph is plotted successively with offset equal to unit duration and the overlapping ordinates are summed up we get ordinates of '*S*' hydrograph. At this stage it is necessary to know how many times the unit hydrograph have to be successively plotted. As a thumb rule the unit hydrograph may be added *T*/*t* times with offset equal to unit duration of the unit hydrograph.

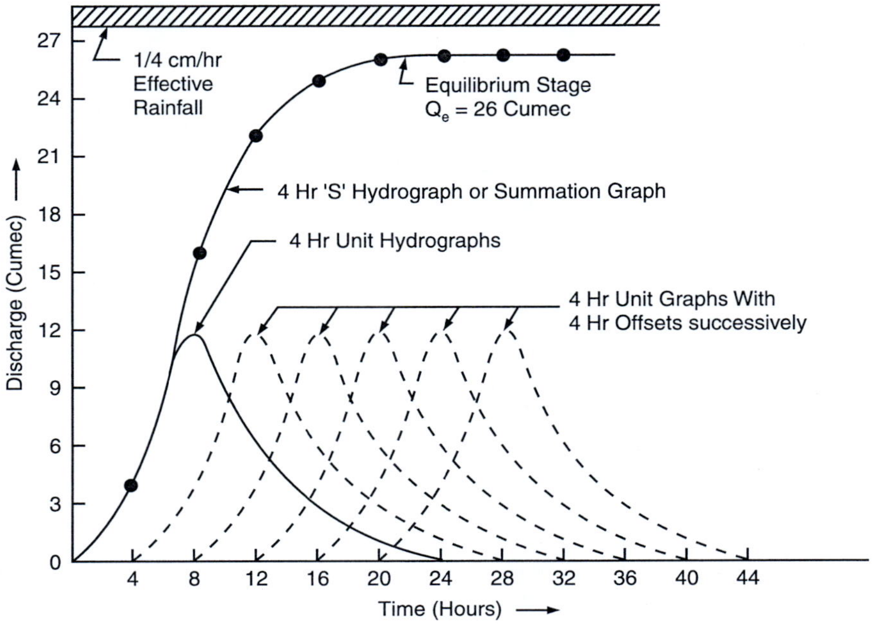

Fig. 4.13. Construction of 'S' hydrograph (Problem 4)

where T is time base of unit hydrograph;

and t is unit duration of unit hydrograph.

The '*S*' hydrograph ordinate at equilibrium stage has the maximum value because after this point the ordinate has same value which is equal to rate of effective rainfall. The value of maximum ordinate of '*S*' hydrograph (say Q_e) at equilibrium stage can be calculated as follows:

We know that rate of direct runoff which is also equal to rate of effective rainfall is such that 1 cm runoff takes place in unit duration '*t*'.

$$\therefore \quad Q_e = \frac{\text{Area of basin} \times 1\,\text{cm}}{t}$$

$$= 1000 \times 1000 \times \text{A} \times \frac{1}{100} \times \frac{1}{60 \times 60 \times t} \text{ cumec} = \frac{2.78\,\text{A}}{t} \text{ cumec}$$

where A is area of the drainage basin in km^2;

and t is unit duration of unit hydrograph in hours.

Now since 1 cm direct runoff or effective rainfall occurs in t hours $1/t$ gives intensity of effective rainfall.

If intensity of effective rainfall in cm/hr is denoted by I

$$Q_e = 2.78 \text{ A I cumec}$$

Use of '*S*' hydrograph: The '*S*' hydrograph provides convenient method of deriving a unit hydrograph of any duration from the available duration and it may be either long or short duration, integral or fractional multiple of the available unit hydrograph duration.

Following example makes the procedure of derivation of '*S*' hydrograph clear.

PROBLEM 4.4. The ordinates of a 4 hour unit hydrograph are given below. Using the principle of superposition construct a '*S*' hydrograph and calculate the discharge at equilibrium stage and the time of its occurrence from the beginning of direct runoff.

Time hr	0	4	8	12	16	20	24
cumec	0	4	12	6	3	1	0

Solution: The '*S*' hydrograph can be obtained by arranging the 4 hr unit hydrographs successively with 4 hour offset and adding the overlaps. The number of times such unit hydrograph addition required is given by ratio T/t.

From the given data T = 24 hrs and t = 4 hours.

Since $T/t = 6$; The unit hydrograph needs plotting at least 6 times shown in Table 4.9 and Fig. 4.13.

Table 4.9: Derivation of 'S' hydrograph

(all values in cumec unless specified)

Time (hr)	*ordinates of 4 hr unit hydrograph*	*ordinates of 4 hr unit hyd. with 4 hr. offset successively*						*ordinates of 4 hr S hydrograph*
(1)	*(2)*	*(3)*	*(4)*	*(5)*	*(6)*	*(7)*	*(8)*	*Σ(2) to (8)*
0	0	—	—	—	—	—	—	0
4	4	0	—	—	—	—	—	4
8	12	4	0	—	—	—	—	16
12	6	12	4	0	—	—	—	22
16	3	6	12	4	0	—	—	25
20	1	3	6	12	4	0	—	26*
24	0	1	3	6	12	4	0	26
28		0	1	3	6	12	4	26
32			0	1	3	6	12	& so on
36				0	1	3	6	
40					0	1	3	
44						0	1	

✍ **Note:** Sometimes summation of various overlapping ordinates does not produce smooth '*S*' curve. In such cases smoothening can be done by drawing average curve and adopting those values for final '*S*' hydrograph. In the present case it is not necessary.

From Table 4.9 it may be seen that the discharge at equilibrium stage is 26 cumec and it occurs at 20th hour from the beginning of effective rainfall.

4.16. OTHER DURATION HYDROGRAPH BY 'S' HYDROGRAPH METHOD

Other duration unit hydrograph can be derived by adopting following step-by-step procedure:

(*i*) If instead of giving '*S*' hydrograph ordinates only unit hydrograph ordinates of a particular duration are given then it becomes necessary to first construct '*S*' hydrograph. This can be done by successively adding the unit hydrograph ordinates arranged with offset equal to unit duration. The method to be followed is described earlier.

(*ii*) Let us assume that '*S*' hydrograph is available with such continuous effective rainfall that 1 cm of effective rainfall occurs in a duration of t_1 hours, *i.e.*, '*S*' hydrograph is obtained from t_1 hour unit hydrograph. The intensity of effective rainfall I is therefore $1/t_1$ cm per hour. Plot this '*S*' hydrograph ordinates against time on a graph and tabulate.

(*iii*) Offset or advance the position of ordinates of '*S*' hydrograph by a period (say t_2 hours) for which duration unit hydrograph is to be obtained. This is called offset '*S*' hydrograph. Tabulate ordinates of offset '*S*' hydrograph.

(*iv*) Find the difference between the ordinates of the two '*S*' hydrographs and plot the difference hydrograph. The difference hydrograph represents runoff occurring due to t_2 hours effective rainfall occurring at a uniform intensity I (*i.e.*, $1/t_1$ cm per hour). Thus the hydrograph gives $I \times t_2$ or $\frac{t_2}{t_1}$ cm runoff.

(*v*) To obtain unit hydrograph for t_2 hour duration (which should give 1 cm runoff) divide the ordinates of difference hydrograph by $I \times t_2$ or $\frac{t_2}{t_1}$. Obtain the ordinates and plot the unit hydrograph.

PROBLEM 4.5. Using '*S*' hydrograph ordinates given in Problem 4.4 derive a unit hydrograph for 6 hours duration.

Solution:

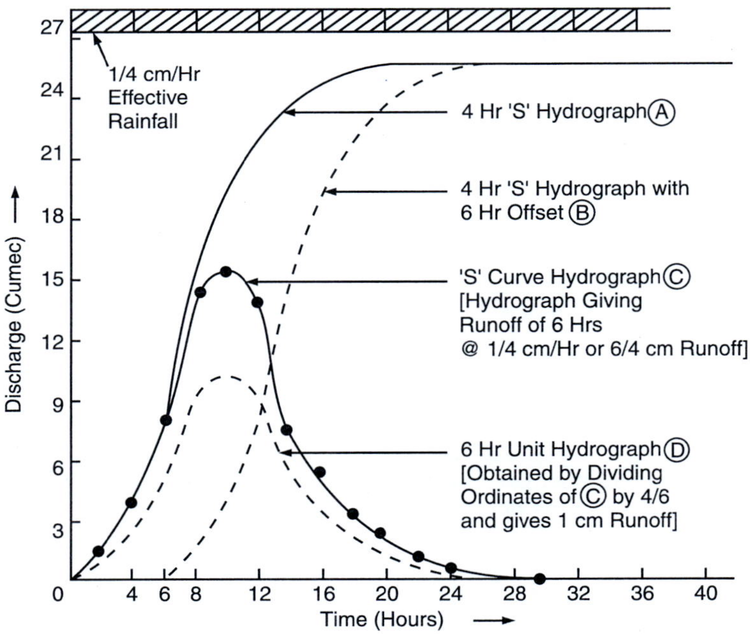

Fig. 4.14. Other duration unit hydrograph (Problem 5)

The ordinates of 4 hour '*S*' hydrograph may be read at 2 hour interval to facilitate calculations. The ordinates of 4 hour '*S*' hydrograph and offset 'S' hydrograph with 6 hours offset may be plotted and tabulated as shown in Fig. 4.14 and Table 4.10 respectively. Find the difference in two '*S*' hydrograph ordinates which gives a hydrograph representing $6 \times \frac{1}{4}$ cm runoff. Divide ordinates of this hydrograph by 6/4 (t_2/t_1) to get unit hydrograph ordinates of 6 hr (t_2 hr) duration. Here t_1 hr is duration of given unit hydrograph and t_2 hr is duration of desired unit hydrograph. The calculations can be done as shown in Table 4.10.

Table 4.10: Derivation of other duration unit hydrograph

(all values in cumec unless specified)

Time (hr)	*Ordinates of 4 hr 'S' Hydrograph*	*Ordinates of offset 'S' Hydrograph*	*Difference Hydrograph (2) – (3)*	*Ordinates of 6 hr. unit Hydrograph* $(4) \div \left(\frac{6}{4}\right)$
(1)	(2)	(3)	(4)	(5)
0	0	—	0	0
2	1.5	—	1.5	1
4	4	—	4	2.67
6	8	0	8	5.33
8	16	1.5	14.5	9.67
10	19.5	4	15.5	10.33
12	22	8	14	9.33
14	23.5	16	7.5	5
16	25	19.5	5.5	3.67
18	25.5	22	3.5	2.33
20	26	23.5	2.5	1.67
22	26	25	1	0.67
24	26	25.5	0.5	0.33
26	26	26	0	0
28	26	26	0	0
30	26	26	0	0
32	26	26	0	0

4.17. CONSTRUCTION OF UNIT HYDROGRAPH FROM A COMPLEX HYDROGRAPH

Many times it becomes difficult to select an ideal single peak isolated rain storm to derive unit hydrograph. In such cases it becomes necessary to analyse complex direct runoff hydrograph to derive unit hydrograph. The analysis is done in step-by-step manner using the principle of superposition as outlined below.

(*i*) Select a storm in such a way that the whole storm can be divided into few consecutive effective rainfall bursts of same unit duration. Let us assume that in the present case three such periods generate the complex hydrograph.

(*ii*) Separate out the three peaks of complex hydrograph and the base flow using the method described in para 4.10. Find out amount of direct runoff either by using graph or planimeter during each rain burst.

(*iii*) Calculate effective rainfall occurring in each rain burst by hyetograph technique and ϕ index. Let the effective rainfall values during the three unit durations be R_1, R_2 and R_3 cm

(*iv*) Measure the direct runoff ordinates of the complex storm at the interval of unit duration. Let the ordinates be Q_1, Q_2, Q_3, Q_4 ..... say Q_n.

(*v*) Assume that ordinates of unit hydrograph are U_1, U_2, U_3, U_4, U_5 etc.

(*vi*) It may be seen from Fig. 4.15 that the complex hydrograph of direct runoff is generated by superposition of different single peaked direct runoff hydrographs. Each direct runoff hydrograph can be reduced to corresponding unit hydrograph spaced at suitable time lag.

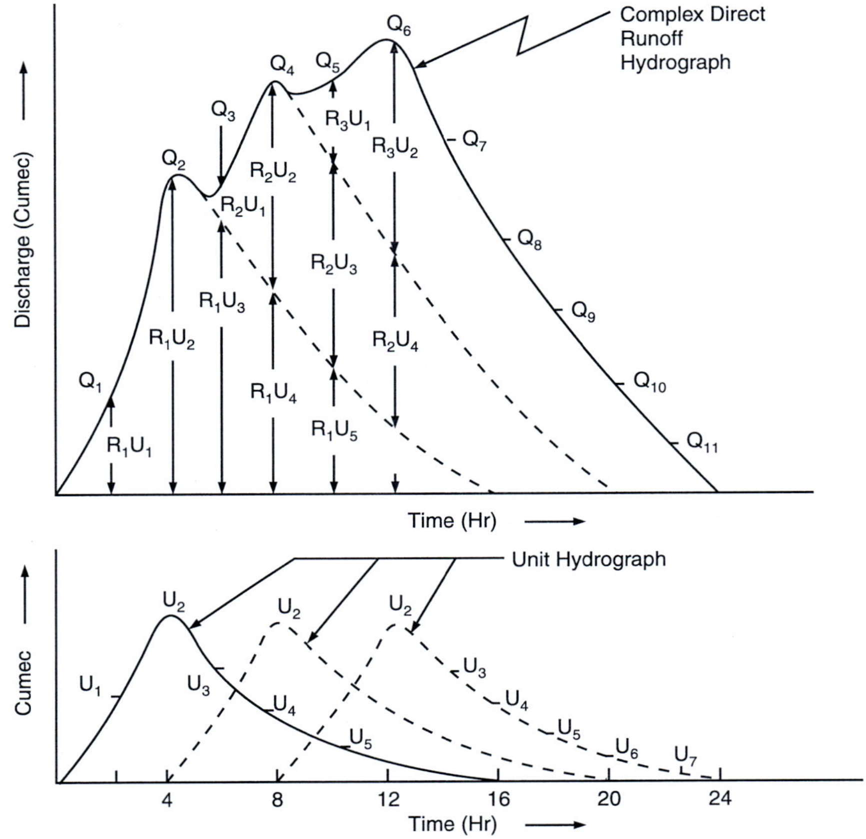

Fig. 4.15. Unit hydrograph from complex hydrograph

Also we know that unit hydrograph ordinates are obtained by dividing direct runoff hydrdgraph by effective rainfall or direct runoff expressed in cm. It may also be ssen from Fig. 4.15. that upto the peak of the first simple hydrograph there is no overlapping ordinate. Therefore knowing Q and R values and following step-by-step procedure unit hydrograph ordinates (u) can be calculated. For example:

There is no overlap for the first two ordinates in the present case as shown in Fig. 4.15.

$\therefore \qquad Q_1 = R_1U_1 \quad \text{and} \quad Q_2 = R_1U_2$

Since Q_1, Q_2 and R_1 are known U_1 and U_2 can be calculated.

Now from third ordinate second duration direct runoff hydrograph begins

$\therefore \qquad Q_3 = R_1U_3 + R_2U_1$

We know Q_3, R_1, R_2 so U_1 and U_3 can be calculated similarly

$$Q_4 = R_1U_4 + R_2U_2$$

We know Q_4, R_1, R_2, so U_2, and U_4 can be calculated. Proceeding further

$$Q_5 = R_1U_5 + R_2U_3 + R_3U_1$$

In this manner equating the successive ordinates of complex direct runoff hydrograph with the sum of the values of various overlapping simple hydrograph, ordinates of unit hydrograph can be completely calculated. Problem 4.6 solved below makes the procedure clear.

PROBLEM 4.6. The direct runoff ordinates of a complex storm made up of 3 effective rainfall bursts are given below.

Time (hr)	0	2	4	6	8	10	12	14	16	18	20	22	24
Hydrograph ordinate (cumec)	0	6	16	22	38	23	31	54	32	18	9	3	0

This hydrograph is generated by three 4 hour effective rainfalls having 1 cm, 2 cm and 3 cm effective rainfalls respectively. Second rain burst follows the first. However the third burst starts after a break of two hours from the end of the second burst. Using complex hydrograph analysis method plot and derive 4 hour unit hydrograph. If the area of the drainage basin is 30.25 km² check validity of the derived unit hydrograph.

Solution: Step 1. Let the ordinates of complex hydrograph of direct runoff be Q_0, Q_1, Q_2, Q_3, Q_4, Q_5, Q_6, Q_7, Q_8, Q_9, Q_{10}, Q_{11}, Q_{12} respectively. Separate out the three peaks of complex hydrograph as shown by dotted lines in Fig. 4.15. Thus we now have three simple direct runoff hydrographs.

Step 2. Also $R_1 = 1$ cm; $R_2 = 2$ cm and $R_3 = 3$ cm

R_1 starts from time 0 hr

R_2 starts from time 4 hr (*i.e.*, 4 hr lag)

R_3 starts from time 10 hr (*i.e.*, 8 + 2 hr lag)

Step 3. Referring Fig. 4.16.

at $Q_0 = 0 \quad U_0 = 0$ ...(0)

At 2 hr $Q_1 = R_1 \times U_1$ or $6 = 1 \times U_1$

∴ $U_1 = 6$ cumec ...(1)

At 4 hr $Q_2 = R_1 \times U_2$ or $16 = 1 \times U_2$

∴ $U_2 = 16$ cumec ...(2)

At 6 hr $Q_3 = R_1 \times U_3 + R_2 \times U_1$

or $22 = 1 \times U_3 + 2 \times 6$

∴ $U_3 = 10$ cumec

At 8 hr $Q_4 = R_1 \times U_4 + R_2 \times U_2$ ...(3)

or $38 = 1 \times U_4 + 2 \times 16$

∴ $U_4 = 6$ cumec ...(4)

At 10 hr $Q_5 = R_1 \times U_5 + R_2 \times U_3 + R_3 \times U_0$

or $23 = 1 \times U_5 + 2 \times 10 + 3 \times 0$

∴ $U_5 = 3$ cumec ...(5)

At 12 hr $Q_6 = R_1 \times U_6 + R_2 \times U_4 + R_3 \times U_1$

or $31 = 1 \times U_6 + 2 \times 6 + 3 \times 6$

∴ $U_6 = 1$ cumec ...(6)

At 14 hr $Q_7 = R_1 \times U_7 + R_2 \times U_5 + R_3 \times U_2$

or $54 = 1 \times U_7 + 2 \times 3 + 3 \times 16$

∴ $U_7 = 0$ ...(7)

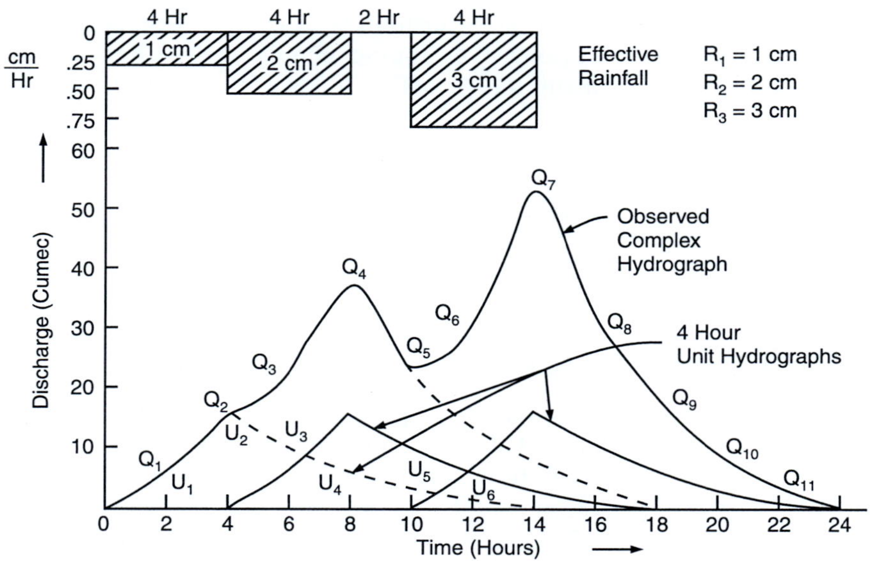

Fig. 4.16. Problem 6

Thus ordinate of 4 hour unit hydrograph are

Time (hr)	0	2	4	6	8	10	12	14	Σ
Ordinate cumec	0	6	16	10	6	3	1	0	42

The total runoff given by unit hydrograph should be 1 cm if the ordinates of unit hydrograph are correctly calculated.

$$\text{Direct runoff from unit hydrograph} = \frac{\Sigma \text{ordinates} \times \text{time interval}}{\text{basin area}}$$

$$= \frac{42 \times 2 \times 60 \times 60}{30.25 \times 1000 \times 1000} \times 100 \text{ cm} = \frac{42 \times 2 \times 36}{30.25 \times 100} = 1 \text{ cm}$$

Hence unit hydrograph has been correctly derived.

4.18. SNYDER'S SYNTHETIC UNIT HYDROGRAPH OR DERIVATION OF UNIT HYDROGRAPH FOR UNGAUGED BASIN

Where adequate rainfall-runoff data is not available unit hydrograph can be derived based on **known physical characteristics** of the basin.

In this method a correlation analysis is made between 3 unit hydrograph parameters, namely:

(*i*) peak flow;
(*ii*) Lag time; and
(*iii*) base time;

and 3 basin characteristics namely:

(*i*) size of basin area;
(*ii*) slope of basin (area-elevation curve); and
(*iii*) number of major storms in basin.

Snyder analysed a large number of hydrographs from drainage basin in the Appalachian Mountain region in the USA and developed following group of equations:

$$t_p = C_t\,(L.L_c)^{0.3} \quad ...(1)$$

$$t_r = \frac{t_p}{5.5} \quad ...(2)$$

$$q_p = \frac{640\,C_p}{t_p} \quad ...(3)$$

$$T = 3 + \frac{t_p}{8} = 3\left(\frac{t_p}{24}\right) \quad ...(4)$$

$$q_{pR} = \frac{640\,C_p}{t_{pR}} \quad ...(5)$$

$$t_{pR} = t_p + \left(\frac{t_R - t_r}{4}\right) \quad ...(6)$$

where t_p = lag time from midpoint of effective rainfall duration t_r to peak of a unit hydrograph (UH) in hrs.

t_r = standard duration of effective rainfall, in hrs. (It is a standard duration with such a minimum value below which further decrease would have little or no affect on the basin lag).

q_p = peak discharge per unit drainage area of UH for standard duration t_r, in cfs/sq. mile.

T = base length of UH in days (for a standard duration of t_r hours).

t_R = duration of effective rainfall other than standard duration t_r, adopted in specific study in hr.

t_{pR} = lag time from midpoint of duration t_R to peak of UH in hr.

q_{pR} = peak discharge per unit drainage area of UH for duration t_R in cfs/sq mile.

L_c = river mileage from the station to centre of gravity of the drainage area in miles.

L = river mileage from the station to the u/s limits of the drainage area in miles (measured along main channel).

C_t and C_p = coefficients depending upon units used and basin characteristics.

Important Note: All coefficient have to be verified for hydrological area under study before they are accepted.

The C_t value of Snyder have been found to vary considerably depending upon the topography, geology and climates *S*. Linsley, Paulhus and Kohler have another expression involving basin slope 'S'.

$$t_p = C_t\left(\frac{L.L_c}{\sqrt{S}}\right)^n$$

where $n = 0.38$; $C_t = 1.2$ for mountanous area; $C_t = 0.72$ for foot hills area; $C_t = 0.35$ for valley areas.

Equation 4 generally gieves long base length for small basin areas because it may include the effect of subsurface runoff (The base length largely depends on the method of base flow separation).

To know the shape of UH more definitely two more widths, *i.e.*, at 75% and 50% of peak discahrge of UH are given by empirical equations developed by US Army Corps of Engineers. They are

$$W_{75} = \frac{440}{(q_{pR})^{1.08}} \quad \text{(in hours)} \quad ...(7)$$

$$W_{50} = \frac{770}{(q_{pR})^{1.08}} \quad \text{(in hours)} \quad ...(8)$$

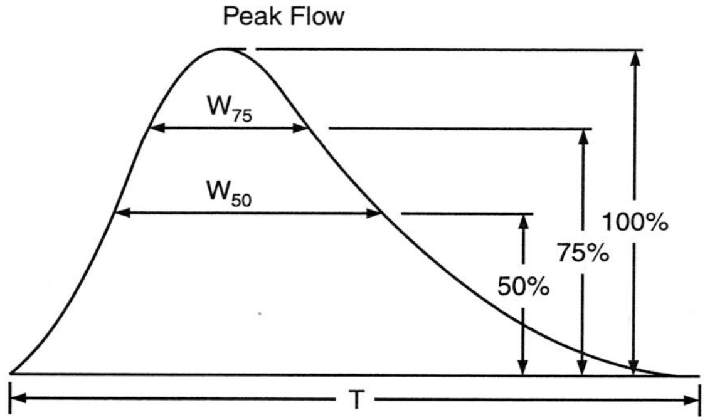

Fig. 4.17. Shape of unit hydrograph

It may be seen that to know the UH for any basin it is first necessary to know coefficient C_p and C_t for that basin. Naturally this can be done by finding these values for a gauged basin in a hydrologically similar region. Then these values of C_t and C_p can be adopted for a basin under consideration and unit hydrograph (UH) derived. The steps involved are enumerated below.

1. Analyse Rainfall-runoff records of a gauged basin located within same hydrologically similar region (of which ungauged basin is a part) and derive unit hydrograph. Determine its unit duration t_R, lag-time t_{pR} and peak discharge per unit area (q_{pR}). This is so because it may not be the standard duration, lag and 'q' for known basin also,
2. Measure L and L_c from the map of gauged basin. (The centroid of the basin is located by cutting out a stiff pattern of the basin and intersecting plumb lines drawn from different rotations of the pattern)
3. Compute C_t and C_p as follows:
   (*i*) Assume $t_{pR} = t_p$ and compute t_r by eqn. (2).
   (*ii*) If computed t_r happens to be equal to or close to t_R assume $q_{pR} = q_p$ and compute C_t and C_p by equations (1) and (3).
   (*iii*) If computed t_r does not equal to t_R, use t_R, t_{pR} and equation (2) for t_r in equation (6) and compute t_p.
   (*iv*) Then compute C_t by equation (1) and C_p by equation (5).
4. Use these values of C_t and C_p to derive unit hydrograph of a desired duration t_R for homogenous ungauged basin as follows :
5. First find L and L_c from map of drainage basin. Then use equations (1) and (2) to calculate t_p and t_r.
6. Use equation (6) to calculate t_{pR}.
7. Use equation (5) to calculate q_{pR}.
8. Calculate peak discharge for UH by multiplying by drainage area of the ungauged basin.
9. Find base length T using equation (4). (Alternatively find ratio of peaking time to base length for gauged basin UH and use same ratio.)
10. Calculate W_{75} and W_{50} using equations (7) and (8).

PROBLEM ON UNGAUGED BASIN

PROBLEM 4.7. Derive a 3 hour unit hydrograph for ungauged drainage basin 'A' having 14765 sq. miles catchment area from a 9 hour unit hydrograph developed for an adjacent drainage basin 'B'. The adjacent drainage basin is hydrologically homogeneous with ungauged drainage basin 'A'. The gauged drainage basin 'B' has following data available:

Drainage basin 'B'

$$\text{Catchment area} = 1750 \text{ sq. miles}$$
$$L = 198.8 \text{ miles}$$
$$L_c = 123.8 \text{ miles}$$
$$t_R = 9 \text{ hours}$$
$$Q_{pR} = 35300 \text{ cusec}$$
$$t_{pR} = 13.5 \text{ hours}$$
$$q_{pR} = \frac{Q_{pR}}{A} = \frac{35300}{1750} = 20.17 \text{ cusec/sq.mile}$$

For ungauged drainage basin 'A' following data is available:

$$\text{Catchment area} = 1476.5 \text{ sq miles}$$
$$L = 177.5 \text{ miles}$$
$$L_c = 115 \text{ miles}$$
$$t_R = 3 \text{ hours (given)}$$

Solution: Step 1: Using data of basin 'B' C_t and C_p is to be calculated.

Assuming $t_{pR} = t_p$ from equation (2)

$$t_r = \frac{t_p}{5.5} = \frac{13.5}{5.5} = 2.46 \text{ hours}$$

Since this value of t_r is not equal to t_R (= 9 hrs) t_p is to be computed from equation (6).

$$t_{pR} = t_p + \left(\frac{t_R - t_r}{4}\right)$$

$$13.5 = t_p + \frac{(9-2.46)}{4}$$

$$t_p = 11.87 \text{ hours}$$

Using the lag equation (1)

$$t_p = C_t\,(L.L_c)^{0.3}$$

or

$$11.87 = C_t\,(198.8 \times 123.8)^{0.3}$$

$$11.87 = C_t \times 20.76 = 0.57$$

C_p can be calculated from equation (5)

$$q_{pR} = \frac{640 C_p}{t_{pR}}$$

$$20.17 = \frac{640\, C_p}{13.5}$$

$$\therefore \quad C_p = 0.42$$

Since basin '*A*' and '*B*' are homogeneous C_t and C_p values of '*B*' basin may be used for '*A*' basin also.

Step 2: For basin '*A*' using values of C_t, C_p, L, L_c, t_R and t_p, t_r, t_{pR}, q_{pR} can be calculated.

From equation (1) $\quad t_p = C_t(L.L_c)^{0.3} = 0.57(177.5 \times 115)^{0.3} = 0.57 \times 19.63 = 11.19$ hours

From equation (2) $\quad t_r = \dfrac{t_p}{5.5} = \dfrac{11.19}{5.5} = 2$ hours

From equation (6) $\quad t_{pR} = t_p + \dfrac{(t_R - t_r)}{4} = 11.19 + \dfrac{(3-2)}{4} = 11.44$ say 11.5 hours

From equation (5) $q_{pR} = \dfrac{640 \times C_p}{t_{pR}} = \dfrac{640 \times 0.42}{11.5} = 23.4$ cusec/sq.mile.

$\therefore$ Peak discharge of 3 hour unit hydrograph for basin 'A' is

$$23.4 \times 1476.5 = 34550 \text{ cusec}$$

Step 3: Time base of the 3 hr unit hydrograph can be obtained from equation (4)

$$T = 3 + \frac{t_p}{8} = 3 + \frac{11.87}{8} = 3 + 1.48 = 4.48 \text{ days}$$

or $$T = 107.5 \text{ hours}$$

Step 4: To determine the shape of 3 hr. unit hydrograph from equations (7) and (8) width of unit hydrograph at 75% and 50% of peak discharge can be calculated.

$$W_{75} = \frac{440}{(q_{pR})^{1.08}} = \frac{440}{(23.4)^{1.08}} = 14.5 \text{ hours}$$

and $$W_{50} = \frac{770}{(q_{pR})^{1.08}} = \frac{770}{(23.4)^{1.08}} = 25.3 \text{ hours}$$

4.19. SPECIAL TYPES OF UNIT HYDROGRAPHS

To facilitate correct and quick analysis of runoff hydrographs various special types of unit hydrographs have been devised. They are:

(i) Normal or average unit hydrograph (Av. UH)

(ii) Dimensionless unit hydrograph

(iii) Instantaneous unit hydrograph (IUH).

4.19.1. Average Unit Hydrograph

To obtain normal or average unit hydrograph for a basin several storms are taken and unit hydrographs plotted for each of them. Obviously these unit hydrographs will be of different unit durations. They are then reduced to suitable unit duration by 'S' hydrograph method. All such unit hydrographs of equal duration are then plotted on the same sheet overlapping each other with their peaks on the same ordinate. The unit hydrographs with wide peaks are neglected. Using the plottings of other unit hydrographs a mean curve is drawn which gives normal or average unit hydrograph for that basin. This unit hydrograph may be tested by reproducing an observed flood by applying corresponding effective rainfall. Such a unit hydrograph is considered quite adequate for all design floods except probable maximum flood.

4.19.2. Dimensionless Unit Hydrograph

It is a unit hydrograph for which the discharge is expressed by the ratio of discharge to peak discharge and the time is expressed by the ratio of time to lag time.

Knowing the peak discharge and lag time for the duration of effective rainfall the unit hydrograph can be estimated. The peak discharge q_p and lag time may be estimated by Snyder method.

Advantages

1. It eliminates effects of basin size and shape. It facilitates comparison of unit hydrographs of basin of different sizes and shapes.
2. It is excellent means of averaging unit hydrographs.

4.19.3. Instantaneous Unit Hydrograph

If the duration of the effective rainfall becomes infinitesimally small, the resulting unit hydrograph is called instantaneous unit hydrograph (IUH) words, for an IUH the effective rainfall is applied to the drainage basin in zero time. Of course this is only fictitious situation and a concept to be used in hydrograph analysis.

QUESTIONS FOR REVISION

1. (*a*) What are various components of runoff ?
   (*b*) What are the various factors which affect runoff ?
2. (*a*) Define hydrograph; and
   (*b*) Analyse the annual runoff hydrograph of a river.
3. Describe the various components of a simple hydrograph resulted from a storm rainfall.
4. (*a*) Define unit hydrograph; List three different possible uses of the unit hydrograph theory.
   (*b*) Give various assumptions on which unit hydrograph theory is based and their limitations.
5. Write short notes on the following:
   (*i*) 'S' Hydrograph
   (*ii*) Instantaneous unit hydrograph
   (*iii*) Peaking time
   (*iv*) Dimensionless unit hydrograph
   (*v*) Hyetograph
   (*vi*) W and ϕ indexes.
6. Describe how Snyder's synthetic unit hydrograph is derived.
7. (*a*) How the base flow is separated from the total surface runoff hydrograph to obtain direct runoff?
   (*b*) Briefly explain how a unit hydrograph is derived by analysing rainfall and runoff data.
8. The 4 hour unit hydrograph ordinates are given below which has been developed for a drainage basin having 550 km^2 area.

Time (hr)	0	1	2	3	4	5	6	7	8	9	10
Discharge (cumec)	0	11	71	124	170	198	172	147	127	107	90

Time (hr)	11	12	13	14	15	16	17	18	19	20	21
Discharge (cumec)	76	62	51	40	31	27	17	11	5	3	0

A uniform storm of 4 hr duration with intensity of 11 mm/hr is followed after a 2 hour break by a further uniform intensity storm of 2 hr. duration and an intensity of 6 mm/hr. Rain loss is estimated to be 1 mm/hr during both storms. Base flow may be assumed to be 25 cumec uniformly.

Compute the peak discharge, total volume of direct runoff and the time of occurrence of the peak.

Hint for solution:

(*i*) Alter the 4 hour unit hydrograph to 2 hour unit hydrograph by 'S' hydrograph method.
(*ii*) Plot three rain bursts of 2 hr. duration with their time of occurrence.
(*iii*) Considering the quantity of effective rainfall obtain three separate hydrographs for each storm with proper offsets.
(*iv*) Using principle of superposition add up the ordinates of three overlapping hydrographs to get total hydrograph of direct runoff.
(*v*) Find the answers required.

□□□

5

Peak Flow and its Estimation

5.1. PEAK FLOW

As a result of rain storms of various durations and melting of snow, the rivers carry relatively high flows. In a every day language they are called floods. These flood flows many times overtop the banks and inundate the surrounding area and damage agricultural produce and valuable property. Also they create problems for construction, operation or safety of the hydraulic structures constructed across the river. There are three important parameters of high flows, namely, water levels attained, discharge rates and duration and volume of flow. The knowledge of all the three aspects is equally important for planning and design of hydraulic structures like spillways, dams, bridges, cross drainage works etc., because they temporarily exert additional force on the structures. With the knowledge of the magnitude of these aspects it is possible to design the structures suitably to counteract and contain the forces generated by high flows without causing damage to various installations.

Definition of peak flow or maximum rate of runoff: Right from the beginning of rain the surface runoff starts contributing its proper share to the natural drainage. Suppose an outlet of the catchment located at the terminal is under consideration. In the early stage of rain storm only part of the catchment of the point under consideration receives rainfall and hence, the rate of flow is less. But as the time passes more and more part of the catchment contributors runoff to the point. After reaching a maximum value, the contribution again decreases in the reverse manner after the rain stops. The process can be well understood with the help of the hydrograph of that storm as shown in Fig. 5.1.

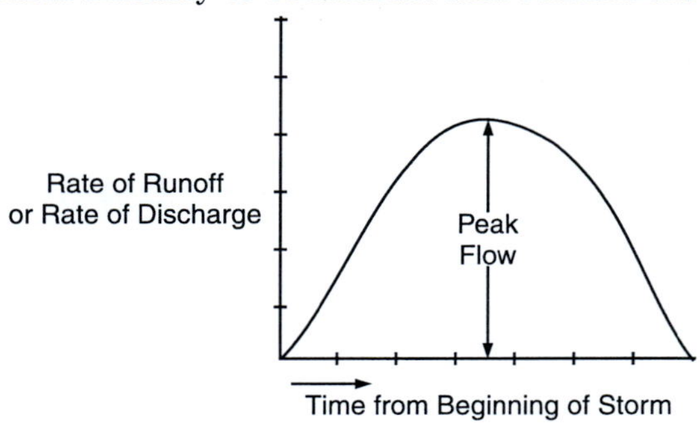

Fig. 5.1. Hydrograph of a storm

The maximum rate of discharge during the period of runoff caused by a storm is called a "peak flow".

The information about peak flows and associated volume of runoff during various floods that have been observed in the past can be complied by the analysis of the observed discharge data and runoff hydrographs. The estimate of peak flow that is likely to be experienced in future is very important for design of structures to be built across the river. The estimation of peak flow and volume of runoff can be done using various empirical formulae, unit hydrograph theory or from probability of occurrence of floods based on frequency concept. The methods of estimation of runoff is given in chapter IV. In this chapter methods of computing peak flows are dealt with.

5.2. FACTORS INFLUENCING PEAK FLOW

Factors which influence the peak flow can be divided in two categories:

- basin characteristics, and
- storm characteristics.

In basin characteristics we can group the following:

1. **Size and shape of a basin :** It is already mentioned that fan shaped catchment gives more quantity of runoff. At this stage it can also be stated that fan shaped catchment gives bigger peak. Reason for it is, small tributaries of fan shaped catchment meet main drainage almost at the same place to give maximum rate of discharge. Conversely fern shaped catchment gives smaller peak.
2. **Geographical location of a basin:** If an area is hilly or on the slopes of hills, peak will be bigger.
3. **Orientation of a basin with respect to storm pattern:** If storm pattern is such that it covers the catchment fully, Fig. 5.2 (*a*), the runoff and consequent peak flow will be bigger.
4. **Topography of a basin:** If the slope of a basin is steep then peak flow will be high.
5. **Geology of a basin:** If the basin is such that it has top crest of rocky layer, peak flow will be high as losses will be less. If there are fissures or cracks in rocky bed then losses will be more and peak flow will be less.

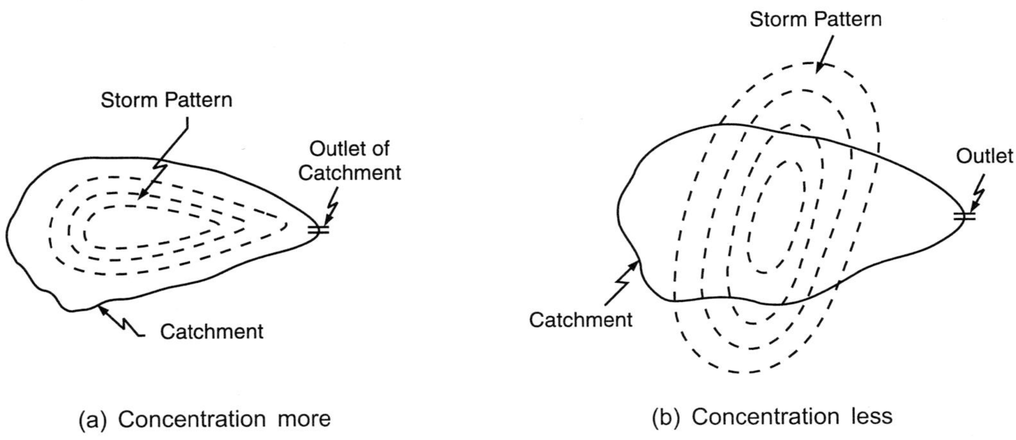

Fig. 5.2. Orientation of a basin with respect to a storm

6. **Type of vegetal cover:** If the basin is covered with big trees evaporation and absorption losses will be less and peak will be high. If basin is covered with grass and small bushes it will obstruct runoff and peak may be low.
7. **Extent of surface detention:** If there are local depressions in the basin the runoff water will be intercepted and peak flow will be reduced.

In storm characteristics we can include the following points:

1. **Intensity of rainfall:** If rainfall per hour is high the peak flow will also be high.
2. **Duration of a storm:** If storm lasts for longer period peak flow also be prolonged.
3. **Distribution pattern of a storm:** If storm is such that it covers whole basin area peak flow will be bigger.
4. **Storm direction:** If storm or rain proceeds from the outlet inwards (Fig. 5.3 (*b*)), the rain that has occurred near the outlet will pass through the outlet before the water from the farthest area reaches the outlet. Whereas if a storm advances towards the outlet then by the time the rainfall at the farthest point reaches outlet, storm also reaches the outlet. Now local rainfall and runoff from farthest area contributes water to that point simultaneously to give high peak flow.

(a) More Peak Flow (b) Less Peak Flow

Fig. 5.3. Effect of storm direction on peak flow

5.3. ESTIMATION OF PEAK FLOW

Estimation of probable maximum peak flow or flood is the first step in planning for flood regulation and protection. There are various methods of estimating the highest peak flow. The mention of following methods is essential at this stage:

- Empirical formulae
- Envelope curves
- Rational method
- Unit hydrograph method
- Frequency analysis

5.4. EMPIRICAL FORMULAE

In this method area of a basin or a catchment is considered mainly. All other factors which influence peak flow are merged in a constant. A general equation may be written in the form

$$Q = CA^n$$

where Q is peak flow or rate of maximum discharge
C is a constant for the catchment
A is area of the catchment and n is an index

The constant for a catchment is arrived at, after taking following factors into account:

(*a*) Basin characteristics

(*i*) area,

(*ii*) shape, and

(*iii*) slope.

(*b*) Storm characteristics

(*i*) intensity,

(*ii*) duration,

(*iii*) distribution.

Limitations

1. This method does not take frequency of flood into consideration.
2. This method cannot be applied universally.
3. Fixing of constant is very difficult and exact theory can not be put forth for its selection.

However, they give fairly accurate idea about the peak flow for the catchments they represent. Some important empirical formulae are mentioned below.

(*i*) **Dicken's formula:** It was formerly adopted only in Northern India but now it can be used in most of the States in India after proper modification of the constant.

$$Q = C.M^{3/4}$$

where Q is discharge in m^3/sec.

M is area of catchment in km^2.

C is a constant.

According to the area of catchment and amount of rainfall, C varies from 11.37 to 22.04 as given in Table 5.1.

Table 5.1.

Region	*Value of C*
North India	11.37
Central India	11.77—19.28
Western India	22.04

(*ii*) **Ryve's formula:** This formula is used only in Southern India.

$$Q = C.M^{2/3}$$

C = 6.74 for areas within 24 km from coast.

= 8.45 for areas within 24 —161 km from coast.

= 10.1 for limited hilly areas.

In worst cases it is found that value of C goes up to 40.5.

(*iii*) **The Inglis Formula:** This formula is used only in Maharashtra. Here three different cases are taken into consideration.

(*a*) For small areas only (It is also applicable for fan shaped catchment).

$$Q = 123.2\sqrt{A}$$

(*b*) For areas between 160 to 1000 km^2

$$Q = 123.2\sqrt{A} - 2.62(A - 259)$$

(*c*) For large areas $Q = \dfrac{123.2A}{\sqrt{A + 10.36}}$

In all the equations A is area in km^2.

(iv) Nawab Ali Nawaz's formula

$$Q = C\left(\frac{A}{2.59}\right)(a - b.\log A)$$

where a, b and C are constants

$$a = 0.993 \text{ and } b = \frac{1}{14}$$

C = 59.5 for North India and 48.1 for south India.

(v) Khosla's formula: It is a very rational formula. It is based on the equation

$P = R + L$ Or

$R = P - L$ where R is runoff

P is rainfall and L is loss.

According to Mr. Khosla main factor influencing losses is temperature. He expressed losses in mean temperature T_m in Fahrenhite. On the right hand side formula in CGS system is given.

$L = \frac{T_m - 32}{9.5}$ here L is in inches.	$L = 4.82\ T_m$ here L is in mm
$R = P - \frac{T_m - 32}{9.5}$	and T_m in centigrade.

The abovementioned formula of losses cannot be used for the values of temperature less than 40°F. For lower temperatures he gave following Table 5.2.

Table 5.2.

T_m°F	40	30	20	10	0
L **inches**	0.84	0.70	0.60	0.50	0.40

Khosla formula is useful in assessing water potential of a basin in a river valley project.

(*vi*) **Besson's formula:** This formula is very rational and can be used in any case.

$$Q_m = \frac{P_m \times Q_r}{P_r}$$

where Q_m is peak flow expected.

Q_r is some observed peak flow.

P_r is observed rainfall.

P_m is expected rainfall.

5.5. ENVELOPE CURVE

It is another method of estimation of peak flow. It is based on the assumption that highest known peak flow per unit area registered in the past in one basin in a region may occur in future in another basin in the same region or a region possessing similar hydrologic characteristics. A graph is constructed by plotting the highest peak flows observed per unit area of the catchment against their catchment areas in the region. The points obtained on the graph are joined by an envelope curve. The curve once constructed can be used to calculate the probable maximum peak flow for any basin in that region.

This method was given formerly by Creager Justin and Hinds in USA. The equation to the curve was of the type :

$$q = C.A^n.$$

where q represents the peak flow per unit area

A represents the catchment area

C is a constant, and

n is some index.

By multiplying both sides of the above equation by area of the basin 'A', we get

$$Q = C.A^{n+1}$$

where Q represents the peak flow.

Kanwar Sain and Karpov have developed two envelope curves to suit Indian conditions as shown in Fig. 5.4. One curve has been developed for the rivers in South India and the other for Northern and Central Indian rivers.

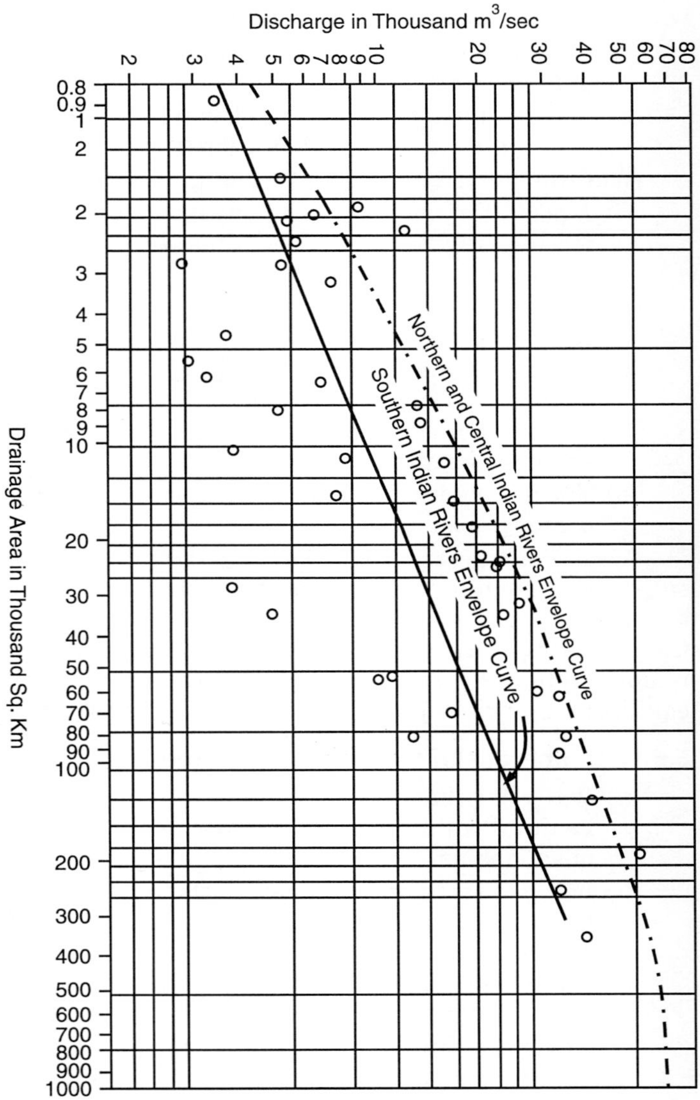

Fig. 5.4. Envelope curves for river basins of India

5.6. RATIONAL METHOD

This method is also based on the principle of the relationship between rainfall and runoff and hence can be considered to be similar to empirical method. It is, however, called rational method because the units of the quantities used are approximately numerically consistent. This method has become popular because of its simplicity. The formula is expressed as follows:

$$Q = PIA$$

where Q is peak discharge in cumec

P is runoff coefficient which depends on the characteristics of the catchment area. It is a ratio of runoff: rainfall. (P values are given later).

I is the intensity of rainfall in m/sec for the duration at least equal to "time of concentration".

and A is area of the catchment in m^2.

Time of concentration: It is the time taken by the rain water falling at the remotest point of the drainage basin to reach the discharge measurement point. It is given by the formula

$$t_c = 0.000324 \frac{L^{0.77}}{S^{0.385}}$$

where t_c is time of concentration in hours,

L is length of the drainage basin in m measured along river channel upto the farthest point on the periphery of the basin,

S is average slope of the basin from the farthest point to the discharge measuring point under consideration.

Assumptions: The rational formula is given on the following assumptions:

(*i*) A peak flow is produced on any drainage basin by a rainfall intensity which continues for a period equal to the time of concentration of the flow at the point under consideration.

(*ii*) The peak flow resulting from any rainfall intensity attains maximum value when the rainfall intensity lasts for the time equal to or greater than the time of concentration.

(*iii*) The maximum peak flow resulting from long duration rainfall intensity as mentioned above is its simple fraction.

(*iv*) The coefficient of runoff is same for all storms of varying frequencies on a given drainage basin.

(*v*) The frequency of peak flow is same as that of the rainfall intensity for a given drainage basin.

While defining the peak flow it is already mentioned that when rainfall continues for such long enough time that all portions of the drainage area simultaneously contribute runoff to an outlet peak flow is reached. Obviously the rainfall must continue till water falling at the farthest point also reaches the discharge measurement point. If the rainfall occurs at uniform rate right from the beginning the time of concentration will be equal to the time of equilibrium when effective rainfall equals direct runoff.

5.6.1. Limitations of the Rational Method

(*i*) It is clear that as the extent of the catchment area increases all assumptions cannot be fulfilled. Hence, for large catchment areas the utility of rational formula is questionable.

(*ii*) For very large and complex catchment areas before the water reaches outlet from the farthest point if the rainfall ceases then there is no possibility of whole catchment contributing its share of runoff to the outlet simultaneously. In such cases the lag time of peak flow is smaller than the time of concentration. In the above circumstances the rational formula does not give maximum peak flow.

Obviously the rational formula is applicable for small and simple drainage basins for which time of concentration is nearly equal to the lag time of the peak flow.

(*iii*) It is seen that rational formula gives better results for paved areas with drainages having fixed and stable dimensions. Therefore, it is popularly used for urban areas and small catchments only when detailed study of the problem is not justified. (The catchment area best suited is of the order of 50 to 100 ha). Since flood records are not available for small areas this method is found convenient.

(*iv*) The choice and selection of value of (P) the runoff coefficient is the most subjective thing and requires good judgement. Otherwise it is likely to introduce substantial inaccuracy.

Refinement of rational method: As a refinement sometimes the drainage basin is divided into zones by contours. Each zone is so selected that the time of concentration of each zone is same. Each zone is then assigned appropriate value of (P) the runoff coefficient depending upon the imperviousness of the area. The total discharge is taken as summation of discharges from various zones. Using this value of total discharge average runoff coefficient for the drainage basin can be worked out.

Table 5.3: Typical values of the runoff coefficient 'P'

Basin characteristics	*Runoff coefficient*
Sandy soil covered with bushes heavily	0.10
Sandy soil covered with bushes lightly	0.20
Gravel paving and bare soil	0.15 to 0.30
Loamy soils largely cultivated	0.40
Asphalt macadam paving	0.50
Concrete paving and flat roof surfaces	0.80
Clayey soils stiff and bare, steep rock bare; Pitched roof surfaces	0.90

PROBLEM 5.1. The areas of the small drainage basin is 500 ha. Using rational formula and making use of the following data calculate peak flow:

The catchment area is under different land use and value of 'P' for various categories is as follows:

Land use category	*Area (ha)*	*runoff coeff.*
Wooded steep rocky land	**130**	**0.8**
Lightly covered plateau land	**60**	**0.7**
Lightly covered clayey soil	**110**	**0.5**
Cultivated loamy soil	**90**	**0.3**
Sandy soil covered sparsely	**110**	**0.2**

The rain storm continued for 5 hours and gave 30 cm of rainfall during this period. The farthest point from the drainage outlet is 10 km away and the difference of elevation between the locations is 100 m.

Solution: Step 1. Time of concentration is given by formula

$$t_c = 0.000324\frac{(L)^{0.77}}{S^{0.385}} = 0.000324\frac{(10,000)^{0.77}}{(0.01)^{0.385}} = 2.3 \text{ hours.}$$

Since t_c is less than the duration of rain storm maximum peak flow will be given by the rational formula.

Step 2. In the example different '*P*' values are given for different land use classes. So weighted average value of '*P*' can be calculated as below

$$\begin{aligned} \frac{A \times P}{130 \times 0.8} &= \frac{AP}{104} \\ 60 \times 0.7 &= 42 \\ 110 \times 0.5 &= 55 \\ 90 \times 0.3 &= 27 \\ 110 \times 0.2 &= 22 \\ &\Sigma\ 250 \end{aligned}$$

Average $$P = \frac{250}{500} = 0.5$$

Step 3. Putting values in rational formula

$$Q = \text{P.I.A.} = 0.5 \times \frac{0.3}{5 \times 60 \times 60} \times 500 \times 10^4 = \frac{0.15}{36} \times 10^4 = 41.6 \text{ cumec}$$

5.7. UNIT HYDROGRAPH METHOD

In the last chapter it is already mentioned that the largest ordinate of the unit hydrograph multiplied by the effective rainfall (in cm) occurring in the unit duration gives the peak flow. To this amount base flow may also be added to get total peak flow. The method is fully explained and examples solved in the last chapter to make the procedure clear. In case of ungauged basins Snyder's Snythetic unit hydrograph may be developed to estimate the peak flow.

5.8. FREQUENCY ANALYSIS

Definition of frequency analysis: Frequency analysis is a method which involves study and analysis of past records (historical data) of hydrologic events to predict the future probabilities (chances) of occurrence. It is based on the assumption that the past data are indicative of the future.

Frequency analysis is done to estimate various things like annual runoff variations, frequencies of floods, droughts, rainfall etc. In other words the primary objective of the frequency analysis of hydrologic data (say flood events) is to determine the recurrence interval of the hydrologic event of a given magnitude.

For such analysis so called probability curves have been used. Given the observed data (for example maximum discharges for estimating maximum flood, average annual discharges for annual variations etc.) the task is to find a theoretical curve the ordinates of which will coincide with those observed. Good agreement of a theoretical curve with an empirical one ensures that the extrapolation can be rightly done.

When stream flood records of sufficient length and reliability are available they may yield satisfactory estimates. The accuracy of the estimates reduces with the degree of extrapolation. It is considered by some that extrapolation may be done only upto double the period for which data is available. For example, to get a 100 year flood 50 years record is necessary. However, insufficiency of recorded data makes it obligatory to use short term data to predict 1000 and 10,000 year floods also.

Frequency analysis is a method which involves statistical analysis of recorded data to estimate flood magnitude of a specified frequency. It, therefore, requires knowledge of statistics to clearly appreciate the methods of frequency analysis.

5.9. DEFINITIONS OF STATISTICAL TERMS

Before proceeding with the methods of frequency analysis it is useful to understand the definitions of some basic statistical terms and parameters commonly used in frequency analysis.

(*i*) **Population:** In statistics the whole collection of objects under consideration is called a population. It may be finite or infinite collection. It includes complete set of values that the object under consideration has taken in the part and/or can or will take in the future.

(*ii*) **Sample:** A sub-population drawn from a larger population is called a sample. The size of the sample is the number of units it contains. If a sample is representative of a population important conclusions about the population can be drawn from analysis of the sample.

(*iii*) **Random Sample:** When every unit in a population has an equal chance of being chosen the sample is called a random sample.

(*iv*) **Variable:** It represents one or more characteristics of a segment of a population. It has at least one property which can be measured and numerically defined. In hydrology, for example, rain intensity, rate of evaporation, stream discharge are hydrologic variables (but only terms evaporation, rainfall, runoff etc, are hydrologic phenomenon). There are two types of variables, namely, continuous and discrete or discontinuous.

Continuous variable can be measured along a continuous scale. For example, a hydrograph presents the stage or discharge of a stream as a continuous variable. Also depth of rainfall is a continuous variable.

On the other hand discrete variable is one which can be counted. For example, number of rainy days in a week or a month is a discrete variable.

(*v*) **Variate:** The value of a variable is called variate.

For example, in hydrology variables like rainfall, annual flood peak, minimum flow is usually designated by X whereas, the variate which represents individual observation is usually designated by x.

(*vi*) **Array:** An array is an arrangement of raw numerical data in ascending of descending order of magnitude.

(*vii*) **Range of data:** The difference between the largest and smallest number is called the range of the data.

(*viii*) **Time series:** It is a sequence of values arranged in order of their occurrence.

(*ix*) **Frequency:** It is the number of occurrences of an item in the sample. It is also called absolute frequency.

(*x*) **Probability:** It is also called relative frequency. Probability is the number of occurrences of a variate divided by the total number of occurrences. It is designated by 'P'. The total probability for all variates is equal to unity, *i.e.*, $\Sigma P = 1$. *The probability cannot be less than zero and more than one.* Probability of zero means an impossible event whereas probability of one means a sure event. For example, there are 10 variates of stream discharge values with discharges ranging as shown in Table 5.4 below. (It is called grouped data)

Table 5.4.

Class interval (cumec)	*Frequency*	*Probability or (relative frequency)*
< 100	1	$\frac{1}{10}$
100 – 200	2	$\frac{2}{10}$
200 – 300	4	$\frac{4}{10}$
300 – 400	2	$\frac{2}{10}$
> 400	1	$\frac{1}{10}$
Σ	10	$\frac{10}{10} = 1$

(*xi*) **Frequency distribution:** It is a (pattern of distribution) graph obtained by plotting the number of occurrences or the frequency against the variate as the abscissa (Fig. 5.5).

Frequency distribution can be plotted as block values or as point values with class mark as abscissa and absolute, relative or % frequency as ordinate.

If the adjacent points are joined by straight lines a *frequency polygon* is obtained. The polygon converges to the *frequency distribution curve* as the number of observations and classes increase indefinitely.

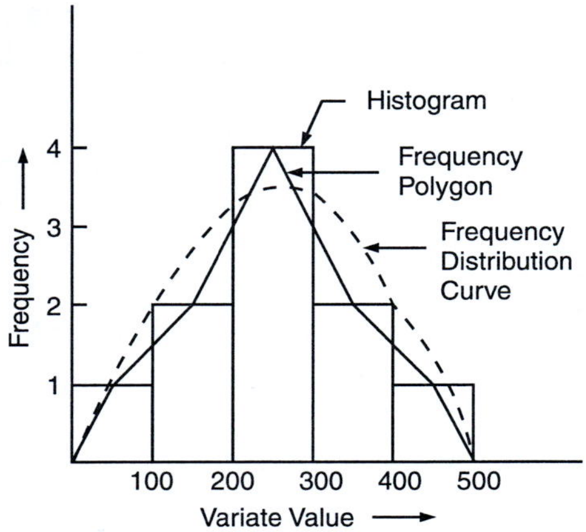

Fig. 5.5. Frequency distribution

(*xii*) **Probability distribution:** It is a pattern of distribution obtained by plotting relative frequency (per cent frequency) against variate as abscissa.

It may be mentioned here that the ordinates of the frequency distribution and its corresponding probability distribution are proportional to each other.

(*xiii*) **Cumulative or mass frequency diagram:** For grouped data (as given in a table above) the cumulative frequency of variates not exceeding a given value is the sum of all frequencies less than or equal to the given value.

The cumulative probability curve is plotted on rectangular coordinates by taking cumulative probability on the ordinate and magnitude of the variate on the abscissa. This type of curve is shown in Fig. 5.6.

The cumulative probability represents the probability that the random variable has a value equal to or less than certain assigned value say x. Then the cumulative probability is designated as

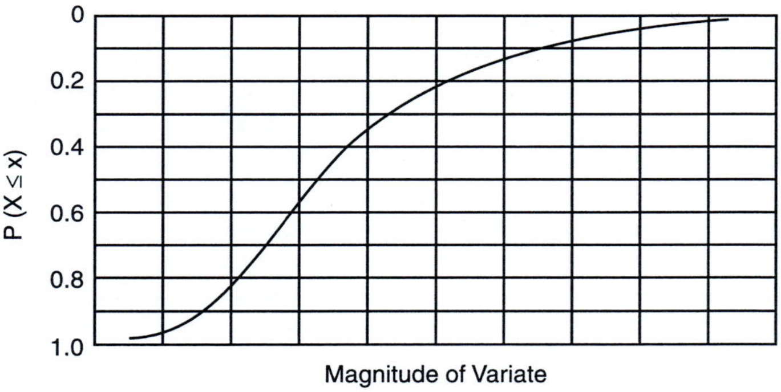

Fig. 5.6. Cumulative probability curve

$P\ (X \le x)$

Since the total probability is unity. The probability of being equal to or greater than x is given by

$P(X \ge x) = 1 - P(X \le x)$

In order to linearize the distribution a probability paper is used (Fig. 5.7). It is a specially ruled paper on which the ordinate and/or abscissa scales are so designed that the distribution plots more nearly on a straight line. It permits better definition of the upper and lower parts of the frequency curve and facilitates extrapolation.

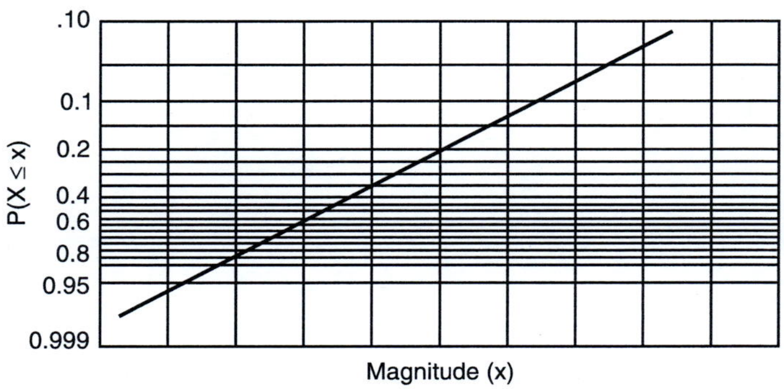

Fig. 5.7. Cumulative probability curve on probability paper

For finding out cumulative probability of the ungrouped data (which normally is the case) the values of the variable are arranged in ascending or descending order of magnitude.

For example, there are 50 observations which form ungrouped sample. Each observation has same (relative frequency) probability = 1/50 or 2%. The data is arranged in say descending order of magnitude.

The smallest value will have cumulative relative frequency of exceedence (equal to or greater than, *i.e.*, $\geq$) of 100%, the next smallest 98% and so on and it will end up with 2% for the maximum observed value.

Conversely since all the sample observations have values equal to or less than (*i.e.*, $\leq$) the maximum value we can assign to maximum value a cumulative relative frequency of non-exceeding of 100%, to next higher 98% and so on.

5.10. RECURRENCE INTERVAL OR RETURN PERIOD

Return period or recurrence interval is the average interval, in years, between events which equal or exceed a given magnitude. It is usually designated by a letter 'T'.

For example, a flood which can be expected to occur, on an average, once every 50 years is called the 50, years flood. The return period being 50 years. It may, however, be clearly understood that the concept of period does not imply that the event of any given magnitude will occur at constant or even approximately constant interval of n years. It only indicates average frequency of occurrence of an event over a long period of years. Say in a long period of 1000 years there will be (1000 ÷ 50 = 20) floods of equal to or larger than the 50 years flood. Such floods may occur at any interval of time regular or even in consecutive years.

The cumulative probability curve can also be plotted on the probability paper by calculating the return period of the given magnitude of a variable (say flood). When for every individual observed flood event the return period is calculated we can say that *plotting position* has been assigned to each event (Plotting position of any point on the graph is known when both the coordinates, magnitude and cumulative probability in this case, are available).

It may be shown that cumulative probability is nothing but inverse of return period (T).

If a hydrologic variable (like flood) equal to or greater than ($\geq$)x occurs once in T years, the probability $P(X \geq x)$ is equal to 1 in T years.

or $$P(X \geq x) = \frac{1}{T} \quad \text{...(5.1)}$$

where X is hydrologic variable

x is selected magnitude

T is return period or recurrence interval

and P is probability

The equation (5.1) above can be generalised as follows:

$$P(X \geq x) = \frac{1}{T} = \frac{m}{n}$$

where m is rank or order of the event when arranged in descending magnitude.

and n is total number of items in sample (in this case annual floods).

Various empirical formulae are available for calculating return period 'T' and therefore plotting positions. They are given in the Table 5.5.

Formula (*iii*) in Table 5 is most commonly used.

$$T = \frac{n+1}{m} \quad \text{and} \quad \therefore\ P = \frac{m}{n+1}$$

Table 5.5. Return period formulae

Name of formula	*Formula for 'T'* or $\left[\frac{1}{P(X \geq x)}\right]$	*Remark*
(*i*) California formula	$\frac{n}{m}$	It is oldest formula of all (1923)
(*ii*) Hazen formula	$\frac{n}{m-0.5}$	
(*iii*) Weibull formula or Gumbel's equation	$\frac{n+1}{m}$	Good for external and normal distribution so commonly used
(*iv*) Bernard and Bos-Levenbach formula	$\frac{n+0.4}{m-0.3}$	Gives close result with Weibull formula. Commonly used in Russia.

5.11. DEFINITIONS OF STATISTICAL PARAMETERS

Characteristics of a statistical distribution, for example that of probability or frequency distribution may be described by statistical parameters. The important once are briefly described below.

5.12. MEASURES (DESCRIPTORS) OF CENTRAL TENDENCY

Sample observations or population individuals usually cluster around some centrally located value. The parameters which describe the central tendency of a statistical distribution are various averages. The important among them are mean, median and mode.

(*i*) **Mean:** The more commonly used arithmetic mean is usually referred to simply as the mean. It gives an idea of the general magnitude of the items. It is designated by $\bar{x}$.

$$\bar{x} = \frac{\Sigma x}{n}$$

where x is the variable

and n is total number of observations.

Arithmetic mean is a good measure when departures of values are not great. In hydrology there are many occasions when a mean becomes meaningless due to the presence of extreme high or low values of a variable in the sample. The arithmetic mean of the sample is then not representative of the population mean.

(*ii*) **Median:** Median is the middle value of X or the variate which divides the cumulative frequencies in two equal portions.

It is already mentioned that cumulative frequency diagram has a range of frequencies from 0 to 100%. So the *median* marks 50% frequency.

The median divides the set of observations into two numerically equal groups. Thus, the number of observations (values) above and below the median are the same.

The median is used when the distribution is extremely skew. Here, median provides better indication, particularly for continuous variable because all variates greater or less than median always occur half the time.

(*iii*) **Mode:** The variate which corresponds to the largest ordinate of a frequency curve is called a mode.

Or

It is the value of the variable with maximum frequency. In a distribution of continuous variables mode is the variate which has maximum probability density.

For example : There are rainfall depths in cm in ascending order for 8 years as follows:

10, 11, 12, 12, 14, 17, 18

The $$mean\ \bar{x} = \frac{\Sigma x}{n} = \frac{110}{8} = 13.75 \text{ cm}$$

The *median* is mean of 4th and 5th observations because the number of observations is even

$$\therefore \qquad \text{Median} = \frac{12+14}{2} = 13 \text{ cm}$$

The mode is = 12 cm

5.13. MEASURES (DESCRIPTORS) OF VARIABILITY

The mean indicates general order of magnitude of a set of data. It is also necessary to know the extent to which the items vary from the mean. The important parameters representing variability or dispersion of a distribution are mean deviation, standard deviation, variance and coefficient of variation.

(*i*) **Mean deviation:** The mean of the absolute deviations of the values from their mean is called mean deviation. It is represented as

$$\text{Mean deviation} = \frac{\Sigma |x - \bar{x}|}{n}$$

(*ii*) **Standard deviation:** It is the square root of the mean squared deviation of individual measurements from their mean. An unbiased estimate of this parameter from the sample is given by

$$S = \sqrt{\frac{\Sigma(x - \bar{x})^2}{n-1}}$$

or $$= \sqrt{\frac{n}{n-1}(\overline{x^2} - \bar{x}^2)}$$

where $$\overline{x^2} = \frac{\Sigma x^2}{n} \text{ (It is called mean of squares)}$$

$\bar{x}^2$ is squared mean

S is standard division.

and n is number of individuals in the sample.

Standard deviation as a variability is most adaptable to statistical analysis.

(*iii*) **Variance:** It is nothing but square of the standard deviation.

$$\text{Variance} = S^2$$

(*iv*) **Coefficient of variation:** It is denoted by letter C_v. It is standard deviation divided by the mean.

$$\therefore \qquad C_v = \frac{S}{\bar{x}}$$

It can be defined as a measure of the relative variation of a variable. Since it is dimensionless it is widely used in hydrology particularly as a regionalisation parameter.

5.14. SHAPES OF FREQUENCY CURVES (DISTRIBUTION) AND MEASURES OR DESCRIPTORS OF SKEWNESS

(*i*) **Symmetrical distribution:** It is that type of frequency curve Fig. 5.8 (*a*), for which

$$\text{mean} = \text{median} = \text{mode}$$

(*ii*) **Skewed distribution:** It is also called asymmetrical distribution. For this type of frequency distribution mean, median and mode do not have same value (Fig. 5.8 (*b*) and (*c*)). Skewness is + ve or – ve depending upon location of the mode with respect to the mean.

In a positively skewed distribution there is a long tail on the right and the mean is on the right of the mode (Fig. 5.8 (*b*)).

In a negatively skewed distribution there is a long tail on the left and the mean is on the left of the mode (Fig. 5.8 (*c*)).

Most hydrological variables are asymmetrically distributed. The important parameters are skewness and coefficient of skewness.

(*iii*) **Skewness:** It is degree of asymmetry or departure from symmetry of a distribution. An unbiased estimate of skewness from the sample is given by the following relationship:

$$\alpha = \frac{n}{(n-1)(n-2)} \Sigma(x - \bar{x})^3$$

or

$$\alpha = \frac{n^2}{(n-1)(n-2)} \left(\overline{x^3} - 3\overline{x^2}.\bar{x} + 2\bar{x}^3\right)$$

where

$$\overline{x^3} = \frac{\Sigma x^3}{n}$$

and α is skewness.

All other notations have same meaning as already defined earlier.

(*iv*) **Coefficient of skewness:**

$$C_s = \frac{\alpha}{S^3}$$

This parameter is also dimensionless and is greatly used as C_v.

For symmetrical distribution $C_s = 0$.

When $C_s > 0$ distribution is skewed to the right (with a long tail on the right side).

When $C_s < 0$ distribution is skewed to the left (with a long tail on the left side).

5.15. PROBABILITY DISTRIBUTIONS

There are many probability distributions that have been found to be useful for hydrologic frequency analysis. Some of the important ones are briefly mentioned below.

1. Discrete Probability distributions: The binomial distribution and Poisson distribution are the two main types under this category. They can be applied to the probabilities of occurrence and non-occurrence of rare events in hydrology.

2. Continuous distributions: Normal distribution coming under this category is a symmetrical, bell shaped, continuous distribution theoretically representing Gaussian law of errors. (Gauss suggested that variate value observed for a continuous variable is combination of the true value + an "error term"). In this distribution mean = median = mode. Normal distribution implies continuous variate values covering a range from – ∞ to + ∞. The great merit of a continuous distribution is that it enables interpolation and extrapolation of variate values other than those observed.

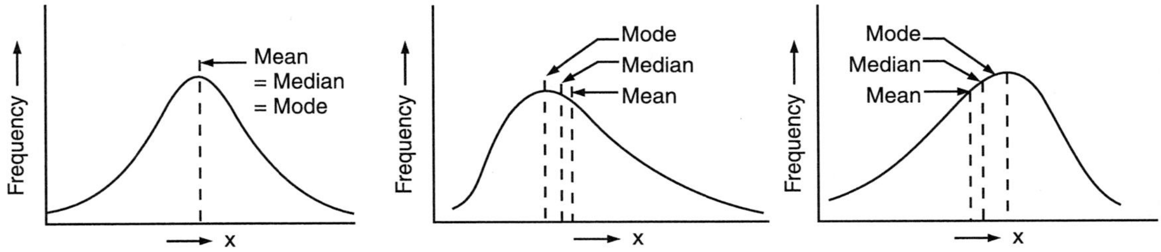

(a) Symmetrical distribution (b) Positively skewed distribution (c) Negatively skewed distribution

Fig. 5.8. Shapes of frequency curves

Yearly mean discharge of a perennial stream might be thought of as composed of average yearly flow over a long period plus a variation term (analogous to the error term). This does not, however, imply that annual flows of perennial streams are normally distributed. Certain features of non-normal populations have been shown to have close affinities to the normal. For a number of hydrologic variables the logarithms of the variates are seen to be approximately normally distributed. The variates are then said to be log normally distributed. The log-normal distribution requires that the variate is essentially positive and greater than zero. In log-normal distribution variates are replaced by their logarithmic values.

3. Pearson's distributions: Mr. K. Pearson stated that the characteristic of the frequency distribution is such that generally it starts at zero, rises to a maximum and then falls again to a low frequency or to zero but often at different rates. He developed 12 types of probability functions that virtually fit any distribution.

Pearson's type III function has been widely used to fit the empirical distribution of flood flows.

Now as per recommendations of the Hydrology Committee of the Water Resources Council, USA for flood peak discharges the current practice is to transform the data to their logarithms and then to compute the statistical parameters. Owing to this transformation the method is called Log-Pearson type III method.

4. Distribution of extreme values: This distribution was first proposed by *Gumbel* for *analysis of flood frequencies* and hence it is called **Gumbel's method** also. He considered a flood as the extreme value of the 365 daily flows. According to the theory of extreme values, the annual largest values of a number of years of record will approach a definite pattern of frequency distribution. Thus annual maximum flood constitutes a series which can be fitted to type I extremal distribution. (Similarly type III extremal distribution can be used for drought frequency analysis).

The external-value law assumes a constant skewness. The variate of a given recurrence interval, therefore, theoretically depends on the coefficient of variation and the mean.

Specially prepared external probability paper with non-uniform probability scale is used to linearise the distribution or frequency curve so that the plotted data can be analysed for extrapolation or comparison purposes. The paper is known as Gumbel-Powell probability paper or type—I extremal probability paper.

Annual flood peaks may also be plotted on log-extremal probability paper which is same as the one mentioned above except that the variate scale is divided logarithmically. The log extremal paper is always used for drought frequency analysis.

For flood frequency studies log-normal probability law as well as extreme-value law have been extensively used. From theoretical point of view Mr. Chow has shown that the type — I extremal distribution is practically a special case of the log-normal distribution when $C_v = 0.364$ and $C_s = 1.139$.

5.16. FREQUENCY ANALYSIS BY GUMBEL METHOD

General principle of frequency analysis can be stated as below:

As a simple method, frequencies (or probabilities), $P(X \geq x)$, of the observed flood peaks could be calculated. The curve of probabilities versus flood peaks ($f\ V_s\ .\ x$) is then plotted on log-probability paper and a smooth curve is fitted covering all points. By extrapolation of the curve extreme values could be obtained.

Since the observed data is usually short it may not represent the population and hence we can not entirely keep reliance on the curve obtained from observed data.

Now considering that recorded data constitutes a random sample of their parent population a theoretical frequency distribution suitable for the data could be fitted.

Once the distribution is fitted properly to the observed data extrapolation to calculate required probabilities can be easily done.

The Gumbel method of frequency analysis is based on extreme value distribution and uses frequency factors developed for theoretical distribution. The method utilises general equation given for hydrologic frequency analysis which is stated as below.

$$x = \bar{x} + \Delta x \qquad \ldots(i)$$

where x is magnitude of flood of some given probability (P) or return period (T)

$\bar{x}$ is mean of floods on record

Δx is departure of variate from the mean.

Δx depends on dispersion characteristics, recurrence interval (T) and other statistical parameters. It can be expressed as

$$\Delta x = S\ K$$

where S is standard deviation of the sample

and K is frequency factor

Thus, equation (i) above can be expressed as

$$x = \bar{x} + K\ S$$

Table 5.6 gives theoretically derived values of the frequency factor K for various sample sizes and return periods.

Various steps involved in frequency analysis by the Gumbel method are as follows:

(i) List and arrange annual floods (x) in descending order of magnitude.

(ii) Assign rank 'm', $m = 1$ for highest value and so on.

(iii) Calculate return period (T) and/or probability of exceedence (P) by equations $\frac{n+1}{m}$ and $\frac{m}{n+1}$ respectively. These values together with respective flood magnitude give plotting positions.

(iv) Using tabular form calculate x^2 and Σx and Σx^2.

(v) Now calculate mean $\bar{x}$; squared mean $\bar{x}^2$; mean of squares $\overline{x^2}$ and standard deviation S.

(*vi*) From the Table 5.6 of frequency factors for Gumbel method read K values for desired return periods (T) and the available sample size.

(*vii*) Using relation $x = \bar{x} + \text{KS}$ calculate flood values for various return periods.

(*viii*) Using the extreme value probability paper plot the x values against respective return periods or P values and join the points to obtain the required frequency curve.

> ✍ **Note.** (*a*) For calculating x, we require respective T; $\bar{x}$; S; K; KS; and ($\bar{x}$ + KS) values. If only one value of flood event (x) of return period T is required it can be worked out straight away. But if number of values of (x) and respective (T) are required to plot theoretical frequency curve it is better to perform calculation in a tabular form. Provide separate columns for T, $\bar{x}$, S, K, KS and ($\bar{x}$ + KS) to facilitate computation.
> (*b*) Normally it is not necessary to plot the observed data on the same plot where theoretical frequency line is plotted. Yet this may be done for the purpose of comparison purpose, *i.e.*, to see how closely the derived frequency line fits to the observed data.

Table 5.6. Frequency factors for Gumbel method

Sample size	*Return priod (T) in years*										
n	*5*	*10*	*15*	*20*	*25*	*30*	*50*	*60*	*75*	*100*	*1,000*
15	0.967	1.703	2.117	2.410	2.632	2.823	3.321	3.501	3.721	4.005	6.265
20	0.919	1.625	2.023	2.302	2.517	2.690	3.179	3.352	3.563	3.836	6.006
25	0.888	1.575	1.963	2.235	2.444	2.614	3.088	3.257	3.463	3.729	5.842
30	0.866	1.541	1.922	2.18	2.393	2.560	3.026	3.191	3.393	3.653	5.727
35	0.851	1.516	1.891	2.152	2.354	2.520	2.979	3.142	3.341	3.598	
40	0.838	1.495	1.866	2.126	2.326	2.489	2.943	3.104	3.301	3.554	5.576
45	0.829	1.478	1.847	2.104	2.303	2.464	2.913	3.078	3.268	3.520	
50	0.820	1.466	1.831	2.086	2.283	2.443	2.889	3.027	3.241	3.491	5.478
55	0.813	1.455	1.818	2.071	2.267	2.426	2.869	3.027	3.219	3.467	
60	0.807	1.455	1.818	2.071	2.267	2.426	2.869	3.008	3.219	3.467	
65	0.801	1.446	1.806	2.059	2.253	2.411	2.852	2.992	3.200	3.446	
70	0.797	1.437	1.796	2.048	2.241	2.398	2.837	2.979	3.183	3.429	
75	0.792	1.430	1.788	2.038	2.230	2.387	2.824	2.9.67	3.169	3.413	5.359
80	0.788	1.423	1.780	2.029	2.220	2.377	2.812	2.956	3.155	3.400	
85	0.785	1.413	1.767	2.020	2.212	2.368	2.802	2.946	3.145	3.387	
90	0.782	1.409	1.762	2.007	2.205	2.353	2.785	2.938	3.125	3.367	
95	0.780	1.405	1.757	2.002	2.193	2.347	2.777	2.930	3.116	3.357	
100	0.779	1.401	1.172	1.998	2.187	2.341	2.770	2.922	3.109	3.349	5.261

Following problem makes the procedure clear.

PROBLEM ON GUMBEL METHOD

PROBLEM 5.2. The annual flood series for a river is available for 21 years. The observed flood peaks are as given below. Calculate the 100 year flood using Gumbel's method and plot the theoretical frequency curve obtained by using frequency factor and compare it with the frequency curve of observed data.

Year	*Flood peak (cumec)*	*Year*	*Flood peak (cumec)*
1962	2960	73	7700
63	6400	74	5910
64	4800	75	4520
65	18,120	76	9120
66	7980	77	8240
67	5880	78	6500
68	12,010	79	8390
69	7990	80	9570
70	8780	81	11,020
71	17,950	82	5120
72	10,820		

Solution: Following the steps mentioned above the flood data can be arranged in descending order in Table 5.7. Rank can be assigned as shown in column 3 and T, $P(X \geq x)$ and x^2 calculated in subsequent columns.

Table 5.7. Flood frequency analysis by Gumbel Method (n = 21 years, period from 1962 to 1982)

Year	*Annual peak discharge (x) in 100 cumec arranged in descending order*	*Order Number m*	*Plotting*	*Positions*	x^2
			Return period $T = \frac{n+1}{m}$ *(years)*	*Probability* $P(X \geq x) = \frac{m}{n+1}$ *(per cent)*	
1	2	3	4	5	6
1965	181.2	1	22.00	4.54	32833.44
1971	179.5	2	11.00	9.09	32220.25
1968	120.1	3	7.33	13.63	14424.01
1981	110.2	4	5.50	18.18	12144.04
1972	108.2	5	4.40	22.72	11707.24
1980	95.7	6	3.66	27.27	9158.49
1976	91.4	7	3.14	31.81	8353.96
1970	87.8	8	2.75	36.36	7708.84
1979	83.9	9	2.44	40.91	7039.21
1977	82.4	10	2.20	45.55	6789.76
1969	79.9	11	2.00	50.00	6384.01
1973	77.0	12	1.83	54.54	5929.00
1966	70.8	13	1.69	59.09	5012.64
1978	65.0	14	1.57	63.63	4225.00
1963	64.0	15	1.46	68.18	4096.06
1974	59.1	16	1.37	72.72	3492.81
1967	58.8	17	1.29	77.27	3457.44
1982	51.2	18	1.22	81.81	2621.44
1964	48.0	19	1.15	86.36	2304.00
1975	45.2	20	1.10	90.91	2043.04
1962	29.6	21	1.05	95.45	876.16
	Σx = 1789.0				Σx^2 = 182820.78

From the Table 5.7

Mean $\bar{x} = \frac{\Sigma x}{n} = \frac{1789}{21} = 85.19$

Squared mean $\bar{x}^2 = (85.19)^2 = 7257.34$

Mean of squares $\overline{x^2} = \frac{\Sigma x^2}{n} = \frac{182820.78}{21} = 3705.75$

Now standard deviation $S = \sqrt{\frac{n}{n-1}\left(\overline{x^2} - \bar{x}^2\right)} = \sqrt{\frac{21}{21-1}(3705.75 - 7257.34)} = 39.00$

Now, using equation $x = \bar{x} + KS$ and adopting values of $\bar{x}$ and S from above and different K and T values from Table 5.6 flood flows (*i.e.*, x values) of various return periods can be calculated as shown in Table 5.8.

Table 5.8.

Return Period T in years	$\bar{x}$	***S***	***K (From table 5.6)***	***KS***	***Flood flow in (00 cumec)*** $x = \bar{x} + KS$
(1)	*(2)*	*(3)*	*(4)*	*(5) = 3 × 4*	*(6) = 2 + 5*
5	85.19	39.00	0.913	35.61	120.80
10	85.19	39.00	1.615	62.98	148.17
20	85.19	39.00	2.289	89.27	174.46
50	85.19	39.00	3.161	123.28	208.47
100	85.19	39.00	3.815	148.78	233.97

From Table 5.8, 100 year flood comes out to be 23,397 say 23400 cumec.

Using extreme value probability paper (Fig. 5.9) flood (x values) flows of column 6 from Table 5.8 are plotted against Return period (T) of column 1 of the same table.

Plotted points are joined to obtain a straight line shown firm in Fig. 5.9.

To compare the fitting of this line to observed data, on the same graph (x values) observed flood flows from column 2 of Table 5.7 are plotted against return period (T) values from column 4 of the same table. It may be seen that on the whole observed data fits the frequency curve satisfactorily. Hence, selected distribution is satisfactory.

5.17. FREQUENCY ANALYSIS BY CURVE FITTING

The method described in paragraph 5.16 above is based on the assumption that observed data follows the theoretical distribution to be fitted and will exhibit a straight line on the probability paper designed for the distribution. In as much as nature does not strictly obey the theoretical laws, the logical solution is to plot the observed data at determined plotting positions on a suitable probability paper and to fit a best-fit curve to the plotted points. It can be done either graphically or mathematically.

5.17.1. Graphical curve-fitting Procedure

In a simple graphical curve fitting procedure the observed floods are plotted on a probability paper and a best-fit curve drawn by "eye" through the points. Log-normal probability paper and extreme value probability paper are commonly used for the purpose. In the case of former, the plotting position of the individual flood of the annual series is found by the formula $P = m/(n + 1)$ where P is the exceedance probability, m the order of magnitude of a given flood in an array of observed floods and n the number of years. If extreme value probability paper, also called Gumbel paper is used, the plotting positions of the floods are found by formula $T = (n + 1)/m$, where T is the return period in years (Fig. 5.9).

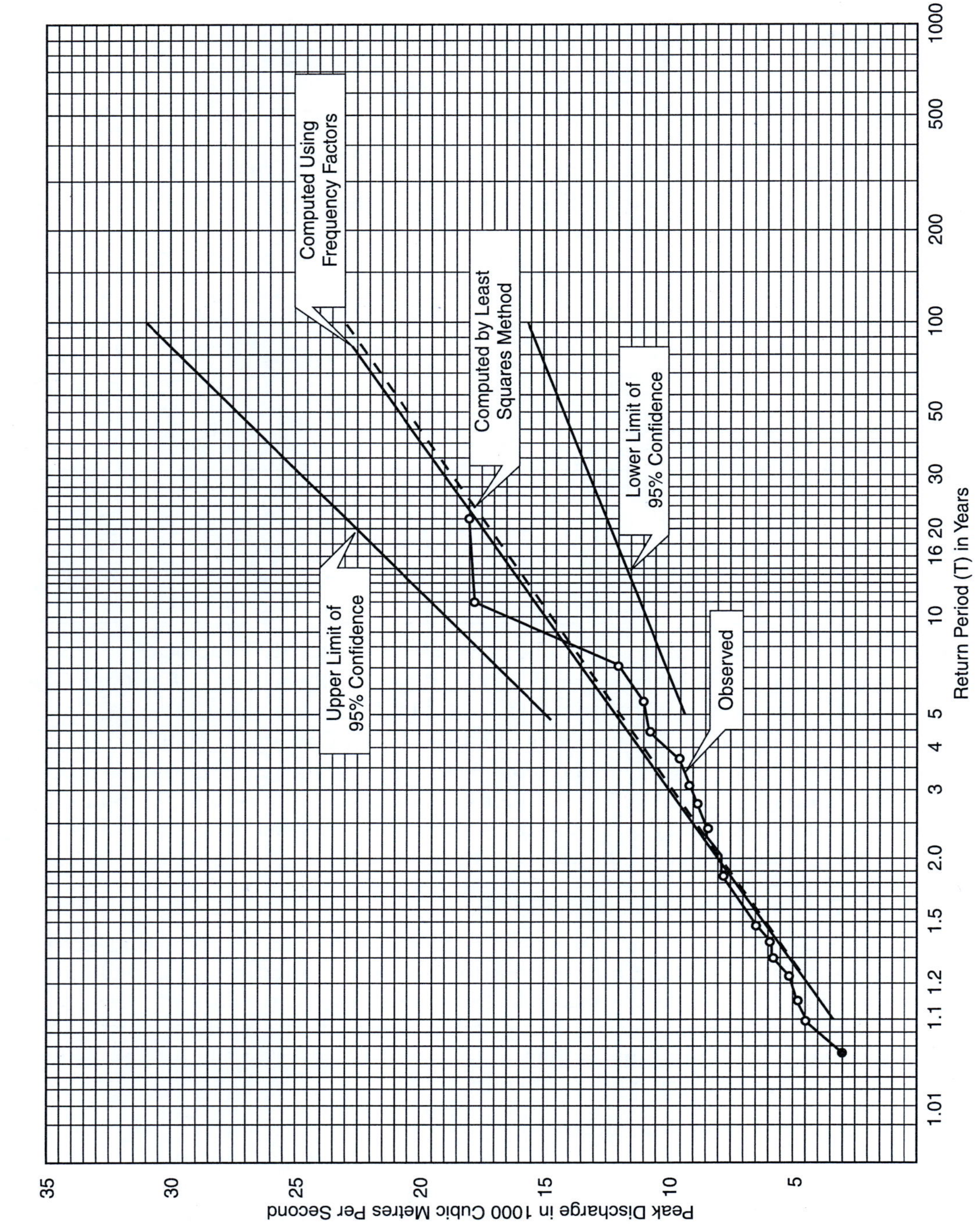

Fig. 5.9. Extreme value probability paper

5.17.2. Mathematical Curve Fitting Methods

To avoid the subjective errors in graphical fitting, curve fitting is done mathematically. Three methods are available for this purpose; the method of moments, the method of least squares and the method of maximum likelihood. The last method gives the best estimates but it is usually very complicated for practical application. The method of least squares gives a better overall fit than the method of moments and involves relatively less computations and therefore is commonly adopted. A brief outline of the principle of least squares and a procedure for fitting Gumbel's distribution using this principle are described below.

In Fig. 5.10 for a given value of x, say x_1, there will be a difference between the value of y_1 and the corresponding value as determined from the curve. This difference (indicated as D in the figure) or the departure may be positive, negative or zero. A measure of the *goodness of fit*' of the curve to the given data is provided by the sum of the squares of departures. If this is small the fit is good and if large it is bad. The least square line approximating the set of points (x_1, y_1), (x_2, y_2), (x_3, y_3), ...(x^n, y^n) has the equation $y = A + Bx$ where the constants A and B are determined by solving simultaneously the equations

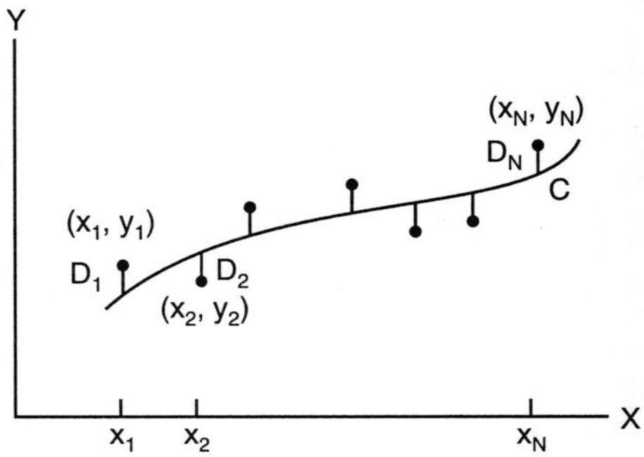

Fig. 5.10. Principle of method of least squares

$$\Sigma y = A.n + B\Sigma x$$

and $$\Sigma xy = A\Sigma x + B\Sigma x$$

which are called normal equations for the least square line. From these equations the constants A and B can be found out as

$$A = \overline{y} - B\overline{x} \text{ and } B = \frac{(\Sigma xy - n\,\overline{x}.\overline{y})}{(\Sigma x^2 - n\,\overline{x}^2)}$$

where $$\overline{x} = \Sigma x/n \quad \text{and} \quad \overline{y} = \frac{\Sigma y}{n}$$

Tables 5.9 and 5.10 show the computations (using data of problem 2) for fitting Gumbel's law (as adopted by Ven Te Chow) by the above method. The law is expressed as

$$y = A + B \log_{10} \log_{10} \frac{T}{T-1}$$

where y is the flood with a return period T.

The step-by-step procedure adopted is given below:

(*i*) Rank the observed floods (y) of the annual series in descending order.

(*ii*) Compute T-values for each of y-values by using relation

$$T = \frac{n+1}{m}$$

(*iii*) Compute x values where $x = \log_{10} \log_{10} \dfrac{T}{T-1}$ for all the items.

(*iv*) Compute the product xy and x^2 for all the items.

(*v*) Find out summations Σx, Σy, Σx^2 and xy and substitute these values in the normal equations to obtain parameters A and B of the least square line.

(*vi*) Plot the fitted equation of line on extreme value probability paper after computing a few values of y for selected T-values. This is the required frequency line.

(*vii*) To judge the goodness of fit the observed data are also plotted on the same paper.

Figure 5.9 shows the best fit line and the observed plotted on an extreme value probability paper.

Table 5.9: Fitting of Gumbel's law y = A + B log10 log10 T/T–1 to annual floods of problem 1 by method of least squares

Year	*Annual peak discharge (y) (00 cumec)*	*m*	$T = \frac{n+1}{m}$	$x = \log_{10} \cdot \log_{10} \frac{T}{T-1}$	*x.y*	x^2
1965	181.2	1	22.00	– 1.6946	– 307.0615	2.87167
1971	179.5	2	11.00	– 1.3831	– 248.2664	1.91297
1968	120.1	3	7.33	– 1.1961	– 143.6516	1.43066
1981	110.2	4	5.50	– 1.0597	– 116.7789	1.12296
1972	108.2	5	4.40	– 0.9509	– 102.8873	0.90421
1980	95.7	6	3.66	– 0.8592	– 82.2254	0.73822
1976	91.4	7	3.14	– 0.7790	– 71.2006	0.60684
1970	87.8	8	2.75	– 0.7071	– 62.0833	0.49999
1979	83.9	9	2.44	– 0.6412	– 53.7966	0.41114
1977	82.4	10	2.20	– 0.5796	– 47.7590	0.33594
1969	79.9	11	2.00	– 0.5214	– 41.6598	0.27186
1973	77.0	12	1.83	– 0.4654	– 35.8358	0.21659
1966	70.8	13	1.69	– 0.4110	– 29.0988	0.16892
1978	65.0	14	1.57	– 0.3572	– 23.2180	0.12759
1963	64.0	15	1.46	– 0.3034	– 19.4176	0.09205
1974	59.1	16	1.37	– 0.2485	– 14.6863	0.06175
1967	58.8	17	1.29	– 0.1915	– 11.2602	0.03667
1982	51.2	18	1.22	– 0.1306	– 6.6867	0.01706
1964	48.0	19	1.15	– 0.0628	– 3.0144	0.00394
1975	45.2	20	1.10	0.0176	– 0.7955	0.00310
1962	29.6	21	1.05	0.1279	– 3.7858	0.01636
	Σ 1789.0			– 12.3968	– 1416.0069	11.85049

From the Table 5.9

$$\bar{y} = \frac{\Sigma y}{n} = 85.19 \qquad \bar{x} = \frac{\Sigma x}{n} = -0.59302$$

$$B = \frac{\Sigma xy - n\,\bar{x}\,\bar{y}}{(\Sigma x^2 - n\bar{x}^2)} = 79.41 \qquad \therefore\ \bar{x}^2 = 0.34848$$

and $$A = \bar{y} - B\bar{x} = 38.31$$

$\therefore$ Line of best fit $y = 38.31 - 79.41 \log_{10} \log_{10} \frac{T}{T-1}$

Now, adopting various values of T corresponding values of (y), *i.e.*, estimated flood flows are calculated as shown in Table 5.10.

Table 5.10.

Return period (T) in years	$x \log_{10} \log_{10} \frac{T}{T-1}$	*Estimated flood flow (y) (00 cumec)*
10	– 1.3395	144.68
25	– 1.7513	177.38
100	– 2.3600	225.71

5.18. DESIGN FLOOD AND DESIGN FLOOD HYDROGRAPH

5.18.1. Definition of Design Flood

Design flood is the value of the instantaneous peak discharge adopted for the design of a particular project or any of its structures. In addition to the considerations of the flood characteristics, frequencies and potentiality of the contributing drainage area above the structure, social, economic and other non-hydrologic considerations which are likely to have influence are considered in deriving a design flood. The term "design flood" is used to denote the maximum flood flow that could be passed without damage or serious threat to the stability of engineering structures. It is therefore, necessary to specify volume and period of flood flow that may flow down in conjunction with the peak flow. Thus, design flood hydrograph is also fully worked out along with peak flow to facilitate design.

The Probable Maximum Flood (PMF) is the flood that would result from the most severe combination of critical meteorological and hydrologic conditions considered physically possible in the region under consideration.

As already mentioned the peakflow is estimated either by empirical formulae, envelope curves, unit hydrograph method, rational method or flood frequency analysis. Then considering social and economic factors the value of the peak flow is increased by applying a safety factor.

For major structures like big dams it is not sufficient to consider design flood only. It calls for estimates of extreme flood like Probable Maximum Flood (PMF). In deriving PMF attempt is made to analyse the causative factors responsible for the production of flood with a view to arriving at a most critical combination of meteorological and hydrologic factors. Among the factors producing severe floods the important ones are the following:

- Occurrence of very intense storm.
- Centering of the storm over the catchment so as to produce maximum runoff.
- Critical time sequence and movement of storm.
- Occurrence of the storm at a time when the drainage basin is already saturated by preceding rainfall.

5.18.2. Computation of Design Flood Hydrograph

Design flood hydrograph for drainage basins having less than 5000 km^2 area can be computed by adopting unit hydrograph method. The steps involved in its derivation could be summarised as follows:

(*i*) Select or derive a critical storm which is likely to occur in the region.

(*ii*) Derive an average or design unit hydrograph either using runoff data or synthetically.

(*iii*) Calculate the amount of effective rainfall that may result from the selected critical or design storm.

(*iv*) Obtain the rainfall amounts for time intervals equal to the unit duration of the design unit hydrograph from the depth duration curve of the selected storm.

(*v*) Adjust the sequence of increments of effective rainfall so that it gives most critical conditions of runoff. (Such adjustment is a must for deriving PMF, however, for design flood calculation only, considering storm characteristics of the area conservative sequence is taken).

(*vi*) Apply the effective rainfall increments with their adjusted sequence to the design unit hydrograph to obtain the total direct runoff hydrograph.

(*vi*) Add expected base flow during the design storm to the total direct runoff hydrograph to obtain the design flood hydrograph of total runoff.

In the case of drainage basins larger than 5000 km^2 and in those smaller basins also where different parts of the basin have widely different runoff characteristics, the total drainage area is sub-divided into sub-areas. Then for each sub-area separate unit hydrographs are derived. From them flood hydrographs are determined for various sub-areas adopting the above mentioned procedure. The sub-area flood hydrographs are then routed down the stream channel to the project site to finally compute the design flood hydrograph there.

QUESTIONS FOR REVISION

1. (*a*) Why is it important to estimate peak flow from a basin?
   (*b*) What factors affect the peak flow?
2. What is rational formula? What are its uses and limitations?
3. Define the following:
   (*i*) Return period
   (*ii*) Standard deviation
   (*iii*) Frequency distribution
   (*iv*) Design flood.
4. (*a*) What do you understand by frequency analysis?
   (*b*) Explain Gumbel method frequency distribution briefly.
5. List various empirical formulae used in estimation of peak flows. Mention their limitations.

6

Irrigation Development in India

6.1. NECESSITY FOR IRRIGATION

There is no single requirement of plant life which is more vital than provision of water. Adequate quantities of water should be readily available within the root-zone of all kinds of plant life. Such water if not present in the soil naturally it may be applied by irrigation or derived directly from the rainfall during the crop season. Even today rainfall is beyond the control of a man. It is estimated that one-third of the earth's surface receives less than 250 mm of yearly rainfall and that another one-third receives only 250 to 500 mm of yearly rainfall. Even in remaining areas rainfall is received within few months during the year. It is essential that for good results rainfall process fulfils following characteristics:

(*a*) Rainfall should be sufficient to make good moisture deficiency in the root-zone.

(*b*) Rainfall frequency should be such as to replenish soil moisture deficiency in time so that plants do not suffer from drought.

(*c*) Rainfall intensity should be low to permit the soil to absorb water to the maximum extent.

It shows that we cannot afford to depend exclusively on the rainfall. Necessity for adopting some method of irrigation is four-fold.

Firstly, when the seasonal rainfall is less than the minimum required for satisfactory crop growth. Obviously **irrigation is needed when rainfall is insufficient**. Such situation prevails in semi-arid areas of the country where unless some assured source of water is created agriculture is likely to fail year after year (*e.g.*, Rayalaseema region of Andhra Pradesh).

Secondly, every crop requires a certain quantity of water after fixed intervals of time till the crop matures. It is well known that rainfall cannot supply water at fixed intervals of time in measured quantities. Naturally **irrigation is required when rainfall is unevenly distributed** according to the crop requirements. This phenomenon is common in the semi-arid areas lying out of the path of prevailing rain bearing winds. (*e.g.*, Southern part of Rajasthan).

Thirdly, some crops (for example, sugar cane) require regular supply for long time which rainfall can not provide. Thus, **irrigation is needed to give regular supply** for long duration (*e.g.*, areas east of Western ghat in Maharashtra).

Fourthly, areas with nominal rainfall are always exposed to drought conditions. Arid regions, rain shadow areas come under this category. **Irrigation makes it possible to cultivate lands where rain usually fails** (*e.g.*, Rajasthan desert).

All these factors have made provision of irrigation facility a real necessity. It is befitting that irrigation is given pivotal status in agricultural development programmes of the country.

6.2. IRRIGATION IN ANCIENT INDIA

Irrigation is an art practised in our country from time immemorial. Its existence could be traced back to centuries before the Christian era. The ancient scriptures such as Riga Veda, Atharva Veda, Ramayana, Mahabharat and Puranas contain many references to the value and practice of irrigation. The scriptures laid down construction of wells, canals and tanks as part of the duty of the rulers to the public. The old traditions were maintained by various rulers throughout the history. The famous historian Megasthenes the Greek Ambassador, in his records dating 300 B.C. had written that the country was under irrigation and was prosperous and growing double crops.

In the ancient days, irrigation works in our country comprised dug wells, small tanks and diversion works on perennial rivers. The grand anicut (weir), constructed with stone laid in clay, across the Cauvery river is one of the notable ancient irrigation works constructed by the great Chola ruler in 200 A.D. The earliest canals were mostly of inunudation type which utilised the monsoon flows for a period of three to four months when the water level in rivers remained high enough to permit flow of water into the canals. The western and eastern Yamuna canals constructed by the King Ferozshah Tughlak in 1355 A.D. and the Bari-Doab canal constructed by the Mughal Emperor Shahjahan are the notable canals constructed in northern India in early days.

6.3. IRRIGATION DEVELOPMENT DURING BRITISH RULE

These canals were subsequently remodelled, renovated and converted into perennial canals in the first half of the nineteenth century during the British period. This was followed by a large number of diversion works with extensive canal systems. Important among them are the Upper Ganga Canal, the Upper Bari-Doab Canal, the Godavari delta system, the Krishna delta system, the Sirhind canal on the Sutluj river which were constructed towards the middle of the nineteenth century. Then started an era of storage reservoirs. The storage dams like the Khadakwasala near Pune with the Mutha canal system and the Periyar dam in the erstwhile Travancore State to divert the west flowing Periyar waters towards the east were constructed in the later half of the nineteenth century.

During the closing decades of the nineteenth century the country was stalked by successive famines. It led to setting up of a series of Famine Commissions. As a follow up, schemes like the Betwa canal in Uttar Pradesh, the Nira left bank canal in Maharashtra and the Rushikulya system in Orissa were executed. At the turn of the twentieth century First Irrigation Commision was set up to take stock of the existing irrigation facility and to make recommendations for its furtherance. As a result of the findings of the Irrigation Commission special attention was given to the development of irrigation in the country. The Godavari canal, the Pravara canal and the Nira right bank canal in Maharashtra, the Sarda canal in Uttar Pradesh and the Gang canal in Rajashtan were some of the important works constructed in the pre-independence period of the twentieth century. Three important reservoir projects namely, the Mettur reservoir in Tamil Nadu, the Krishnaraj Sagar in Karnataka and Nizam Sagar in Andhra Pradesh were also completed during this period. In addition, many other big and small irrigation works like those mentioned above and also tanks and wells were constructed. Table 6.1 gives the statewise list of some important irrigation works, benefitting about 2 lakh hectares and above, completed before independence.

If one looks at the irrigation works completed during British rule it will appear that mostly diversion works were constructed on the main rivers. These works were constructed in short period at low cost and gave immediate relief. The copious river flows were directly diverted on to the fields by constructing canal network and vast fertile areas were brought under irrigation. Obviously extent of irrigation depended on the adequacy of river flows. In good rainfall years extensive irrigation was possible while in bad years irrigation was limited to availability of flows. It, however, provided food security to the population during famines to some extent. Very few storage dams were constructed during this period. Krishna Raj Sagar in Karnataka, Khadakwasala in Maharashtra, Periyar dam in Kerala and Mettur in Tamil Nadu are noteworthy.

Table 6.1. Irrigation Works Completed before Independence

Important irrigation works	*Year of completion*	*Irrigation benefits (lakh ha)*
Andhra Pradesh		
(*i*) Godavari delta system	1890	5.58
(*ii*) Krishna delta system	1898	4.42
Bihar		
(*iii*) Sone canal system	1874	3.47
Haryana		
(*iv*) Western Yamuna canal system	1820	4.31
Punjab		
(*v*) Upper Bari Doab canal	1859	3.35
(*vi*) Sirhind canal	1873	6.00
Rajasthan		
(*vii*) Gang canal	1927	3.04
Tamil Nadu		
(*viii*) Cauvery delta system	1889	5.05
Uttar Pradesh		
(*ix*) Upper Gang canal system	1856	6.99
(*x*) Lower Gang canal system	1880	5.28
(*xi*) Eastern Yamuna canal system	1830	1.91
(*xii*) Sarda canal system	1926	6.12

6.4. IRRIGATION DEVELOPMENT AFTER INDEPENDENCE OF INDIA

Immediately before partition of the country net area irrigated in the Indian sub-continent was the largest in the world. The country, however, received a severe blow in the field of irrigation when it lost nearly 30 per cent of its best irrigated areas as a result of the partition of the country in 1947. To off-set the drastic reduction in irrigated areas great emphasis was placed on irrigation when an era of planning began in 1951. During the last six decades over 2200 big and small irrigation projects have been taken up. They include various types of dams, barrages and canals. Nearly 1400 of them have already been completed by the end of Tenth Plan which ended in March 2007. The remaining have been carried over to Eleventh Plan which started from April 1, 2007 and will continue for a five year period ending March 31, 2012.

Physical progress of irrigation over the period has been quite impressive. At the beginning of planned era in 1951 total number of large dams (height 15 m and above) was about 280.

However, after 1951 the number of large dams rose rapidly and now there are more than 4500 dams, majority of which have already been completed.

Some important projects of various States with ultimate irrigation potential more than 2 lakh hectares and which are either completed since Independence or are under various stages of construction are given in Table 6.2.

Table 6.3 gives development of Irrigation at a glance during different periods, starting from preplan era upto end of tenth plan *i.e.*, (2002-07). From the data presented in Table 6.3 it can be seen that irrigation potential at the beginning of First Plan was 22.6 M.ha which increased to 123.3 M.ha by end of Tenth Plan. Out of this 42.3 M.ha is from major and medium schemes and the remaining 81 M.ha from minor schemes. The percentage of utilisation of total created potential in the beginning was 90 to 95% upto 1980. There after lag between potential created and utilized went on increasing. From the Ninth plan the utilisation percentage is around 74%.

Table 6.2. Important Post-independence Projects of India

State	*Project*	*Ultimate irrigation potential (lakh ha)*
Andhra Pradesh	Nagarjuna Sagar	8.62
	Pochampad	2.67
Bihar	Gandak	15.07
	Kosi	7.48
Gujarat	Mahi-Kadana	2.75
	Ukai-Kakrapar	3.31
	Sardar Sarovar	18.75
Haryana	Western Yamuna Canal extension	2.48
Karnataka	Tungabhadra	5.04
	Ghatprabha	3.17
	Upper Krishna	4.25
Madhya Pradesh	Chambal	5.03
	Tawa	3.32
	Bansagar	2.49
	Bargi	2.45
Maharashtra	Jayakwadi	2.77
Orissa	Hirakud	2.51
	Mahanadi Delta	5.63
Punjab	Bhakara	14.60
	Beas (with Haryana)	3.20
	Thein	2.48
	Pong Dam (with Rajasthan)	16.50
Rajasthan	Rajasthan canal	12.58
Uttar Pradesh	Sarda Sahayak	15.82
Uttarakhand	Ramganga	5.91
	Tehri	2.70
	Lakhwar Vyasi	9.11
West Bengal	Damodar Valley	5.15
	Kangsabati	4.01
	Mayurakshi	2.51
	Teesta barrage	3.42

Maximum annual irrigation under any project is called irrigation potential of the project.

Table 6.3. Planwise and Schemewise (Cumulative) Irrigation Potential Created and Utilised

(Figures in M.ha)

Period	*Schemewise Cumulative Irr. Potential*						*% Utilisation*	*GIA as per Land Stat.*
	Major and Med.		*Minor*		*Total*			
	P	*U*	*P*	*U*	*P*	*U*		
1	**2**	**3**	**4**	**5**	**6**	**7**	**8**	**9**
Preplan (upto 1951)	9.7	9.7	12.9	12.9	22.6	22.6	100	22.6
End of 1st plan (1951-56)	12.2	11.0	14.0	14.0	26.2	25.0	95.4	25.0
End of 2nd plan (1961-66)	14.3	13.1	14.7	14.7	29.0	27.8	95.6	27.8
End of 3rd plan (1961-66)	16.6	15.2	17.0	17.0	33.6	32.2	95.9	30.9
End of Annual Plan (1966-69)	18.1	16.8	19.0	19.0	37.1	35.8	96.4	35.5
End of 4th plan (1969-74)	20.7	18.7	23.4	23.4	44.1	42.1	95.4	40.3
End of 5th plan (1974-78)	24.7	21.2	27.3	27.3	52.0	48.5	93.2	46.1
End of Annual Plan (1978-80)	26.6	22.6	30.0	30.0	56.6	52.6	93.0	49.2
End of 6th plan (1980-85)	27.7	23.6	37.5	35.2	65.2	58.8	87.1	54.5
End of 7th plan (1985-90)	29.9	25.5	46.6	43.1	76.5	68.6	89.6	61.9
End of Annual Plan (1990-92)	30.7	26.3	50.4	46.6	81.1	72.9	89.8	65.7
End of 8th Plan (1992-97)	32.9	27.9	62.5	48.3	95.4	76.2	79.9	76.0
End of 9th Plan (1997-02)	37.0	31.0	75.4	52.8	112.4	83.8	74.6	78.7
End of 10th Plan (2002-07)	42.3	34.4	81.0	56.7	123.3	91.1	73.9	82.6

Note: (*i*) GIA is gross irrigated area as per land use statistics.

(*ii*) P is potential created; U-potential utilised.

(*iii*) Percentage utilisation may not tally exactly due to rounding off.

6.5. TYPES OF IRRIGATION

Different types of irrigation can be best understood from the following schematic diagram:

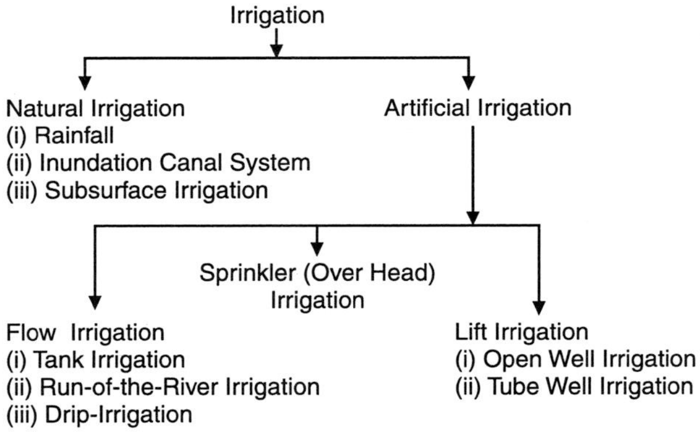

Irrigation is a process in which water flows and spreads over the land to nourish the plant life to give maximum crop output.

When the object of irrigation is fulfilled without constructing any engineering work of significant magnitude it can be said to be a natural type of irrigation. In a very broad sense natural rainfall may be included in natural type of irrigation.

When flood occurs in the river surrounding area is submerged under water. This flood water can be utilised for irrigation purposes, just by giving proper direction to the flood water. It is nothing but inundation canal system of irrigation. As such we can include inundation canal system in natural type of irrigation.

In some places soil formation is such that sub-strata is porous and the leakages from water bodies like tanks, lakes, canal flow under ground and directly fulfil water need of the crops by capillary action which are sown lower down the slopes. In a very broad sense this can be called natural irrigation.

In artificial irrigation, fields are irrigated after constructing properly designed and located hydraulic structures, for example, headworks, outlets etc. In turn artificial irrigation can be subdivided into three subheads depending upon the source of water supply. When water is lifted up from the source and then taken in channels, for example, from open wells and tube wells it is called lift irrigation. In flow irrigation water flows in channels directly from the temporary or permanent storage of water, for example, from a river or from a tank respectively. Flow irrigation may be either perennial (there is flow in the channels continuously for all the year), or non-perennial.

In areas where evaporation and percolation losses are very high from land surfaces and also soils may be saline, water is directly applied at the roots of the plants by means of pipes with perforations or pipes fitted with drip nozzles. The perforations are so adjusted that water comes out drop-by-drop at the root of the plant and directly fulfils crop water requirement. It is called drip irrigation. This system of irrigation gives atmost economy in water use and is very useful in arid areas.

The system in which water is spread over the fields by means of network of pipes fitted with spray nozzles at their open ends is called overhead or sprinkler irrigation. This system is just like rainfall created artificially. The sprinkler irrigation is most useful in the following situations where:

(*i*) cost of water is very high and the water has to be used very economically;
(*ii*) land is liable to be waterlogged easily through excessive use of water;
(*iii*) water use is to be controlled fully; and
(*iv*) topography does not permit surface spreading of irrigation water.

6.6. ADVANTAGES AND DISADVANTAGES OF IRRIGATION

Irrigation schemes are always introduced for welfare of the people of any particular locality but every region has its own climatic, topographic and other characteristics due to which introduction of this artificial flow of water in any particular region may disturb the natural balance to produce some ill-effects. Advantages and disadvantages of irrigation can be summarised as given below :

Advantages of Irrigation

1. For proper nourishment of crops certain amount of water is required. If rainfall is insufficient there will be deficiency in fulfilment of water requirement. Irrigation tries to remove this deficiency caused due to inadequate rainfall. Thus, **irrigation comes to rescue** in dry years.
2. Irrigation improves the yield of crops and **makes people prosperous**. The living standards of the people is thereby improved.
3. Irrigation also **adds to the wealth** of the country in two ways. Firstly as bumper crops are produced due to irrigation it makes country self-sufficient in food requirements. Secondly as the irrigation water is taxed when it is supplied to the cultivators, it adds to the revenue.
4. Irrigation makes it **possible to grow cash crops** which give good returns to the cultivators than the ordinary crops they might have grown in absence of irrigation. Fruit gardens, sugarcane, potato, tobacco etc., are the cash crops.
5. Sometimes large irrigation channels can be used as **a means of communication**.
6. The falls which come across the irrigation channels can be utilised for **producing hydroelectric power**.
7. Domestic advantages should not be overlooked. Irrigation **facilitates bathing, cattle watering** etc., and improves freshwater circulation.
8. Irrigation **improves the groundwater storage** as water lost due to seepage adds to the groundwater storage.
9. Along the banks of large irrigation channels plantation can be successfully done which not only helps introducing social forestry but also improves environmental status of the region.
10. New irrigation works are started at the time of famines to provide employment to a large number of population. These works are called famine works or relief works.
11. When watering facility is provided to a barren land, the value of this land gets appreciated.

Disadvantages of Irrigation

1. Excessive seepage and leakage of water forms marshes and ponds all along the channels. The marshes and the ponds in course of time become the colonies of the mosquito, which **gives rise to a disease like malaria**.
2. Excessive seepage into the ground raises the water-table and this in turn completely saturates the crop root-zone. It **causes waterlogging** of that area.
3. It lowers the temperature and **makes the locality damp** due to the presence of irrigation water.

4. Under irrigation canal system valuable residential and industrial land is lost.
5. Initial cost of irrigation project is very high and thereby the cultivators have to pay more taxes in the form of levy.
6. Irrigation works become obstacles in the way of free drainage of water during rainy season and thus results in submerging standing crops and even villages.

6.7. IRRIGATION METHODS

Water courses bring irrigation water to the fields for irrigation purposes. The next problem is how the application of water to the fields can be accomplished economically and efficiently for maximum crop production. It is essential to adopt some scientific method to solve this problem.

It should be remembered that water cannot be applied by same method to each field. There are various methods in practice in different parts of the country. The methods differ even from field to field in the same tract. Some obvious reasons for adopting various methods are mentioned below:

(*i*) Soil formation may not be the same.
(*ii*) Topographical features of the tracts may vary.
(*iii*) Mode of water supply may be different.
(*iv*) Crops to be grown may not be the same.
(*v*) Lastly the customs in the tract may be different.

Irrigation water can be applied to the fields by three different methods. They are **surface irrigation, drip irrigation** and **overhead irrigation**. These methods are further subdivided as mentioned below:

(1) Surface Irrigation

Flooting method
(*a*) Wild or free flooding
(*b*) Border strip method
(*c*) Check flooding
(*d*) Basin method

Furrow Irrigation
(*e*) Deep-furrows
(*f*) Corrugations or shallow furrows

(2) Drip Irrigation

(*g*) Drip or trickle irrigation

(3) Over-head Irrigation

(*h*) Fixed pipe sprinkler
(*i*) Portable sprinkler

6.8. SURFACE IRRIGATION

In this method water flows and spreads over the surface of the land. Varied quantities of water are allowed on the fields at different times. Hence, flow of water under surface irrigation comes under unsteady flow. As a result it is very difficult to understand the hydraulics of surface irrigation. However, suitable and efficient surface irrigation method can be adopted after taking into consideration various factors which are involved in the hydraulics of surface irrigation. They are:

(*i*) Surface slope of the field,
(*ii*) Roughness of the field surface,
(*iii*) Depth of water to be applied,

(*iv*) Length of run and time required,
(*v*) Size and shape of water-course,
(*vi*) Discharge of the water-course, and
(*vii*) Field resistance to erosion.

If the surface irrigation method is correctly chosen it fulfils following requirements:

(*a*) It helps in storing required amount of water in the root-zone-depth.
(*b*) It minimizes the wastage of irrigation water from the field in the form of run-off water.
(*c*) It reduces the soil erosion to minimum.
(*d*) It helps applying uniform application of water to the fields.
(*e*) Amount of manual labour required is minimum.
(*f*) It is best suited to the size of the field and at the same time it uses minimum land for making ditches, furrows, strips, etc.
(*g*) It does not prevent use of machinery for land preparation, cultivation, harvesting, etc.

6.9. WILD OR FREE FLOODING

Flooding method of irrigation is in use since centuries. Flooding method consists in applying the water by flooding the land of rather smooth and flat topography. In modern irrigation practice several flooding methods have been developed.

In free flooding method water is applied to the land from field ditches without any check or guidance to the flow. The method is very clear from Fig. 6.1.

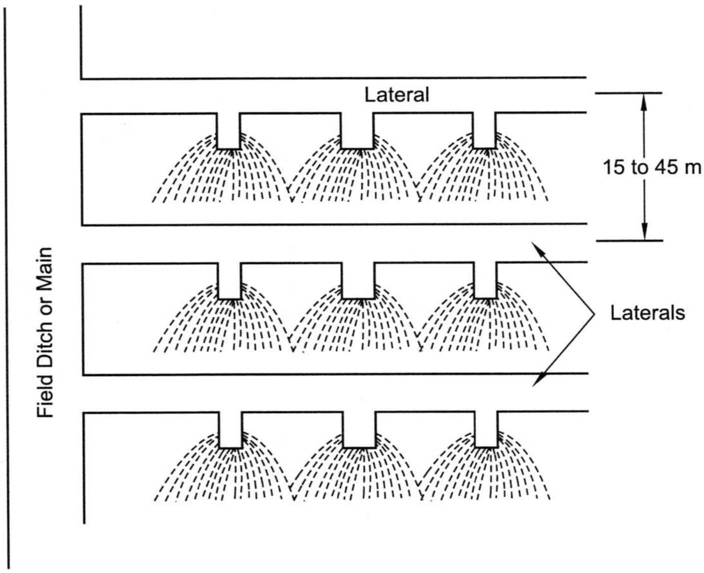

Fig. 6.1. (a) Free-flooding method for erodable soils

From the main or field ditch, laterals are taken across the fields at a spacing of 15 m to 45 m. The laterals run along the contours. The water flows out through the openings in laterals to flood the field. On the other side of the field a drain ditch is provided to take away excess water. To divert the water from the main to the laterals generally earthen dams are used. Sometimes steel shutters may also be used.

This method is very useful for newly established farms where making furrows, etc., is very costly. This method is cheap and can be successfully used where water supply is in plenty. This method is also well suited for the fields with very irregular surface which makes adoption of other methods difficult. Disadvantage of this method is, as there is no perfect control over the flow of water it is difficult to attain high efficiency. Sometimes the flow of water over the soil is too rapid to fulfil soil moisture deficiency. On the other hand sometimes water is retained on the field for a very long time and consequently the water is lost in infiltration or deep percolation.

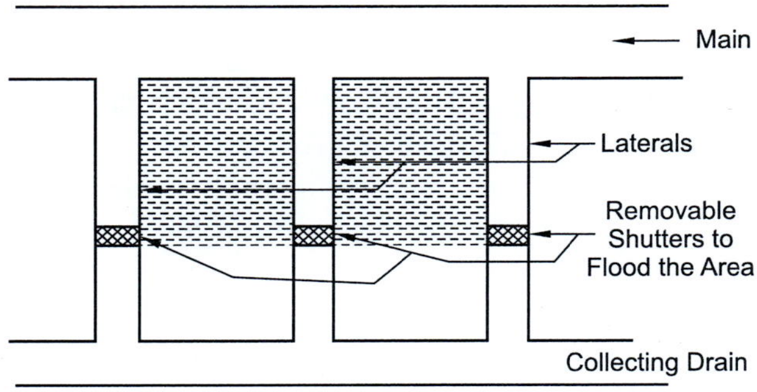

Fig. 6.1. (b) Free-flooding method for non-erodable soils

This method should be adopted only when other flooding methods are impossible.

6.10. BORDER STRIP METHOD

In this method a field is divided into number of strips. The width of strip varies from 10 to 15 metres and length varies from 90 m to 400 m. Strips are separated by low embankments or levees. The arrangement is as shown in Fig. 6.2.

The water is diverted from the field channel into the strips. The water flows slowly towards lower end, wetting the soil as it advances.

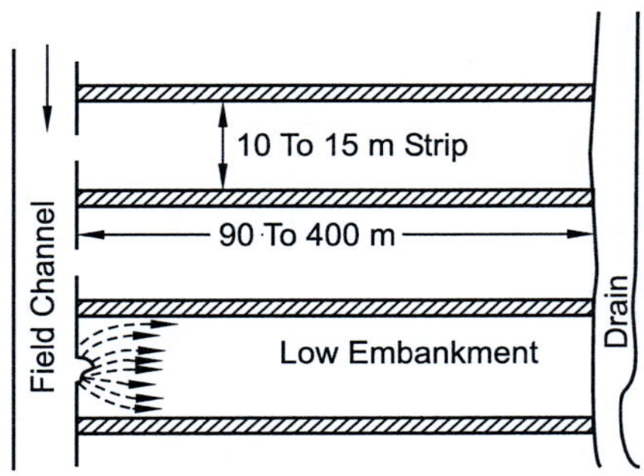

Fig. 6.2. Border strip method

The surface between two embankments should essentially be level. It helps in covering the entire width of the strip. There is a general surface slope from opening to the lower end. The surface slope from 2 to 4 m/1000 m is best suited. When the slope is steeper, special arrangement is made to prevent erosion of soil.

In this method it is possible to maintain more discharges successfully. The discharge may vary from 0.015 to 0.30 cumec depending upon the kind of soil, nature of crop, size of strips etc. This method is suitable on the fields where soil is sufficiently previous to absorb the water as it advances. In other words the soil should have wide variation in texture.

To divert water into the strip and to cutoff the water supply after the whole strip is flooded a gate is provided in the field channel at the head of each strip. Extra water is generally removed from the strip by means of a collecting drain. It is provided at the other end.

For economic and efficient use following points should be kept in mind:

1. Short strips may be more economical, the optimum distance is 90 metres.
2. A large rate of application for short time will result in more uniform and economical water use.
3. Larger the absorption rate larger the size of stream required and shorter the strip length.

6.11. CHECK FLOODING

In this method, relatively level plots are enclosed by small levees or embankments. Irrigation water enters the closed area and subsequently floods it. Check flooding method is very suitable for soils having high permeability. The reason is that the water quickly spreads over the entire area before it goes deep, below the root zone depths, into the ground and joins the water table. Thus, the water loss due to infiltration is prevented or reduced.

It is best adopted for heavy soils also. In heavy soils water absoption rates are low. The land can be kept flooded for longer time till sufficient water infiltrates into the ground to fulfil soil-moisture deficiency.

This method can be subdivided into two groups depending upon the mode of levee construction. The subdivisions are the following:

1. Rectangular checks: In this system the levees are constructed to enclose rectangular area as shown in Fig. 6.3.

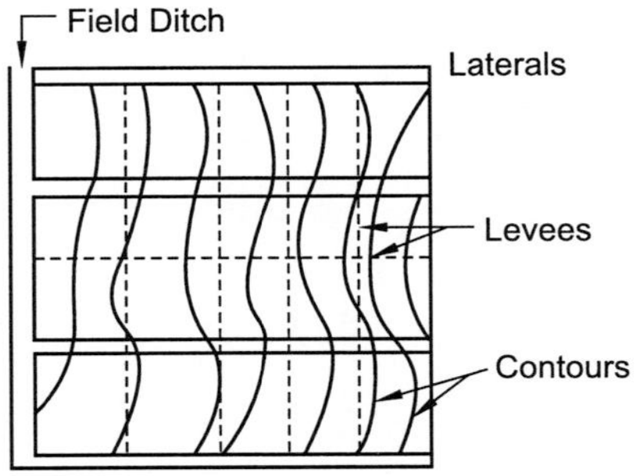

Fig. 6.3. Rectangular check method

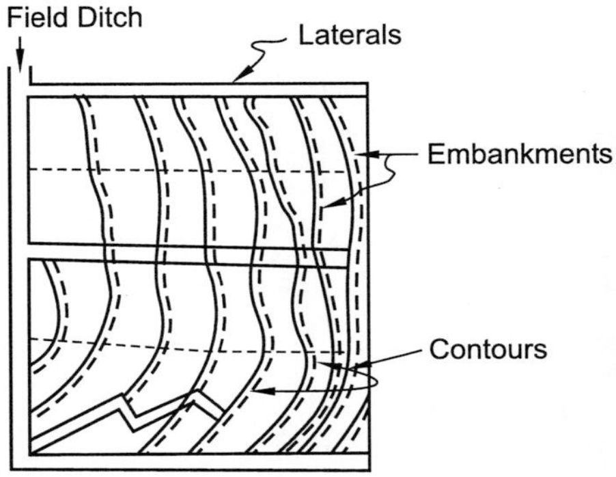

Fig. 6.4. Contour check method

2. Contour checks: In this system checks are prepared by constructing embankments or levees along the contours having vertical intervals of about 10 to 15 cm. Cross levees are constructed sometimes at convenient places as shown in Fig. 6.4.

In check method 0.2 to 0.8 hectares area is better. The embankment should be about 25 to 30 cm high and the width at the base should not be more than 2.5 m. Otherwise the embankments or levees may put hindrance to farm machinery.

This method is very useful in leaching the soil and reducing the salinity. It is generally used for grain and fodder crops.

6.12. FURROW IRRIGATION

When the crops are grown and planted in rows this method is best suited. Most of the crops are grown and planted in rows and thus this is the most commonly used method. In this method, unlike flooding, wetting is done of only a part of the field. The area wetted varies from 1/2 to 1/5 of total area over which crops are grown. Thus the losses due to evaporation, deep percolation, etc., are reduced. In this method water just flows through the furrows or small ditches and during that time the water enters the soil and the plant roots derive their proper share from the soil.

The longitudinal slope of a furrow may be from 2 to 10 m/1000 m. Steeper slopes may also be adopted, but then it is necessary to see that the water is not overflowing the banks of furrows. Length of furrows may be upto 800 m for field crops, about 30 m or less for gardens. However, no rigid limit can be fixed in any case.

The spacing of furrows varies according to the nature of the crop. For row crops, corn, onions, potatoes, etc., the spacing is determined by the spacing of the plant rows. Thus one irrigation furrow being provided for each row.

Spacing of furrows may be from 1 to 2 meters apart in case of orchard irrigation.

When furrows are 8 to 12 cm deep they are called corrugations or shallow furrows. Depth of deep furrows may vary from 20 to 30 cm. The advantage of deep furrow is that in initial stages the roots are not damaged. Deep furrow irrigation system is generally used for sugar cane, orchards, and some field crops also. Deep furrows are best suited in soils with low permeability.

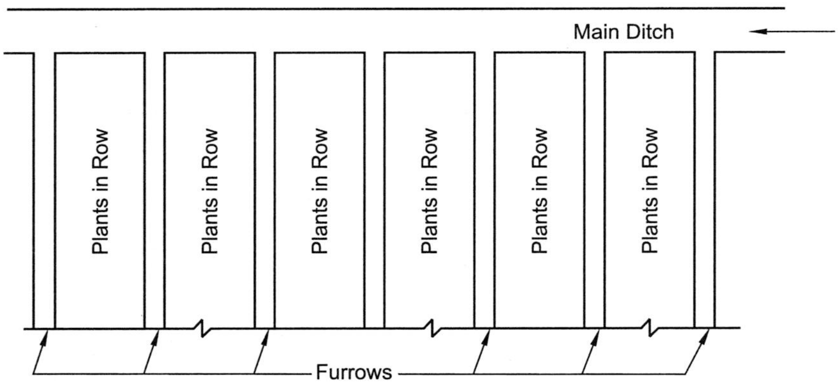

Fig. 6.5 Furrow irrigation

The water is distributed to the furrows from the field channel or main ditch. Figure 6.5 shows the arrangement clearly.

In the side of field or main ditch small openings are made, and through the openings water flows into one or more furrows. Small adjustable gates may be provided to control the supply of water from the field ditch to the furrows.

Furrow irrigation has distinct advantages over other methods. They are :

(*i*) As the area wetted is just $\frac{1}{2}$ to $\frac{1}{5}$ of the cropped area of the field, puddling and crusting of the soil is minimum.

(*ii*) Also, loss of water due to deep percolation and evaporation is restricted.

(*iii*) Furrows do not put hindrance in use of field machinery or other farming methods.

(*iv*) In this method plants in their early tender age are not damaged by flow of water.

(*v*) By laying the furrows along the contours, across the slope of land, soil erosion can be minimised.

(*vi*) Furrow making is a simple and cheap method and working expenses are also nominal.

(*vii*) Land between the rows of plants is utilized to construct furrows therefore, useful irrigable land is not wasted.

6.13. OVERHEAD OR SPRINKLER IRRIGATION

In this method an attempt is made to simulate natural rainfall. Irrigation water is applied to the land in the form of a spray. This method is also known as sprinkler irrigation, Fig. 6.6.

Sprinklers can be used on all soil types of any topography. According to the equipment and procedure used the sprinkler method may fall in fixed type or portable type. The sprinkler irrigation may also be subdivided according to the functions it has to perform, namely:

(*i*) the main irrigation system,

(*ii*) the supplementary irrigation system, and

(*iii*) the protective irrigation system.

The sprinkler irrigation system is in use since 1920 A.D. in some advanced countries. In India this method has come into use since 1950. It is mostly adopted in tea and coffee gardens. But the time has come to explore the possibilities of using this method on large scale. Experiments have already been undertaken on this aspect at various research centres in the country. There is great scope and likelihood of this system becoming popular in our country in the coming years. By introduction of spray irrigation about 35 per cent of water can be saved which is otherwise wasted in surface methods. Following conditions favour implementation of sprinkler irrigation:

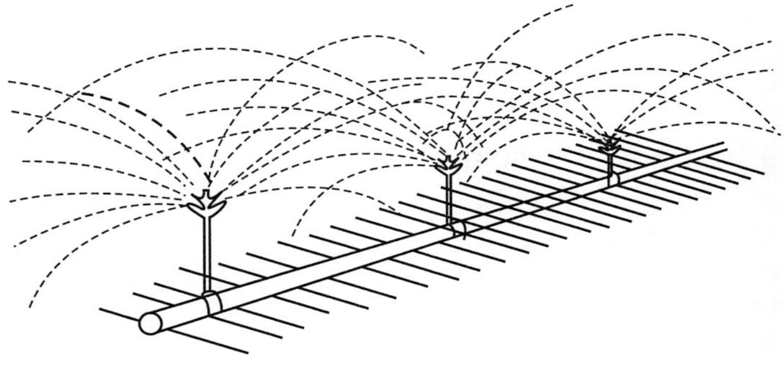

Fig. 6.6. Sprinkler irrigation

1. When the soil is too porous for good distribution by surface irrigation.
2. When there are fields with uneven surface.
3. When soil is easily erodable.
4. When water supply is just sufficient for crop growth.

There are three general type of the sprayers. They are fixed nozzles attached to the pipe, perforated pipe and rotating sprayers.

Perforated fixed pipe sprinklers: In this method parallel pipes are installed at a suitable spacing (say 15 m) and supported on the posts. The water is discharged through the perforations at right angles to the pipe line. The water jets emerge out of the pipe line at an angle of 135° to the vertical. With this inclination it is possible to irrigate the whole strip of 15 m width. Sometimes nozzles can be fixed to the pipes. The water passes through pipes under pressure.

Portable sprinklers: They are mainly used for irrigation in orchards and nurseries. They consist of underground main pipe lines, portable lateral flexible pipelines and sprinklers. Pumping plant is kept at one point for the whole system.

Advantages of the sprinkler irrigation:

1. There is no erosion of soil.
2. Fertilizers are used economically since they can be injected through the water.
3. Water can be applied uniformly and in controlled manner as required for light irrigation for nursery etc.
4. This method is natural and helpful in sowing stage.
5. This method can be used on any type of field.
6. There is good saving of water.

Disadvantages of sprinkler irrigation:

1. Initial cost of the system is rather high.
2. Any cost of power to provide pressure must be added to irrigation chargers. It makes this costly and therefore unpopular.
3. Wind interferes with the distribution pattern. It reduces the spreading rate and in turn the efficiency. Under high temperatures and strong winds heavy evaporation loss takes place thereby offsetting the saving in water.

4. Perforations in the pipe may choke up.
5. Labour is required to move the pipes and plant.
6. Water supply has to be free from sediment to avoid choking of nozzles etc.

6.14. DRIP OR TRICKLE IRRIGATION

It is a latest advancement over other methods. The name of the method itself implies water saving. In this method irrigation water is conveyed on the surface in 12 to 16 mm diameter tubings fed from large feeder pipes. The water is allowed to drip or trickle slowly through the nozzle or orifices at practically zero pressure. In this way the soil in the root-zone of crops is constantly kept wet. By using this method crops can be grown successfully over the saline lands also. This method has been found to be of great value in reclaiming and developing desert and arid areas. The main drawback of this method is its high cost. But with growing realisation of the value of water this method has been introduced in other countries of the world particularly in desert areas. The method is still in initial stages of development in our country.

Main advantages of drip irrigation are following:

(*i*) It helps in optimum utilisation of irrigation water by reducing percolation and evaporation losses on one hand and by maintaining appropriate water content in the root-zone of plants.

(*ii*) There is no chance of land getting waterlogged and thereby becoming saline or alkaline.

(*iii*) Crop yield is substantially increased.

(*iv*) It makes possible to go cash crops.

(*v*) The fields do not get infested with weeds and pest due to non-availability of excess water.

(*vi*) It helps in economical use of fertilizers since they are applied along with irrigation water in solution with it.

(*vii*) The fields do not get eroded or degraded since there is no excessive use of water on the fields.

QUESTIONS FOR REVISION

1. What is the necessity for Irrigation?
2. Name important projects of your region executed before Independence as well as after Independence and give their locations?
3. (*a*) What are disadvantages of irrigation?
   (*b*) In how many ways irrigation benefits the country?
4. Describe comparatively various methods of applying water to the land in general.
5. What is the flooding method of irrigation? In how many ways it can be accomplished?
6. Write short notes on the following:
   (*i*) Drip irrigation,
   (*ii*) Portable sprinklers,
   (*iii*) Furrow irrigation.
7. Describe in detail sprinkler irrigation method and enumerate its advantages and disadvantages.

7

Water Requirement of Crops

7.1. FUNCTIONS OF IRRIGATION WATER

There is no other vital requirement of crop than water. In absence of water plant life cannot survive even if remaining conditions are most favourable for crop growth.

The water is absorbed by the plant roots from the available soil-moisture storage. Then the absorbed water is circulated through the plant structure. Finally it goes to the atmosphere in the form of vapour by the process of transpiration.

The water is a solvent for the nutriment required for plant growth. Water forms the solution of nourishing food and more precisely this solution is absorbed by the plant roots. Thus it is very clear at this stage that water acts as a nutriment carrier. After each part of the plant structure takes up nourishing food from the solution, water leaves the plant structure and goes to atmospheric vapour storage.

Water is not only the nutriment carrier. It also performs other duties. Some salts present in soil react to produce nourishing food products only in the presence of water. Water when added to the soil externally wets the soil and when temperature change takes place water tries to neutralize the unfavourable change by exposing itself to the change before the soil gets affected. Thus it keeps the temperature of the soil-controlled.

In short, following important functions are performed by irrigation water:

- It acts as a solvent for nutrient.
- It circulates through plant structure and supplies moisture and nutriment.
- It provides medium for reaction of salts present in the soil to be subsequently absorbed by the roots.
- It keeps the soil temperature under permissible limits.
- It dilutes the concentration or washes down the harmful salts from the soil mass.

7.2. WATER NEED OF CROPS

Crops need water right from the time of sowing to maturity continuously. The rate of use of water is not, however, same for all crops. The rate of use varies with the kind of crop grown, time taken by the crop to mature, the temperature, and weather conditions. It is already mentioned that the root zone acts as a store house for available water for crop use. When irrigation water is provided to the land, quantity of the water stored in the available zone of soil-moisture should be sufficient for crop growth for a period varying from few days to several weeks. It is actually the period between two consecutive waterings. The length of the period between two waterings is dependent on the capacity of the particular soil to hold the available water and the type of crop grown on that soil.

Optimum irrigation requirement of a crop depends upon the following three factors:

- Soil factors affecting storage and transmissivity of water;
- plant factors affecting water absorption, transmission within plants and transpiration; and
- climate factors affecting transmission, vapour pressure gradient and energy for evapotranspiration.

Method of obtaining optimum irrigation water use is further discussed while dealing with (δ) delta or depth of water to be applied to crops.

7.3. FACTORS AFFECTING IRRIGATION WATER REQUIREMENT

Recent experiments have shown that productivity of a given quantity of water can be doubled if scientific water management technique is adopted. The scientific water management comprises various activities like proper tillage operations and land levelling, correct estimation of water requirement of crops and also irrigability characteristic of soil, most suitable cropping pattern and adoption of correct method of irrigation apart from efficient and economic storage and conveyance practices. The water management thus involves perfect knowledge of soil-water-crop relationship under any given climatic condition as well as various engineering aspects of irrigation from the storage down to the fields. The estimation of irrigation water requirement of a crop assumes a pivotal role in the scientific water management technique.

Many factors affect the amount of irrigation water required by a plant. The effect of these factors vary from place-to-place and also fluctuate from time-to-time at the same place. The important factors which influence the irrigation water requirement are the following:

- Amount and rate of precipitation during a crop season.
- Temperature variations in any particular locality greatly affect the rate of consumptive use.
- Hours of day-light also increase evaporation.
- Water requirement of a plant varies for different stages of growth during the crop season.
- Evaporation and transpiration depends upon humidity of a locality.
- Evaporation of water is greatly affected by movement of wind. Hot, dry winds increase evaporation.
- The growing season influences seasonal use of water by plants.
- Quality of water used has also minor effect on the consumptive use.
- Soil characteristics like fertility, permeability, pH value etc., also affect the consumptive use to some extent.

7.4. DEFINITIONS OF WATER BALANCE PARAMETERS

Before proceeding with the subject it is useful to understand following terms:

Water Requirement of Crop: It is the total amount of water required at the field head to mature a crop without regard to the source of water. Obviously, it includes the amount of water needed to meet:

- losses through evaporation;
- losses through transpiration;

- plant metabolism needs;
- application losses; and
- special needs.

It, however, does not include transit losses. Water requirement of crops can be expressed mathematically as

$$WR = CU + \text{Application losses} + \text{Special needs}$$

Alternatively, in relation to source it can also be expressed as

$$WR = IR + ER + S$$

where WR is water requirement of a crop
CU is consumptive use of water
IR is irrigation requirement of a crop at field head
ER is effective rainfall
S is soil moisture contribution

All quantities are expressed as depth in cm or volumetrically according to convenience.

Consumptive use of water: It is a matter of common experience that water losses are simultaneously taking place from the soil adjacent to the plant and through transpiration from the plant structure. Hence the term evapo-transpiration is commonly used to denote combined water loss. It is designated by abbreviation ET also. In addition, some water is utilized in the plant metabolism. The consumptive use of water is the sum total of ET and water used up in plant metabolism. But since water used in metabolism is negligible, for all practical purposes—

$$ET = CU \quad \text{where } ET \text{ is evapo-transpiration.}$$

Application losses: The amount of water required to replenish the soil moisture deficit back to field capacity (FC) in the root-zone of a crop cannot be applied exactly by means of irrigation. Some loss of water is, therefore, unavoidable under field conditions. These unavoidable losses are called application losses.

Water applied in excess of field capacity are avoidable losses. Such losses include surface run off and percolation losses beyond crop root-zone. In scientific management of water occurrence of such losses is to be prevented.

Special needs: Depending upon field conditions and soil characteristics extra water is needed to meet purposes like leaching of excess salts, puddling, pre-planting (paleo) irrigation, etc.

Irrigation Requirement (IR): It is the depth of water exclusive of precipitation, stored soil moisture or groundwater that is required to meet evapo-transpiration needs of a crop.

It can be expressed as $IR = WR - (ER + S)$

It is also known as *net irrigation water requirement* (NIR). Irrigation requirement of a field will be IR plus transit losses in the farm distribution system.

Similarly, irrigation requirement at head of an outlet will constitute farm irrigation requirement (FIR) plus losses taking place below outlet.

Gross irrigation requirement (GIR) will, therefore, be net irrigation requirement plus all losses in transit upto the point where water is drawn in the conveyance system from the storage or by diversion.

Effective Rainfall (ER): It is that portion of precipitation which has fallen during the base period of a crop and is available to meet the evapo-transpiration need of the crop. In a broad sense effective rainfall is the portion of rainfall that has not lost by surface run-off or through deep percolation below the root zone. On certain problematic soils rainwater is used in leaching of salts and accounts for ER.

Carryover soil moisture (S): It is the moisture stored in the root zone depth in the soil before the crop is planted. This moisture is also available for meeting the evapo-transpiration needs of a plant. Sometimes shallow groundwater also contributes to crop growth.

7.5. MEASUREMENT OF PARAMETERS

The irrigation requirement can be determined directly by conducting field experiments based on the modern concept of soil-water-plant relationships.

(*i*) **Soil moisture sampling method:** To measure evapo-transperation soil-moisture sampling can be done on irrigated field where depth of groundwater is sufficiently below and does not influence soil moisture fluctuations in the root zone. Soil samples are taken before and after each irrigation to determine moisture content by usual laboratory methods. The loss of water in noted time is calculated. The consumptive use can be calculated for various periods ranging from interval between two consecutive waterings to whole crop season as required.

(*ii*) **By field experiment:** In this method irrigation water is applied to a selected plot in such a manner that there is neither deep percolation nor surface run-off. This method is repeated on number of plots under different conditions of water supply. Yield obtained from different plots is plotted against the water applied. The most profitable yields are taken as the ones for which water has been applied consumptively in optimum manner. The results show that for every crop the yield increases rapidly with water supplied upto a certain point. Thereafter there is break in the curve and yield starts decreasing even though water supply is increased. This transition point is taken to be consumptive use for that crop.

(*iii*) **By use of Lysimeters:** The field lysimeters of a appropriate size can be utilized to determine various parameters like IR, ER, S, run off, percolation losses etc. Lysimeters are blocks of soil in a metal or plastic containers sunk in the ground. Sometimes the bottom is made pervious. The crop is grown in the tank. By taking the soil samples at fixed intervals loss of water can be measured to calculate parameters like evaporation, IR, ER, S, etc. The aim being to determine loss of water to achieve satisfactory growth of crop. The size of lysimeter may be from 1 to 3 m in dia and 2 to 3 m in depth. The installation of lysimeter, however, needs great care apart from sufficient guard area to overcome clothes line effect. Larger the size of the lysimeter closer will be the results to actual field condition.

(*iv*) **By analysis of basic equation:** The irrigation requirement can also be estimated on the basis of values of WR, ER and S. For estimating WR it is necessary to measure ET, application losses and special needs. In practice, application losses and special needs can be measured directly.

(*a*) The effective rainfall can be determined directly by using field lysimeters as already referred above. It can also be estimated by computing the soil moisture deficit in the root zone before an anticipated rain. Any rain in excess of this deficit would be ineffective. This, however, presume that rain has occurred at an intensity equal or lower than infiltration rate of the soil.

The estimation of ER is based on estimate of ET or potential evapo-transpiration (PET) values derived by using open pan evaporimeter or from any suitable formulae. The data on pan evaporation and other climatological factors may be obtained from the nearby meteorological observatories for working out ET estimates. The average ET values for different crops in different seasons may also be obtained from nearby research stations or may actually be determined in a sample field to be used in similar areas.

(*b*) There is no simple method to determine or estimate the contributions from the soil profile moisture under all situations. The contribution of soil profile moisture in crop growth depends on the capacity of different soils to hold water. The difference between soil moisture available at sowing and that remained unused at harvest gives the actual amount of soil moisture contributed to WR of a crop. This determination is possible in areas where groundwater table is far below the crop root zone.

In places where water table is near or within the root zone considerable amount of water is contributed to WR of a crop directly from the water-table or capillary fringe. Under certain conditions this amount completely satisfies the water requirement (WR) of crop and under worse condition may even necessitate drainage arrangements. Basic studies are, therefore, needed for different places to ascertain soil profile moisture contribution under different water table conditions.

(*v*) **By use of formulae:** As already mentioned the evapo-transpiration (ET) can be measured directly by soil moisture depletion studies by conducting field experiments. Alternatively the ET values can be estimated indirectly from potential evapo-transpiration (PET) values calculated by using evaporimeters or by using various formulae based on thermodynamic principles. Most of the formulae produced relate to the conditions where those were developed. Also PET occurs only when there is ample water (so that evapo-transpiration is not restricted for want of soil water) and a large area of land is completely covered with actively growing green vegetation. Such conditions are not normally met with in practice except in fully grown wet paddy fields or for a short period after a rain or irrigation in other crops. It, therefore, requires conversion of estimated PET values to actual ET values by adopting a factor determined for local conditions. Therefore, the empirical formulae developed elsewhere should not be used directly.

The following are some of the commonly used equations for computation of consumptive use:

- Blaney-Criddle equation
- Hargreaves class A pan evaporation formula
- Penman formula.

7.6. BLANEY-CRIDDLE EQUATION

It gives monthly consumptive use by a relation in terms of temperature and day-light hours as follows:

$$C_u = k.f$$

where C_u is monthly consumptive use in cm

k is a crop factor

It is determined experimentally for each crop under certain environmental conditions.

f is monthly consumptive use factor.

and $$f = \frac{P}{40}[1.8\,t + 32]$$

P is percentage of daylight hours occurring during the period. It is taken from sunshine table.

t is mean monthly temperature in °C.

The weakness of this equation is that it does not consider factors like wind velocity and humidity on which consumptive use depends.

7.7. HARGREAVES CLASS A PAN EVAPORATION FORMULA

It gives consumptive use as a function of pan evaporation. The formula is in the form:

$$C_u \text{ or } E_t = KE_p$$

where E_t or C_u is consumptive use;

E_p is class A pan evaporation;

and K is consumptive use coefficient.

K is different for different crops and depends upon several climatic factors and needs to be determined experimentally. Value of K for some crops in India is given in Table 7.1.

Table 7.1. Consumptive Use Coefficient K for Some Indian Crops (To be Used in the Equation $E_t = KE_p$)

% of crop growing Season	*Wheat (Ludhiana)*	*Wheat (Pune)*	*Cotton (Pune)*	*Maize (Ludhiana)*
0	0.14	0.30	0.22	0.40
5	0.17	0.40	0.22	0.42
10	0.23	0.51	0.23	0.47
15	0.33	0.62	0.24	0.54
20	0.45	0.73	0.26	0.63
25	0.60	0.84	0.35	0.75
30	0.72	0.92	0.58	0.85
35	0.81	0.96	0.80	0.96
40	0.88	1.10	0.95	1.04
45	0.90	1.10	1.03	1.07
50	0.91	1.00	1.08	1.09
55	0.90	0.91	1.08	1.10
60	0.89	0.80	1.07	1.11
65	0.86	0.65	1.05	1.10
70	0.83	0.51	1.00	1.07
75	0.80	0.40	0.93	1.04
80	0.76	0.30	0.85	1.00
85	0.71	0.20	0.73	0.97
90	0.65	0.12	0.62	0.89
95	0.58	0.10	0.50	0.81
100	0.51	0.10	0.40	0.70
Seasonal value of K	**0.61**	**0.61**	**0.68**	**0.86**

E_p is generally calculated by making use of Christiansen formula for obtaining better results.

It states that, $$E_p = 0.459R \cdot C_t \cdot C_w \cdot C_h \cdot C_s \cdot C_e \text{ cm}$$

Table 7.2. Mean Monthly Values of Extra-terrestrial Radiation Expressed as Equivalent Evaporation at 20°C, in mm per Month of Water*. (Christiansen, 1968)

Latitude (degree)	*Jan.*	*Feb. 28/29 days*		*Mar.*	*Apr.*	*May*	*June*	*July*	*Aug.*	*Sept.*	*Oct.*	*Nov.*	*Dec.*
1	**2**	**3**	**4**	**5**	**6**	**7**	**8**	**9**	**10**	**11**	**12**	**13**	**14**
North 60	45	99	102	217	337	458	496	486	391	260	144	58	30
50	117	167	173	286	384	483	503	501	430	317	221	131	97
45	154	201	208	318	404	491	505	505	445	343	257	168	132
40	191	234	242	348	421	497	505	508	458	367	291	204	169
35	229	265	275	374	435	500	503	508	468	388	322	240	207
30	267	295	306	398	446	500	498	505	474	407	350	274	245
25	304	323	334	419	455	498	490	500	478	422	376	307	283
20	339	348	361	437	460	491	479	491	478	435	400	339	320
15	372	371	384	451	461	482	465	479	475	442	422	368	356
10	402	391	405	462	459	469	448	464	468	448	441	395	390
5	429	408	423	470	453	454	428	446	458	450	457	420	421
Equator	453	422	437	473	445	435	406	424	445	449	468	441	450

* Christiansen computed the values in inches from data of Napier Shaw. These values are multiplied by 25.4 and reported here in mm.

where R = Extra-terrestrial radiation (cm), to be determined from the table (Refer Table 7.2)

C_t = Coefficient for temperature, determined from the expression:

$C_t = 0.393 + 0.02796\, T_c + 0.0001189\, T_c^2$ (T_c is mean temp. in °C)

C_w = Coefficient for wind velocity, given by

$C_w = 0.708 + 0.0034\, W - 0.0000038\, W^2$

(W is mean wind velocity in km/day at 0.6 m above ground surface)

C_h= Coefficient for relative humidity given by

$C_h = 1.250 - 0.0087\, H + 0.75 \times 10^{-4}\, H^2 - 0.83 \times 10^{-8}\, H^4$

(H is the mean % relative humidity at noon or average relative for 11 and 18 hours)

C_s = Coefficient for per cent of possible sunshine and is given by

$C_s = 0.542 + 0.008S - 0.78 \times 10^{-4}\, S^2 + 0.62 \times 10^{-6}\, S^3$

(S is mean sunshine percentage)

C_e = coefficient of elevation, given by

$C_e = 0.97 + 0.00984\, E$ (E is elevation in 100 metres)

7.8. PENMAN FORMULA

It is gives consumptive use or potential evapo-transpiration. The formulae based on energy radiation concept and aerodynamic principles such as the one developed by Penman give reliable PET values. It requires data on large number of weather parameters. In 1975, Doorenbos and Pruit gave a modified Penman method for estimating PET values on the basis of extensive study of climatic and measured grass evapo-transpiration data from various research stations in the world. The method with fair accuracy gives reference crop ET value. Necessary tables for doing computations have also been prepared by them.

7.9. RADIATION

At this state it is useful to understand phenomenon of radiation that is taking place. From the sun two types of radiation is received by the earth. They are short wave and long wave radiation.

Net radiation (R_n) with which we are concerned is the difference between all radiation that is coming from the sun (R_a) and all that goes out. Outgoing radiation is the sum total of four items. They are :

(*a*) While amount of radiation received at the top of the atmosphere is, R_a ; part of it gets absorbed in the atmosphere during its passage to the earth. Radiation gets absorbed due to clouds present in the atmosphere. Earth actually receives 'R_s'.

(*b*) Part of the radiation (R_s) gets directly reflected back to the atmosphere from the earth and crop cover. Reflection 'δ' depends on the extent of crop cover and wetness of the adjacent exposed soil surface. What remains is net short wave solar radiation 'R_{ns}'. Therefore, $R_{ns} = (1 - \delta).R_s$.

(*c*) In addition further loss of radiation occurs at the earth's surface. Part of the absorbed shortwave energy is radiated back by the earth to the atmosphere as long wave radiation.

(*d*) Fourthly, part of incoming long wave radiation also goes back to the atmosphere. Actually outgoing long wave radiation is more than the incoming long wave radiation because part of the absorbed short wave radiation also goes back as long wave radiation

Table 7.3. (a) Values of Weighting Factor W for the Effect of Radiation on ET at Different Temperatures and Altitudes

Altitude m	*Temperature °C* 2	4	6	8	10	12	14	16	18	20	22	24	26	28	30	32	34	36	38	40
	Values of W																			
0	.43	.46	.49	.52	.55	.58	.61	.64	.66	.68	.71	.73	.75	.77	.78	.80	.82	.83	.84	.85
500	.44	.48	.51	.54	.57	.60	.62	.65	.67	.70	.72	.74	.76	.78	.79	.81	.82	.84	.85	.86
1000	.46	.49	.52	.55	.58	.61	.64	.66	.69	.71	.73	.75	.77	.79	.80	.82	.83	.85	.86	.87
2000	.49	.52	.55	.58	.61	.64	.66	.69	.71	.73	.75	.77	.79	.81	.82	.84	.85	.86	.87	.88
3000	.52	.55	.58	.61	.64	.66	.69	.71	.73	.75	.77	.79	.81	.82	.84	.85	.86	.87	.88	.89
4000	.54	.58	.61	.64	.66	.69	.71	.73	.75	.77	.79	.81	.82	.84	.85	.86	.87	.89	.90	.90

Table 7.3. (b) Values of Weighting Factor (1 – W) for the Effect of Wind and Humidity on ET at Different Temperatures and Altitudes

(1 – W) at altitude m	*Temperature °C* 2	4	6	8	10	12	14	16	18	20	22	24	26	28	30	32	34	36	38	40
0	0.57	.54	.51	.48	.45	.42	.39	.36	.34	.32	.29	.27	.25	.23*	.22	.20	.19	.17	.16	.15
500	.56	.52	.49	.46	.43	.40	.38	.35	.33	.30	.28	.26	.24	.22	.21	.19	.18	.16	.15	.14
1000	.54	.51	.48	.45	.42	.39	.36	.34	.31	.29	.27	.25	.23	.21	.20	.18	.17	.15	.14	.13
2000	.51	.48	.45	.42	.39	.36	.34	.31	.29	.27	.25	.23	.21	.19	.18	.16	.15	.14	.12	.12
3000	.48	.45	.42	.39	.36	.34	.31	.29	.27	.25	.23	.21	.19	.18	.16	.15	.14	.13	.12	.11
4000	.46	.42	.39	.36	.34	.31	.29	.27	.25	.23	.21	.19	.18	.16	.15	.14	.13	.12	.11	.10

Table 7.4. Extra-terrestrial Radiation (Ra)
(Expressed in equivalent Evaporation in mm/day)

Northern Hemisphere

Lat	Jan.	Feb.	Mar.	Apr.	May	June	July	Aug.	Sep.	Oct.	Nov.	Dec.
50	3.8	6.1	9.4	12.7	15.8	17.1	16.4	14.1	10.9	7.4	4.5	3.2
48	4.3	6.6	9.8	13.0	15.9	17.2	16.5	14.3	11.2	7.8	5.0	3.7
46	4.9	7.1	10.2	13.3	16.0	17.2	16.6	14.5	11.5	8.3	5.5	4.3
44	5.3	7.6	10.6	13.7	16.1	17.2	16.6	14.7	11.9	8.7	6.0	4.7
42	5.9	8.1	11.0	14.0	16.2	17.3	16.7	15.0	12.2	9.1	6.5	5.2
40	6.4	8.6	11.4	14.3	16.4	17.3	16.7	15.2	12.5	9.6	7.0	5.7
38	6.9	9.0	1.8	14.5	16.4	17.2	16.7	15.3	12.8	10.0	7.5	6.1
36	7.4	9.4	12.1	14.7	16.4	17.2	16.7	15.4	13.1	10.6	8.0	6.6
34	7.9	9.8	12.4	14.8	16.5	17.1	16.8	15.5	13.4	10.8	8.5	7.2
32	8.3	10.2	12.8	15.0	16.5	17.0	16.8	15.6	13.6	11.2	9.0	7.8
30	8.8	10.7	13.1	15.2	16.5	17.0	16.8	15.7	13.9	11.6	9.5	8.3
28	9.3	11.1	13.4	15.3	16.5	16.8	16.7	15.7	14.1	12.0	9.9	8.8
26	9.8	11.5	13.7	15.3	16.4	16.7	16.6	15.7	14.3	12.3	10.3	9.3
24	10.2	11.9	13.9	15.4	16.4	16.6	16.5	15.8	14.5	12.6	10.7	9.7
22	10.7	12.3	14.2	15.5	16.3	16.4	16.4	15.8	14.6	13.0	11.1	10.2
20	11.2	12.7	14.4	15.6	16.3	16.4	16.4	15.9	14.8	13.3	11.6	10.7
18	11.6	13.0	14.6	15.6	16.1	16.1	16.1	15.8	14.9	13.6	12.0	11.1
16	12.0	13.3	14.7	15.6	16.0	15.9	15.9	15.7	15.0	13.9	12.4	11.6
14	12.4	13.6	14.9	15.7	15.8	15.7	15.7	15.7	15.1	14.1	12.8	12.0
12	12.8	13.9	15.1	15.7	15.7	15.5	15.5	15.6	15.2	14.4	13.3	12.5
10	13.2	14.2	15.3	15.7	15.5	15.3	15.3	15.5	15.3	14.7	13.6	12.9
8	13.6	14.5	15.3	15.6	15.3	15.0	15.1	15.4	15.3	14.8	13.9	13.3
4	14.3	15.0	15.5	15.5	14.9	14.4	14.6	15.1	15.3	15.1	14.5	14.1
2	14.7	15.3	15.6	15.3	14.6	14.2	14.3	14.9	15.3	15.3	14.8	14.4
0	15.0	15.5	15.7	15.3	14.4	13.9	14.1	14.8	15.3	15.4	15.1	14.8

Table 7.5. Mean Daily Maximum Possible Duration of Bright Sunshine Hours (N) for Different Months and Latitudes.

Northern Lats.	*Jan.*	*Feb.*	*march*	*April*	*May*	*June*	*July*	*Aug.*	*Sept.*	*Oct.*	*Nov.*	*Dec.*
Southern Lats.	*July*	*Aug.*	*Sept.*	*oct.*	*Nov.*	*Dec.*	*Jan.*	*Feb.*	*March*	*April*	*May*	*June*
50°	8.5	10.1	11.8	13.8	15.4	16.3	15.9	14.5	12.7	10.8	9.1	8.1
48°	8.8	10.2	11.8	13.6	15.2	16.0	15.6	14.3	12.6	10.9	9.3	8.3
46°	9.1	10.4	11.9	13.5	14.9	15.7	15.4	14.2	12.6	10.9	9.5	8.7
44°	9.3	10.5	11.9	13.4	14.7	15.4	15.2	14.0	12.6	11.0	9.7	8.9
42°	9.4	10.6	11.9	13.4	14.6	15.2	14.9	13.9	12.9	11.1	9.8	9.1
40°	9.6	10.7	11.9	13.3	14.4	15.0	14.7	13.7	12.5	11.2	10.0	9.3
35°	10.1	11.0	11.9	13.1	14.0	14.5	14.3	13.5	12.4	11.3	10.3	9.8
30°	10.4	11.1	12.0	12.9	13.6	14.0	13.9	13.2	12.4	11.5	10.6	10.2
25°	10.7	11.3	12.0	12.7	13.3	13.7	13.5	13.0	12.3	11.6	10.9	10.6
20°	10.0	11.5	12.0	12.6	13.1	13.3	13.2	12.8	12.3	11.7	11.2	10.9
15°	11.3	11.6	12.0	12.5	12.8	13.0	12.9	12.6	12.2	11.8	11.4	11.2
10°	11.6	11.8	12.0	12.3	12.6	12.7	12.6	12.4	12.1	11.8	11.6	11.5
5°	11.8	11.9	12.0	12.2	12.3	12.4	12.3	12.3	12.1	12.0	11.9	11.8
0°	12.1	12.1	12.1	12.1	12.1	12.1	12.1	12.1	12.1	12.1	12.1	12.1

Table 7.6. Correction for Temperature f(t) on Long Wave Radiation R_{nl}

t°C	*0*	*2*	*4*	*6*	*8*	*10*	*12*	*14*	*16*	*18*	*20*	*22*	*24*	*26*	*28*	*30*	*32*	*36*
$f(t) = \sigma TK^4$	11.0	11.4	11.7	12.0	12.4	12.7	13.1	13.5	13.8	14.2	14.6	15.0	15.4	15.9	16.3	16.7	17.2	18.1

Table 7.7. Correction for Vapour $f(e_d)$ on Long Wave Radiation R_{nl}

e_d *mbar*	*6*	*8*	*10*	*12*	*14*	*16*	*18*	*20*	*22*	*24*	*26*	*28*	*30*	*32*	*34*	*36*	*38*	*40*
$f(e_d) =$ $0.34 - 0.444\sqrt{e_d}$	.23	.22	.20	.19	.18	.16	.16	.14	.13	.12	.12	.11	.10	.09	.08	.08	.07	.06

Table 7.8. Correction for the Ratio Actual to Maximum Bright Sunshine Hours f(n/N) on Long Wave Radiation R_{nl}

n/N	*0*	*.05*	*.10*	*.15*	*.20*	*.25*	*.30*	*.35*	*.40*	*.45*	*.50*	*.55*	*.60*	*.65*	*.70*	*.75*	*.80*	*.85*	*.90*	*.95*	*1.0*
$f(n/N)$ = 0.1 + 0.9 n/N	.10	.15	.19	.24	.28	.33	.37	.42	.46	.51	.55	.60	.64	.69	.73	.78	.82	.87	.91	.96	1.0

Table 7.9. Values of Function f(u) = 0.27 $\left(1+\frac{u_2}{100}\right)$ for Wind run at a 2 m height* in km/day

Wind km/day	*0*	*10*	*20*	*30*	*40*	*50*	*60*	*70*	*80*	*90*
	–	0.30	0.32	0.35	0.38	0.41	0.43	0.46	0.49	0.51
100	0.54	0.57	0.59	0.62	0.65	0.67	0.70	0.73	0.76	0.78
200	0.81	0.84	0.86	0.89	0.92	0.94	0.97	1.00	1.03	1.05
300	1.08	1.11	1.13	1.16	1.19	1.21	1.24	1.27	1.30	1.32
400	1.35	1.38	1.40	1.43	1.46	1.39	1.51	1.54	1.57	1.59
500	1.62	1.65	1.67	1.70	1.73	1.76	1.78	1.81	1.84	1.80
600	1.89	1.92	1.94	1.97	2.00	2.02	2.05	2.08	2.11	2.15
700	2.16	2.19	2.21	2.24	2.27	2.29	2.32	2.35	2.38	2.40
800	2.43	2.46	2.48	2.51	2.54	2.56	2.59	2.62	2.65	2.65
900	2.70									

* where wind data are not collected at 2 m height the appropriate corrections for wind measurements taken at different heights are given below:

Measurement height, m	0.5	1.0	1.5	2.0	3.0	4.0	5.0	6.0	10.0
Correction factor	1.35	1.15	1.06	1.00	0.93	0.88	0.85	0.83	0.77

from the earth. The difference between outgoing and incoming long wave radiation is called net long wave radiation 'R_{nl}'. Since outgoing long wave radiation is greater than incoming long wave radiation R_{nl} represents net energy loss. Therefore, mathematically:

Net radiation = (Net solar radiation) – (net long wave radiation)

or
$$R_n = R_{ns} - R_{nl}$$
$$= R_s(1 - \delta) - R_{nl}$$

Radiation is expressed in different ways. When converted into heat radiation it can be expressed as energy required to evaporate water from an open surface with which we are concerned in the present context. In such a situation it is expressed as equivalent evaporation in mm/day.

7.10. MODIFIED PENMAN FORMULA FOR EVAPO-TRANSPIRATION

The modified Penman formula can be stated as

$$\text{ET} = C\,\{W \,.\, R_n + (1 - W) \,.\, f(u) \,.\, (e_a - e_d)\}$$

In the above equation:

ET is reference crop evapo-transpiration in mm/day;

C is adjustment factor to compensate for the day and night weather effects (Table 7.11)

$W \,.\, R_n$ represents radiation term; and

$(1 - W) \,.\, f(u) \,.\, (e_a - e_d)$ represents aerodynamic term.

The meaning of various notations used in the formula is as follows:

W is temperature and elevation related weighting factor for the effect of radiation on ET (Refer Table 7.3(*a*))

R_n is net radiation $(R_{ns} - R_{nl})$ in mm/day

R_{ns} is the net incoming shortwave solar radiation $= R_a\,(1 - \alpha)\left(0.25 + \frac{0.5\,n}{N}\right)$

R_a is extraterrestrial radiation expressed in equivalent evaporation in mm/day (Refer Table 7.4), which is received at the top of the atmosphere

R_s is solar radiation that reaches the earth after part of R_a is absorbed in the atmosphere. It is given by relation $R_s = R_a\left(0.25 + \frac{0.5\,n}{N}\right)$

n is actual duration of bright sunshine.

N is maximum possible duration of bright sunshine (Refer Table 7.5)

α is reflection coefficient of surface = 0.25 for most of the crops.

R_{nl} is the net long wave radiation $= f(t) \,.\, f(e_d) \cdot f(n/N)$

(Correction factors for temperature t, mean actual vapour pressure e_d and n/N are given in Tables 7.6, 7.7, 7.8 respectively).

$(1 - W)$ is a temperature and elevation related weighting factor and the effect of wind humidity on ET. (Refer Table 7.3(*b*))

$f(u)$ is a wind related function (Refer Table 7.9)

e_a is saturation vapour pressure in *m* bar at the mean air temperature in degrees centigrade (Refer Table 7.10)

e_d is mean actual vapour pressure of air in *m* bar $= \frac{e_a \times \text{RH mean}}{100}$; where *RH* is relative humidity

Table 7.10. Saturation Vapour Pressure (e_a) in m Bar (as a Function of Mean air Temperature in °C)

Temp. (°C)	e_a (*m bar*)	*Temp* (°C)	e_a (*m bar*)
0	6.1	21	24.9
1	6.6	22	26.4
2	7.1	23	28.1
3	7.6	24	29.8
4	8.1	25	31.7
5	8.7	26	33.6
6	9.4	27	35.7
7	10.0	28	37.8
8	10.7	29	40.1
9	11.5	30	42.4
10	12.3	31	44.9
11	13.1	32	47.6
12	14.0	33	50.3
13	15.0	34	53.2
14	16.1	35	56.2
15	17.0	36	59.4
16	18.2	37	62.8
17	19.4	38	66.3
18	20.6	39	69.9
19	22.0		
20	23.4		

Table 7.11. Adjustment Factor (*C*) in Penman Equation

	RH_{max} = 30%				RH_{max} = 60%				RH_{max} = 90%			
R_s *mm/day*	3	6	9	12	3	6	9	12	3	6	9	12
U_{day} *m/sec*					U_{day}/U_{night} = 4.0							
0	.86	.90	1.00	1.00	.96	.98	1.05	1.05	1.02	1.06	1.10	1.10
3	.79	.84	.92	.97	.92	1.00	1.11	1.19	.99	1.10	1.27	1.32
6	.68	.77	.87	.93	.85	.96	1.11	1.19	.94	1.10	1.26	1.33
9	.55	.65	.78	.90	.76	.88	1.02	1.14	.88	1.01	1.16	1.27
					U_{day}/U_{night} = 3.0							
0	.86	.90	1.00	1.00	.96	.98	1.05	1.05	1.02	1.06	1.10	1.10
3	.76	.81	.88	.94	.87	.96	1.06	1.12	.94	1.04	1.18	1.28
6	.61	.68	.81	.88	.77	.88	1.02	1.10	.86	1.01	1.15	1.22
9	.46	.56	.72	.82	.67	.79	.88	1.05	.78	.92	1.06	1.18
					U_{day}/U_{night} = 2.0							
0	.86	.90	1.00	1.00	.96	.98	1.05	1.05	1.02	1.06	1.10	1.10
3	.69	.76	.85	.92	.83	.91	.99	1.05	.89	.98	1.10	1.14
6	.53	.61	.74	.84	.70	.80	.94	1.02	.79	.92	1.05	1.12
9	.37	.48	.65	.76	.59	.70	.84	.95	.71	.81	.96	1.06
					U_{day}/U_{night} = 1.0							
0	.86	.90	1.00	1.00	.96	.98	1.05	1.05	1.02	1.06	1.10	1.10
3	.64	.71	.82	.89	.78	.86	.94	.99	.85	.92	1.01	1.05
6	.43	.53	.68	.79	.62	.70	.84	.93	.72	.82	.95	1.00
9	.27	.41	.59	.70	.50	.60	.75	.87	.62	.72	.87	.96

The adjustment weather factor '*C*' and wind related function *f*(*u*) are the important variations made in the modified Penman formula. This method gives very satisfactory value of crop ET. Computation technique is given in a solved problem subsequently.

7.11. COMPUTATION OF NET IRRIGATION REQUIREMENT FROM THE CONSUMPTIVE USE

The net irrigation requirement of a crop is determined by using step-by-step procedure. The whole crop period is divided in suitable parts, say in 10-day periods, from the time of preparing the fields for sowing to the maturity period. Then for each period the per cent of growing season elapsed has been worked out. For each period and corresponding per cent of grown season the value of consumptive use coefficient '*k*' is calculated. It is nothing but a ratio of Evapo-transpiration and Pan-evaporation values. It is based on the fact that there is definite relationship between pan evaporation and evapo-transpiration. It is of course necessary to before hand workout in the research centres such relationships for various crops in different localities and in different growing seasons. By multiplying pan evaporation value with the consumptive use coefficient the consumptive use is calculated. Finally the net irrigation requirement during various periods is determined by subtracting effective rainfall from the consumptive use.

The example given in Table 7.12 will make the procedure clear.

Theoretically Net irrigation requirement for any period is estimated by subtracting from consumptive use the amount of moisture contributed by such natural sources as: (*i*) effective rainfall, (*ii*) carry over soil moisture, and (*iii*) groundwater. The later two are, however, difficult to estimate. Also, their contribution is not generally significant and these are not dealt with here. Once the net irrigation requirement is calculated the efficiency of irrigation at various point is calculated to find out total irrigation water requirement at say outlet, head works storage reservoir, etc.

Table 7.12. Computation of Net Irrigation Requirement for Paddy (in cm)

Period	*No. of Days*	*Mid-point*	*% GS*	*ET/Pe = k*	*Pe*	*CU*	*ER*	*NIR*
July 15 – 20	6	3	3	1.00	4.3	4.3	5.8	–
July 21 – 31	11	9	8	1.00	8.6	8.6	27.0	–
Aug. 1 – 10	10	22	20	1.00	8.6	8.6	17.0	–
Aug. 11 – 20	10	32	30	1.15	8.6	9.9	3.3	5.6
Aug. 21 – 31	11	43	40	1.20	8.0	9.6	1.5	8.1
Sept. 1 – 10	10	53	49	1.30	10.0	8.9	11.0	–
Sept. 11 – 20	10	63	58	1.30	10.0	8.9	7.4	1.5
Sept. 21 – 20	10	73	67	1.20	9.3	8.9	2.3	6.6
Oct. 1 – 10	10	83	76	1.15	8.9	8.9	0.5	8.4
Oct. 11 – 30	10	93	85	1.00	8.6	8.6	6.1	2.5
Oct. 21 – 31	11	103	95	0.80	10.1	8.6	–	8.6
Total	**109 days**							

Note: %GS = Per cent growing season; ET = Evapo-transpiration;
Pe = Pan-evaporation; CU = Consumptive use;
ER = Effective rainfall; NIR = Net irrigation requirement;
k = Consumptive use coefficient; ET = CU

7.12. IRRIGATION SCHEDULE OR FREQUENCY OF IRRIGATION

In order to prepare schedule of irrigation it is necessary to decide the time interval between successive irrigations. The interval should be as large as possible without detriment to growth, quality and yield of crops. Also, it has to be ensured that available water resources are utilised efficiently in crop production. As mentioned in earlier part the water requirement of crop is affected by soil, plant and climatic factors. The schedule of irrigation has therefore to be prepared on the basis of these factors.

(*i*) **Soil factor:** Available soil moisture content at which irrigation should be applied is a good criteria as it indicates moisture content present in the soil and its availability for plant growth. Soil moisture can be considered extending full root zone of the crop or at least 60 cm from ground level because most of the roots are concentrated in these layers. The limit to which the soil moisture content is allowed to be depleted is determined by field experiments. For cereals and grain crops around 50 to 60 per cent of available soil moisture may be allowed to be depleted for good crop growth. The field experiments for soil moisture determination require extra labour and time.

(*ii*) **Plant factor:** Plants have distinct growth stages. For example, for wheat crop various growth stages are crown root initiation, tillering, jointing, flowering, milk and dough stage etc. Certain stages of crop growth are more critical than others so far as demand of water is concerned. While following criterion of critical stages it is necessary to apply irrigation sufficiently before the particular stage of growth is reached so that optimum moisture conditions prevail during that stage.

(*iii*) **Climatic factor:** At a place climatic conditions influence the evapo-transpiration. Knowledge of amount of evapo-transpiration during various intervals serve as a best criterion to know the soil water balance in the field. Evapo-transpiration can be measured directly by lysimeters or by indirect methods as already explained.

In addition to the above factors timeliness, adequacy and cost of irrigation water supplies greatly influence the irrigation scheduling of crops. Other considerations like pest control, temperature control (*e.g.,* by maintaining submerged in case of rice), field preparation and harvesting etc., also influence irrigation schedule.

Considering the availability of water resources and cultivated land two approaches could be mentioned for irrigation scheduling. They are:

- Where water resources are abundant as compared to cultivated land the aim should be to obtain **maximum yield per unit area of land**.
- Where water resources are limited as compared to cultivated land the aim should be to obtain **maximum production per unit volume of water used**.

Based on the various research data available as mentioned above in respect of irrigation requirement of crop in various regions a chart showing dates and quantity of irrigation is prepared. This serves as a guide to cultivators and irrigation engineers who control water supplies. In course of crop season if water is not sufficient to meet optimum requirements or in case rainfall occurs beyond calculations the schedule of irrigation can be adjusted suitably to achieve best results.

7.13. IRRIGATION EFFICIENCY

The objective of every water manager is to use irrigation water efficiently. The water can be said to have been utilized efficiently when the supply put in is utilized to the maximum without allowing the losses to occur. In general term, efficiency is a ratio of output to input expressed as percentage. Similarly irrigation efficiency is also expressed as a percentage and is a ratio of irrigation water utilized to the water supplied. The irrigation efficiency inversely depends on the extent of water loss occurring in the system. The losses and therefore irrigation efficiency depends on the following points:

- Application losses including evaporation, deep percolating and surface runoff.
- Intake characteristics of soils.
- Topographical features of irrigation area.
- Climatic characteristic of the region.
- Irrigation method.
- Tail water recovery system.
- Adequacy of design and construction of irrigation network and other hydraulic structures.

It is already explained earlier that water gets lost by different processes right from storage to the field. As a result efficiency of water utilization also varies for different points on irrigation water distribution system. Since irrigation water requirement is different at different locations in the irrigation system, for example, net irrigation requirement, irrigation requirement at field head, irrigation requirement at outlet and so on, efficiency of irrigation can be calculated separately for different locations. In general, on farm irrigation efficiency is highest whereas gross irrigation efficiency which takes into account various losses is the lowest.

Popularly following two types of irrigation efficiencies are calculated:

(*i*) **Water application efficiency:** It is the ratio of quantity of water stored in the crop root zone to the quantity of water applied or delivered on the field.

$$\text{Application efficiency} = \frac{\text{NIR}}{\text{FIR}} \times 100$$

where NIR is net irrigation requirement, and
FIR is field irrigation requirement

(*ii*) **Water conveyance efficiency:** It is the ratio of water applied or delivered on the field to water diverted from the river or the reservoir for supplying it to irrigation land.

$$\text{Water conveyance efficiency} = \frac{\text{FIR}}{\text{GIR}} \times 100$$

FIR is field irrigation requirement, and
GIR is gross irrigation requirement.

Similarly various other types of efficiencies can be calculated.

They are:

(*iii*) **Water storage efficiency:** It is the ratio of the water stored in the crop root zone during irrigation to water needed in the root zone prior to irrigation. It is expressed as below:

$$\eta = \frac{\text{Water stored in root zone}}{(\text{Field capacity} - \text{Available moisture})}$$

(*iv*) **Water use efficiency:** It is the ratio of the water put to beneficial use (it may include water used for leaching of soil in some cases) to the water delivered both measured in same unit.

SOLVED PROBLEMS

Problem 7.1. The climatic conditions existing in a certain area at latitude 25° N during *rabi* season are given below. Wheat crop is grown for which crop factor may be taken to be 0.75. Making use of Blaney Criddle equation determine:

(a) Consumptive use of crop;
(b) Net irrigation requirement (assume leaching requirement to be nil); and
(c) Field irrigation requirement (assume water application efficiency to be 80%).

Month	*Average month temp. (°C)*	*Monthly % (P) of day time hours of the year*	*Effective rainfall (cm)*
Nov.	19	7.39	1.5
Dec.	15	7.41	1.2
Jan.	13	7.53	2.0
Feb.	14	7.15	—

> **Note:** Mean daily percentage of annual day time hours for a given month and latitude is calculated from accompanying table.

Solution: Formula used is: $C_u = K \Sigma f$ ... (1)

where $f = \dfrac{P}{40}(1.8\, t + 32)$

Step 1: Values of f are worked out and tabulated in a table below.
Monthwise values of t, P, R_e and f

Month	*t* (°C)	*P* %	*Re* (cm)	$f = \dfrac{P}{40}(1.8\,t + 32)$
Nov.	19	7.39	1.5	12.2
Dec.	15	7.41	1.2	10.9
Jan.	13	7.53	2.0	10.4
Feb.	14	7.15	—	10.2
Σ			4.7	43.7

Step 2: Putting values in equation (1)

$$C_u = 0.75 \times 43.7$$

Consumptive use = 32.78 cm

Step 3: $\text{NIR} = C_u - R_e = 32.78 - 4.7$

Net irrigation requirement = 28.08 cm

Step 4: $\text{FIR} = \dfrac{\text{NIR}}{\eta} = \dfrac{28.08}{0.8}$

Field irrigation requirement = 35.1 cm

Problem 7.2. **Making use of following information and Charistiansen method calculate pan evaporation and consumptive use for the month of April:**

(i)	**Consumptive use coefficient**	**0.8**
(ii)	**Latitude of location**	**20°N**
(iii)	**Mean temperature**	**30°C**
(iv)	**Mean wind velocity at 0.5 m above ground**	**180 km/day**

(*v*) Mean relative humidity	**38%**
(*vi*) Mean sunshine per cent	**85%**
(*vii*) Elevation of the location	**400 m**
(*viii*) Extra-terrestrial radiation from Table 7.2	**46 cm**

Solution: Pan evaporation is given by equation

$$E_p = 0.459\ R \cdot C_t \cdot C_w \cdot C_h \cdot C_s \cdot C_e$$

Step 1: We have to solve for each of the coefficients C_t, C_w, C_h, C_s, C_e from simplified and non-dimensional equations.

Step 2: $C_t = 0.393 + 0.02796\ T_c + 0.0001189\ T_c^2$

$= 0.392 + 0.02796 \times 30 + 0.0001189\ (30)^2$

$= 0.393 + 0.8388 + 0.1070 = 1.3388$

Step 3: $C_w = 0.708 + 0.0034\ W - 0.0000038\ W^2$

$= 0.708 + 0.0034 \times 180 - 0.0000038\ (180)^2$

$= 0.708 + 0.612 - 0.12312 = 1.1969$

Step 4: $C_h = 1.250 - 0.87 \times 10^{-2} \times H + 0.75 \times 10^{-4} \times H^2 - 0.83 \times 10^{-8} \times H^4$

$= 1.250 - 0.3306 + 0.1083 - 0.1731$

$= 1.3583 - 0.5037 = 0.8546$

Step 5: $C_s = 0.542 - 0.8 \times 10^{-2} \times S - 0.78 \times 10^{-4} \times (S)^2 + 0.62 \times 10^{-6} \times (S)^3$

$= 0.542 + 0.8 \times 10^2 \times 85 - 0.78 \times 10^{-4} \times (85)^2 + 0.62 \times 10^{-6} \times (85)^3$

$= 0.542 + 0.68 - 0.5636 + 0.3808 = 1.603 - 0.564 = 1.039$

Step 6: $C_e = 0.97 + 0.00984\ E = 0.97 + 0.00984 \times 4 = 0.97 + 0.039 = 1.009$

Step 7: Putting values in Christihansen equation

$$E_p = 0.459(46)\ (1.3388)\ (1.1969)\ (0.8546)\ (1.039)\ (1.009) = 30.31 \text{ cm.}$$

Step 8: Consumptive use $= K \cdot E_p = 0.8 \times 30.31 = 24.25$ cm.

Problem 7.3. Using modified Penman formula calculate the potential evapo-transpiration for the reference crop with the following data:

1. Particulars of place – (*a*) location 28° N
(*b*) altitude 230 m

2. Period under consideration	**....**	**....**	**First ten days of August**
3. Mean air temperature	**....**	**....**	**29°C**
4. Mean relative humidity	**....**	**....**	**88.5**
5. Mean bright sunshine hours	**....**	**....**	**3.42**
6. Wind speed at 3 m height in km/day	**....**	**....**	**145.2**
7. Reflection coefficient (α)	**....**	**....**	**0.25**

Solution: Formula used is

$$\text{ET} = C(\text{W}\ R_n + (1 - W) \cdot f(u) \cdot (e_a - e_d) \qquad \text{... (1)}$$

Step 1: Values of various parameters to be used

$W = 0.78$ (from Table 7.3(*a*))

$\alpha = 0.25$ (reflection coefficient given data)

$R_a = 15.7$ mm/day (from Table 7.4)

$N = 13.1$ hours (from Table 7.5)

$e_a = 40.1$ m bars (from Table 7.10)

$$e_d = \frac{e_a \times RH \text{ mean}}{100} = \frac{40.1 \times 88.5}{100} = 35.5$$

U = wind speed at 2 m height

= 145.2 × 0.93 (from Table 7.9) = 135 km /day

Step 2: Solving for radiation term $W\, R_n$

$$R_n = (R_{ns} - R_{nl}) \qquad \text{... (2)}$$

$$R_{ns} = R_a\,(1 - \alpha)\left(0.25 + \frac{0.5n}{N}\right)$$

$$= 15.7\,(1 - 0.25)\left(0.25 + \frac{0.5 \times 3.42}{13.1}\right) = 15.7\,(0.75)\,(0.38) = 4.47$$

$$R_{nl} = f(t)\,.\,f(e_d)\,.\,f(n/N) \qquad \text{... (3)}$$

Using Tables 7.6, 7.7 and 7.8

$f(t)$ = 16.5

$f(e_d)$ = 0.08

$f(n/N)$ = 0.34

Putting values in equation (3)

R_{nl} = 16.5 × 0.08 × 0.34 = 0.45

Putting values in equation (2)

R_n = (4.47 – 0.45) = 4.02

Radiation term $W \cdot R_n$ = 0.78 × 4.02 = 3.13

Step 3: Solving for aerodynamic term $(1 - W) \cdot f(u) \cdot (e_a - e_d)$

$(1 - W)$ = (1 – 0.78) = 0.22

$f(u)$ = 0.635 (from Table 7.9)

$(e_a - e_d)$ = 40.1 – 35.5 = 4.6

Aerodynamic term

$(1 - W) \cdot f(u) \cdot (e_a - e_d)$ = 0.22 × 0.635 × 4.6 = 0.64

Step 4: $ET = C\,\{3.13 + 0.64\}$

Step 5: Value of C can be calculated from Table 7.11.

Since data is not given assume.

$$RH_{\max} = 90\% \qquad \text{... (1)}$$

Solar radiation is given by formula

$$R_s = \left(0.25 + 0.5\frac{n}{N}\right) \cdot R_a$$

$$= \left(0.25 + 0.5\frac{3.42}{13.1}\right) \cdot 15.7$$

$$= 15.7\,(0.38) = 5.97 \text{ (say 6)} \qquad \text{... (2)}$$

$$U_{\text{day}} = \frac{145.2 \times 1000}{24 \times 60 \times 60} = 1.68 \text{ m/sec} \qquad \text{... (3)}$$

Using Table 7.11 by interpolation [$U_{\text{day}}/U_{\text{night}}$ = 2 assumed]

C = 1.02

Step 6: E_T = 1.02 (3.78) = 3.9 mm/day.

Problem 7.4. Find out frequency of irrigation for a crop grown on clay-loam soil were groundwater table is far below the root zone and contribution from other sources like existing soil moisture and rainfall is negligible. Use following data:

(*i*) Depth of root zone = 1.5 m

(*ii*) Field capacity = 25%

(*iii*) Permanent wilting point = 13%

(*iv*) Daily evapo-transpiration or consumptive use by crop = 15 mm

(*v*) Dry density of soil = 1.5 g/cm^3

Only 60% of available soil moisture is to be depleted.

Solution: Step 1: Available soil moisture in cm

$$= \frac{v \times d}{v_w} (FC - PWP)\ 100$$

$$= \frac{1.5 \times 1.5}{1} (25\% - 13\%)\ 100$$

$$= 2.25 \left(\frac{12}{100}\right) \times 100 = 27 \text{ cm.}$$

Step 2: Readily available moisture for consumptive use

$$= 60\% \text{ of } 27 \text{ cm}$$

$$= 27 \times .6 = 16.2 \text{ cm}$$

Step 3: Interval between to irrigation application

$$= \frac{\text{Readily available moisture}}{\text{Daily consumptive use}}$$

$$= \frac{16.2}{1.5} = 10.8 \text{ say } 11 \text{ days.}$$

7.14. CLIMATIC SEASONS

It is already mentioned that as we go towards south of the country the climate becomes hotter and hotter. In southern part of India there is no marked change in temperature and for all the twelve months of a year the climate is hot. On the contrary if we come towards north there is marked change from summer to winter. For the central part of India also there is marked change.

Broadly for southern and Central India four seasons may be recognized. Spring season ranges in February and March, Summer from April to July, autumn season in August and September and winter from October to January. Whereas for Northern India spring season ranges in March and April, summer from May to August, autumn season in September and October and winter from November to February.

From the above explanation it can be inferred that there is a lag of about one month in the beginning and end of the seasons as we come towards north. The range of each season cannot be taken in a very rigid sense for all the areas of the region.

7.15. CROP SEASONS AND CROPS OF VARIOUS SEASONS

Taking natural seasons into account together with the number of times different crops can be grown in a year on the same land, the agriculturists have given the following crop seasons for various parts of India in a broad sense:

1. For **Southern India** the weather conditions are moderate neither touching extreme, hot or extreme cold. As a result three crop seasons can be recognized. They are Hot weather **crop season**, **rabi season** and **kharif season**. The hot weather crop season ranges from February to May and the crops which can sustain heat and humid conditions are grown. Kharif season ranges from June to October. The crops are sown in the beginning of the monsoon and are harvested in autumn. Rabi season ranges from October to March. The crops are sown in the beginning of winter and are harvested in spring.
2. For **Northern India** there are two distinct crop seasons in a year. They are **rabi season** and **kharif season**. Rabi season starts in the month of November and ends in the month of March. While kharif season starts in June and ends in October. As a matter of fact the crops of one season may extend in other season depending upon the prevailing climatic conditions of the area.

Apart from these there are some crops which take comparatively longer time for maturing. Cotton takes 8 months for maturing. It is, therefore, known as a **eight months**' crop. Crop season extends from June to February.

Sugar cane takes full 12 months for maturing. It is, therefore, known as a **perennial crop**. Garden crops also come under this category.

Crops of various seasons: A crop is said to pertain to a particular season when the crop extends from the time of sowing to the time of harvesting in that season mostly.

The following is the division of crops according to the two main seasons **rabi** and **kharif** over which the crops extend.

The extent of cultivated areas under the Principal crops in India is given in brackets. The areas are given in million hectares in Table 7.13.

Table 7.13. Area under Principal Crops

(Figure in Mha)

Rabi crops	*Kharif crops*	*Other crops*
Wheat (26.48)	Rice (43.66)	Tea (0.38)
Gram (6.90)	Great millet (8.67)	Coffee (0.08)
Rapeseed and Mustard (7.28)	Spiked millet (9.58)	Sugarcane (4.20)
Linseed (1.02)	Maize (7.59)	Cotton (8.68)
Castor (0.56)	Arhar (3.58)	
	Pulses (4.09)	Tobacco (0.37)
	(Urd, Mung, Moth, Kulthi etc.)	
Barley (1.23)	Jute (0.81)	Potato (0.84)
Other pulses (7.88)		
(Peas, Masur, etc.)	Seasame (1.83)	
Onion (0.26)	Groundnut (6.74)	

(Figures relate to 2005-06)

Cash crops: There is one more classification of crop which has nothing to do with the season in which it is grown. There are certain crops which a farmer mostly grows for selling in the market to meet his financial requirements. They are called cash crops. Wheat, rice, various millets, maize are used by the farmers to run their household and to pay wages to workers as food for work. All other crops listed in Table 7.13 above and many others like various vegetables could be termed as cash crops.

Sometimes a cash crop is sown on the some field together with a food crop (for example, wheat and mustard) then the field is said to have a **mixed crop**.

When on the same field two crops are grown one after the other in one calender year the farmer is said to have grown **double crop**.

7.16. CROP ROTATION

The method of crop rotation is most natural and at the same time economical. The crop rotation means simply changing the crop to be grown every year on the same field. Under lying principle is that each crop requires different nutrients and in different proportions. By changing a crop every year in a way soil is given rest as the same nutrients are not used every year. At the same time cultivated land is not kept idle.

In this method only precaution to be taken is to select the crops properly for rotation. So far as crop rotation is concerned there is no distinction between rabi and kharif crops. Rabi crops can be brought in rotation with kharif crops. For this best and quick results manure may also be added to the field.

For rotation some crop groups are given below to order of sequences.

- Wheat — Great millet — Gram.
- Rice — Gram.
- Cotton — Wheat — Gram.
- Cotton — Wheat — Sugar cane
- Cotton — Great millet — Gram.

From the above information on the crop rotation the following three inferences can be drawn:

1. In most of the crop rotation groups gram is the crop which completes a cycle. The reason for gram being last is rotation is, it is a leguminous crop and when sown on the field gives nitrogen to the soil to make a rich.
2. Crop rotation keeps the soil free from soil diseases. If the same crop is grown on the field it will make soil poorer at the same time crop may impart some peculiar diseases of its own to the soil.
3. Each crop has different depth of root-zone. Naturally maximum possible usage of the available soil moisture of a field can be made by growing different crops, some of which have shallow roots and others deep roots.

7.17. CROP CALENDAR

A crop calendar showing typical sowing and harvesting time for principal crops together with total water requirement for each, number of waterings required and an average assessment of crop yield in meter tonnes per hectare is given in Table 7.14.

Table 7.14. Crop Calendar

Crop season and crops	*Sowing period*	*Harvesting period*	*Average (D) in cm*	*Number of waterings*	*Average yield in metric tons/ha*
Rabi Crops					
1. Wheat	Oct. – Nov.	Mar. – Apr.	40	5 – 6	2.0
2. Gram	Oct.	Jan.	30	3 – 4	1.0
3. Potato	Oct.	Feb.	75	10	1.2
4. Mustard	Sep.	Feb.	45	3 – 4	0.8
5. Tobacco	Oct.	March	60	6 — 8	1.0
Kharif Crops					
1. Rice	Jul. – Aug.	Dec. – Jan.	125	6 — 8	2.0
2. Maize	June – Jul.	Sep. – Oct.	45	5 — 6	1.5
3. Spiked millet (Bajra)	June – Jul.	Sep. – Oct.	30	3 — 4	1.0
4. Great millet (Jwar)	June – Jul.	Sep. – Oct.	30	3 — 4	1.0
5. Groundnut	June – Jul.	Nov. – Dec.	30	3 — 4	1.0
6. Pulses	Jul. – Aug.	Oct. – Nov.	30	3 — 4	0.8
7. Sesame	Jul. – Aug.	Oct. – Nov.	45	4 — 5	0.8
Eight Month Crop					
1. Cotton	June – Aug.	Jan. – Mar.	55	6 — 8	0.5
Perennial Crop					
Sugar cane	Feb. – Mar.	Dec. – Mar.	160	12 — 14	2.5

QUESTIONS FOR REVISION

1. Write short notes on the following:
   1. Crop rotation
   2. Crop calendar
   3. Frequency of irrigation
2. (*a*) What do you understand by consumptive use of water ? How it is different from evapo-transpiration ?
   (*b*) What are the factors affecting irrigation water requirement of a crop ?
   (*c*) How evapo-transpiration is determined ?
3. (*a*) What are principal rabi and kharif crops of India ?
   (*b*) Mention their time of sowing, harvesting, water requirement and number of waterings required in general.

8

Canal Irrigation

8.1. COMPONENTS OF FLOW IRRIGATION PROJECT

When the rainfall is inadequate or ill-distributed and does not meet the water needs of crops throughout the crop season an irrigation project is constructed. Normally during the short period of the rainy season sufficient water flows down the river. To make best use of the river water, flow irrigation scheme is undertaken. It is necessary in such cases to create sufficient storage of water at a suitable place to take the water to the fields under the force of the gravity. Obviously main components of a flow irrigation system are the following:

(*i*) Storage works,

(*ii*) Diversion works,

(*iii*) River training works, and

(*iv*) Distribution system.

A typical sketch of the flow irrigation system is shown in Fig. 8.1.

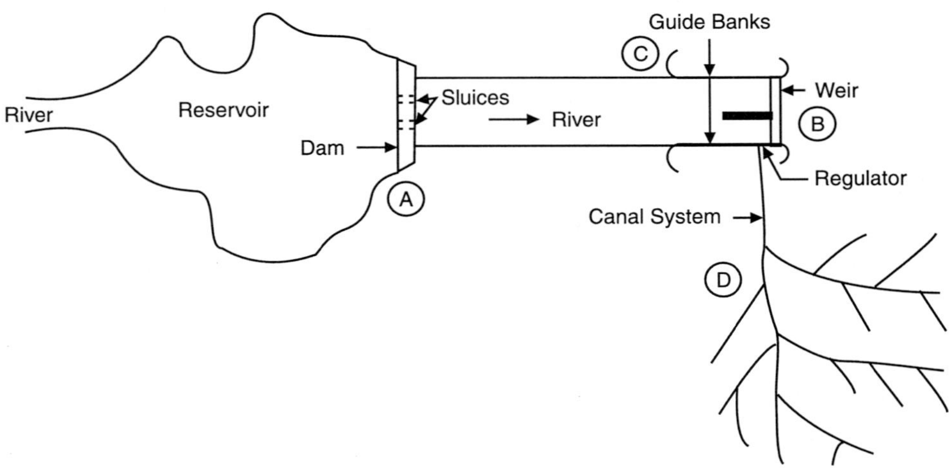

Fig. 8.1. Components of flow irrigation system

Storage work comprises a reservoir created by constructing a dam across the river. In the body of the dam a spillway is provided to pass down the flood flow safely. Also, sluices and outlets are provided to release the water for irrigation purposes.

Generally, a diversion work is provided across the river below the dam in the vicinity of irrigable area. It is also called headworks. A weir or a barrage with its constituent parts raise the water level in the river. A regulator diverts a measured quantity of water into the canal system.

In order to guide the river flow in proper direction and to protect the river course some sort of river training work is constructed at the site of the head works. Commonly constructed training works are guide banks, spurs or groynes and bank revetment. The head works and the river training works are dealt with in separate chapters.

Once the measured quantities are diverted in the canal irrigation water is carried to the fields by a network of big and small irrigation canals. The network of the canals is called as canal system and is made up of main, branch, distributary canals and field channels. The distribution system is dealt with in this chapter.

Also, various other hydraulic structures (*e.g.,* cross drainage works, falls, outlets) are constructed across the canals to carry irrigation water efficiently to the fields. As the name suggests the cross drainage works are constructed to crossover natural drainage. The falls are provided to negotiate steep ground slopes. The outlets are constructed to divert irrigation supply in the field channels. All these regulatory works have been described in different chapters in this book.

8.2. CLASSIFICATION OF FLOW CANALS

It has been already mentioned that canals can be used as a means of communication apart from irrigation purposes. Flow canals can be successfully used for the following purposes:

- Irrigation.
- Navigation or water transport.
- Power generation.

India is an agricultural country and irrigation has been given important place. In India there are many irrigation canals of appreciable magnitude. To achieve economy, power generation can be successfully done form the irrigation canal itself. Some years before canals were constructed to provide irrigation and navigation facilities from the same canal system. For example, Ganga, Western Yamuna and Sirhind canals were constructed to perform both the functions. However, this combination has not proved successful because of the following characteristic differences:

(*i*) The velocity of flow of water in navigation canal is negligible. Then only barges and small ships can move in both direction of flow without experiencing any difficulty. On the contrary velocity of water in the irrigation canal is always higher to keep sand and silt load moving. Naturally when navigation is to be done in the irrigation canal, that too in the direction opposite to flow, process becomes expensive. It needs introduction of powered craft.

(*ii*) When natural falls are present in the irrigation canals, power generation can be done successfully. But the falls of great height practically hinder navigation in irrigation canals. To facilitate navigation, canal locks are to be provided. It may create hinderance in irrigation water management.

(*iii*) Irrigation canals are aligned to avoid big cities and industrial places so far as possible. Whereas navigation canals generally pass through and connect the industrial places.

(*iv*) Navigation canal is aligned to follow low country. This procedure when adopted reduces the cost of construction of the navigation canals. On the other hand irrigation canals are to be located on the ridges of commanded area. It gives good command and gravity flow to the irrigation water.

Power generation and irrigation purposes can be easily and successfully combined on one canal. When hydro-electric power is to be generated it is always better to provide falls of greater depths in irrigation canals whenever possible rather than providing large number of smaller falls.

Classification of Irrigation canals may be done in number of ways. The following are the four main classifications:

I. Classification according to the nature of source of supply and system of distribution. Here irrigation canals are subdivided into:

- permanent canals, and
- inundation canals.

A canal is said to be permanent when its source of supply is sufficiently assured to warrant construction of protective and regulatory works with a regular, permanent and graded channel.

A canal is said to be inundation type when the supply of water to the canal depends on periodical rise in the water level of a river from which the canal takes off. Construction of a permanent regulatory work is very rarely done in this case.

II. Classification based on financial aspect. The subdivision of canals under this head is as follows:

- productive canals, and
- protective canals.

Productive canals are those which are introduced with an idea of recovering all the expenditure, incurred during the construction, in a specified time limit. The following explanation clarifies the meaning:

After full development of the canal system revenue obtained from the cultivators is such that it is more than the maintenance expenditure. This exceeding amount of revenue is so adjusted that it is about 2 per cent of the cost of the canal system. With this rate the total cost of the project will be recovered in about 50 years. Thus the productive canals is that which adds to the wealth of nation directly.

Protective canal is one which is not actually remunerative as the previous type. The construction work of this type of canal and its development may be started during a famine. It helps in giving employment to the cultivators at the time of the famine, thus the protective canals is a sort of relief work.

III. Classification based on the function served by the canal. With the advancement in the field of irrigation engineering and to fulfil multi-objective needs canals are constructed to perform different functions. Such canals are:

- link canals,
- feeder canals, and
- power canals.

All the above three types of canals are in essence carrier channels.

Link canals are those which are constructed to join two rivers. The canals may or may not provide direct irrigation enroute.

Feeder canals convey water from a river or a storage project to feed another water deficient irrigation system. Like link canals these canals also may or may not provide direct irrigation along its run.

Power canals are generally meant to feed the hydro-electric power station. Generally such canals do not provide irrigation before water is utilized in power generation. A tail race channel which leads the water away from the power house may discharge into the river or another canal system for subsequent utilization.

IV. Classification based on the capacity of the canal. Depending upon the size and capacity of the canals different names (*e.g.,* main canal, branch, distributary etc.) are given to them. All such canals, however, form component of a complete canal system and are described subsequently.

8.3. IRRIGATION CANAL ALIGNMENT

It is now clear that irrigation water, in flow type, should reach the fields by gravity. To accomplish this requirement irrigation canal is always aligned in such a way that the water gets proper command over the whole irrigable area. Obviously if the canal follows a watershed or a ridge of the drainage area it will get necessary gravity flow. The watershed or the ridge is a dividing line between two drainage areas. Thus a canal which runs over the ridge, gets command of area on both sides of the ridge. Another advantage of keeping a canal on the ridge is that crossing of drains and canal are completely avoided. The obvious reason is drains do not cross the ridge or the watershed.

So far irrigation flow canals are concerned they are taken off from the river or from the storage reservoirs or tanks created or constructed in the upper or hilly reaches of the river.

River is nothing but a big drain and canal takes off from it. Naturally the head of the canal is in the valley portion and not on the ridge for that section of the catchment. Obviously first thing to be done after take off is to bring the canal on the watershed of the catchment. Canal can be brought to the watershed without much difficulty because the slope given to canal bed is very flat in comparison with the slope of ground in the reach. From Fig. 8.2, it can be visualized very clearly.

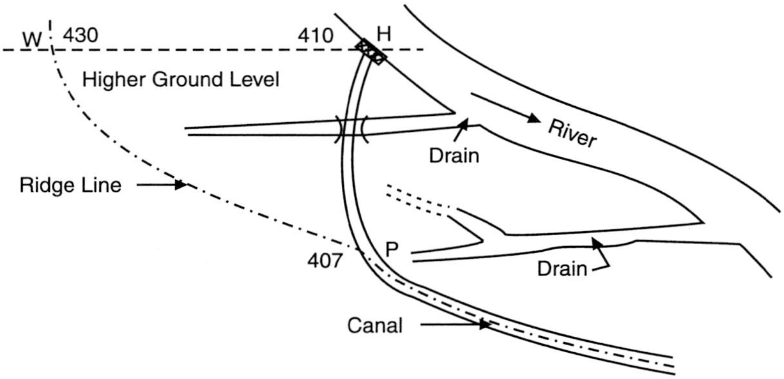

Fig. 8.2. Canal taking off from river and mounting on the ridge of a catchment

The canal takes off from point H on the river. If a cross-section of the catchment area at this point is taken, the level of the river at the head of the canal is say 410, whereas level of point W on the ridge is 430. The canal starting at level 410 is to be taken over the ridge at a suitable point as soon as possible. This is easily accomplished. Point P is at level 407 and lies on the ridge. This is possible because ground level falls very quickly from 430 at W to 407 at P. The distance from H to P is say 9 km. So if the canal is laid with a slope of 1/3000 it is sufficient to bring the canal on the ridge. The value of slope is also standard for laying the canal. Of course selection of point P is a matter of trials. In some cases head works will have to be located according to the conditions of the watershed. Another limiting condition is about crossing of drains. There are various drains starting from watershed and meeting the river. While aligning the canal it is very essential to see that the canal is crossing least number of drains.

The command should not be overlooked in the above process particularly so when there is some important irrigable area above the canal. In the initial stages there is higher ground level on one side of the canal and canal water cannot irrigate that portion of the land by gravity.

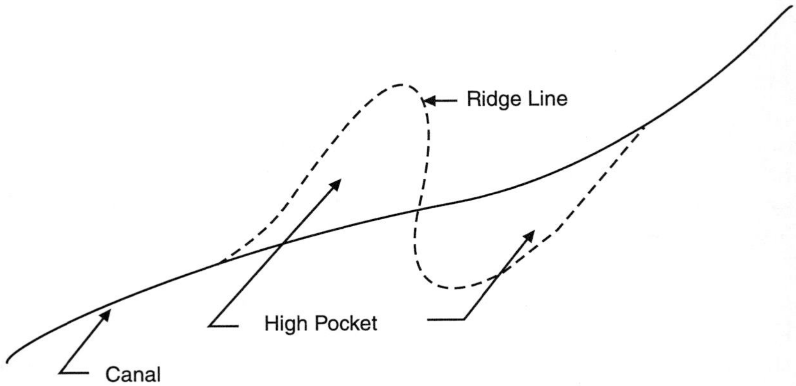

Fig. 8.3. Canal alignment for loop forming ridges

Keeping all these points in view the canal can be very efficiently brought on the ridge. Once the canal is brought on the ridge it is mostly kept on the ridge to fulfil the basic condition of gravity flow. But some times the ridge line may be zigzag and it may take sharp turns forming a loop. The canal alignment in such cases is taken straight, off the ridge line for a very short length and then again it mounts the ridge. In this process saving in the length of the canal is accomplished as shown in Fig. 8.3.

In this process high pockets are left uncommanded. Also drainages of these pockets cross the canal. When all the existing conditions fully justify the canal is aligned straight to reduce the length.

8.4. METHODS OF ALIGNMENT

Uptil now it is assumed that the country is fairly plain. This is not a usual condition, naturally the procedure of keeping a canal on the watershed cannot be applied everywhere. Depending upon the nature of the country various types of alignments are possible which are mentioned below:

- Alignment of contour channels
- Alignment of ridge or watershed channels
- Alignment of side slope channels

In hilly reaches first or third method is used depending upon the situation. When the reach is fairly plain second method is adopted. In the first method the canal is aligned along a falling contour.

Hence, it is named contour channel. The flow of water is generally perpendicular to the slope of the ground. In this type of alignment drainages are not avoided. To reduce the length and subsequently the cost, the channel is not taken strictly along any contour. It follows the contour but at the same time it is taken in such a way that maximum distance is placed between the river and the canal as shown in Fig. 8.4.

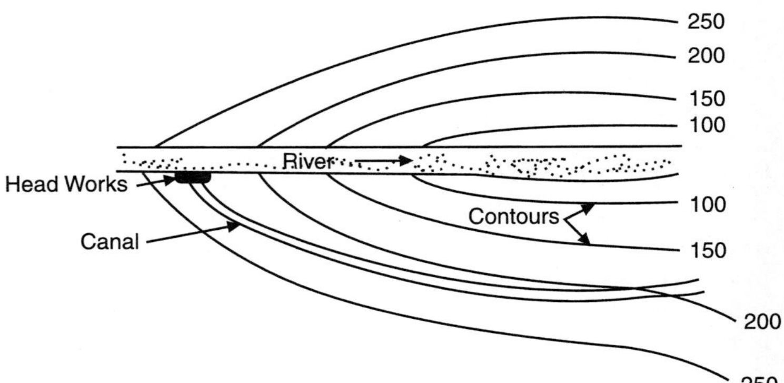

Fig. 8.4. Contour channel

Thus, the command of the canal is increased because on the other side of the canal land is high and flow irrigation is not possible.

In this type when the land is very high on one side it is not necessary to construct the banks on both sides. Generally higher side is left without a bank. This type of canal is called a single bank canal.

Second method of aligning the channel is already explained. The advantage of this method is that cross drainage works are avoided and the area on both sides of the ridge is under command. In this type there are banks on both sides of the canal and hence it may be termed as a double bank canal.

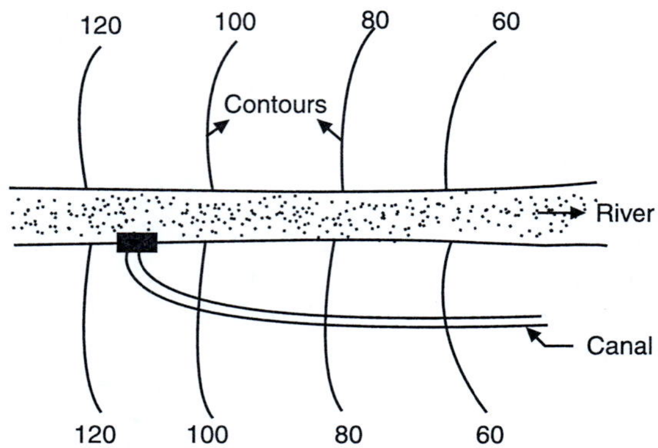

Fig. 8.5. Side slope channel

In the third method canal is aligned at right angles to the contours. Naturally canal runs parallel to the drains or river and hence cross drainage works are avoided to the maximum extent. Fig. 8.5 shows this type of alignment.

8.5. PRECAUTIONS IN CANAL ALIGNMENT

While aligning a canal following points should be considered in general:

- The canal should be aligned on the ridge or in such a way as to obtain maximum command.
- So far as possible the canal alignment should be kept in the centre of the commanded area.
- The canal should be aligned in such a way that the length is minimum possible.
- The alignment should avoid inhabited places, roads, railways, properties, places of worship etc.
- Canal should be taken through the area where subsoil formation is favourable. Water-logged, alkali, saline, rocky soils create troubles.
- The alignment should be straight so far as possible. Where alignment is not straight simple circular curves of large radius should be provided.
- To ensure economy the alignment of the canal should be such that excessive cuttings and fillings are not required. The alignment should not cross hills or depressions.
- While aligning the canal, cost of the land to be acquired should be taken into consideration.
- Cost of the alignment should be in proportion to the total cost of the project.
- The canal should cross minimum number of drainages.

8.6. CANAL CURVES

It is needless to mention that so far as possible straight alignment is not only the shortest but also most efficient in transportation of water. However, sometimes it becomes unavoidable to divert the alignment due to presence of such an important property or structures which cannot be expected to give way to the canal alignment. In such circumstance canal alignment is given curvature. It is then seen that as far as possible the curves are gentle enough to avoid undue scouring on concave side and silting on the convex side of the canal. To make the curvature gentle larger radius is provided. Also larger radius helps in maintaining smooth flow as the discharge increases.

Tables 8.1 and 8.2 show the minimum radius requirement for various capacities of lined and unlined canals.

Table 8.1. Minimum radii for curves on earthen canals

Capacity of canal in m^3/sec	*Minimum Radius in metres*
More than 80	1525
80 – 30	925
30 – 15	610
15 – 3	325
3 – 0.3	150
Less than 0.3	90

Table 8.2. Minimum radii for curves on lined channels

Capacity of canal in m^3/sec	*Minimum Radius in metres*
40 – 70	325
70 – 140	450
140 – 210	600
210 – 275	750
More than 275	925

So far as possible larger radius should be used.

8.7. PARTS OF CANAL SYSTEM

In this type the canal system is designed to carry specific discharge. The section of the channel is properly designed and maintained throughout the year to fulfil the designed requirements.

The source of water supply for this system is also well assured at all times. A permanent canal may be perennial type or non-perennial type depending upon water requirement of crops of a particular tract.

The discharge used in designing the head of a canal system satisfies the irrigation requirements of the commanded area fully. It is called the full supply discharge of the canal there. It is generally designated as FSD. Then the top level of water in the canal will be called full supply level and designated as FSL. While calculating the FSD. it is essential to account for all sorts of losses in conveyance and in distribution. Obviously to reach at a FSD we will have to work back from the tail of the canal system. First net water requirement on the field is found out. Then adding the possible losses, which occur till head of the canal system is reached, FSD can be calculated.

Components of a Permanent Canal System: When the canal takes off from the river the FSD is substantial. As the canal transverses the land it goes on satisfying the water requirement enroute. Naturally when the canal carries smaller amount of water and when small patches of land in various directions are to be irrigated, same section cannot be adopted throughout. Depending upon the capacity of the canal and the function served by that length of canal system various components can be correctly recognized. All the components put together form a full canal system.

Before actually coming to the components it is essential to understand clearly what is meant by headworks of the canal system. The canal takes off from a river. Naturally the river has to be trained at the site of take off. It is also very essential to maintain the water level in the river at a particular fixed elevation to assure constant and continuous supply. The river water carries lot of sand and silt load. It is necessary to allow only clear water into the canal. To accomplish this a still pond is to be created. Also, fixed amount of water is always to be taken in the canal.

To fulfil the above mentioned requirements headworks is constructed. Thus a headwork is an assemblage of various hydraulic structures namely, training works, weir, divide will, head regulator etc.

The various components of a canal system are mentioned below. Figure 8.6 also clearly explains the canal system.

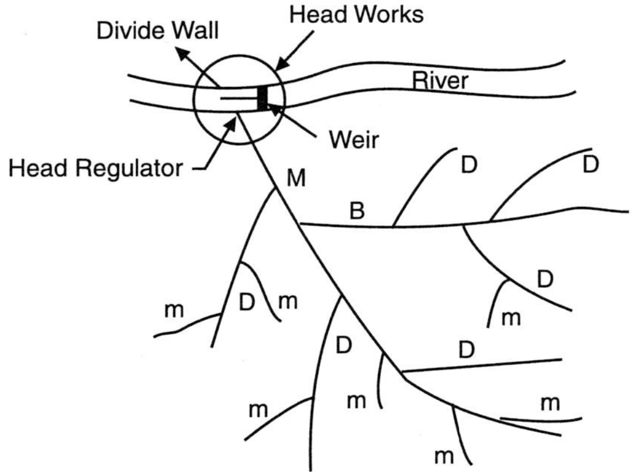

Fig. 8.6. A canal system

Main Canal (M): This canal takes off from the river and derives water through a regulator. For a certain length in the beginning the canal may be in full cutting. Obviously direct flow irrigation from such a channel is not possible as its water surface is below the general ground surface. The main function of the canal is to take the required amount of irrigation water at the head and to deliver it in the canal system below.

Branch Canals (B): After the main canal comes in the tract which is to be irrigated by the canal system, water is distributed over the entire commanded area. For this purpose the main canal is branched to cover the whole area. The bifurcated canals are called branch canals or simply branches. From the branches also direct irrigation is very rarely done. The main function of the branches is to make the water available in different parts of the tract for further distribution. The section of a branch canal is smaller than the main canal.

Distributaries: They can be subdivided into:

- Major distributaries, and
- Minor distributaries or minors.

Major Distributaries (D): They are also named Rajbaha. Major distributaries are smaller in section than the branches. They are taken off mainly from the branches to distribute the water to various parts for direct irrigation purposes. Sometimes major distributaries may take off directly from the main canal. The distributaries provide water to field channels through outlets for irrigation.

Minor Distributaries (m): They are smaller in section than the major distributaries. The main purpose of these distributaries is to reduce the length of the field channels. They are taken off from the major distributaries and sometimes from the branch canals. When the length of the field channels is more than 3 km the minor distributaries are introduced to reduce the length. They are also called minors.

Water courses: They are also named as guls or field channels. They are the channels which take irrigation water to the fields. The water courses derive their supply from the distributaries through outlets. An outlet is a simple and a small irrigation structure constructed on the distributaries. It may be a simple pipe outlet or any other type provided to allow flow of water into the water courses. The amount of water withdrawn through one outlet is in proportion to the area to be irrigated below that point. Table 8.3 gives the ranges of discharges carried by each type of canal.

Table 8.3.

Type	*Discharge capacity in m^3/sec*
Minor	Less than 0.70
Major distributary	Less than 5.50
Branch	More than 5.50
Main canal	Whole discharge of the canal system

Irrigation department is responsible for construction and maintenance of the canal system upto minor distributaries. The field channels are constructed by cultivators themselves as per the alignment sanctioned by the department. Nowadays the field channels are also constructed by irrigation department for cultivators. The job of maintenance of field channels is left entirely to the cultivators. The length of a water course should not exceed 3 km in any case. It generally serves an area of 40 to 50 hectares. The upper limit of discharge in a watercourse ranges between 0.08 to 0.06 m^3/sec.

8.8. LOSSES IN CANALS

An irrigation canal, in principle is to be partly excavated in the nautral ground and partly constructed above the natural ground to hold the required quantity of irrigation water. The canal section is never made hundred per cent watertight. The canal water tries to seep into the soil. Moreover the canals are exposed to the atmosphere at the surface. The water also goes to the atmosphere in the form of vapour. Thus irrigation water, when taken in the canal system for conveying it to the fields, is liable to be lost in various ways. Loss of canal water occurs due to two main reasons, namely:

- Evaporation, and
- Seepage.

Out of these two types of losses, seepage loss is the one which is of most concern. It is observed that in initial stages total losses may amount to 40 per cent of the total discharge of the canal system.

In porous soil formations the seepage loss is exceptionally high. Sometimes, it is possible that the whole supply diverted in the canal may be lost if lining is not done.

After running the canal for a sufficient period the seepage loss is minimised to a considerable extent. The reduction in the loss with time is due to two main reasons.

Firstly, when the subsoil formation below the canal gets saturated the entry of seepage water does not remain free from resistance. The reason for this resistance lies in the fact that the soil pores are already filled with water. Obviously the rate of seepage is lowered.

Secondly in course of time the soil pores also get filled with the silt carried by the canal water. In early stages of introduction of the canal system the canal is run even in rainy season to accelerate the process of silting.

Since the irrigation water carries maximum suspended and bed load during that period the soil pores get filled up in a short period.

It may be mentioned here that evaporation loss cannot be checked whereas seepage loss goes on reducing with the age of the canal system and it can also be checked by adopting suitable measures.

Thus it is clear that while designing the canal section proper allowance has to be made for these losses. Generally, the losses due to evaporation and seepage are combined while making allowance for them. The water loss is generally expressed in m^3/sec per million square metres of the exposed water surface or in m^3/sec per million square metres of the wetted surface. Experience has, however, proved that in actual practice it is somewhat difficult to arrive at the exact amount of loss due to these natural causes.

8.9. LOSS DUE TO EVAPORATION

As canal water is exposed to the atmosphere at the surface, loss due to vapouration is obvious. It is of course true that in most of the cases evaporation loss is not significant. It may range from 0.25 to 1% of the total canal discharge. The rate of loss of water in the process of evaporation depends mainly on the following factors:

- Temperature of the region,
- Prevailing wind velocity of the region,
- Humidity,
- Area of water surface exposed to the atmosphere.

Generally it is considered that the rate of evaporation loss depends mainly on temperature. It is not hundred per cent correct. The rate of loss also equally depends on the velocity of wind which carries vapour from the water surface to the atmosphere. Loss due to evaporation is more for shallow water depths. Many times it is observed that due to above mentioned factors the rate of loss due to evaporation does not differ much for day and night periods.

Thus it can be inferred that the loss due to evaporation is directly dependent on the climatic conditions of the region and hence can not be checked. It also depends directly on the exposed area of the water surface and inversely on the depth of water in the channel.

8.10. LOSS DUE TO SEEPAGE AND ITS MEASUREMENT

The water lost in seepage may find its way finally into the river valley on enters an aquifer where it can be utilized again. But many times the seepage water is not recoverable.

The loss due to seepage is the one which is most significant so far as irrigation water loss from a canal is concerned. The seepage loss depends mainly on the following factors:

- Underground water table conditions,
- Porosity of the soil,
- Physical properties of the canal water for example its temperature and quantity of suspended load carried by the water (turbidity of water),
- Condition of the canal system.

The seepage loss which occurs in an existing canal system or which may occur in the proposed canal system can be estimated by several methods. Some useful methods of calculating seepage loss are mentioned below.

(*a*) **Permeameters and seepage meters:** Permeameters measure permeability of a canal bed or lining. It is no doubt very much different from the seepage rate. In order to determine the seepage rate it is necessary also to know the hydraulic slope causing flow through the canal bed. It is also essential to know the effective permeability of the sides of the canal. The depth of water in permeameter should be approximately equal to the normal depth of water in the canal.

Seepage-meter consists of a metal cylinder. At the top it is dome shaped. A valve is fixed on the dome to remove the entrapped air. A plastic bag is attached to the cylinder by a tube. For calculating seepage rate the plastic bag is filled with water and the cylinder is then pressed slowly into the soil. The remaining space above the cylinder is also filled with water. The whole meter remains below the water surface of the canal. Seepage occured in the cylinder causes a corresponding reduction in the water content of the plastic bag. The rate of reduction of water content gives the seepage rate. The plastic bag helps in maintaining the same hydrostatic pressure on the soil of the cylinder as that surrounding the meter. The area under test is very small. The character of the bed material may not be same everywhere. This method only gives an indication of the order of the seepage rate. For collecting reliable information it is very essential to obtain several readings on the bed and sides of the canal.

(*b*) **Ponding method:** This method involves isolation of a section of a canal by means of temporary cross bunds. The enclosed area is filled with water and the decrease in the volume over a certain period of time is noted. It may then be used to calculate the rate of loss. Proper allowance is made for rainfall and evaporation. The canal can not be used during the period of tests.

(*c*) **Inflow and outflow method:** It is a very simple method. It consists in measuring the quantity of water entering a certain reach and the quantity of water going out of that reach. The difference gives the amount of water lost. The discharge can be measured by flumes, weirs, current-metres or venturimetres. It is necessary to maintain constant water level. Allowance should be made for evaporation loss also.

The seepage loss may occur in two characteristic ways, namely:

- Absorption, and
- Percolation.

(*i*) **Absorption:** When the undergrond water table is at a considerable depth, the water entering the soil is unable to join the saturated zone and wets the subsoil locally immediately below the canal bed (Fig. 8.7).

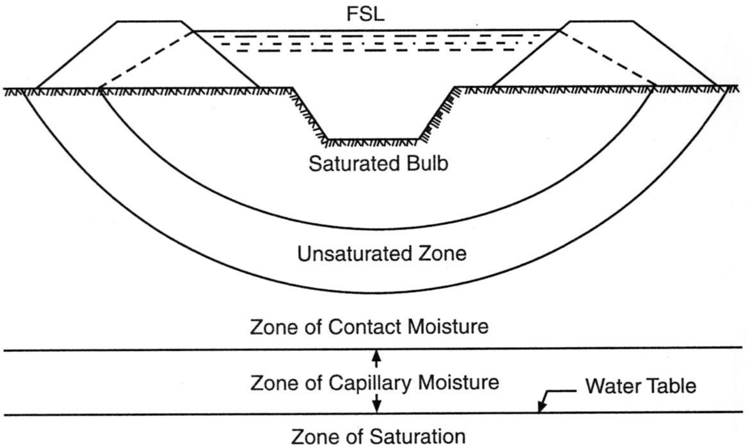

Fig. 8.7. Loss of water due to absorption

The soil layer which is in immediate contact with the channel section is completely saturated due to the absorbed water. It forms a bulb of saturated soil below the channel. The soil layer below the saturated bulb is not fully saturated. Thus the extent of saturation goes on decreasing from the ground level below in the soil with depth. There now exists a zone of unsaturation between the underground saturated zone and the saturated bulb. Thus there is no chance of continued and constant flow from the canal to the groundwater reservoir.

It is observed that loss of due to absorption is more when the canal is in cutting reach. The loss due to absorption may be calculated from an empirical formula given below. It was given primarily for the state of Punjab.

$$P = C\sqrt{D}\,\frac{\text{WL}}{10{,}00{,}000}$$

where P is the loss due to absorption in m^3/sec

C is a constant and generally taken as 1.932

D is the depth of water in a canal in m

W is the width of water surface in m

L is the length of reach in m.

From the above formla it is clear that $C\sqrt{D}$ or $1.932\sqrt{D}$ gives the loss due to absorption in m^3/sec per million m^2 of exposed water surface area.

Another formula which is also used in Punjab is

$$K = 1.905\ Q^{0.0625}$$

where K is absorption loss per million m^2 of wetted perimeter.

and Q is discharge in any reach of a canal in m^3/sec.

This formula holds good for average soil.

In case channel is lined the loss is given by

$$K = 0.467\ Q^{0.056}$$

***(ii)* Percolation:** When the underground water-table is nearer to the natural surface, the water which has entered the subsoil may join the saturated zone or underground reseavoir to maintain a continuous direct flow (Fig. 8.8).

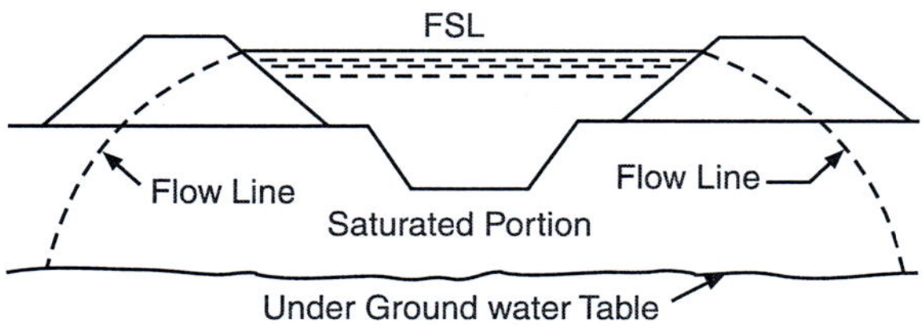

Fig. 8.8. Loss of water due to percolation

Thus the portion between the two extreme flow lines is completely saturated from the canal section to the underground water table. Here water directly flows from the canal to the underground reservior through the soil pores under pressure. The pressure which is responsible for this flow is due to the difference of level between the underground reservoir and the water level in the canal. It can be measured in metres of water depth.

Etchevery and Harding gave the amount of water lost in conveyance, in different types of soils. Table 7.4 gives the values suggested by them.

Table 8.4. Etchevery and Harding given Losses

Type of soil	*Losses*	
	m^3/sq. m of wetted perimeter in 24 hrs	*m^3 per sec/million m^2 of wetted perimeter*
Very gravelly soil	0.92 to 1.8	10.70 to 21.40
Porous gravel soil	0.76 to 0.92	8.95 to 10.70
Sandy gravel soil	0.61 to 0.76	7.00 to 8.95
Loose sandy soil	0.46 to 0.53	5.18 to 6.10
Sandy loam soil	0.30 to 0.46	3.66 to 5.18
Clay loam soil	0.23 to 0.46	2.74 to 3.66
Ordinary clay loam	0.15 to 0.23	1.83 to 2.74
Medium clay loam with hard pan below	0.11 to 0.15	1.22 to 1.83
Imprevious clay loam	0.08 to 0.11	1.92 to 1.22

QUESTIONS FOR REVISION

1. How will you classify flow canals ? Explain elaborately.
2. What are the considerations in aligning a channel ?
   Briefly explain the effect of cross-drainage works on the alignment.
3. Write short notes on the following:
   (*i*) Main canal
   (*ii*) Branch
   (*iii*) Rajbaha
   (*iv*) Minor
   (*v*) Water course
4. Distinguish between the following:
   (*i*) Contour canal and side slope canal.
   (*ii*) Minor and water course.
   (*iii*) Productive and protective irrigation works.
   (*iv*) Inundation canal and perennial canal.
5. How seepage loss in a canal is measured?

❑❑❑

9

Design of Irrigation Canals
(On Alluvial and Non-Alluvial Soils)

9.1. DEFINITIONS

Before proceeding with the subject it is essential to understand following terms:

Hydraulic mean radius (R): It is also called hydraulic mean depth. It is the ratio of wetted cross-sectional area and wetted perimeter.

That is,

$$R = \frac{A}{P_w}$$

where R is hydraulic mean radius.

A is sectional area of flow.

and P_w is wetted perimeter of a channel.

Full Supply Diacharge (FSD): It is the maximum discharge for which the canal is designed. It is the sum of the total irrigation requirement and the various losses in conveyance of water. The level of water in the canal at the FSD is called the full supply level (FSL).

Economical section and balancing depth: In Irrigigation canal water flows under the force of gravity. To achieve this condition the FSL is generally kept above the natural surface level (NSL). Naturally to hold the water in the channel it is partly excavated below the NSL and partly built above the NSL.

For achieving economy the depth of excavation is so arranged that the quantity of the earth excavated from the canal section is just sufficient to construct the banks.

Coefficient of rugosity: It is generally denoted by letter N. This coefficient is mainly a function of:

(*i*) Grain size of bed and bank material,

(*ii*) Bed shear stress, and

(*iii*) Depth of water in a channel.

In short, in addition to grain size it depends on the size of rupples formed on the bed of the channel. Table 9.1 gives the values of N for different types of bed material.

Table 9.1. Values of N

Material		*Coefficient*
Wood	.....	0.013 to 0.0165
Steel	.....	0.0125 to 0.018
Concrete	.....	0.013 to 0.018
Masonry	....	0.02 to 0.036
Earth	.....	0.0225 to 0.035

Channel in Regime: When the channel water flows in "incoherent unlimited alluvium" and when the character of its bed and bank material is same as that of transported material and also when silt charge and silt grade is constant, the channel may be said to be in the regime. It is then called a *regime channel.*

The term "incoherent unlimited alluvium" gives the information about character of the bed material and its formation. It is such a state where the bed is made up of loose grains of small size (silt) and the processes of silting and scouring need no special efforts.

Depending upon the existing conditions at site under observation, regime may be initial or final. The channel is said to be in initial regime condition when it has formed its section only and yet not secured the longitudinal slope. The channel after attaining its section and longitudinal slope, is said to be in **final regime**. Whereas if the channel is protected at bed and sides with some protecting material, there is no possibility of change of section and longitudinal slope. Then the channel is said to be in **permanent regime**. Thus it is clear that permanent and final regime are two different things.

Critical velocity (V_0): When the velocity of water flowing in the channel is such that no silting or scouring is taking place then the velocity is said to be critical.

Table 9.2. Critical velocity

Nature of soil	*Critical velocity in m/sec*
Earth	0.3 to 0.6
Ordinary Murum	0.6 to 0.9
Hard Murum	1.2
Boulders	1.5 to 1.8
Soft Rock	1.8 to 2.4
Hard Rock	more than 3

Naturally if velocity is lowered below critical value silting occurs. On contrary if velocity is increased above critical value scouring occurs.

It is seen that the critical velocity is dependent on the nature of soil formation in which the canal flows. Table 9.2 gives various values of critical velocity for different soil formation.

Critical velocity ratio or CVR $\left(\frac{V}{V_0}\right)$**:** It is generally denoted by '*m*' and is the ratio of actual mean velocity of flow to the critical velocity of the channel. It is clear that for non-silting conditions CVR should be slightly more than one or at least equal to one.

Silt factor: It is a physical constant and generally denoted by letter '*f*'. It is a function of silt grade. It is also dependent on silt charge, density and shape of silt particles. Thus, it is clear that it can not be fixed arbitrarily but actually depends on the characteristics of bed material present at the site under consideration. The silt factor is, however, independent of discharge.

Table 9.3 gives different values of '*f*' for various types of bed materials.

Table 9.3. Values of 'f'

Type	*Diameter of grain in mm*	*Values of 'f'*
Fine silt	0.12	0.60
Sandy silt	0.325	1.00*
Coarse sand	0.725	1.5
Gravel	7	4.75
Boulder	72	15.0

* It is Kennedy's standard silt of Bari Doab Canal in Punjab.

Sediment Load: When water flows in natural or artificial channel it carries soil particles of varying sizes. This extra burden that water is required to carry along with it is called sediment load. Depending upon size of the (soil particles) sediment and the force (turbulence as well as horizontal drag) exerted by water on the sediment the particles move forward either in suspension or along the bed. The particles in suspension are called **suspended sediment load** and those moving along the bed are called bed load. The bed load may move along the bed with intermittent hopping or small jumps close to the bed of the channel.

Threshold velocity: When the water in a channel flows with low velocities it neither has turbulence nor sufficient drag force to carry forward soil particles. When the velocity of flow is slowly increased a stage reaches when the particle on the bed start getting dislodged; immediately afterwards the sediment comes into motion. This stage is called threshold stage and the corresponding velocity is called threshold velocity.

As the velocity of flow keeps increasing the material at the bed starts violent movement and the bed of the channel undergoes numerous changes from rupples to dunes to antidotes and so on.

Tractive force and shear stress or unit tractive force

Tractive force or drag force or shear force is a force exerted by water on the bed material in the direction of flow so that it causes movement of the bed particles. Tractive force is the tangential component of the weight of water flowing in the channel. In uniform flows, the horizontal component of the weight of water per unit length is calculated as follows:

In uniform flow

The weight of water per unit length of channel ..... $= \upsilon_W \times A \times 1$

where υ_W is unit weight of water and A is cross-sectional area of flow.

If the bed slope is S

The horizontal component of this (force) wt. of water $= \upsilon_W \times A \times S$

Since S = Sin of angle made by the bed of the channel with horizontal line.

The mean tractive force $= \upsilon_w AS$

Unit tractive force or shear stress is nothing but tractive force per unit wetted area $(P_W \times 1)$

$$\therefore \qquad \tau = \frac{\upsilon_W AS}{P_W}$$

Since, $$\frac{A}{P_W} = R \text{ (hydraulic mean radius)}$$

$$\tau = \upsilon_W RS$$

Critical tractive force and critical tractive stress

The bed particles in the flowing channel start moving when the tractive force exceeds frictional resistance offered by the particles. The tractive force which just causes the movement of bed particles is called critical tractive force and the corresponding unit tractive force is called critical tractive stress (τ_c). Average critical tractive stress for common sediment types is given below in Table 9.4.

Table 9.4. Critical Tractive Force (τ_c) for common soil types

Sediment type	τ_c *(kN/m²)*
Fine gravel	0.0037
Coarse sand, alluvial soils, silty loam	0.0025
Sandy loam	0.0020
Medium sand	0.0017

9.2. CANALS ON NON-ALLUVIAL SOILS

Here non-alluvial soils are supposed to be stable for the purpose of design of the irrigation canal. The dimensions of the canal can be worked out on the basis of well known hydraulic formulae. The bed slope of the canal may be kept anything. Only consideration is that the velocity of flow should be quite close to the critical velocity for the available soil. Generally side slope of 1½ : 1 (Horizontal : Vertical) in filling and 1 : 1 in cutting is given. Knowing the ratio of bed width B and depth D the dimensions of the canal can be determined uniquely.

The hydraulic formulae commonly used are the following:

1. $$Q = A\,.V$$

where Q is design discharge in m³/sec;

A is cross-sectional area of the canal in m²;

and V is mean velocity of flow in m/sec.

2. Chezy's formula:

$$V = C\sqrt{R.S}$$

where R is hydraulic mean depth in m;

S is bed slope of the canal;

and C is Chezy's constant.

The value of Chezy's 'C' can be calculated from following formulae:

(a) Bazin's formula

$$C = \frac{87}{1+\dfrac{K}{\sqrt{R}}}$$

where R is hydraulic mean radius in m;

and K is roughness constant.

The value of K depends on the character of canal bed and side material. In Table 9.5 values of 'K', roughness constant are given.

Table 9.5. Values of K

Types of Channel	*Bazin's K*
Smooth plaster, planed timber	0.06
Timber, Ashlar, neat brickwork	0.16
Rubble masonry	0.46
Plaster, very smooth earth	0.85
Earth channels in good condition	1.30
Earth channels in bad condition	1.75

(b) Manning's formula

$$C = \frac{R^{1/6}}{N}$$

Putting this value of C in Chezy's formula

$$V = \frac{1}{N} R^{3/3} \cdot S^{1/2}.$$ It is also called Manning formula.

Here N is the coefficient of rugosity.

(c) Ganguillet and Kutter's formula

$$C = \frac{23 + \frac{1}{n} + \frac{0.000155}{S}}{1 + \left(23 + \frac{0.000155}{S}\right)\frac{n}{\sqrt{R}}}$$

where S is bed slope or slope of hydraulic gradient:
and n is a rugosity factor. (It has same values as N)
Table 9.6 gives values of n for various types of channels.

Table 9.6. Values of n

Types of Channel	*Values of n*
Straight earth canals	0.0225
Rocky, with smooth and uniform bed	0.033
Rough strong beds and weeds on earth banks	0.035
Rocky and jogged canals	0.045

Once V is fixed using Chezy's or Manning's formula, sectional area can be determined from fundamental equation $Q = A.V.$

9.3. CANALS ON ALLUVIAL SOILS

If the procedure adopted for the design of canals on non-alluvial soils is studied carefully it can be recognised that no reference is made to the silt load carried by the irrigation water. The principle of design of a canal on the alluvial soil is totally different from that of a canal on non-alluvial soils. Canals on alluvial soil carry appreciable silt and sand load. Silt concentration of the canal water affects the velocity of flow considerably. Hence, the Manning formula cannot be used here to determine the velocity of flow. When the canal water has excess silt load silting occurs in the canal. On the contrary when the water is silt free it picks up the silt from the canal bed and sides. It results in erosion of canal section. Manning's equation or Chezy's equation do not consider this aspect. When silting takes place the channel section is reduced and consequently capacity of the channel is reduced. When scouring occurs firstly the water level is lowered which in turn reduces the command. Secondly, the scoured material is deposited at some other place to disturb the equilibrium conditions there. Taking the problem of silt transportation into account it was necessary to evolve some basis for the design of a stable section with critical velocity. There are two important and most commonly used theories. They are Kennedy's silt theory or transportation theory and Lacey's silt theory.

9.4. KENNEDY'S THEORY (1895)

Upper Bari Doab canal system is one of the oldest irrigation systems. Mr. R.G. Kennedy, Executive Engineer in Punjab, PWD was incharge of this canal system in 1895. The canal was not silt cleared for a long period. He recognised that the canal has attained stable stage and hence he pleaded that the velocity of flow has also attained critical stage. He studied nearly 20 sites on the stable reach of this canal system and finally developed a theory which is known after his name.

Kennedy's silt theory or as it is called the transportation theory is as mentioned below.

Sediment in flowing canals is kept in suspension solely by the vertical components of the constant eddies which can always be observed over the full width of any stream, boiling up gently to the surface. (The reason for the production of eddies is the roughness of the bed).

In order to obtain an expression for the silt supporting power of the stream, it may be safely assumed that the quantity of silt supported is proportional to the width of the bed, all other conditions remaining the same.

It also varies with the velocity of stream V_0. It may be taken proportional to V_0^{n-1}. It is clear that greater the velocity, force of eddies will be greater. The force becomes zero when velocity is zero. Factor 'n' is some index.

According to Kennedy, though depth is the third variable, it could not affect either the number or the force of eddies.

Hence the amount of silt supported in the stream may be expressed by $A \cdot B \cdot V_0 \cdot V_0^{n-1}$.

where, A is some constant;

B is bed width of canal; and

V_0 is the velocity in stable state.

The amount of silt transported is given by

$$A \cdot B \cdot V_0^{n-1} \times V_0 = A \cdot B \cdot V_0^{n}$$

It is essential to recognize here that all the silt sediment is taken to be in suspension. Of course there is no doubt about the fact, that small quantity of heavier silt is carried as a bed load which rolls along the bed. This amount would vary directly as BV_0 instead BV_0^n.

To include rolling silt also, the value of n should be taken less, than it would be, if the suspended silt were alone considered.

Kennedy plotted various graphs between V_0 and depth of flow and finally gave a formula to calculate V_0. The formula is

$$V_0 = C \cdot D^n$$

where

V_0 is critical velocity in m/sec;

D is full supply depth in m;

and C is constant. It depends on the character of silt. Coarser the material greater the value of the constant.

n is some index. It also depends on type of silt.

Table 9.7. Values of *n* for various regions

Type of silt load in the canal water	*Value of n*
Western Ghat fine silt	0.53
In Punjab sandy silt	0.64
Himalayan sandy silt	0.69

For the State of Punjab he gave values of C and n and thus the formula is

$$V_0 = 0.546\ D^{0.64} \quad \text{...(1)}$$

After recognizing the fact that silt grade also plays an important part, he modified the formula (1). The new form is

$$V = 0.546\ \ \text{m} \cdot D^{0.64} \quad \text{...(2)}$$

where m is CVR or $\dfrac{V}{V_0}$

For coarse sand value of m may be taken as 1.1 to 1.2. Whereas for finer material it may be kept between 0.8 and 0.9.

Based on Kennedy theory the formulae of similar nature, $V = C \cdot D^n$, were advanced in various regions.

Table 9.8. Values of *C* and *n* for various regions

Region	*C*	*n*
Upper Bari Doab canal system	0.546	0.64
Lower Chenab canal system	0.518	0.50
Godavari Delta region	0.393	0.55
Krishna Delta region	0.532	0.52

9.5. LIMITATIONS OF KENNEDY'S THEORY

- In the absence of B/D relation the Kennedy theory do not provide easy basis for fixing channel dimensions uniquely.
- Perfect definitions of silt grade and silt charge are not given.
- Complex phenomenon of silt transportation is not fully accounted and only critical velocity ratio (m) concept is considered sufficient.
- There is no provision to decide longitudinal slope under the scope of the theory.
- By use of Kutter's formula inherent limitations therein remain applicable in Kennedy's channel design procedure.

9.6. DESIGN OF IRRIGATION CHANNELS MAKING USE OF THE KENNEDY THEORY

When an irrigation channel is to be designed by the Kennedy theory it is essential to know, FSD (Q), coefficient of rugosity (N), CVR (m), and longitudinal slope of channel (S), before hand. Then making use of the following three equations section can be designed by trials:

- $V = 0.546\ \text{m} \cdot D^{0.64}$
- $Q = A \cdot V$; and
- $V = C\sqrt{RS}$

The procedure of designing may be outlined in the following steps:

- Assume the reasonable full supply depth, D.
- Using equation (1) find out value of V.
- With this value of V, using equation (2) find out A.
- Assuming side slopes and from the knowledge of A and D find out bed width B.
- Calculate R which is ratio of area and wetted perimeter.
- Using equation (3) find out value of actual velocity V.

When the assumed value of D is correct, the value of V in step (*f*) will be same as V calculated in step (*b*), if not assume another suitable value of D and repeat the procedure till both values of velocity come out to be the same.

It may be recognized here that for same values of Q, N and m but with different values of S various channel sections may be designed. It is needless to mention that all of them would not be equally satisfactory. To give some guidance for fixing particular slope (S). Woods has given a table (Table 9.9) on the basis of experience in which he gives suitable B/D ratios for various values of Q, S, N, m. Further, by adopting suitable B/D ratio labour of making trials can be avoided.

Thus when Q, N, m and B/D ratio are given using the above given three formulae viz., Kennedy's equation, continuity equation and Chezy's formula the channel can be designed uniquely without making trials. The procedure of designing may be outlined in the following steps:

- Using equation (1) express V in terms of D only.
- From given B/D ratio and side slopes calculate area A in terms of depth D only (if side slopes are not given take ½ : 1 as side slopes for alluvial tract).
- Using equation (2) obtain another relation between V and D only.
- Solve two equations giving relation between V and D as simultaneous equations and find out the value of D.
- Calculate value of B from known B/D ratio.
- Calculate V from Kennedy's equation (1).
- Using Chezy's equation calculate value of S.

Note: Determine Chezy's constant from Bazin's formula.

Table 9.9. Woods design table for use in Kennedy theory

Discharge (m^3/sec)	*B/D*	*B (m)*	*D (m)*	*Slope 1 in*	*Kutter's N*	*V/V_0*	*V (m/sec)*
0.283	2.9	1.45	0.488	3333	0.0225	0.92	0.344
0.708	3.4	2.21	0.656	3636	0.0225	1.01	0.424
1.415	3.7	3.125	0.839	4000	0.0225	1.00	0.476
2.83	4.2	4.425	1.036	4444	0.0225	1.00	0.555
7.08	4.8	6.705	1.54	4444	0.0225	1.01	0.702
14.15	5.7	9.77	1.725	5000	0.0225	1.01	0.755
28.30	7.6	15.25	1.984	5000	0.0225	1.03	0.882
56.60	11.3	25.48	2.255	5714	0.0225	1.03	0.945
141.50	22.5	56.40	2.50	6666	0.0225	0.98	0.975
283.00	41.0	105.40	2.59	6666	0.0225	1.02	1.01
		110.00	2.68	8000	0.0225	0.98	0.955

Problem 9.1. Design an irrigation canal for the following data:

Use Kennedy's equations.

FS discharge = 14 m³/sec

Bed slope = 1 in 5000

Kutter's N = 0.0225

CVR (m) = 1

Side slope = ½ : 1 (Horizontal : Vertical)

Solution: Assuming full supply depth 1.8 m velocity can be calculated from Kennedy's equation.

$$V = 0.546 \text{ m} \times D^{0.64} = 0.545 \times 1 \times 1.8^{0.64} = 0.795 \text{ m/sec}$$

From equation $Q = AV$

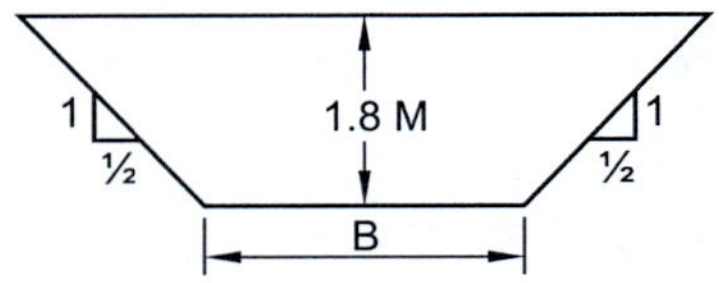

Fig. 9.1. Problem 1

$$A = \frac{14}{0.795} = 17.6 \text{ m}^2$$

$$A = BD + D^2/2 = 17.6$$

or $$1.8\,B + 3.24/2 = 17.6$$

$$B = 8.88 \text{ m}$$

$$R = \frac{A}{P_w} = \frac{17.6}{B + D\sqrt{5}} = \frac{17.6}{8.88 + 1.8\sqrt{5}} = \frac{17.6}{8.88 + 4.025} = 1.365$$

Chezy's constant can be calculated from Kutter's formula.

$$C = \frac{23 + \dfrac{1}{0.0225} + \dfrac{0.000155}{0.0002}}{1 + \left(23 + \dfrac{0.000155}{0.0002}\right)\dfrac{0.0225}{\sqrt{1.365}}} = 46.8$$

Now using Chezy's formula

$$V = 46.8\ \sqrt{1.365 \times 0.0002} = 46.8 \times 0.0165 = 0.772 \text{ m/sec.}$$

Thus this velocity comes out to be slightly less than velocity obtained from Kennedy's equation. Hence, one more trial with smaller depth will give fairly close result.

The same problem can be solved easily and uniquely when B/D ratio is given. From the Table 9.7 for 14.15 m³/sec discharge, the value of B/D ratio is 9.7.

That is $$B = 5.7\ D$$

∴ $$A = 5.7\ D \times D + D^2/2 = 6.2\ D^2$$

Now using equations (1) and (2)

$$V = 0.546 \cdot D^{0.64} \qquad \text{...(A)}$$

and $$14 = 6.2\ D^2 \cdot V \qquad \text{...(B)}$$

Treating equations (A) and (B) as simultaneous value of D can be uniquely determined. Equation (B) may be written as:

$$V = \frac{14}{6.2 \times D^2}$$

Now equating this equation to (A)

$$\frac{14}{6.2 D^2} = 0.546\ D^{0.64}$$

or $$\frac{14}{6.2 \times 0.546} = D^{2.64}$$

or $D^{2.64} = 4.14$

or $D = 1.712$ m

and hence $B = 9.76$ m

Value of required slope may be calculated from Chezy's formula. Of course it is true that Chezy's constant will have to be determined from Bazin's formula.

Problem 9.2. An irrigation canal has been constructed with following parameters:

Full supply discharge = 45 cumec
Bed width = 30 metres
Full slope depth = 1.8 metres
Side slopes = ½ : 1
Bed slope = 1 in 6600
Mannings N = 0.0225
Critical velocity ratio = 1

Check whether the section designed satisfies Kennedy's theory.

Solution: Step 1. Using Kennedy's equation.

$$V = 0.546 \times m \times D^{0.64} = 0.546 \times 1 \times 1.8^{0.64} = 0.546 \times 1.457$$
$$= 0.795 \text{ m/sec} \quad ...(1)$$

Step 2. Area of flow $= BD + D^2/2 = 30 \times 1.8 + \dfrac{(1.8)^2}{2} = 55.62 \text{ m}^2$

Wetted perimeter $= B + \sqrt{5}\,D = 30 + 2.236 \times 1.8 = 34.025$ m

$$R = \frac{A}{P_w} = \frac{55.62}{34.025} = 1.64 \text{ m}$$

Step 3. Chezy's 'C' can be calculated from Kutter's formula:

$$C = \frac{23 + \dfrac{1}{N} + \dfrac{0.000155}{S}}{1 + \left(23 + \dfrac{0.000155}{S}\right)\dfrac{N}{\sqrt{R}}}$$

$$= \frac{23 + \dfrac{1}{0.0225} + \dfrac{0.000155 \times 6600}{1}}{1 + \left(23 + \dfrac{0.000155 \times 6600}{1}\right)\dfrac{0.0225}{\sqrt{1.64}}} = \frac{23 + 44.44 + 1.02}{1 + (24.02)\dfrac{0.0225}{1.28}}$$

$$= \frac{68.46}{1.42} = 48.2$$

Step 4. Velocity from Chezy's formula.

$$V = C\sqrt{RS} = 48.2\sqrt{1.64 \times \frac{1}{6600}} = 48.2\sqrt{0.00025}$$
$$= 48.2 \times 0.0158 = 0.762 \text{ m/sec} \quad ...(2)$$

Step 5. Comparing velocities obtained under steps 1 and 4.

$$0.795 \approx 0.762$$

So the section designed satisfies Kennedy's theory.

9.7. LACEY'S THEORY (1939)

Taking lead from the Kennedy theory Mr. Gerald Lacey undertook detailed study to evolve more scientific method of designing irrigation channels on alluvial soils. He presented revised version of his study in 1939 which is popularly known as Lacey's theory. In this theory, Lacey described in detail concept of regime conditions and rugosity coefficient. The definitions of these terms are already given. It may be seen that for a channel to achieve regime condition following three conditions have to be fulfilled:

- channel should flow uniformly in "incoherent unlimited alluvium" of same character as that transported by the water;
- silt grade and silt charge should be constant; and
- discharge should be constant.

These conditions are vary rarely achieved and are very difficult to maintain in practice. Hence according to Lacey's conception regime conditions may be subdivided as initial and final. The definition of these two terms are already given earlier.

In rivers achievement of initial or final regime is practically impossible. Only in bank full stage or high floods the river may be considered to achieve temporary or quasi-regime. The recognition of this fact can be utilised to deal with the issues concerning scour and floods.

Lacey also states that the silt is kept in suspension solely by force of eddies. But Lacey adds that eddies are not generated on the bed only but at all points on the wetted perimeter. The force of eddies may be taken normal to the sides, (Fig. 9.2).

Obviously the vertical components of forces due to eddies are responsible for keeping the silt in suspension.

Unlike Kennedy, Lacey takes hydraulic mean radius (R) as a variable rather than depth (D). So far as wide channels are concerned there is hardly any difference between R and D. When the channel section is semi-circular there is no base width and sides actually and hence assumption of R as a variable seems to be more logical. From this point of view velocity is no more dependent on D but, rather depends on R. Consequently amount of silt transported is not dependent on the base width of a channel only.

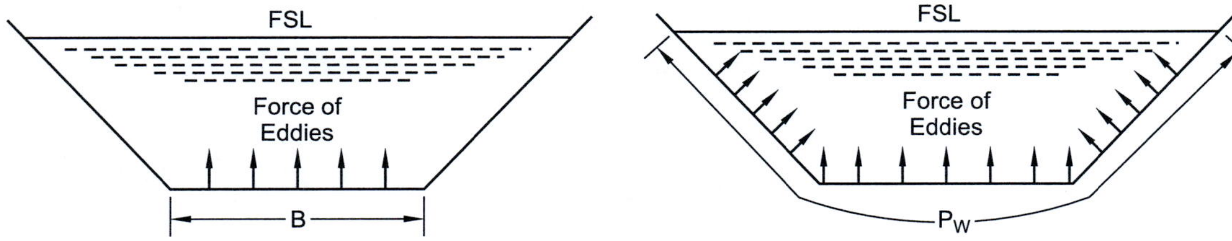

(a) According to Kennedy (b) According to Lacey

Fig. 9.2. Lacey's theory

On the basis of above mentioned arguments Lacey plotted a graph between regime mean velocity (V) and hydraulic mean radius (R) and gave the relationship.

$$V = K \cdot R^{1/2} \qquad ...(1)$$

where K is a constant.

It can be seen here that power of R is a fixed number and needs no alteration to suit different conditions.

Lacey recognised the importance of silt grade in the problem and introduced a concept of function 'f' known as silt factor.

He adjusted the values of 'f' such that it also came under square root sign. Thus it gave scalar conception. Equation (1) is thus modified as

$$V = K \cdot \sqrt{f \cdot R} \qquad \text{...(2)}$$

Kennedy's general equation is

$$V = c \cdot m \cdot D^n \qquad \text{...(3)}$$

Comparing equations (2) and (3)

$$f = m^2$$

It can also be recognized from equation (2) that in regime channels if the mean velocity is same then hydraulic mean radius varies inversely with the silt factor.

Lacey takes silt as a standard silt when silt factor is unity for that silt.

He further states that standard silt is a sandy silt in a regime channel with hydraulic mean radius equal to one metre.

9.8. LACEY'S REGIME EQUATIONS

After study and plotting of large data to justify his theory Lacey gave three fundamental equations from which other equations have been derived for design of irrigation channels. The three fundamental equations are:

$$V = 0.639 \sqrt{fR} \qquad \text{...(1)}$$

where V is regime velocity in m/sec.

$$Af^2 = 141.2\ V^5 \qquad \text{...(2)}$$

$$V = 10.8\ R^{2/3}\ S^{1/3} \qquad \text{...(3)}$$

where S is the slope of water surface.

Equation (3) is called regime flow equation and is of great practical importance. It may be seen that the equation does not contain the term of rugosity coefficient. While adopting similar equations like Manning's or Kutter's equation it is necessary to know the value of coefficient of rugosity (N), the selection of which most of the time remains a matter of experience and many not be reliable especially in case of rivers in floods. Considering that at high flood river flows in quasi-regime the regime flow equation (3) given above can be adopted though it may have some exceptions.

Important equations given by Lacey in his theory are summarised below. As already mentioned first three equation are called fundamental equations on the basis of which other equations have been developed.

$$V = 0.639 \sqrt{f R} \qquad \text{...(4)}$$

$$Af^2 = 141.24\ V^5 \qquad \text{...(5)}$$

$$V = 10.8\ R^{2/3}\ S^{1/3} \qquad \text{...(6)}$$

$$V = \frac{1}{N_a}\ R^{3/4}\ S^{1/2} \qquad \text{...(7)}$$

$$N_a = 0.0225\ f^{1/4} \qquad \text{...(8)}$$

$$V = 0.4382\ (Q\ .\ f^2)^{1/6} \qquad \text{...(9)}$$

$$R = 2.46\ \frac{V^2}{f} \qquad \text{...(10)}$$

$$P_w = 4.825\ \sqrt{Q} \qquad \text{...(11)}$$

$$S = \frac{f^{5/3}}{3316Q^{1/6}} \quad ...(12)$$

$$A = 2.28\ \frac{Q^{5/6}}{f^{1/3}} \quad ...(13)$$

$$f = 1.76\ \sqrt{m_r} \quad ...(14)$$

$$R = 1.35\left(\frac{q^2}{f}\right)^{1/3} \quad ...(15)$$

Although all the above equations are given for regime conditions and are normally, for canals in alluvium, as explained earlier, since the river attains quasi-regime conditions, equations (6), (11) and (15) are very useful in calculating flood discharge, waterway required during floods and scour depth during floods respectively.

When equation (15) is applied to rivers in flood the value of R gives the normal depth of scour. Hence this formula is very useful determining the levels of foundations, vertical cut-offs etc. This formula is popularly called Lacey's scour depth formula.

9.9. LIMITATIONS OF LACEY'S THEORY

- The Lacey's work is based on field observations and empirically derived equations and therefore it cannot be said to be theory in strict sense.
- Regime equations in their derived form cannot be applied universally as they hold good mostly for the regions whose data was taken for study.
- Like Kennedy's theory even though perfect definition of silt grade and silt charge is not given most of the equations are based on the silt factor 'f'.
- In practice regime condition stated by Lacey is very rarely achieved and that too after a long period.
- The field observations have shown limited acceptance of the concept of semi-elliptical section of a regime channel.
- Complex phenomenon of sediment concentration and transport have not been scientifically considered.

9.10. DESIGN OF IRRIGATION CHANNELS MAKING USE OF THE LACEY THEORY

It is already mentioned that the full supply discharge for any canal is always fixed before starting a design. The value of 'f' for particular site may be calculated using equation (11) of article 9.8. or if CVR is given then $f = m^2$.

Thus when Q and f are known design can be done in the following steps:

- Find out V using equation (6) of the Article 9.8.

$$V = 0.4382\ (Q\ .\ f^2)^{1/6}$$

- Calculate value of R using equation (7) of the Article 9.8.

$$R = 2.46\ \frac{V^2}{f}$$

- Calculate wetted perimeter P_w using Lacey's regime perimeter equation $P_w = 4.825\ Q^{1/2}$.
- Calculate cross sectional area A from equation $Q = AV$.

- Assuming side slopes, calculate the full supply depth from A, P_w and R.
- Calculate the longitudinal slope of the channel using equation (9) of the Article 9.8.

$$S = \frac{f^{5/3}}{3316 Q^{1/6}}$$

Problem 9.3. **Design an irrigation canal by Lacey's theory for the following data:**

FSD = 14 m³/sec

f **= 1**

Side slopes = 1/2 : 1 (Horizontal : Vertical)

Coefficient of rugosity, N = 0.0225

Solution: Mean velocity $V = 0.4382\ (14 \times 1^2)^{1/6} = 0.68$ m/sec

Hydraulic mean radius.

$$R = 2.46\ \frac{(0.68)^2}{1} = 1.14 \text{ m}$$

Wetted perimeter, $P_w = 4.825 \times (14)^{1/2} = 18.06$ m

Cross-sectional area, $A = R \times P_w = 1.14 \times 18.06 = 20.55$ m²

From figure

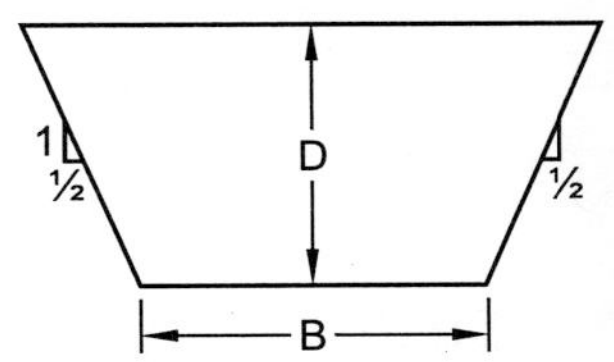

Fig. 9.3. Problem 3

$$A = BD + \frac{D^2}{2} \quad \text{...(1)}$$

$$P_w = B + \sqrt{5}\ D \quad \text{...(2)}$$

Putting values of A and P_w in equations (1) and (2) respectively

$$20.55 = BD + D^2/2$$

$$18.06 = B + \sqrt{5}\ D \quad \text{...(3)}$$

From equation (3)

$$B = 18.06 - \sqrt{5}\ D$$

Now putting value of B in equation (1)

$$20.55 = (18.06 - \sqrt{5}\,D)\ D + D^2/2$$

or $\quad 20.55 = 18.06\ D - 1.73\ D^2$

or $1.73\ D^2 - 18.06\ D + 20.55 = 0$

It is a quadratic equation in D. Hence, solving for D.

$$D = \frac{18.06 \pm \sqrt{(18.06)^2 - 4 \times 1.73 \times 20.55}}{2 \times 1.73}$$

$$D = \frac{18.06 \pm 13.56}{3.46}$$

or $\quad D = 9.13$ and 1.3 m

Taking $\quad D = 9.13$ m

$B = 18.06 - 2.23 \times 9.13 = -2.14$ which is absurd

Hence, taking $\quad D = 1.3$ m

$B = 18.06 - 2.23 \times 1.3 = 15.15$ m

By using the values of B and D hydraulic mean radius can be calculated again to compare it with the value already calculated above using Lacey's equation–

$$R = 2.46 \frac{(0.68)^2}{1}$$

$$R = \frac{A}{P} = \frac{15.15 \times 1.3 + (1.3)^2/2}{15.15 + \sqrt{5} \times 1.3} = \frac{20.54}{18.05} = 1.14 \text{ m}$$

Now slope can be determined from formula

$$S = \frac{f^{5/3}}{3316 \times Q^{1/6}} = \frac{1^{5/3}}{3316 \times (14)^{1/6}} = \frac{1}{3316 \times 1.553} = \frac{1}{5160}$$

Answer : FS depth 1.3 m; bed width 15.15 m; and bed slope 1 in 5160.

Problem 9.4. Using Lecey's theory design a channel section for the following data:
Full supply discharge = 15 m³/sec
Mean diameter of silt particles = 0.33 mm
Side slopes = 1/2 : 1

Solution: Step 1. Silt factor,

$$f = 17.6 \sqrt{m_r} = 1.76 \sqrt{0.33} = 1.76 \times 0.574 = 1.0$$

Step 2. Mean velocity,

$$V = 0.4382 (15 \times 1^2)^{1/6} = 0.69 \text{ m/sec}$$

Step 3. Hydraulic mean radius,

$$R = 2.46 \frac{(0.69)^2}{1} = 1.17 \text{ m}$$

Step 4. Wetted perimeter,

$$P_w = 4.825 \times (15)^{1/2} = 18.69 \text{ m}$$

Step 5. Cross-sectional area

$$A = R \times P_w = 21.86 \text{ m}^2$$

For the given side slopes

$$A = BD + \frac{D^2}{2} \qquad ...(1)$$

and

$$P_w = B + \sqrt{5}\, D \qquad ...(2)$$

Putting values of A and P_w in equations (1) and (2) respectively

$$21.86 = BD + D^2/2$$

$$18.69 = B + \sqrt{5}\, D$$

From equation (2)

$$B = 18.69 - \sqrt{5}\, D$$

Step 6. Putting value of B in equation (1)

$$21.86 = (18.69 - \sqrt{5}\, D)\, D + D^2/2 = 18.69\, D - 1.73\, D^2$$

or $1.73\, D^2 - 18.69\, D + 21.86 = 0$

It is a quadratic equation in D. Hence solving for D.

$$D = \frac{18.69 \pm \sqrt{(18.69)^2 - 4 \times 1.73 \times 21.86}}{2 \times 1.73} = \frac{18.69 \pm 14.07}{3.46}$$

$$\therefore \quad D = 9.47 \text{ m} \quad \text{and} \quad 1.34 \text{ m}$$

Step 7. Taking $D = 9.47$ m

$B = 18.69 - 2.236 \times 9.47 = -2.48$ which is absorb

Hence, taking $D = 1.34$ m

$B = 18.69 - 2.236 \times 1.34 = 15.7$ m

Step 8. These values of B and D can be utilised to calculate value of R.

$$R = \frac{A}{P_w} = \frac{15.7\times1.34+(1.34)^2/2}{15.7+\sqrt{5}\times1.34} = 1.17 \text{ m}$$

This value is equal to value of R calculated in step 3 above.

Step 9. Longitudinal slope

$$S = \frac{f^{5/3}}{3316\times Q^{1/6}} = \frac{1}{3316\times(15)^{1/6}} = \frac{1}{3316\times1.87} = \frac{1}{5210}$$

Answer: FS depth 1.34 m; bed width 15.7 m and longitudinal slope 1 in 5210.

Problem 9.5. Design an irrigation channel using Lacey's regime theory when following data is given:

Mean diameter of silt particles = 0.33 mm

Longitudinal slope of channel = 1 in 6000

Side slopes of the section = 1/2 : 1

Assume that for the given slope maximum discharge the canal has to carry under regime conditions.

Solution: Step 1. Silt factor,

$$f = 1.76\sqrt{m_r} = 1.76\sqrt{0.33} = 1.0$$

Step 2. Using relation between $Q - S$ and f.

$$S = \frac{f^{5/3}}{3316\times Q^{1/6}} = \frac{(1)^{5/3}}{3316\times Q^{1/6}}$$

$$Q = \left(\frac{6000}{3316}\right)^6 = 35.1 \text{ cumec}$$

Step 3. Mean velocity,

$$V = 0.4382\,(35.1 \times 1^2)^{1/6} = 0.793 \text{ m/sec}$$

Step 4. Hydraulic mean radius,

$$R = 2.46\,\frac{(0.793)^2}{1} = 1.54 \text{ m}$$

Step 5. Wetted perimeter,

$$P_w = 4.825\,(35.1)^{1/2} = 28.59 \text{ m}$$

Step 6. Cross-sectional area,

$$A = R \times P_w = 1.54 \times 28.59 = 44.02 \text{ m}^2$$

For given side slopes

$$A = BD + D^2/2 \quad \text{...(1)}$$

and

$$P_w = B + \sqrt{5}\,D \quad \text{...(2)}$$

Putting values of A and P_w in equations (1) and (2) respectively.

$$44.02 = BD + D^2/2$$

$$28.59 = B + \sqrt{5}\,D$$

or

$$B = 28.59 - \sqrt{5}\,D$$

Step 7. Putting value of B in equation (1)

$$44.02 = (28.59 - \sqrt{5}\ D)D + D^2/2 = 28.59\ D - 1.73\ D^2$$

or $1.73\ D^2 - 28.59\ D + 44.02 = 0$

It is a quadratic equation in D. Solving for D

$$D = \frac{28.59 \pm \sqrt{(28.59)^2 - 4\times1.73\times44.02}}{2\times1.73} = \frac{28.59 \pm 22.64}{3.46}$$

$\therefore$ $D = 14.80$ m and 1.72 m

Step 8. Taking $D = 14.80$ m

$B = 28.59 - 14.80 \times \sqrt{5} = -4.50$ which is absurb

Hence, taking $D = 1.72$ m

$B = 28.59 - 1.72\sqrt{5} = 24.75$ m.

Step 9. These values of B and D can be utilised to calculate value of R.

$$R = \frac{A}{P_w} = \frac{24.75\times1.72 + (1.72)^2/2}{24.75 + \sqrt{5}\times1.72} = 1.54 \text{ m}$$

This value of R is equal to the value calculated in step 4 above.

9.11. DESIGN OF IRRIGATION CHANNELS MAKING USE OF GARRET'S DIAGRAMS

Articles 9.6 and 9.10 describe how to design irrigation channels with the help of formulae based on various theories. It has been experienced that every time designing the canal with the help of formulae is time consuming. Hence, the diagrams are evolved to simplify the procedure.

The graphs were originally prepared in British units. After reading the values they can be converted into metric units. Alternatively diagrams can be converted into metric units for direct reading.

Engineer Garret constructed diagrams to provide a graphical procedure of designing irrigation channels. They are based on Kennedy's procedure of canal design and take Kennedy's and Kutter's equation into consideration. The diagrams are available in five sheets to cover a very large range of discharge, slope and coefficient of rugosity. The discharge ranges from 1 to 12,000 cft/sec or 0.03 to 336.6 cumec. The slope ranges from 1 in 100 to 1 in 10,000. The coefficient of rugosity ranges from 0.018 to 0.03. Any sheet may be used to design the channel fully, provided that the range of slope and discharge of that sheet covers the values of slope and discharge under consideration. If the values of S and Q fall outside the range another suitable sheet may be used. In the diagrams discharge and velocity are plotted on abscissa whereas slope and depth are plotted on the left and right ordinates respectively. The discharge lines are curves in full lines and bed-width curves are shown dotted.

The diagrams have been drawn for the coefficient of rugosity of 0.0225. But since the value of N is not always same a nomograph is drawn at the top of each diagram to facilitate adoption of any other value of N. The point of intersection of given values of S and Q is shifted to the left or right as the case may be by equal distance to the left or right as given by new value of N from standard value of N (= 0.0225) on the nomograph at the top of the diagram.

The diagrams are drawn for side slope of 1/2 : 1 (horizontal : vertical). This value of side slope is fixed on the assumption that an irrigation channel subsequently adopts this slope (Fig. 9.4). (Plate 1) gives one sheet of Garret's diagram.

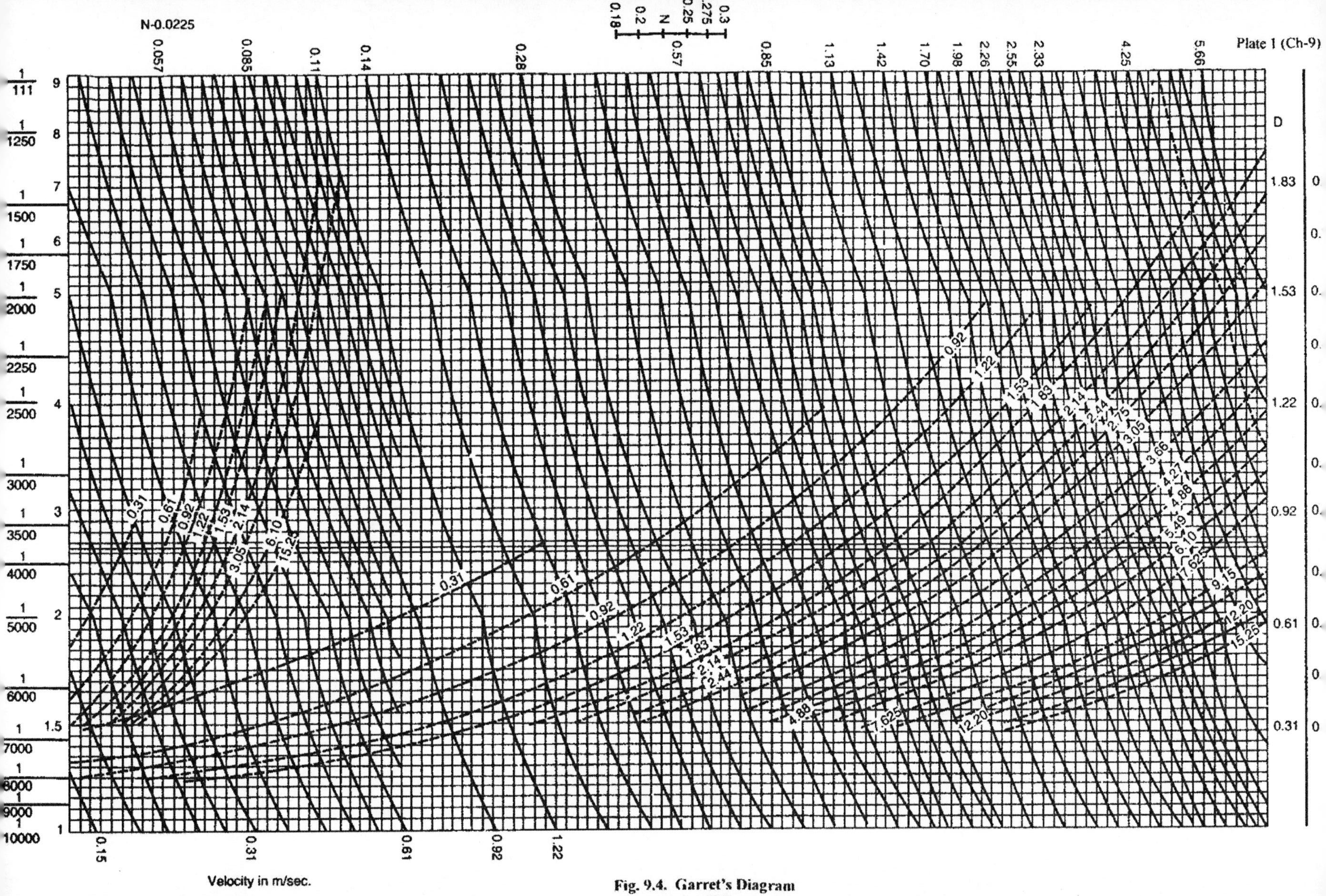

Fig. 9.4. Garret's Diagram

The method of using Garret's diagrams may be outlined in the following steps:

- Find out the point of intersection of given slope and discharge on the diagram.
- Erect a vertical line through this point to cut various bed-width curves.
- Each point of intersection of vertical line and bed-width curve gives a depth on the right hand side ordinate.
- Each depth value gives a separate value of critical velocity (V_0) from the column of V_0 next to depth ordinate.
- For each depth value actual mean velocity (V) is calculated. It may be calculated in two ways.

  Firstly, from equation $V = \dfrac{Q}{A}$, which gives fairly good results.

  Secondly from the same diagram making use of bed width-velocity curves. The procedure of finding V from graph is as follows:

  (*a*) Take a particular pair of B and D and from that depth draw a line parallel to abscissa to cut bed width-velocity curve of appropriate value.

  (*b*) From this point of intersection come vertically downwards to cut the given slope line.

  (*c*) Then from this new point of intersection comedown along the velocity curve to abscissa to get value of V.

  This procedure gives approximate value of V.

- Then calculate $\dfrac{V}{V_0}$ ratios. One for each pair of B and D.

  Thus we get various sets, each consisting of B, D and $\dfrac{V}{V_0}$ values.

- As a last step choose pair of B and D out of all sets which has ratio $\dfrac{V}{V_0}$ unity.

To simplify the procedure all results may be arranged in a tabular form.

Problem 9.6. Design a distributary canal making use of Garret's diagram for the data given below:

FS discharge = 2.26 cumec
Bed slope = 1 in 5000
Kutter's N = 0.0225
CVR (m) = 1
Side slopes = 1/2 : 1 (Horizontal : Vertical)

Verify the answer using Kennedy's equation.

Solution: Refer, plate – 1 Fig. 9.4 Garret's diagram converted in metric units.

(*i*) Find out the point of intersection of given slope $\dfrac{1}{5000}$ and discharge 2.26 m^2/sec, on the graph.

(*ii*) Erect a vertical line through this point to cut various bed-width curves say, 9.15, 6.10 and 10.68 metres.

(*iii*) Each point of intersection of the vertical line and bed width curve gives a depth on right hand side ordinate say, 0.61, 0.76 and 0.55 metres respectively.

(*iv*) Each depth value gives a separate value of critical velocity, V_0 from the column of V_0 next to depth ordinate, say, 0.40, 0.46 and 0.37 m/sec, respectively.

(*v*) For each depth value actual mean velocity, V is calculated from equation $V = Q/A$ where A is area of flow. For side slope 1/2 : 1 $A = BD + D^2/2$.

The results may be tabulated as below for getting correct answer.

Set	*B* Bed width	*D* Depth velocity	V_0 Critical velocity	*V* Actual velocity	$\frac{V}{V_0}$	*Remarks*
1	9.15	0.61	0.40	0.393	0.982	
2	6.10	0.76	0.46	0.46	1	Correct value
3	10.68	0.55	0.37	0.378	1.02	

Second set gives $\frac{V}{V_0}$ equal to unity hence, bed-width and depth of the distributary may be taken as 6.1 m and 0.76 m respectively.

These results can be checked by using Kennedy's equation as follows:

$$V = 0.546 \times 1 \times m \times D^{0.64}$$

Taking $D = 0.76$

$$V = 0.546 \times 0.76^{0.64} = 0.46 \text{ m/sec} \qquad ...(1)$$

$$A = \frac{Q}{V} = \frac{2.26}{0.46} = 4.92 \text{ m}^2$$

$$\therefore \quad 4.92 = BD + D^2/2 = 0.76\,B + \frac{(0.76)^2}{2} = 0.76\,B + 0.29$$

$$\therefore \quad B = 6.1 \text{ metres}$$

$$R = \frac{A}{P_w} = \frac{4.92}{B + D\sqrt{5}} = \frac{4.92}{6.1 + 0.76\sqrt{5}} = 0.63$$

Chezy's constant '*C*' from Kutter's formula.

$$C = \frac{23 + \frac{1}{0.0225} + \frac{0.000155}{0.0002}}{1 + \left(23 + \frac{0.000155}{0.0002}\right)\frac{0.0225}{\sqrt{0.63}}} = 40.7$$

$\therefore$ Velocity V from Chezy's formula is

$$V = 40.7\sqrt{0.63 \times 0.0002} = 0.46 \text{ /sec} \qquad ...(2)$$

Since velocities of (1) and (2) are equal the depth assumed is correct.

$$\therefore \quad B = 6.1 \text{ metres} \quad \text{and} \quad D = 0.76 \text{ metres}.$$

9.12. LACEY'S REGIME DIAGRAMS

These diagrams can be used to design the canal graphically. They are based on Lacey's regime equations. There are two sets of diagrams. First set gives depth and bed-width for the known values of Q and f. Second set gives Lacey's regime slope for the known values of Q and f.

It will not be out of place to mention here that the Lacey's theory gives a fixed value for bed slope. Strictly speaking it cannot be adopted in truest sense. The obvious reason is that the longitudinal slope depends on topography or natural ground levels. But it is true that Lacey's regime slope value provides good basis for selecting bed slope in longitudinal section. Once the bed slope is fixed on the longitudinal section, design of channel may be carried out for known values of Q, N, S, m (= 1). It means that Kennedy's theory or Garret's diagrams may then be used successefully.

9.13. COMPARISON OF KENNEDY'S AND LACEY'S THEORIES

1. The basic concept regarding silt transporation is the same in both the theories. In both the theories it is stated that the silt remains in suspension due to vertical force of eddies.
2. Kennedy assumes that the eddies are generated on the bed only and hence he derives the formula for finding out critical velocity in terms of depth.

   Lacey proposes that regime section is semi-circular ultimately and eddies are generated along the whole wetted perimeter. He drives formula for mean regime velocity in terms of hydraulic mean radius.
3. Lacey states that as the shape of irrigation channel is fixed to a particular geometrical figure (generally trapezoidal) it can not achieve final regime conditions and hence may be said to achieve initial regime.

   Kennedy assumes that when there is neither silting nor scouring the channel is in its regime.
4. Kennedy selects Kutter's formula for designing irrigation channel. But in Kutter's formula value of N is arbitrarily fixed.

   Lacey has not fixed any value arbitrarily.
5. Kennedy has made use of term "CVR" (m) but he did not give any basis for calculating m. He simply states that it depends on silt charge and silt grade.

   Lacey has introduced a term "silt factor" (f). He related f to mean diameter of the bed material and gave basis to calculate f.

   The formula is $f = 1.76\ \sqrt{m_r}$
6. Kennedy gives no clue for calculating longitudinal regime slope. Lacey produced a regime slope formula.
7. Design based on Kennedy's theory can only be achieved after making trials. Of course Woods has simplified the procedure by giving normal design table which provides B/D ratio.

   Lacey gave very important wetted regime perimeter equation

$$P_w = 4.825\ Q^{1/2}$$

He of course admitted that value of constant in the above equation is in no way constant and varies from 4 to 5.8 for regime channels.

9.14. DEFECTS IN KENNEDY'S AND LACEY'S THEORIES

Both the theories lack in the following respects:

1. In both the theories perfect definitions of silt grade and silt charge are not given.
2. It is clear that water loss will be taking place in the channel (say 12 to 15 per cent of total discharge). In both the theories the effect of this loss on the silt distribution is not considered.
3. Even though regime condition is reached the silt grade is never constant due to attrition. No account is made for this reduction in silt grade.

4. Branches, distributaries, outlet take their proper share from the parent canal progressively. None of the two theories mention what will be the effect of this water withdrawal on the regime.

9.15. SEDIMENT TRANSPORT AND ITS DETERMINATION

It is a matter of common experience that the flowing water invariably carries fragmented rock and soil particles with it. Obviously knowledge of sediment transported by water is essential while designing water conveyance systems. As velocity of flow varies on any vertical plane so also distribution of sediment is variable. It may be noted that pattern of sediment distribution is, however, not similar to velocity distribution. The sediment concentration in the lower layers of flow is more than in the upper layers. Roughly average concentration is found to be at 0.6 D. Owing to its weight each particle tries to settle down on the bottom. The eddies generated in turbulent flow are responsible for supporting and keeping the particles in suspension. The sediment transported by water can be divided into two categories, namely:

- Suspended sediment load; and
- bed load.

9.15.1. Suspended Sediment

The soil particles which are transporated by water without coming in contact with the bottom of the channel is called suspended sediment. The particles are kept in suspension by the upward component of turbulent current. It is of course true that some particles drop down on the bed whereas some particles are picked up by the flow. In turbulent flow the rising eddies transfer sediment from heavy sediment concentration bottom layers towards top. On the other hand particles settle down under the force of gravity. Under stable condition the sedimnent transferred upward balances with that fallen on the bottom. Also the weight of suspended sediment exerts additional pressure on the bed of the channel which is in excess of fluid pressure.

The suspended load concentration 'C' at a height 'y' above the bottom can be determined from the known concentration "C_a" at a reference point at height "a" above the bottom. The equation given is

$$\frac{C}{C_a} = \left[\left(\frac{D-y}{y}\right)\times\left(\frac{a}{D-a}\right)\right]^{\frac{w}{KV}}$$

where D is depth of water

w is fall velocity of a grain in still water

K is Von Karman's universal constant = 0.4

V is shear velocity = $\sqrt{\tau_o/\rho}$

ρ is mean density of water

and τ_0 is intensity of shear stress at bottom

Evaluation of total sediment load per metre width of the channel can be made by integrating product of velocity and concentration over the entire depth.

9.15.2. Bed Load

It is that part of sediment which moves along the bottom of the channel. The grains move forward by rolling, sliding or jumping along the bed. The movement of sediment along the bed is mainly due to fluid drag. It is the total tangential component of weight of water in unit length of channel. It is given by expression:

$$\text{fluid drag} = v_w\ A\ S \qquad ...(1)$$

where υ_w = is unit weight of water;
A is cross sectional area; and
S is bed slope

Tractive force is the fluid drag per unit area and is given by dividing A by wetted perimeter P.

Thus, $\tau_0 = \upsilon_w RS$

For wide channels $R \simeq D$

$\therefore$ $\tau_0 = \upsilon_w DS$

When the value of tractive force is such that the grains just start moving it is called critical tractive force and is denoted by term "τ_{cr}".

For wide channels with smooth bed τ_{cr} is given by relation

$$\tau_{cr} = 0.047\ (\upsilon - \upsilon_w)\ d$$

where υ is unit weight of sediment
and d is grain diameter.

Thus it may be seen that the rate of bed load transport is a function of the difference of τ_0 and τ_{cr}. It is of course not so straight forward because with an increase in tractive force the bed forms undergo changes and ripples are formed. These ripples create form resistance and absorb part of tractive force. Two equations generally used in determining rate of bed load transport are given by Meyer-Peter and Einstein.

9.16. MEYER-PETER'S EQUATION

It states that the bed load transported by water in kilogram per metre width is given by equation

$$q_B = 4666 \left\{ \tau_0 \left(\frac{n'}{n} \right)^{3/2} - \tau_{cr} \right\}^{3/2}$$

where q_B is the rate of transport of bed load in kg/m/hr.

τ_o is tractive force intensity on bed in kg/m^2

n' is Manning's coefficient for the grains on a plain bed without ripples. It can be obtained from equation

$$n' = \frac{(Ks)^{1/6}}{76}$$

K_s is effective grain diameter in mm. It equals mean grain diameter for closely spaced uniform grains. It may be taken as d_{65} or the value of the diameter than which 65% of the material is finer for graded sands.

n is actual value of mannings coefficient on the bed with ripples.

τ_{cr} is critical tractive force in kg/m^2

9.17. EINSTEIN'S EQUATION

Einstein adopted satistical approach and derived bed load function for equilibrium rate of bed load transport when number of deposited and scoured particles was same. He equated probability of particle being eroded to the probability that weight of lifted particle is less than submerged weight. In deriving this equation he has made number of assumptions and adopted many experimental coefficients. The probability P of bed particle movement is given by him as

$$P = 1 - \frac{1}{\sqrt{\pi}} \int_{-B_*\psi_*-1/\eta_0}^{B_*\psi_*-1/\eta_0} e^{-12} dt = \frac{A_*\phi_*}{1 - A\phi_*}$$

In the above relation all parameters like ϕ_*, ψ_*, η_0, A_*, B_* are constants. ψ_* is a dimensionless shear parameter whereas ϕ_* is dimensionsless transport parameter.

When the bed material is made up of uniform grain material various parameters reduce to $\phi_* = \phi$ and $\psi_* = \psi$ and so on.

Since the above relationship is cumbersome he further correlated two dimensionless parameters ϕ and ψ as $\phi_* = f(\psi_*)$ for practical purposes.

For uniform bed material the relationship was represented by a curve on semi-logarithmic plot with equation

$$0.465\ \phi = e^{-0.391\ \psi}$$

He gave value of ϕ by following equation:

$$\phi = \frac{q_B}{Gv_w}\left(\frac{1}{G-1}\right)^{1/2}\left(\frac{1}{gd^3}\right)^{1/2}$$

where G is specific gravity of the grains;
d is diameter of grains;
g is acceleration due to gravity
v_w is specific weight of water

Other symbols have the similar meanings already given earlier.

Also, he gave relation for ϕ as

$$\psi = (G - 1)\ \frac{d}{R'S}$$

where R' is hydraulic mean radius that would exist if bed was unrippled. When rugosity coefficient used represents granular roughness only R' can be calculated from Manning's equation.

To simplify the procedure he gave a curve on log-log paper as $\phi_* = f(\psi_*)$ for working use and is given in Fig. 9.5.

Einstein-Brown's Relation: Brown plotted the data on log-log plot and found that all data reduces to a single linear function of the form

$$\phi = \frac{40}{(\psi)^3}$$

This relationship is also found to be useful in calculating bed load transport in certain cases.

Problem 9.7. In a vary wide channel the concentration of suspended load was found to be 500 ppm at 0.4 m above the bed. If the velocity of fall of a grain in still water is 0.04 m/sec and bed slope of the channel is 1 in 4500 determine the suspended load concentration at 0.8 m above the bed of the channel. Take depth of flow as 2 m.

Solution: Step 1. C_a at 0.4 m above bed = 500 ppm = $500 \times 10^{-6} \times 10^3 = 0.5$ kg/m^2

Step 2. Shear velocity $V = \sqrt{\dfrac{\tau_0}{\rho}}$

Since $\tau_0 = v_w\ RS$

$$R \approx D \quad \text{and} \quad \rho = \frac{v}{g}$$

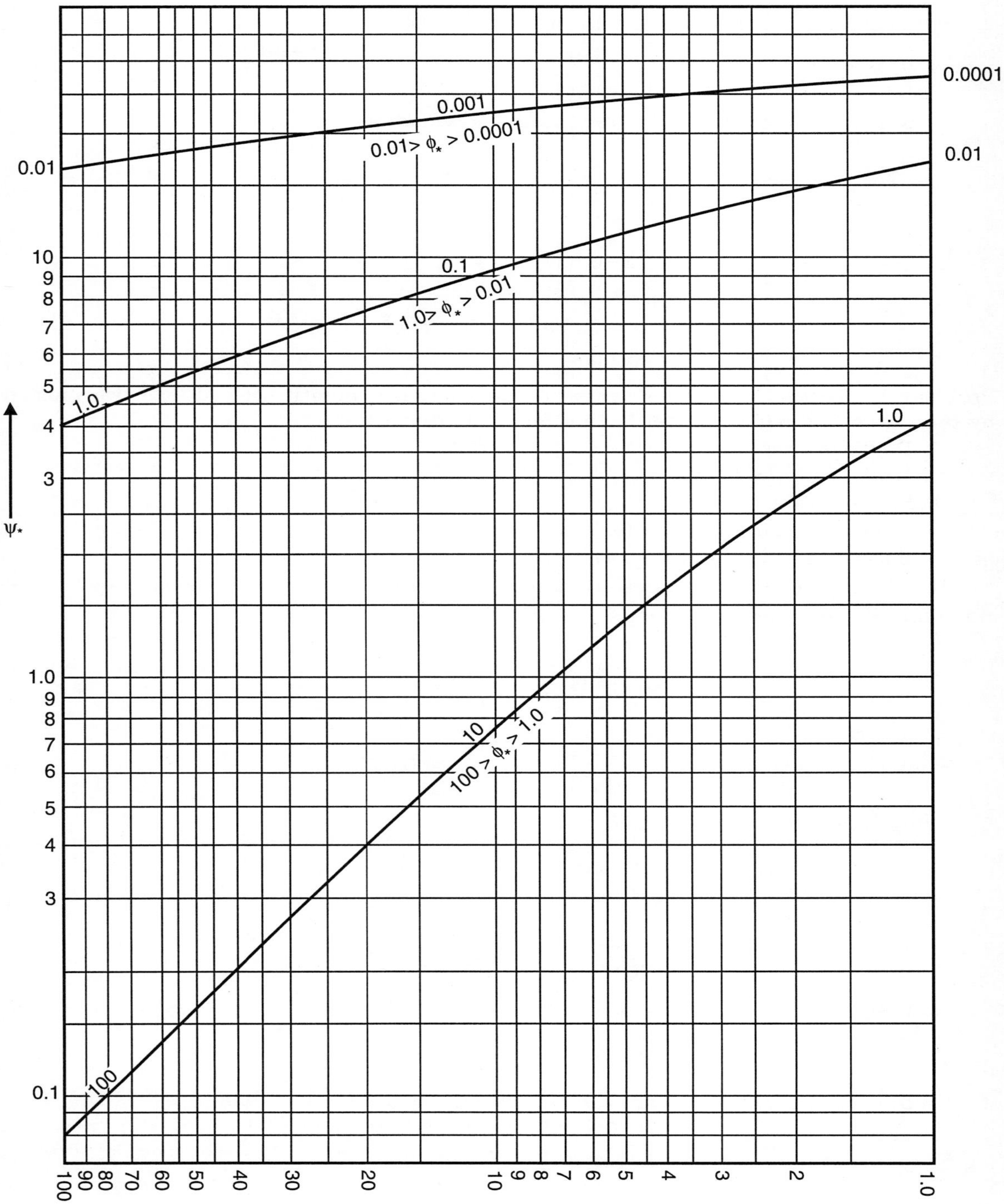

Fig. 9.5. $\phi^* - \psi^*$ curve

$$V = \sqrt{gDS} = \sqrt{\frac{9.81 \times 2}{4500}} = \sqrt{0.00436} = 0.066 \text{ m/sec}$$

Step 3. Von Karman's Universal constant $K = 0.4$

$$\therefore \quad \frac{w}{KV} = \frac{0.04}{0.4 \times 0.066} = 1.515$$

Step 4. Putting all values in suspended load equation

$$\frac{C}{C_a} = \left\{\left(\frac{D-y}{y}\right) \times \left(\frac{a}{D-a}\right)\right\}^{\frac{w}{KV}}$$

$$\frac{C}{0.5} = \left\{\left(\frac{2-0.8}{0.8}\right) \times \left(\frac{0.4}{2-0.4}\right)\right\}^{1.515} = 0.226$$

$\therefore \quad C = 0.5 \times 0.226 = 0.113$ kg/m^3

or $\quad C = 113$ ppm

Problem 9.8. Using Meyer-Peter's equation calculate the quantity of bed load transported by the water in a wide channel having following dimensions:

1. **Depth of flow 4 m**
2. **Bed slope 1 in 4300**
3. **Size of grain 0.3 mm**
4. **Effective grain diameter (Ks) 0.58 mm**
5. **Specific gravity of grain 2.47**

Also determine bed load transported using Einstein's equation and compare the results.

Solution: Meyer-Peter's equation is

$$q_B = 4666 \left[\tau_0 \left(\frac{n'}{n}\right)^{3/2} - \tau_{cr}\right]^{3/2}$$

Step 1. Using relation $n' = \frac{Ks^{1/6}}{76} = \frac{(0.58)^{1/6}}{76} = 0.012$

Using Lacey's equation $f = 1.76\sqrt{m_r}$

$$f = 1.76\sqrt{0.3} \simeq 1.0$$

So $\quad n = 0.0225$

$$\therefore \quad \left(\frac{n'}{n}\right)^{3/2} = \left(\frac{0.012}{0.0225}\right)^{3/2} = 0.39$$

Step 2. $\quad \tau_0 = v_w D S \quad$ Since channel is wide

$$= 1000 \times 4 \times \frac{1}{4300} = 0.93 \text{ kg/m}^2$$

Step 3. $\quad \tau_{cr} = 0.047 (v - v_w)d = 0.047 (2.47 - 1) \times 1000 \times 0.3 \times 10^{-3}$

$$= 0.047 \times 1.47 \times 0.3 = 0.021 \text{ kg/m}^2$$

Step 4. Putting all values in Meyer-Peter's equation.

$$q_B = 4666 \{0.93 \times 0.39 - 0.021\}^{3/2} = 4666 \{0.363 - 0.021\}^{3/2}$$

$$= 4666 \{0.199\} = 929 \text{ kg/m/hr} \quad \text{...(1)}$$

Comparing by Einstein's equation

Step 5. $$\psi = \left(\frac{G-1}{1}\right)\frac{d}{R'S}$$

For wide channel $R = D = 4$ m

Also from Manning's formula

$$\frac{R'}{R} = \left(\frac{n'}{n}\right)^{3/2}$$

$$R' = 4 \times 0.39 = 1.56$$

$$\therefore \quad \psi = \frac{\left(\frac{2.47-1}{1}\right)3\times10^{-4}\times4300}{1.56} = \frac{1.47\times3\times43}{1.56} = \frac{189.63}{1.56} = 1.22$$

Step 6. From Fig. 9.5 for $\psi = 1.22$

$$\phi = 5.5$$

Step 7. Using relation

$$\phi = \frac{q_B}{Gv_w}\left(\frac{1}{G-1}\right)^{1/2}\cdot\left(\frac{1}{gd^3}\right)^{1/2}$$

q_B can be calculated

$$\therefore \quad 5.5 = \frac{q_B}{2.47\times1000\times\sqrt{1.47}\times\sqrt{9.81}\times(0.0003)^{3/2}\times3600}$$

$$= \frac{q_B}{2.47\times1.21\times3.13\times0.0000052\times3600\times1000} = \frac{q_B}{175.24}$$

$$\therefore \quad q_B = 963 \text{ kg/m/hr}$$

Answer: Rate of bed load transport

by Meyer-Peter's method = 929 kg/m/hr

by Einstein's method = 963 kg/m/hr

Problem 9.9. Find out regime slope for a canal with following data:

(*i*) FS discharge = 30 cumec

(*ii*) Bed load concentration = 50 ppm by weight

(*iii*) Average grain size = 0.3 mm

(*iv*) Effective grain diameter = 0.58 mm = (K_s)

(*v*) Specific gravity of grain = 2.67

Use Meyer-Peter's equation.

Solution: Step 1. Bed load transported by water per hour = $50 \times 10^{-6} \times 10^3 \times 30 \times 3600$ = 5400 kg/hour

Step 2. Using Lacey's regime perimeter formula

$$P_w = 4.825\sqrt{30} = 26.43 \text{ m}$$

Against this value of P_w assuming bed width to be 22 m

$$q_B = \frac{5400}{22} = 245.5 \text{ kg/m/hr}$$

Step 3. $$\tau_{cr} = 0.047\,(v - v_w)d = 0.047\,(2.67 - 1) \times 10^3 \times 0.3 \times 10^{-3} = 0.024 \text{ kg/m}^2$$

Step 4. $$n' = \frac{Ks^{1/6}}{76} = \frac{(0.58)^{1/6}}{76} = 0.012$$

Using Lacey's equation, $f = 1.76\sqrt{m_r} = 1.76\sqrt{0.3} \simeq 1.0$

$\therefore \quad n = 0.0225$

The ratio, $$\left(\frac{n'}{n}\right)^{3/2} = \left(\frac{0.012}{0.0225}\right)^{3/2} = 0.39$$

Step 5. $\tau_0 = v_w\, R\, S = 1000\, RS$

Step 6. Putting all values in Meyer-Peter's equation

$$245.5 = 4666\,\{1000 \times R \times S \times 0.39 - 0.024\}^{3/2} \quad ...(i)$$

Step 7. Assuming Area = $26 \times R$

From Manning's formula

$$Q = \frac{1}{0.0225} \cdot R^{2/3}\, S^{1/2}\, A$$

$$30 = \frac{1}{0.0225} \times R^{2/3} \times S^{1/2} \times 26 \times R \quad ...(ii)$$

Step 8. Simplifying (*i*) and (*ii*)

$$R\,S = 4.2 \times 10^{-4} \quad ...(iii)$$

$$R^{10/3}\,S = 6.74 \times 10^{-4} \quad ...(iv)$$

Step 9. Treating (*iii*) and (*iv*) as simultaneous equations and solving for R and S.

$R = 1.23$ and

$S = 0.000343$ (1 in 2900)

Step 10. Assuming slide slope to be1/2 : 1 (Horizontal : Vertical) B and D can be calculated as follows:

$P_w = 26.4$ m as already worked out

$A = P_w \times R = 26.4 \times 1.23 = 32.5\ \text{m}^2$

$$32.5 = BD + D^2/2 \quad ...(v)$$

$$26.4 = B + \sqrt{5}\, D \quad ...(vi)$$

Solving them as simultaneous equations

Bed width = 23.38 m and

Depth = 1.35 m

9.18. UTILITY OF SEDIMENT TRANSPORT EQUATIONS

It is already mentioned that in an unlined channel where velocities are such that channel section does not undergo any change the channel section may be called stable or rigid. A rigid channel can be designed using Chezy's and Manning's formulae. In practice, however such channels are rare. The channel bed material of unlined canal normally becomes mobile with changing flow conditions. When the bed material starts moving various complexities arise. They are:

- With the movement of sediment, bed form starts changing, it increases boundary resistance to flow.
- As the soil particles get dislodged from the bed and sides, extra energy is required by the flow of water to carry the sediment particles.

LONGITUDINAL SECTION

Scale = HOR-1cm=1600m
VER-1cm=45m

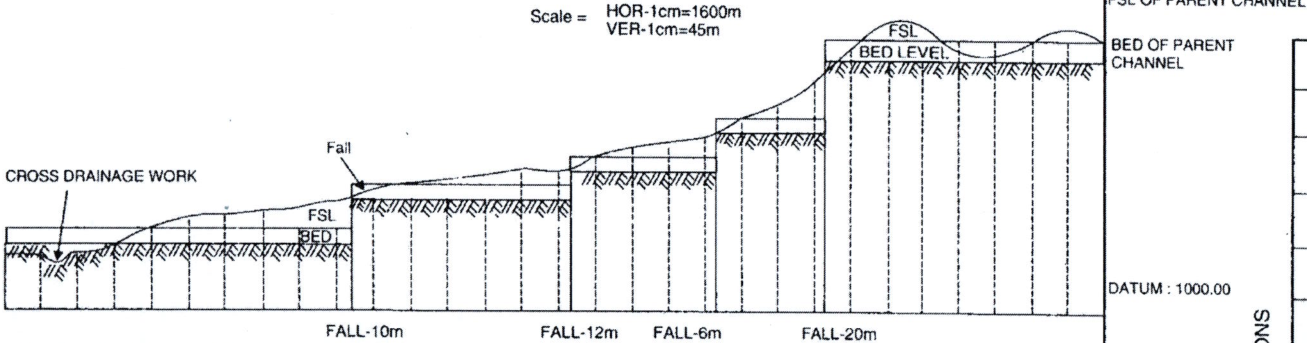

R.D.	N.S.L.	F.S.L.	BED RL.
30000	1148.00	1151.75	1148.05
29000	1148.00	1151.99	1148.30
28000	1148.00	1152.05	1148.35
27000	1150.00	1152.20	1148.50
26000	1153.40	1152.35	1148.65
25000	1153.00	1152.50	1148.80
24000	1154.00	1152.65	1148.95
23000	1155.00	1152.80	1149.10
22000	1157.00	1152.95	1149.25
21000	1160.00	1153.10	1149.40
20000	1162.00	1163.20	1159.50
19000	1164.00	1163.35	1159.65
18000	1165.00	1163.50	1159.80
17000	1168.00	1163.65	1159.95
16000	1170.00	1163.80	1160.10
15000	1172.00	1163.95	1160.25
14000	1174.00	1176.10	1172.40
13000	1176.30	1176.25	1172.55
12000	1176.50	1176.40	1172.70
11000	1177.50	1176.55	1172.85
10000	1181.00	1182.70	1179.00
9000	1185.00	1182.85	1179.15
8000	1195.00	1202.00	1198.30
7000	1207.00	1202.15	1198.45
6000	1207.00	1202.30	1198.60
5000	1207.00	1202.45	1198.75
4000	1201.00	1202.60	1198.90
3000	1202.00	1202.75	1199.05
2000	1202.20	1202.90	1199.20
1000	1203.80	1203.05	1199.35
0000	1203.50	1203.20	1199.50

R.D. (reach)	LAND WIDTH	VELOCITY	SLOPE	LACE Ys 'f'	OUTLETS	FS DISCHARGE	BED WIDTH	FS DEPTH
30000–25000	50m	1m/Sec.	0.15/1000	1.0	12 cm, 15 cm, 12 cm	21m³/Sec	16 m	2.44 m
24000–20000	50 m	1m/Sec.	0.15/1000	1.0	12 cm, 12 cm, 15 cm, 15 cm	22.5 m³/Sec	16.5 m	260 m
19000–15000	52 m	1m/Sec.	0.15/1000	1.0	12 cm, 12 cm, 15 cm, 15 cm	23 m³/Sec	18.5 m	275 m
14000–10000	55m	1m/Sec.	0.15/1000	1.0	12 cm, 15 cm, 12 cm, 12 cm	24 m³/Sec	17 m	3.00 m
9000–5000	55m	1m/Sec.	0.15/1000	1.0	12 cm, 45 cm, 15 cm	25m³/Sec	175 m	3.40 m
4000–0000	55m	1m/Sec.	0.15/1000	1.0	12 cm, 15 cm, 12 cm	27m³/Sec	18 m	3.70 m

Fig.: 9.6. Longitudinal section of a canal

Plate 2 (Ch-9)

SCHEDULE OF AREA STATISTICS AND CHANNEL DIMENSIONS

1	2	3	4	5	6	7	8	9	10	11	12	13	14	15	16	17	18	19
			AREA TO BE IRRIGATED								CHANNEL DIMENSIONS							
BELOW Km	GROSS COMMANDED AREA	CULTURABLE COMMANDED ASEA	RABI	PERENNIAL	KHARIF	OUTLET DISCHARGE FACTOR	CUTLET DISCHARGE	LOSSES IN REACH	TOTAL LOSSESE	TOTAL DISCHARGE	BED SLOPE	BED WIDTH	F.S. DEPTH	F.S. DISCHARE	HEIGHT OF BANKS	WIDTH OF BANKS	VELOCITY	V/vo or *f*
	Hectares	Hectares	Ha	Ha	Ha	m³/Sec	m³/Sec	m³/Sec	m³/Sec	m³/Sec	I	m	m	m³/Sec	m	m	m³/Sec	I
1																		
2																		
3																		
4																		
5																		

- Suspended sediment generated by turbulance in flow further increases resistance to the flow.
- Effect of change of shape of channel.
- Effect of non-uniformity of sediment sizes.
- Effect of variation in channel discharge.

All these issues have not been fully investigated and incorporated in the analytical resistance equations developed so far. Hence accurate mathematical method of designing irrigation canals is not forthcoming. In India, therefore, mostly Lacey's equations are used to design the irrigation canals and has been found quite adequate. It is so because these equations are based on long observations in regions of India where canals were under operation for a very long period.

9.19. LONGITUDINAL SECTION OF IRRIGATION CHANNEL

After desining the irrigation channel, it is essential to draw longitudinal and cross-section at suitable points for proper execution of work and for giving full information at a glance.

First step in a construction programme is to mark an alignment on the contour plan or on shajara sheet. The alignment is then scaled generally in km and its fractions. That is starting from the head of the canal, reduced distances are marked on the alignment and the natural surface levels (NSL) are interpolated for each section.

The scales generally selected for drawing longitudinal section are :

$$1 \text{ cm} = 160 \text{ m as horizontal scale}$$

and

$$1 \text{ cm} = 0.6 \text{ m as vertical scale}$$

Of course it is true that the scales may be changed according to the nature of topography.

The plotting of longitudinal section (*L* section) may be done in the following successive steps. (Fig. 9.6. Plate 2) shows in typical *L* Section of an irrigation channel. The section is not drawn to scale in order to include all details in one sheet.

1. The first step in plotting *L* section is to plot the natural surface on the sheet as per scales selected. Convenient datum level may be selected for the purpose arbitrarily.
2. For reference the FSL of a source or a parent channel and its bed level is also plotted.
3. Next step is to plot the full supply line of the channel. While plotting FSL due consideration should be given to the facts mentioned below. Starting from a take-off point:
   (*a*) When an irrigation channel takes off from the parent channel or a source sufficient working head should be provided. Consideration should be given to: (*i*) silting of the canal, (*ii*) Losses at entrance, (*iii*) future expansion programme of irrigation. For this purpose the FSL of the channel should be fixed at least about 15 cm below the FSL of the parent channel or source. For very big canals working head upto 50 cm is provided at off-take point.
   (*b*) So far as possible the FSL should always be above NSL to give gravity flow. Of course this condition need not be satisfied too strictly because then it may become necessary to run the irrigation canal in filling which may in turn increase the cost and losses. Generally FSL is kept 15 to 30 cm above NSL As the irrigation channel runs over ridge this allowance is sufficient.
   (*c*) The FSL should be fixed in such a way that the canal is mostly in balancing depth. As then the earth excavated from the channel section may be used to construct the banks.

Since payment for the earthwork is made only once either for cutting or for filling, it makes the project economical.

(*d*) It is alread mentioned that though Lacey's equation gives slope for regime channels the general NSL fixes the value of bed slope. Best way is to obtain the bed slope first from Lacey's regime equation for slope and then modify it to suit the site conditions.

(*e*) When it is seen that NSL falls much faster than the slope of the full supply line then it becomes necessary to provide falls. When the falls are provided then FSL downstream of the fall may be carried below NSL for a short length. As then irrigation of land downstream of a fall may be done from outlets situated upstream of the fall.

It is needless to stress that none of the above consideration should be taken as a binding principle in fixing FS line.

Depending upon the conditions prevailing at the site golden mean should be adopted judiciously.

Thus FS line is marked finally on the *L* section.

4. As a fourth step a table is prepared which gives full information about the area to be irrigated per km in different seasons, discharges and outlet factors and channel dimensions. The table is called a **schedule of area statistics and channel dimensions.** The table is generally prepared on the same sheet where *L* section is drawn. Complete columns of the table are shown in Fig. 9.6, in a tabular form. This makes the procedure of filling the table very clear.
5. The information from this table is then filled in the rows below the *L* section as shown in Fig. 9.6. Thus for each km length of the channel relevant information is filled in the respective rows.
6. Knowing the FS depth, a bed line may be plotted on the sheet. The bed line is parallel to FS line and distance between the two lines at any point is equal to the FS depth there.
7. Position of outlets, datum line and land widths are also shown in *L* section.

Thus, the *L* section of an irrigation channel can be drawn completely to give detailed information.

9.20. CROSS-SECTIONS OF IRRIGATION CHANNELS

The *L*-section does not provide the information about the actual shape of the canal at any site. Hence, it becomes necessary to plot the cross-sections for typical sites. For example,

- Canal in full cutting.
- Canal in full filling.
- Canal partially in cutting and partially in filling.

The irrigation canals convey water to the irrigable area in such a way that the water level in the canal mostly remains above the natural ground level. It helps in maintaining gravity flow. The ground level along the selected alignment is never uniform. At one place the ground level may be very high and at some other place it may be low. Sometimes the slope of the ground may be very steep when compared with the canal slope. In order to maintain the water level in the canal above the field level the canal may have to be constructed either in full cutting or in full embankment or in partial cutting and partial embankment. A canal is said to be in full cutting when the natural surface level along the alignment is above the water level of the canal. A canal is said to be in full embankment when the natural

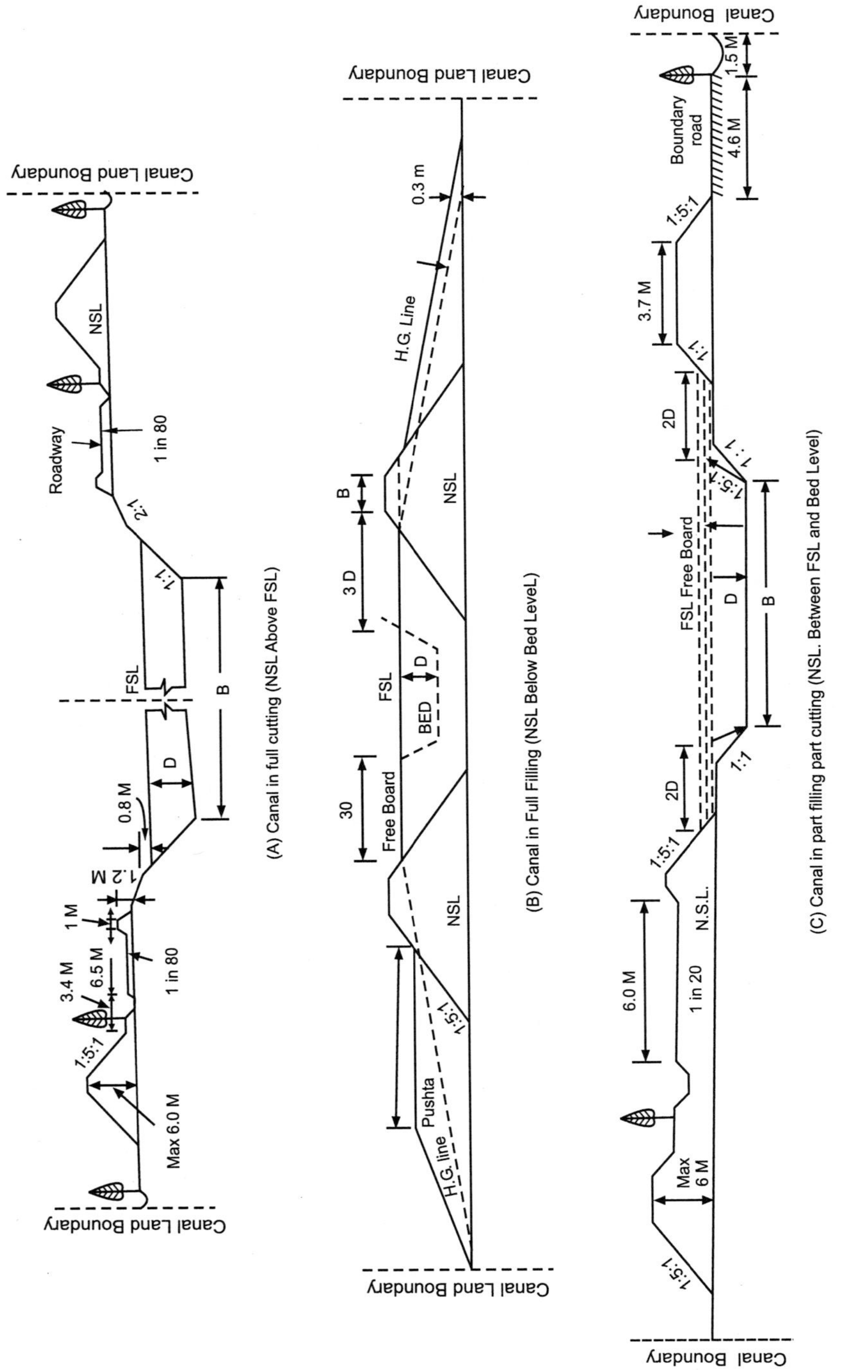

(A) Canal in full cutting (NSL Above FSL)

(B) Canal in Full Filling (NSL Below Bed LeveL)

(C) Canal in part filling part cutting (NSL. Between FSL and Bed Level)

Fig. 9.7. Typical cross-sections of unlined main canal

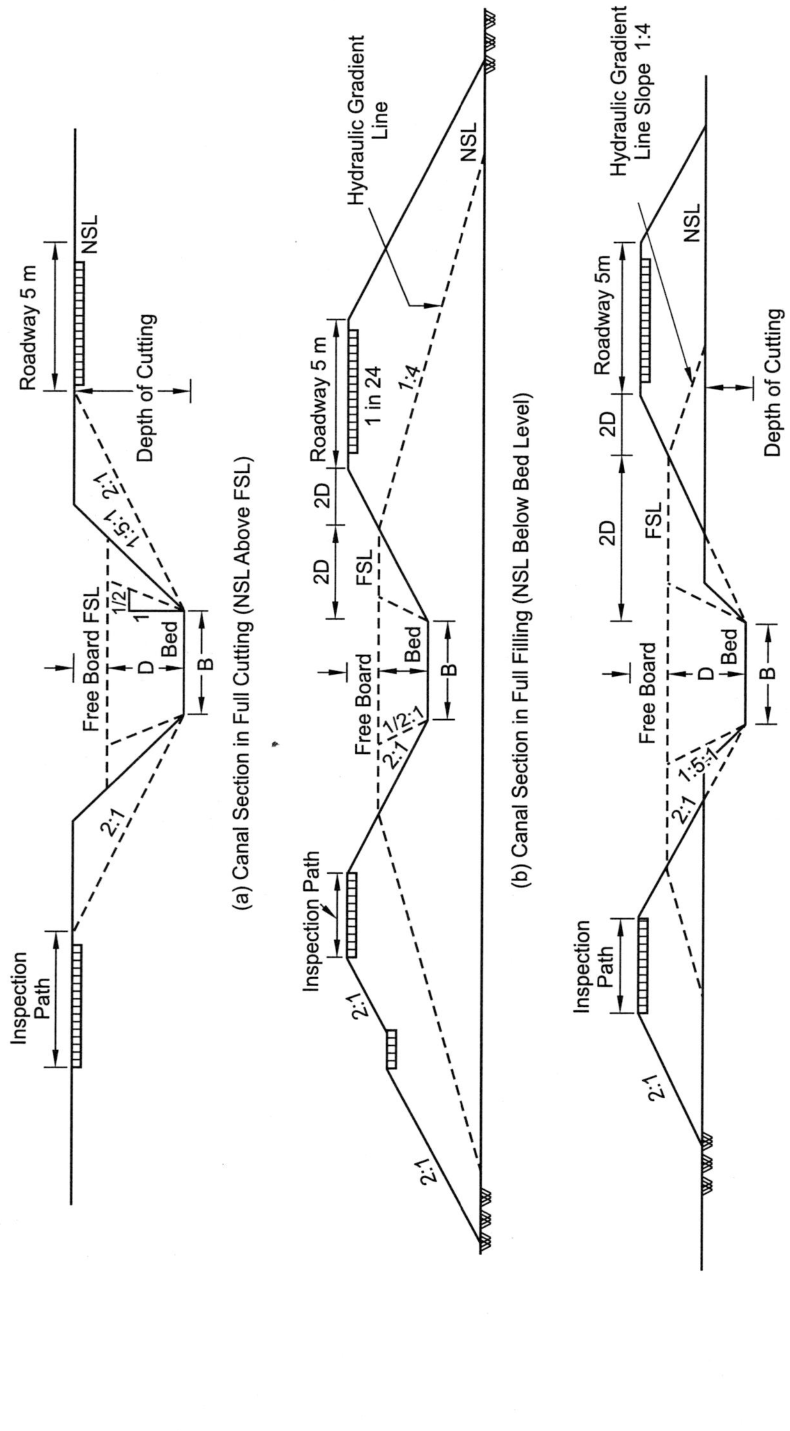

(a) Canal Section in Full Cutting (NSL Above FSL)

(b) Canal Section in Full Filling (NSL Below Bed Level)

(c) Canal Section in Partial cutting (NSL in Between FSL and Bed Level)

Fig. 9.8. Typical cross-sections of unlined distributary canal

surface level along the alignment is below bed level of the canal. When the natural surface level comes beteen the water level and the bed level of the canal, a canal is said to be in part embankment and part cutting. In principle the construction features and various components of all irrigation canals (viz. main canal, distributary canal etc.) are common. But since the main canal has to contain large discharge its section is bigger. Obviously the main canal sections of various type involve provision of almost all component parts. In case of a distributary the sections are rather simple because it carris small discharge. Figures 9.7 and 9.8 show the typical cross-sections of an unlined main canal, and unlined distributary canal respectively.

A canal cross-section consists of following main parts (Fig. 9.9).

- Main channel section.
- Berms
- Banks
- Dowels
- Roadways
- Drains.

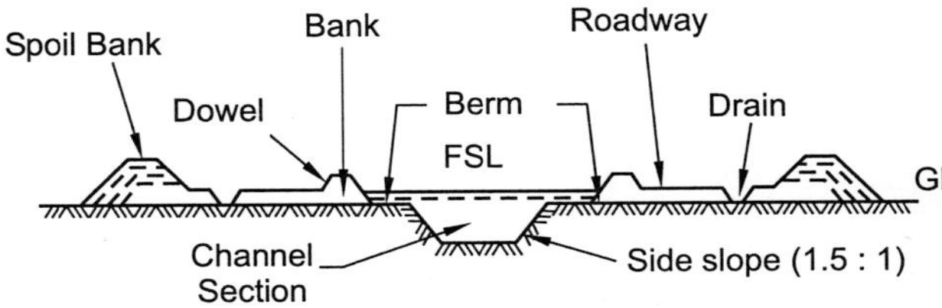

Fig. 9.9. Typical cross-section of canal showing component parts

QUESTIONS FOR REVISION

1. Explain Kennedy's silt theory in brief. How irrigation channel can be designed by this theory?
2. Lacey's theory is an improvement over Kennedy's theory. Explain.
3. What are the defects in Kennedy's and Lacey's theories?
4. (*a*) What factors are to be taken in consideration while fixing FSL in *L*-section?
   (*b*) What are general considerations for canal alignment?
5. Draw three typical cross-sections for irrigation channels :
   (*a*) Full cutting
   (*b*) Full filling
   (*c*) Partially in cutting and partially in filling.

10

Lining of Irrigation Canals
(Design of Lined Canals)

10.1. PURPOSE OF CANAL LINING

Implementation of irrigation scheme which includes extensive distribution system is an expensive proposition. Hence it is very necessary to ensure that large losses do not occur in transit due to seepage. In practice, however, considerable loss of water takes place due to seepage in earthen canals. This loss is unavoidable unless the canal is lined.

An irrigation canal is said to be lined when the bed and the sides of the canal cross-section are protected with imprevious or fairly impervious material of sufficient strength.

Thus main aim of lining the irrigation canals with impermeable material is to stop seepage, thereby saving valuable irrigation water. When canal is lined considerable additional area can be commanded with the help of the saving resulted from lining the canals. When a canal is to be lined the canal may be built to much smaller cross-section than an unlined one as there is practically no loss in the lined canals due to seepage. Condition of canal is also improved a lot and as a result it requires less maintenance. Experience has shown that lined canals check waterlogging in the areas adjoining the canals.

Existing camals may also be lined, but then the construction work may interrupt the irrigation supplies. The canal in head reach does not need lining. The obvious reason is the seepage water is not lost totally. It goes back to the river. However, sometimes it may become necessary to line the canal in the head reach also. Then the maintenance of the canal becomes an important problem as the waters table in the head reach is usually high and water behind the lining may create unstable conditions.

The lining of existing or proposed irrigation canal is justified economically when the cost of the lining can be recovered from the increased monetary benefits in a reasonably short period.

10.2. ADVANTAGES OF CANAL LINING

It will not be an exaggeration to say that lining has advantages to its credit than disadvantages. It is of course true that as other structures it may also create troubles due to faulty design, construction and maintenance procedures. But the benefits derived from canal lining are so great that the disadvantages may be overlooked. The main benefits derived by lining the canal are mentioned below.

1. **Canal lining controls seepage:** It is already mentioned that the seepage losses in unlined canal system may amount to 40% of the discharge. Obviously if all the

seepage losses could be prevented then it is possible to irrigate 40 more hectares per every 60 hectares now under irrigation. It is needless to say that this additional command is free of cost so far as cost of the water is concerned.

2. **Canal linings increase capacity of canals:** When the canal is lined the bed and sides and rendered smooth. Obviously resistance to flow is less and consequently velocity of water is more. The general formula for velocity of flow in channels $V = C\sqrt{RS}$ makes it more clear. Value of C in the Chezy formula is given by Kutter's formula.

$$C = \frac{23 + \frac{1}{N} + \frac{0.000155}{S}}{1 + \left(23 + \frac{0.000155}{S}\right)\frac{N}{\sqrt{R}}}$$

From the study of above mentioned formula it is clear that C is inversely proportional to N (coefficient of rugosity). Obviously when N is small the value C will be more. For unlined canal average value of N is 0.0225 whereas for concrete lined canal N is 0.014. As a result higher velocities are maintained in lined canals. Thus water discharging capacity of the canal is also increased.

3. **Canal lining is an important anti-waterlogging measure:** When canals are lined excessive seepage is prevented and in turn it prevents addition of water to the unders ground reservoir. Thus it avoids deterioration of cultivable land by water-logging.

4. **Canal linings improve command:** It is already mentioned that lining makes it possible to maintain high velocity in the canal. Generally lined canals are dsigned with normal value of silt factor (f = 1.2) and maximum permissible velocity (from 1.8 m/sec to 2.5 m/sec). For these values of f and V steepest practicable longitudinal slope may be adopted. Even then this slope will be flatter than the slope of unlined canal. Thus this flat slope raises the FSL in general and brings the high patches of land under command which have not been irrigated before. Another advantage of maintaining high velocities can be recognised from the fact that lined canals require smaller sections to convey same amount of water as that carried by unlined canals.

5. **Canal linings increase available head for power generation:** When the power canals are lined the slope adopted is very flat. Then it prevents loss of head due to excess slope. As a result available head at power house is increased.

6. **Canal linings make the canal section stable:** When the sides and the bed of a canal are protected with sufficiently strong material a canal may be said to be in permanent regime. It protects the canal section against the forces which may try to change the regime conditions.

7. **Canal linings prevent bank erosion and breaches:** For unlined canals occurrence of breaches and bank erosion are problems of much concern. But in lined canals they crete no problems as the sides are protected.

8. **Canal linings assure economical water distribution:** When canals and field channels are lined the water available may be distributed easily and economically to the cultivators as planned. Experience has shown that lining and constructing permanent turnouts may reduce labour requirement for water distribution by about 50 percent.

9. **Canal linings reduce maintenance costs:** As weed growth is totally prevented in lined canals a major portion of this class of maintenance is eliminated.

Since the canal is lined erosion of bed and sides is totally checked. Moreover velocity is so designed that silting does not occur. This eliminates considerably the annual maintenance expenditure necessary in unlined canals for removal of silt.

The lining is generally designed to resist the water pressue behind the lining. Backfill is also sufficiently compacted to avoid settlement. Thus maintenance required for minor repairs is also substantially reduced. Experience has shown that maintenance cost of a lined canal is about 25 per cent that of a similar unlined canal.

10.3. DRAWBACKS OF CANAL LINING

Though the benefits derived from the lined canal are so great there are some disadvantages also. They are the following:

sym

- Once the lining is damaged it is difficult to repair it.
- The lined canals are constructed without berms. Moving vehicles, pedestrains, animals are liable to fall in the canal directly in the absence of safety berms.
- The lined canal is a permanent structure hence it is difficult to shift the outlets often.
- Lining of a canal can be done only after incurring heavy expenditure.

10.4. ECONOMICS OF CANAL LINING

Though lining has advantages to its credit a lined canal costs 2 to 2.5 times as much as an unlined canal. Hence, it is very essential to make sure before lining the canal that the cost incurred on lining is recoverable during the life time of the canal. While considering the economics of a canal lining it is necessary to evaluate the actual benefit and to compensate them with the cost of the lining. If the value of the annual benefits exceeds the annual cost of the lining then construction of the canal lining may be considered to be a good investment. The indirect benefits may be considered as an extra addition to saving which will provide sound economy to a lining project.

The annual cost of canal lining includes:

- Annual interest charge on the average investment during the life time of the lining.
- annual depreciation chages.

Let

P be the first cost of lining per km

r be the rate of interest

n be the estimated useful life of canal in years (before renewal of canal is required)

d be the anuual depreciation

i be the average annual interest charge.

Then

$$\text{Annual depreciation charge, } d = \frac{P}{n} = \frac{\text{First cost}}{\text{Useful life}}$$

and

$$\text{Average annual interest charge, } i = \frac{P}{2} \,.\, r = \text{average cost} \times \text{rate of interest}$$

The obvious reason for taking average cost is that the first cost is reduced every year due to the process of charging depreciation. The reduction in the cost each year will be equal to the annual depreciation charge.

Then average annual cost of lining per km is

$$d + i = \frac{P}{n} + \frac{P}{2} \cdot r \qquad \text{...(1)}$$

The annual benefits actually derived by lining the canal include:

- Saving in seepage loss
- Saving in maintenance and operation cost
- Other additional benefits.

Let

A be the saving in seepage loss per km length of lined canal

B be the saving in maintenance and operation cost per km

C be the additional benefits.

Then total annual benefits per km $= A + B + C$ ...(2)

For economical justification annual benefits in (2) should be sufficiently more than annual cost in (1).

It may be mentioned here specially that if lining is planned before constructing the canal much greater saving would be possible. Thus because of greater overall savings, the need for lining the canal should be weighted carefully well in advance and if required and considered proper should be included in the original construction programme.

10.5. TYPES OF LINING

The canals can be made fairly watertight by lining the canal section with various materials. The materials which are commonly used for lining are cement, bricks, puddle clay, stone blocks, sodium carbonate, asphalt and road oils. The various types of linings vary in first cost, construction procedure. They are described below.

10.6. CEMENT CONCRETE LINING

There are various types of lining in which cement is used for lining the canal.

This term is reserved for either plain or reinforced plastic concrete made with cement, coarse and fine aggregate and water. Experience has shown that this type of lining is very durable. It is capable of reducing the losses due to seepage by about 90 to 95 per cent.

No general rule can be stated for determining the thickness of the concrete linings. However, proper thickness of the lining should be adopted to fulfil two basic requirements, namely:

- Water-tightness, and
- Sufficient structural strength to resist cracking.

It has been proved by experience that for small canals plain concrete lining of 4 to 6 cm thickness is satisfactory. While for big canals lining may be provided upto thickness of 15 centimetres.

To relieve the canal of various unaccounted forces it is essential to limit the side slopes. The slopes adopted should be such that for the backfill the angle of internal fiction is not exceeded. Then the backfill will be stable and it will not exert any pressure on the lining. Thus the side slopes are limited to a range of 1.25 : 1 to 1.5 : 1. Sometimes side slopes as steep as 1 : 1 may be adopted.

Figure 10.1 gives the details of a concrete lining.

Construction of concrete lining: The first phase in constuction of concrete lining is preparation of subgrade. Ordinally subgrade should be sufficiently porous to permit drainage.

When subgrade is made up of the soil like clay in which great volume changes occur due to moisture it becomes essential to control moisture content of the soil.

Secondly subgrade should be sufficiently compact. When the lining is to be placed against a trimmed subrade in undisturbed soil no compaction is needed. But when the canal is in filling, the banks need compaction. The compaction can be done successfully at optimum moisture content with a sheep-foot roller.

It is seen that when concrete is placed on the subgrade directly the subgrade absorbs the water from the concrete. As a result the concrete is rendered weak due to moisture deficiency. Then curing is not satisfactory and concrete becomes porous.

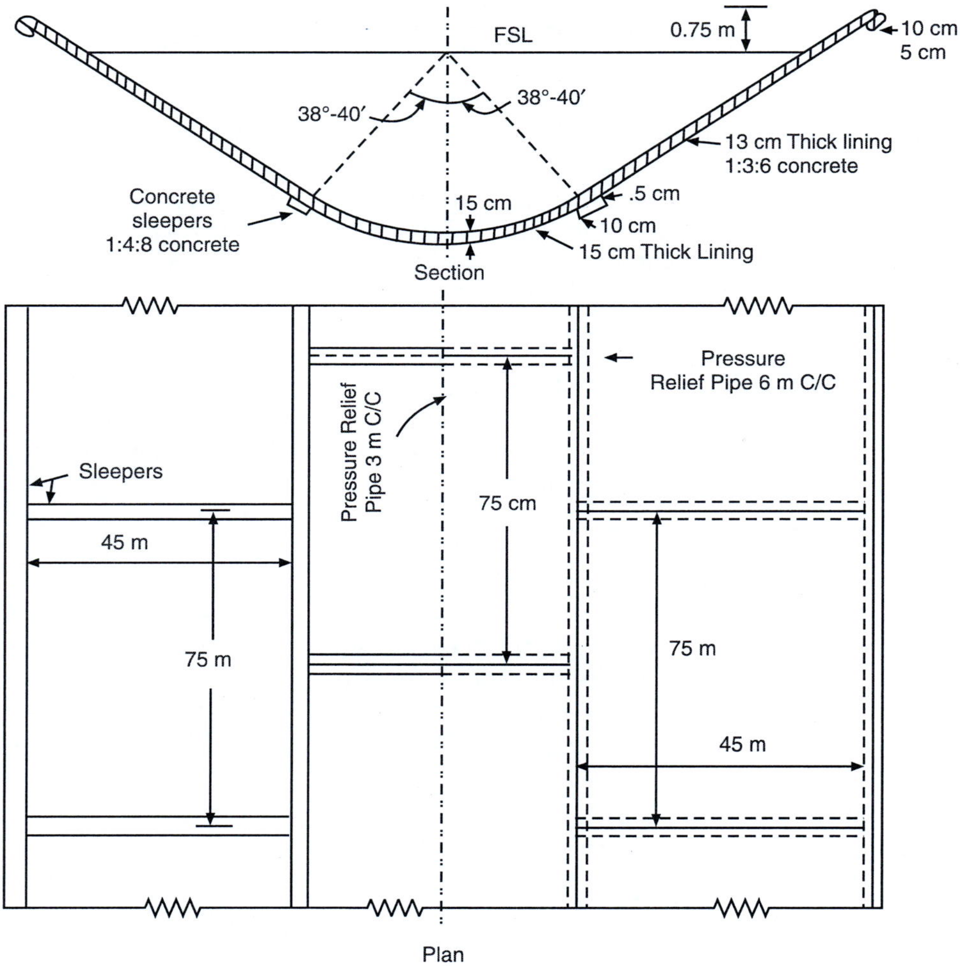

Fig. 10.1. Details of concrete lining

Obviously it is very essential to see that when the subgrade is made of very dry soil it should be sufficiently wetted before placing the concrete. Even this is not sufficient in many cases. Then other methods also are used as a supplement. They are:

(*a*) **Laying a base coat of 1 : 4 cement-sand slurry:** The slurry of 1 : 4 cement-sand may be applied to the subgrade to give a 3 mm thick base layer. This slurry is generally placed few hours before placing the concrete.

(*b*) **Spreading oil paper on the subgrade:** Before laying the concrete oil paper is spread on the subgrade. It has been experienced that it gives good results.

(*c*) **Spraying of crude oil on the subgrade:** Before placing the concrete crude oil may be sprinkled on the subgrade to give a good coating.

(*d*) **Applying 1 : 6 cement plaster on the subgrade:** Initially subgrade is plastered with 1 : 6 cement plaster before palacing the concrete. It is not necessary to make the surface of the plaster smooth.

It has been experienced that the method of applying cement-sand slurry is cheapest.

The second phase in construction of lining is actual placing of concrete. Mixing of concrete may be done either at site or at centrally located batching and mixing plant. Proportion of the concrete may be 1 : 3 : 6 or 1 : 5 : 6 or 1 : 4 : 8 depending upon the requirements. The concrete may be placed by hand or by machinery. When hand methods are used temporary form work is constructed and placed transversely across the canal. The size of the form work should be equal to the size of the blocks to be provided. Blocks are generally laid alternately. The time interval between two adjacent placements should be about 7 days. For big canals concrete may be laid making use of machinery. The machines travel on the rails. The rails are laid on the banks of canal. The trimining operation for subgrade may also be done with the help of a machine travelling on the same rails.

The third phase is curing of concrete and is of great importance. Proper curing may be accomplished in two ways. Firstly by covering the fresh lining with moist earth. Secondly, the concrete may be kept wet continuously for about 5 days.

In large canals, on the side slopes ladders are provided at suitale intervals along the length of the lined canals. The ladder serve dual purpose. Firstly, it helps during inspection of the dry canal. Secondly, it provides easy means to come out of the running canal safely to fallen human beings.

The ladders are provided on both banks, say about 250 to 300 m apart. On two banks they are arranged in a straggered manner. When the canal enters any underground structure, the ladders are invariably povided about 30 metres upstream of the entry point of the canal,

The ladder is made of smooth round mild steel bars, and the ladder rungs are constructed in the canal lining. The steel rungs are galvanised or coated with coal tar after installation to avoid its rusting.

10.7. OTHER TYPES OF LINING USING CEMENT

1. **Shotcrete lining:** It is a type of lining in which slurry of portland cement, processed sand and pure water is applied pneumatically through the nozles on the surface of the subgrade. Ordinarily a mixture of cement and sand (1 : 4) is used. The jet of cement mortar slurry is shot at the subgrade and hence this type of lining is called shotcrete. The thickness of this type of lining generally varies from 2.5 cm to 6.5 cm. This type of lining gives good working rate irrespective of the nature or condition of the subgrade. It works equally successfully on smooth, uneven or cracked subgrade surfaces. It is of course true that for irregular and cracked subgrade surfaces the amount required is more. This type is rich in cement and hence its cost is high. Satisfactory curing of shotcrete is very important requirement for successful lining. This type of lining has been found to be most successful.
2. **Precast concrete lining:** This type of lining is constructed with precast concrete slabs. The slabs are manufactured at a suitably located central place. The slabs may then be taken to the site at the time of construction. The size of concrete slab should be such that it can be handled by one or two men. The size may be 50 cm × 30 cm.

The thickness of the slab may range from 5 to 6.5 cm. The blocks are manufactured with some interlocking arrangement at ends.

The provision of a suitable joint ensures continuity of the lining. The slabs are then laid on a well compacted subgrade. The joints are sealed afterwards with asphalt or cement grout to prevent leakage.

3. **Cement mortar lining:** In this type of lining it is very essential to have well graded sand. The sand should range from fine to coarse to meet the requirements of durability and appearance. The amount of cement required is more and hence the cost is also more. The thickness of this type of lining may vary from 9 mm to 38 mm. The method of construction is similar to that of concrete lining. This type is not durable hence it can be used only in conjuction with some other protective material. To give an example it may be sandwiched beteen two layers of bricks when later is used as a lining material. It has been proved that 25 mm thick cement mortar layer reduces seepage by about 75 per cent.
4. **Soil cement lining:** Some times cement may be mixed with the water and locally available soil. The soil before using should be thoroughly analysed in laboratories by conducting various tests. The water soil and cement is mixed to get a workable mixture. After spreading this mixture on the subgrade it is compacted to attain maximum density. This type of lining may be constructed with a travelling mixer with a slip form. It has been experienced that this type of lining can be constructed rapidly and has a low cost.

10.8. BRICK LINING

The canal is said to be lined with bricks when the sides and bed are protected with brick surfacing laid in cement mortar. It has been experienced that the brick lining generally fails due to:

(*a*) **Spilling of water:** It may be checked by providing sufficient freeboard.

(*b*) **Settlement of weak subgrade:** It may be prevented by giving adequate compaction to the back fill.

(*c*) **Pressure of water behind lining:** The failure of lining may be prevented by providing properly designed drainage system in the banks.

To make the lining successful the lining may be constructed of two layers of bricks laid flat in mortar. The first layer is laid on a 12 mm layer of 1 : 6 cement mortar. This 12 mm thick layer of cement mortar is spread on the properly compacted and wetted subgrade. Then on the top of first brick layer, a 12 mm layer of 1 : 3 cement mortar is given. It is allowed to cure for two days. Finally second layer of bricks in mortar is laid on the top. Thus a 12 mm layer of rich cement mortar (1 : 3) is sandwiched between two brick layers and this layer is actually responsble to make the lining practically watertight.

The size of the bricks generally used is 30 cm × 15 cm × 6½ cm. However, this size of bricks is not standard. To give additional stability to the lining, masonry may be reinforced by using mild steel bars. This reinforcing steel is generally laid in the cement mortar layer sandwiched between the two brick layers.

The bricks used for lining should be manufactured with good earth. Good earth is one which contains about 10 to 20 per cent clay and in which calcium carbonate is less than 2 percent and salt content is not more than 0.3 per cent. The bricks should be thoroughly wetted with water before use.

The brick lining has been successfully done on various canals in Punjab, for example, Haveli canal and Bhakra canal. Sarda canal in Uttar Pradesh is also lined with bricks. The lining of Haveli canal damaged at many places within a year after its construction. The main reason for failure was that the lining was not designed and constructed to fulfil the safety requirements strictly. (Improper compaction, defective material and insufficient freeboard were the main reasons.)

10.9. MISCELLANEOUS TYPES OF CANAL LINING

(*i*) **Clay puddle lining:** The canal may also be lined with a clay puddle. It is already mentioned that the puddle clay is fairly imprevious when properly pugged and saturated with water. It can be laid on the subgrade to form a fairly watertight coating. The thick layer of puddle clay is then protected with 30 cm thick layer of silty soil. It has been seen that it prevents about 80% seepage loss.

(*ii*) **Lining with stone masonry:** This type of lining is constructed with dressed stone blocks laid in mortar. Properly dressed stones are not available in nature. Irregular stone blocks are dressed and chipped off as per requirement. It makes the type costly. When roughly dressed stones are used for lining, the surface is rendered rough which may put lot of resistance to fow. Technically the coefficient of rugosity will be higher. Thus the stone lining is limited to the situation where loss of head is not an important consideration and where stones are available at moderate cost.

(*iii*) **Plastic lining:** As a modern technique use of plastics in canal lining holds good promise. There are three types of plastic membranes which are used in canal lining, namely:

- Low density polyethylene (LDPE)
- High molecular high density polythene (HM); and
- Polyvinyl chloride (PVC)

The plastic as a material for canal lining offers certain characteristic advantages like negligible weight, easy for handling, spreading and transport, immune to chemical action and speedy construction. The plastic film is spread on the prepared subgrade of the canal. To anchor the membrane on the banks '*V*' trenches are provided. The film is then covered with protective soil cover. The plastic films are available like cloth in 3 m width. The thickness of plastic membrane varies according to its quality, for example, LDPE, HM and PVC qualities are used with 250, 100 and 15 micron thickness respectively. The plastic sheets can be welded together at site to increase the width. Considering factors like initial capital cost, installation cost and effectiveness plastic lining will go a long way in reducing loss of water in the water conveyence system.

10.10. SELEECTION OF TYPE OF LINING

The selection of a particular type of lining should be based on the following considerations:

1. **Functional success:** It is already mentioned that the canals are lined to prevent seepage loss. Obviously the type which gives maximum required watertightness to the canal section should be chosen.
2. **Economic consideration:** It is very essential to make sure that the type selected is such that the benefits derived from it are sufficient to balance the annual cost of the lining with safe margin.
3. **Structural stability:** The type of lining selected should be sufficiently strong to resist the damaging forces. Lining should be sufficiently flexible to allow moderate settlement of subgrade without cracking.

4. **Hydraulic efficiency:** The lining selected should give smooth finish to the surface. If the surface is rough the coefficient of rugosity will be high. It reduces the velocity of flow and consequently discharge capacity of the canal is reduced.
5. **Durability:** The type of lining selected should have sufficient working or useful life. The lining should be resistant to wearing, weathering and chemical attack.

10.11. DESIGN OF LINED CANALS

The lined canals are not designed making use of Lacey or Kennedy Theory because the section is rigid. Generally Manning's equation is used in design.

To carry a certain discharge number of channel sections may be designed with different bed widths and side slopes. But it is clear that each section is not equally good for the purpose. The section to be adopted should be economical and at the same time it should be functionally efficient.

It has been found that the most suitable cross-section of a lined canal is a circular section with sloping sides. That is, the bed is not flat but it is an arc of a circle. This arc is tangential to the sloping sides.

Side slopes: The side slope is selected in such a way that it nearly equals the angle of repose of the soil in the subgrade. Care is taken to ensure that no earth pressure is exerted on the back of the lining.

From the knowledge of hydraulics it is clear that the section is economical when cross-sectional area is maximum for minimum wetted perimeter. This condition is achieved when the centre of an arc lies at FSL of the canal. This section is also efficient in the sense that as the velocity of flow is higher silt carrying is also higher than a wide and shallow section. Thus the problem of silting is completely eliminated and functioning is efficient.

It may be mentioned here that such section with circular bed may be designed upto a discharge of 85 m^3/sec. When the discharge is more than 85 m^3/sec the section best suited is one with a flat bed and sloping sides with rounded corners. This section is certainly better than trapezoidal section, because it is more stable and economical to construct.

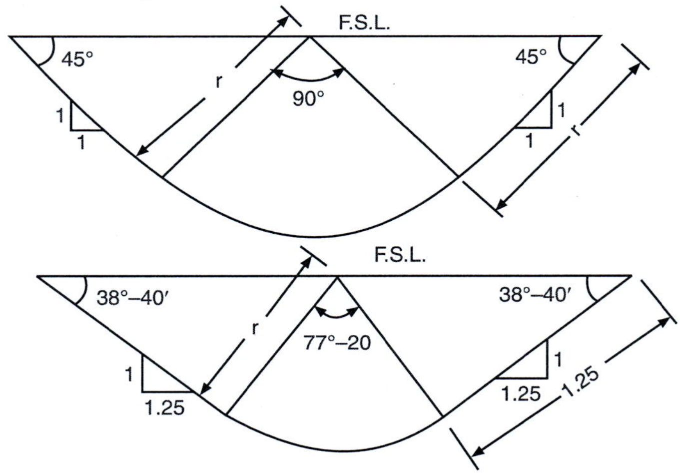

Fig. 10.2. Design of lined canals

Dimensions of a canal section with circular bed may be obtained from the equations given below. Figure 10.2 shows the canal sections of this type with two standard side slopes.

When $r = 3.6$ m or less, side slopes may be taken 1 : 1

When $r > 3.6$ m side slopes may be taken 1.25 : 1

Velocity of flow: Mean velocity of flow may be calculated using Manning's formula.

$$V = \frac{1}{N} \cdot R^{2/3} \cdot S^{1/2}$$

where N is coefficient of rugosity and may be taken as 0.018

S is the slope of bed and expressed in fall of bed in m in 10,000 m length

For the given value of N formula may be reduced to

$$V = 0.556\ R^{2/3} \cdot S^{1/2}$$

where V is velocity of flow in m/sec;

R is hydraulic mean radius in m;

and S is bed slope expressed in metres/metre length,

since $(10{,}000)^{1/2}$ is merged in the constant 0.556.

Following Table 10.1 gives the sectional data for the two side slopes

Table 10.1.

Sectional data	***Side slopes***	
	1 : 1	***1.25 : 1***
Sectional area	$1.785\ r^2$	$1.925\ r^2$
Wetted perimeter	$3.57\ r$	$3.85\ r$
Hydraulic mean radius (R)	$0.5\ r$	$0.5\ r$
FS discharge $Q = A \cdot V$	$0.624\ r^{8/3} \cdot S^{1/2}$	$0.674\ r^{8/3} \cdot S^{1/2}$
Lacey's silt factor $f = 2.46\ \dfrac{V^2}{R}$	$0.603\ r^{1/3} \cdot S$	$0.603\ r^{1/3} \cdot S$

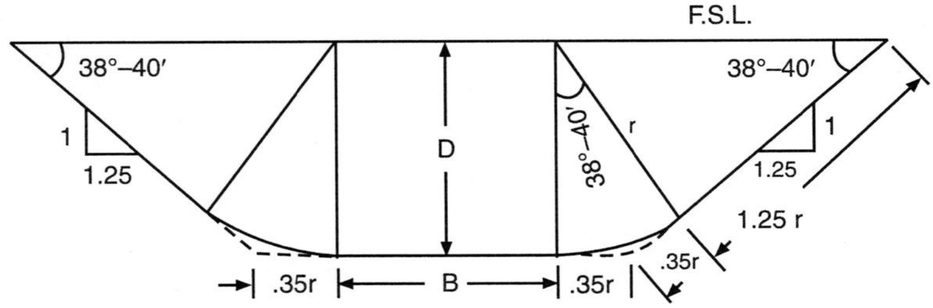

(a)

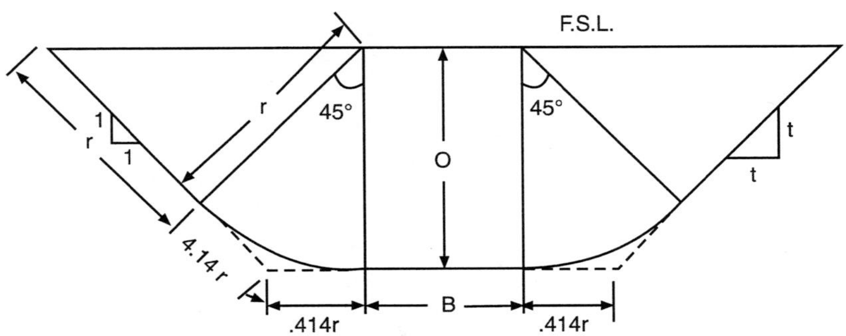

(b) Design of lined canals

Fig. 10.3.

Dimensions of a canal section with flat bottom may be obtained from the equations given below. Figure 10.3 shows the cross-sections of this type with two standard side slopes. It may be mentioned here that generally maximum velocity of flow is recommended to be 2 m/sec and Lacey's '*f*' is limited to 1.2. When *r* exceeds 55 m or when discharge is more than 85 m^3/sec, the canal should be designed with a flat bottom with fillets at corners. The Indian Standard 4745 in fact does not suggest lined canal with circular bottom.

Velocity of flow: Mean velocity of flow may be calculated from Manning's formula for value of $N = 0.018$.

Then $$V = 0.556\ R^{2/3} \cdot S^{1/2}$$

Following Table 10.2 gives sectional data for the two side slopes:

It may be remembered that lined canal could be given steep slope to achieve recommended velocity of 2 m/sec and value of '*f*' of about 1.2. Critical velocity ratio is not applicable to lined canals. But to avoid possibility of sitting CVR should be aimed at more than unity.

Coefficient of rugosity (*N*): In general practice for lined canal average value of *N* may be taken as 0.018. For different types of linings the value of '*N*' varies. The values given for straight channels in Indian Standard 4745 are iven in Table 10.3. When the alignment is not straight loss of head increases and a small increase in the value of '*N*' may be made to allow for additional loss of energy.

Freeboard: Freeboard is measured from the (FSL) full supply level to the top of lining. For lined canals having less than 10 cumec discharge 0.6 m free board is recommended. For bigger lined canals freeboard not less than 0.75 m is generally provided.

Bank widths: The Indian Standard recommended following values for bank widths for main and branch canals:

Main canal in cutting and filling = 8.0 m

Branch canal in cutting = 6.5 m

Branch canal in filling, left bank = 6.5 m

Right bank = 5.0 m

Table 10.2.

Sectional data	*Side slopes*	
	1 : 1	*1.25 : 1*
Sectional area	$(K + 1.785)\ D^2$	$(K + 1.925)\ D^2$
Wetted perimeter	$(K + 3.57)\ D$	$(K + 3.85)\ D$
$C = \dfrac{R}{D}$	$\dfrac{(K+1.785)}{(K+3.57)}$	$\dfrac{(K+1.925)}{(K+3.85)}$
Hydraulic mean radius (*R*)	*C.D*	*C.D*
FS Discharge $Q = A.V$	$(K + 1.785) \times 0.556\ C^{2/3}.\ D^{8/3} . S^{1/2}$	$(K + 1.925) \times 0.556\ C^{2/3}\ D^{8/3} . S^{1/2}$
Lacey's silt factor $f = 2.46\ \dfrac{V^2}{R}$	$0.763\ C^{1/3} \cdot D^{1/3} \cdot S$	$0.763\ C^{1/3} \cdot D^{1/3} \cdot S$
K	*B/D*	*B/D*

Table 10.3.

S.No.	*Surface Characteristics*	*Value of N*
1.	Concrete with surface as indicated below:	
	(*a*) Formed, no finish	0.013 – 0.017
	(*b*) Trovel finish	0.012 – 0.014
	(*c*) Float finish	0.013 – 0.015
	(*d*) Float finish some gravel on bottom	0.015 – 0.017
	(*e*) Gunite, good section	0.016 – 0.019
	(*f*) Gunite, wavy section	0.018 – 0.022
2.	Concrete bottom float finished sides as indicated below:	
	(*a*) Dressed stone in mortar	0.015 – 0.017
	(*b*) Random stone in mortar	0.017 – 0.020
	(*c*) Cement rubble masonry	0.020 – 0.025
	(*d*) Cement rubble masonry plastered	0.016 – 0.020
	(*e*) Dry rubble (rip rap)	0.020 – 0.030
3.	Gravel bottom sides as indicated below:	
	(*a*) Formed concrete	0.017 – 0.020
	(*b*) Random stone in mortar	0.020 – 0.023
	(*c*) Dry rubble (rip rap)	0.023 – 0.033
4.	Brick	0.014 – 0.017
5.	Asphalt	
	(*a*) Smooth	0.013
	(*b*) Rough	0.016
6.	Wood Planed Clean	0.011 – 0.013
7.	Concrete lined excavated rock with:	
	(*a*) Good section	0.017 – 0.020
	(*b*) Irregular section	0.020 – 0.027

Suitable provision of dowel, roadway, catch water drain, under drainage has to be made for every lined canal. Figure 10.4 shows three typical cross-sections of the lined canal in which provision of various components is ilustrated. The canal sections shown have trapezoidal shape. For circular bottomed canal arrangement will be similar. It may be noted that angle θ shown in the section varies with the side slopes adopted. θ for side slope 1 : 1 has 45° value and for 1.25 : 1 is 38° 40′ as shown in Fig. 10.3. Maximum height of the spoil bank is limited to 6 m when the excavation is done by machinery. In case work is accomplished manually maximum height is to be restricted to 4 m only. Also when a canal section in filling involves more than 10.5 m filling each section shall be designed and tested for stability.

DESIGN PROBLEMS

Problem 10.1. Design an irrigation lined canal to carry a discharge of 34 m³/sec. The mean diameter of the average soil particles is 0.464 mm. Assume side slopes 1.25 : 1 and width zero.

Solution: Lacey's silt factor, $f = 1.76\sqrt{m_r} = 1.76\sqrt{0.464} = 1.2$

Taking value of $N = 0.018$

$\therefore$ Mean velocity of flow $V = 0.556 \cdot R^{2/3} \cdot S^{1/2}$

$$Q = 0.674\, r^{8/3} \cdot S^{1/2} \quad \text{(from Table 10.1)}$$

$$\therefore \quad 34 = 0.674\, r^{8/3} \cdot S^{1/2} \quad ...(1)$$

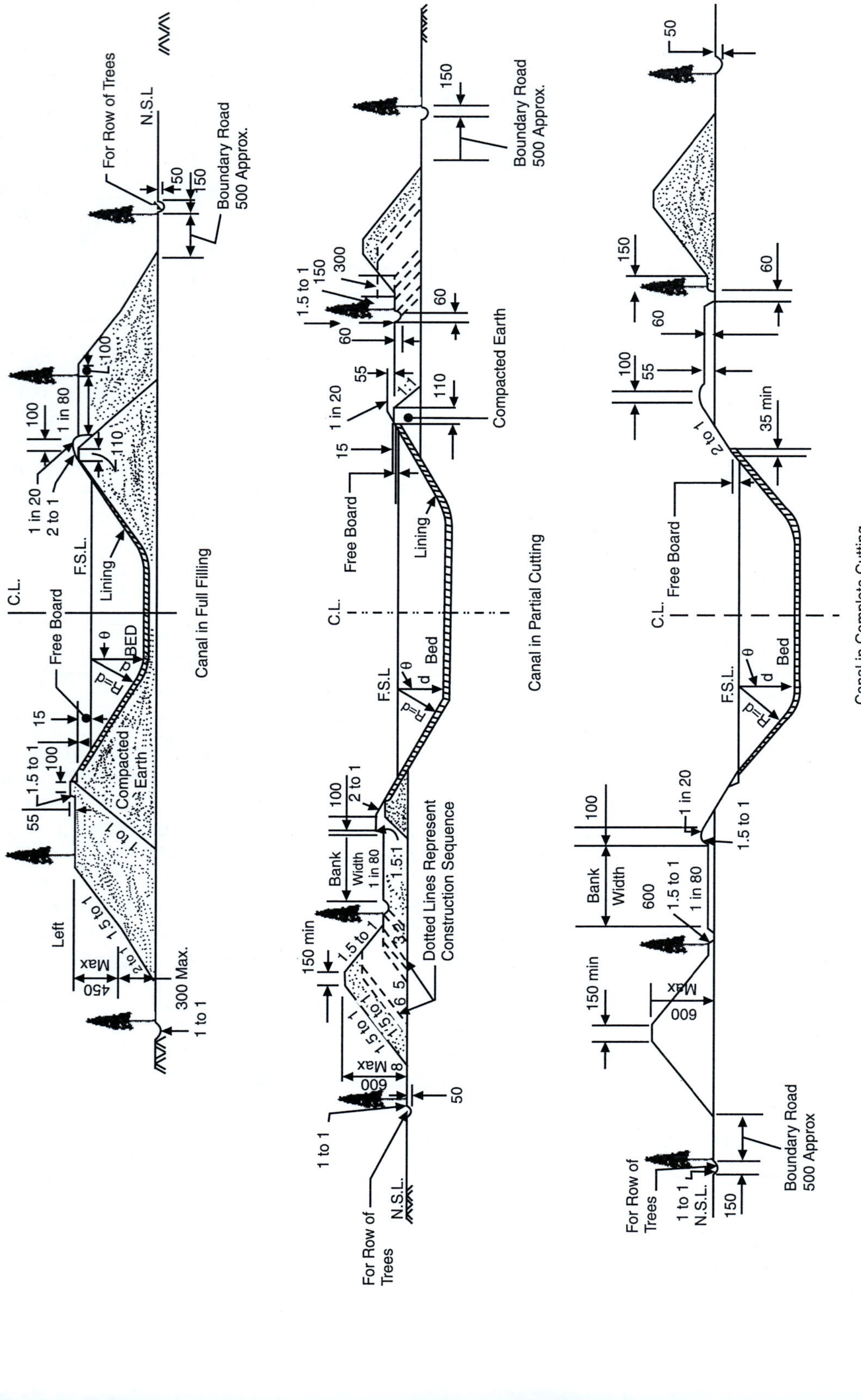

Note: All dimensions are in centimetres.

Fig. 10.4. Typical Cross Sections of Lined Canal

$$f = 0.603\ r^{1/3} \cdot S$$

$$\therefore \quad 1.2 = 0.603\ r^{1/3} \cdot S$$

or $$S = \frac{1.2}{0.603} \times \frac{1}{r^{1/3}}$$

$$= \frac{1.99}{r^{1/3}} \qquad \ldots(2)$$

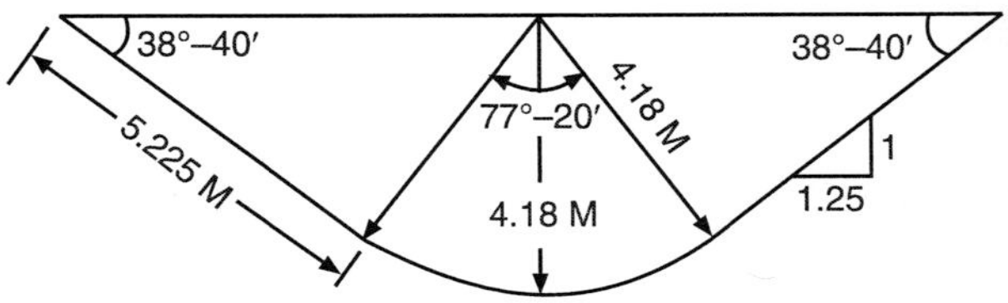

Fig. 10.5. Problem 1

Substituting value of S in equation (1)

$$34 = 0.674\ r^{8/3}\left(\frac{1.99}{r^{1/3}}\right)^{\frac{1}{2}} = 0.674 \times 1.41 \times r^{(8/3\ -\ 1/6)} = 0.95 \times r^{2.5}$$

or $$r^{2.5} = 35.8$$

$\therefore$ Radius $$r = 4.\ 18\ \text{m}$$

Hence, from equation (2)

Slope, $$S = \frac{1.99}{(4.18)^{1/3}} = \frac{1.99}{1.61} = 1.24\ \text{m in } 10{,}000\ \text{m}$$

Hydraulic mean radius, $R = 0.5\ r = \dfrac{4.18}{2} = 2.09$

$$V = 0.556 \times 2.09^{2/3} \times 1.24^{1/2}$$

$$= 0.556 \times 1.635 \times 1.114 = 1.012\ \text{m/sec}$$

Answers. Radius of a canal section 4.18 m. Longitudinal slope 1.24 in 10,000, velocity of flow 1.012 m/sec.

Problem 10.2. Design a lined canal to carry 100 cumec discharge with the following data:

(i) Angle of repose of the soil = 45°

***(ii)* Lacey's silt factor = 1.2**

***(iii)* B/D ratio = 3**

***(iv)* Value of N = 0.018**

Solution: Step 1: Since the angle of repose of the soil is given as 45° side slope of 1 : 1 may be adopted for the canal.

Step 2: For the side slope 1 : 1 using generalised sectional data given in Table 10.2.

$$A = (K + 1.785)\ D^2$$

$$P = (K + 3.57)\ D$$

and $$V = 0.556\ R^{2/3}\ S^{1/2}$$

where $$K = B/D = 3$$

$$\therefore \quad \frac{A}{P} = \left(\frac{4.785}{6.57}\right) D = R$$

So $$R = 0.728\ D \qquad \ldots (1)$$

Step 3: $$Q = A.\ V = (4.785) \cdot D^2 \cdot 0.556\ (0.728\ D)^{2/3} \cdot S^{1/2}$$

$$= 4.785 \times 0.556 \times 0.81\ D^{8/3}\ .S^{1/2}$$

$$100 = 2.15\ D^{8/3} \cdot S^{1/2} \qquad \ldots (2)$$

Step 4: Also $$f = 0.763\ R^{1/3} \cdot S = 0.763\ (0.728)^{1/3}.\ \text{D}^{1/3}\ .S = 0.763 \times 0.9\ D^{1/3}\ .\ S$$

$$\therefore \quad 1.2 = 0.686\ D^{1/3} \cdot S$$

or $$S = \frac{1.75}{D^{1/3}} \quad \text{... (3)}$$

Step 5: Substituting value of S in equation (2) above

$$100 = 2.15\ D^{8/3}.\left(\frac{1.75}{D^{1/2}}\right)^{\frac{1}{2}} = 2.15 \times 1.32 \times D^{(8/3\ -\ 1/6)} = 2.84\ D^{2.5}$$

$$D^{2.5} = 40$$

$$D = 4.37 \text{ m}$$

Since $$B/D = 3$$

$$B = 13.11 \text{ m}$$

Step 6: From relation (3) above

$$S = \frac{1.75}{(4.37)^{1/3}}$$

$$S = 1.06 \text{ m in } 10{,}000 \text{ m}$$

$$A = 91.4 \text{ m}^2$$

$$V = 1.09 \text{ m/sec}$$

Answer: Depth 4.37 m; Bed width 13.11 m; side slopes 1 : 1 and longitudinal slope 1 in 9434 or 1.06 m in 10,000 m.

Problem 10.3. In the particular area water table is at an average depth of 4 m below ground level.

Design a lined canal to carry 100 cumec discharge so that there is no uplift on the bed of the canal.

The velocity of flow may be kept to be 1.6 m/sec and side slope of 1 : 1 may be adopted.

Solution: Step 1: Since the uplift is to be avoided the depth of the canal should be less than 4 m below ground level. Let us take depth of flow as 3.5 m.

Step 2: $$Q = A{\cdot}V$$

$$\therefore \quad A = \frac{100}{1.6} = 62.5 \text{ m}^2$$

Step 3: From Table 10.2.

$$A = (K + 1.785)\ D^2$$

where $$K = B/D$$

$$62.5 = (K + 1.785)\ (3.5)^2$$

$$62.5 = 12.25\ K + 21.87$$

$$\therefore \quad K = \frac{40.63}{12.25} = 3.3$$

$$\therefore \quad B = 3.3 \times 3.5 = 11.55 \text{ m}$$

Step 4: From Table 10.2

$$P = (K + 3.57)\ D = (3.3 + 3.57)\ 3.5 = 6.87 \times 3.7 = 24.04 \text{ m}$$

Step 5: $$R = \frac{A}{P} = \frac{63.5}{24.04} = 2.6$$

Step 6: Using relation, $V = 0.556\ R^{2/3}\ .\ S^{1/2}$

$$1.6 = 0.556 \times (2.6)^{2/3} \cdot S^{1/2} = 0.556 \times 1.89 \times S^{1/2} = 1.051\ S^{1/2}$$

$$S^{1/2} = 1.52$$

$$\therefore \quad S = 2.32 \text{ m}$$

∴ Slope is 2.32 m in 10,000 m

or Slope is 1 in 4300

Answer: Depth 3.5 m; bed width 11.55 m; side slopes 1 : 1 and longitudinal slope 1 in 4300.

QUESTIONS FOR REVISION

1. Why irrigation canals need lining? Explain.
2. What do you understand by lining of canals ?
3. Mention various advantages of lining the canal.
4. Explain various types of lining in which cement is used.
5. Write short notes on:
   (*a*) Clay puddle lining
   (*b*) Plastic lining
6. How canal is lined with bricks?
7. What are the properties that determine the suitability of a particular type of lining?

11

Waterlogging and its Remedies

(Including Drainage and Conjunctive use of Surface and Groundwater)

11.1. DEFINITION OF WATERLOGGING

It is already emphasized that water is the most important requirement of plant life. It is also mentioned that water acts as a nutrient carrier. It may now be cleared here that it is correct only upto a certain limit. For clear understanding some knowledge of chemistry is very helpful. It will be sufficient to state here that nutrients are produced in the soil only when the soil mass is amply ventilated or more specifically, aerated. Thus plant roots require proper air circulation also in addition to the presence of water. Hence, soil in crop root-zone should be such as to provide free circulation facilities to air and water at the same time.

Although development of irrigation has increased agricultural production, it has also caused adverse effect in the form of waterlogging and associated problem of soil salinity/ alkalinity in many irrigation commands. The problem of waterlogging can be mitigated to a large extent by efficient water management and by adopting suitable preventive measures. However, in spite of best efforts the problem of waterlogging has surfaced in many irrigation project commands. It has, therefore, become essential to reclaim such areas by remedial measures so as to ensure optimum agricultural production from all these commands.

When the conditions are so created that the crop root-zone gets deprived of proper aeration due to the presence of excessive moisture or water content, the tract is said to be waterlogged.

To create such conditions it is not always necessary that under groundwater table should enter the crop root-zone. Sometimes even if water table is below the root-zone depth the capillary water zone may extend in the root-zone depth and makes the air circulation impossible by filling the pores in the soil.

The waterlogging may be defined as rendering the soil unproductive and infertile due to excessive moisture and creation of anaerobic conditions.

The phenomenon of waterlogging can be best understood with the help of a hydrologic equation, which states that

$$\text{Inflow} = \text{Outflow} + \text{Storage}$$

Here inflow represents that amount of water which enters the subsoil in various processes. It includes seepage from the canals, infiltration of rainwater, percolation from irrigated fields and subsoil flow. Thus although it is loss or us, it represents the amount of water flowing into the soil.

The term outflow represents mainly evaporation from soil, transpiration from plants and underground drainage of the tract.

The term storage represents the change in the groundwater reservoir.

11.2. CAUSES OF WATERLOGGING

After studying the phenomenon of waterlogging in the light of hydrologic equation main factors which help in raising the water-table may be recognised correctly. They are:

- Inadequate drainage of over-land run-off increases the rate of percolation and in turn helps in raising the watertable.
- The water from rivers may infiltrate into the soil.
- Seepage of water from earthen canals also adds significant quantity of water to the underground reservoir continuously.
- Sometimes subsoil does not permit free flow of subsoil water which may accentuate the process of raising the water table.
- Irrigation water is used to flood the fields. If it is used in excess it may help appreciably in raising the watertable. Good drainage facility is very essential.

11.3. EFFECTS OF WATERLOGGING

The waterlogging affects the land in various ways. The various after effects are the following:

1. **Creation of anaerobic condition in the crop root-zone:** When the aeration of the soil is satisfactory bacteriological activities produce the required nitrates from the nitrogenous compounds present in the soil. It helps the crop growth. Excessive moisture content creates anaerobic condition in the soil. The plant roots do not get the required nourishing food or nutrients. As a result crop growth is badly affected.
2. **Growth of water loving wild plants:** When the soil is waterlogged water loving wild plant life grows abundantly. The growth of wild plants totally prevent the growth of useful crops.
3. **Impossibility of tillage operations:** Waterlogged fields cannot be tilled properly. The reason is that the soil contains excessive moisture content and it does not give proper tilth.
4. **Accumulation of harmful salts:** The upward water movement brings the toxic salts in the crop root-zone. Excess accumulation of these salts may turn the soil alkaline. It may hamper the crop growth.
5. **Lowering of soil temperature:** The presence of excessive moisture content lowers the temperature of the soil. In low temperature the bacteriological activities are retarded which affects the crop growth badly.
6. **Reduction in time of maturity:** Untimely maturity of the crops is the characteristic of waterlogged lands. Due to this shortening of crop period the crop yield is reduced considerably.

11.4. DETECTION OF WATERLOGGING

From the subject matter discussed above it is clear that the waterlogging is indicated when the ground water reservoir goes on building up continuously. When the storage starts building up in the intital stages the crop growth is actually increased because more water is made available for the crop growth. But after some time the waters table rises very high and the land gets waterlogged. Finally the land is rendered unproductive and infertile.

The problem of waterlogging develops in its full form slowly. Therefore its early detection is possible by keeping a close watch over the yields and also on the variations in the groundwater level.

A comparative reduction in crop yields in spite of irrigation and fertilisation and early maturity of crops indicate the symptoms of waterlogging. Also when harmful salts start appearing on the fields as white incrustation or deposit it indicates that waterlogging is likely to follow. In worst cases the water-table rises so high and close to the ground surface that the fields turn into swamps and marshes.

The best way of keeping watch over the problem of waterlogging is by observing variations in the groundwater level. It can be done by measuring the depth of water levels at regular interval in the wells dug in the area. Continuous high water levels indicate that the groundwater storage is building up which may create waterlogging in the area.

11.5. SOLUTION TO THE PROBLEM OF WATERLOGGING

The problem of waterlogging may be attacked on two fronts. First is preventive measures, which keep the land free from waterlogging. Secondly curative measures may be adopted to reclaim the waterlogged area. But in principle both measures aim at reducing the inflow and augmenting the outflow from the underground reservoir.

11.5.1. Preventive Measures

Preventive measures include the following:

(*a*) **Controlling the loss of water due to seepage from the canals:** The seepage loss may be reduced by adopting various measures for example:

- **By lowering the FSL of the canal:** Loss may be due to percolation or absorption but when FSL is lowered the loss is reduced to sufficient extent. It is course essential to see that while lowering the FSL command is not sacrificed.
- **By lining the canal section:** When the canal section is made fairly watertight by providing lining the seepage loss is reduced to quite a good extent.
- **By introducing intercepting drains:** They are generally constructed parallel to the canal. They give exceptionally good results for the reach where the canal runs in high embankments.

(*b*) **Preventing the loss of water due to percolation from field channels and fields:** The percolation loss can be renoed by using water more ecomically. It may also be affected by keeping intensity of irrigation low. Then only small portion of the irrigable tract is flooded and consequently the percolation loss takes place only on the limited area. It keeps the water-table sufficiently low.

(*c*) **Augmentation of outflow and prevention of inflow:** It may be accomplished by introducing artificial open and underground drainage grid. It may also be achieved by improving the flow conditions of existing natural drainages.

(*d*) **Quick disposal of rainwater:** Quick removal of rainwater by surface or open drains is a very effective method of preventing the rise in water table and consequent waterlogging of the tract. It is needless to state that the rainwater removed is net reduction in inflow.

11.5.2. Curative Measures

Curative measures include the following:

(*a*) **Installation of lift irrigation systems:** When a lift irrigation project in the form of a tube well irrigation system is introduced in the waterlogged area the water table gets lowered sufficiently. It is found to be very sucesseful method of reclaiming waterlogged land. Thus a combination of a canal system and a supplementry tube well irrigation system may be considered to be most successful and efficient irrigation scheme. Of course it is true that it will create some complications while assessing the charges for irrigation water. (The canal water being cheaper than tube well water).

(*b*) **Implementation of drainage schemes:** The waterlogged area may be reclaimed by introducing overland and underground drainage schemes.

11.6. EXTENT OR WATERLOGGED AREA

In our country waterlogging is a problem of great concern. It is estimated that total area of waterlogged land is 86.92 lakh hectares. It includes area in irrigation commands as well as other area oustide the command. While the areas in the irrigation command get waterlogged due to rise in water table as a direct consequence of inadequate drainage, other areas get waterlogged due to inundation, as consequence of flooding for long durations. The States mainly affected and the extent of area rendered infertile and unproductive are given in Table 11.1.

About 48 lakh ha are estimated to be affected by salinity and 25 lakh ha by alkalinity. Saline soils include 10 lakh ha in arid and semi-arid regions of Rajasthan and Gujarat and 14 lakh ha in black cotton soils. The alkali problem is mainly in Punjab, Haryana and Uttar Pradesh.

Table 11.1. Extent of waterlogged area

State	*Water-logged area in lakh ha*	
	Total	*In irrigated commands*
Andhra Pradesh	3.39	2.66
Assam	4.50	N.R.
Bihar	7.07	6.20
Gujarat	4.84	1.73
Haryana	6.20	2.49
Jammu and Kashmir	0.10	0.02
Kerala	0.61	0.12
Karnataka	0.25	0.25
Madhya Pradesh	0.73	0.73
Maharashtra	1.11	0.15
Orissa	1.96	1.96
Punjab	10.90	2.00
Rajasthan	3.48	1.80
Tamil Nadu	0.18	0.16
Uttar Pradesh	19.80	4.30
West Bengal	21.80	N.R.
Total	**86.92**	**24.57**

Steps are being taken to reclaim the waterlogged land in the country. The steps taken to reclaim such areas include implementation of drainage schemes, provision of deep drains, excavation of new channels and improvement of existing ones, construction of sluices with marginal embankment and installation of tube wells.

The spread of conjunctive use of groundwater with that of surface water especially in Punjab, Haryana and parts of Uttar Pradesh has substantially lowered the groundwater table and helped in containing waterlogging/salinity.

Summarising the most effective and efficient anti-waterlogging measures are:

- Lining of channels (main canal, branches and field channels).
- Provision of surface drains for drainage of rainwater; and
- Implemenation of tube well projects both extensive and local.

Other methods are of local importance and hence benefit derived from them is limited.

11.7. DRAINAGE OF LAND

By its very nature irrigation creates periodically saturated conditions of upper layers of soil formation. Over a long period where intensive irrigation is practised even deep soil layers tend to become saturated and consequently underground water table rises in absence of adequate drainage facilities. Thus knowledge of drainage engineering is very essential to achieve balance of the hydrologic equation (Inflow = outflow + storage). Drainage of culturable lands is equally essential for humid and arid regions. Adequate lowering of the water-table by drainage is a first and basic necessity in any irrigated tract.

Drains have been defined as devices both natural and artificial, which can be helpful in removing water from any particular area. Consequently their functions are diverse. Thus drains are required to:

- relieve waterlogged areas,
- remove flood water,
- drain off surplus rainwater not required in the normal growth of the agricultural crops, and
- drain off ponds and marshes.

11.8. BENEFITS OF DRAINAGE

It is already mentioned that drainage is the first essential in reclaimation of waterlogged soils. Drainage scheme when implemented properly improves the soil structure and increases productivity of the soil. So far as irrigated lands are concerned following benefits may be achieved from adequate drainage scheme.

- It facilitates early ploughing and in turn early sowing of crops. As a result crop period is increased to derive maximum crop yield.
- It actually extends the crop root-zone. Thus more soil moisture is made available for crop growth.
- It maintains higher soil temperatures. Thus as a result soil is kept warmer. Soils which are waterlogged take more time for warming up. The reason is, waterlogged soil require more heat to raise the temperature of a given volume of water by 1°C then to raise the temperature of same volume of air by 1°C.
- It helps in maintaing proper aeration of upper soil layers. The aeration and higher temperature increase the bacteriological activities in the soil. Thus more and more nutrients are made available to the plants.
- In the process of draining the land, harmful salts are leached off.
- It also improves the sanitary condition and makes surounding more and more attractive and gay.

11.9. CLASSIFICATION OF DRAINS

The drains may be artificial or natural. Drains are termed artificial when they are constructed after proper consideration of existing conditions and functions to be served. Artificial drains are generally constructed to dispose off surplus water quickly, before it gets absorbed deep into the soil. Rivers and valley lines between the two ridges are examples of natural drains. The irrigation channels usually run on the ridges and the lowest valley lines between the two ridges usually form natural drains.

Another significant classification of drains is: (*i*) surface drain, and (*ii*) subsurface drains.

11.9.1. Surface Drains

These are open to the atmosphere. This category can be resolved into following types:

- **Storm water drains:** Their primary function is to drain off excess rain water. They are constructed with an aim to carry the flood flow from the catchments under their command.
- **Seepage drains:** They are generally constructed in the tracts which get irrigation water from canals. Seepage water makes appreciable contribution to the under ground reservoir. As a result water-table rises and crop root-zone is filled with excess water. Then the plant roots get deprived of air. To reduce this contribution to the underground resevoir these drains are constructed. Seepage drains take away the subsoil water to some suitably located outfall. The drains are smaller in size than storm-water drains. Thus these drains help in maintaining free air circulation in root-zone depths.
- **Storm-cum-seepage drains:** They perform both of the above mentioned functions. In the rainy season they carry storm water. But most of the time they serve the purpose of seepage drains. Thus it is very essential to determine the capacity judicially.

11.9.2. Sub-surface Drains

These are tile or pipe drains laid in the permeable stratum below under ground water-table. The sub-surface drains are further classified according to the functions served by each.

- **Relief drains:** The function served by relief drains is similar to that of seepage drains. They relieve the saturated soil of its excess water content which is contributed to the subsoil by various inflow processes, for example, percolation, infiltration, subsoil flow etc.
- **Carrier drains:** They may be supposed to be main drians of which relief drains are tributaries. The carrier drains collect the water from relief drains and carry that water towards outfall. Obviously the size of carrier drains is bigger, say 45 cm diameter. It is of course true that side-by-side these drains also relieve the soil of its excess water.
- **Intercepting drains:** They are generally aligned in a direction parallel to existing canal. There is always seepage taking place to the adjoining low land from the running canal. Intercepting drains check and collect this seepage flow and finally collected water is taken out of the tract to a suitable outfall. Thus aim of these drains is to intercept the seepage water before it joins the underground water-table.

11.10. SURFACE DRAINS AND THEIR DESIGN

11.10.1. Alignment of Surface Drains

Following points should be given due consideration in marking an alignment for drains:

Firstly, alignment should follow a natural drainage line, that is the lowest contour in the valley. To reduce the cost of drainage scheme the length of drains should be minimum. It can be achieved by taking alignment straight rather than zig zag.

Secondly, alignment of drain should not pass through ponds or marshes. The reason is that such drain may act as feeder line to the marsh and then pond will go on expanding.

The solution to the situation is to align the drain clear off the pond. To drain off the pond a small escape drain may be constructed to join the pond with the main drain.

Thirdly, so far as possible drains should not cross irrigation canals. The obvious reason is then some expensive structure will have to be constructed at the crossing point. It increases the cost of drainage scheme.

11.10.2. Design of drains

- **Capacity of drains:** Drains should be designed to carry the maximum anticipated flood efficiently. In Punjab drains are designed for a maximum flood capacity of 0.05 cumec per km^2 of the catchment area in the canal irrigated tract.
- **Velocity:** The velocity of the drain water should be such that the ditch is kept clean by the flow. In other words ditch should be self-cleaning for the designed velocity. It should also be seen that no scouring of bed and sides occur. Chezy's and Manning's formula give good basis for determing velocity.

Etcheverry has given maximum values of mean velocities safe against erosion. These values are given in Table 11.2.

It is experienced that average velocity of 0.6 to 1 m/sec prevents silt deposition.

Table 11.2. Safe velocities in drains

Material	*Mean velocity*
Hard rock	3 to 4.5
Soft sedimentary rock	1.8 to 2.5
Coarse gravel	1.5 to 1.8
Stiff clay soil	1.25 to 1.5
Clay loam	1 to 1.1
Sandy loam	0.75 to 0.8
Coarse sand	0.45 to 0.6
Very light loose sand	0.3 to 0.45

Side slopes: Side slopes to be adopoted depend on the type of the soil formation in which drain is dug. Table 11.3 gives the values of side slopes to be adopted for various formations.

Table 11.3. Side slopes of drains

Soil Type	*Side Slopes (Horizontal : Veritical)*
Rock	Vertical
Loose rock and hard soil	1/2 : 1
Clay soil	1 : 1
Sandy loam	2 : 1
Very sandy soil	3 : 1

Longitudinal slope: Longitudinal slope to be given to the drains is governed by the general slope of natural ground. Of course slope should be fixed in correlation to the permissible velocity. An efficient drain is one which is so designed as not to produce velocity which may induce either silting or scouring.

Economy and efficiency should be the main consideration in designing the drain. The section should be designed in such a way that it will carry maximum discharge for given amount of excavation.

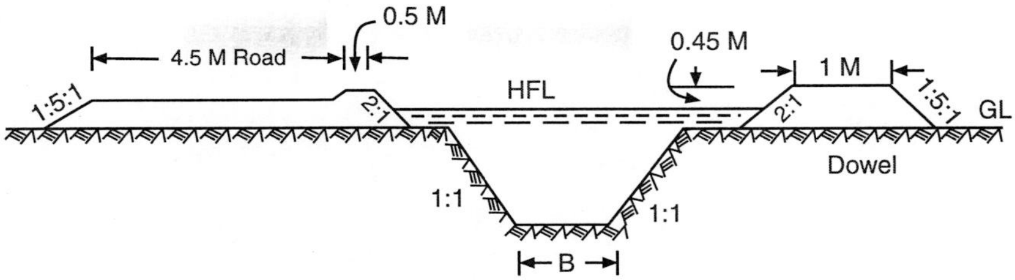

Fig. 11.1. Typical cross-section of a surface drain

Layout: Surface drains are generally laid to follow natural depressions and drainage lines (Fig. 11.1).

11.11. SUBSURFACE DRAINS AND THEIR DESIGN

When the depth of surface drains increases, the surface drainage scheme becomes uneconomical. Then sub-surface drainage scheme in the form of tile drains may be implemented. The depth at which tile drains are to be laid depends upon the level upto which water table is to be lowered. The tile line is laid about 0.6 m below the predetermined level up to which waters table is to be lowered.

The aim of sub-surface drainage scheme is to lower the underground waters table suffciently below the natural ground level so that the plant roots having deeper root-zone get proper aeration. It is used for the crops which have root-zone as far as 1 to 1.25 m below the natural ground level.

Alignment: It is very essential to study the subsoil formation and to prepare **hydroisobath** and **murum isobath** for the area under consideration. Hydro-isobath is an imaginary line which joins the points of similar depth of the underground waters table below the ground surface. To make it clear, 3 metre hydroisobath is a line (contour) which indicates the points where the underground water table is 3 metres below the ground surface. Hydro-isobath is also defined as a line of flow of the underground water.

Similarly murum isobath is a contour of similar depth of top layers of murum below the ground surface. Broadly it can be said to show the configuration of the layer of murum.

Knowing the positions of hydro-isobath and murum isobath drain can be provided accurately below the ground level. Relief drain are provided in the permeable soil stratum (Murum layer) below the water-table.

Longitudinal slope: Generally the slope given is 0.1 m per 300 m length of drainage line. Steeper slopes may necessitate deep digging towards tail whereas flat slope may require greater care in laying of tiles.

Flow in the drains: The velocity of flow in tile drains is calculated from the formula

$$V = 92.87\ R^{2/3} \cdot S^{1/2}$$

Then Q = A.V is used to determine the discharge. Section to be provided is kept liberal as the catchment area of the drain cannot be determined accurately.

11.12. CONSTRUCTION AND LAYOUT OF SUB-SURFACE DRAINS

It is already mentioned that the tile drains are laid 0.6 m below the level up to which water table is to be lowered. The tile drains are circular pipes made of vitrified clay. The trench is excavated in the ground up to the required depth and then the tile line is laid on 15 cm sand bed, Fig. 11.2.

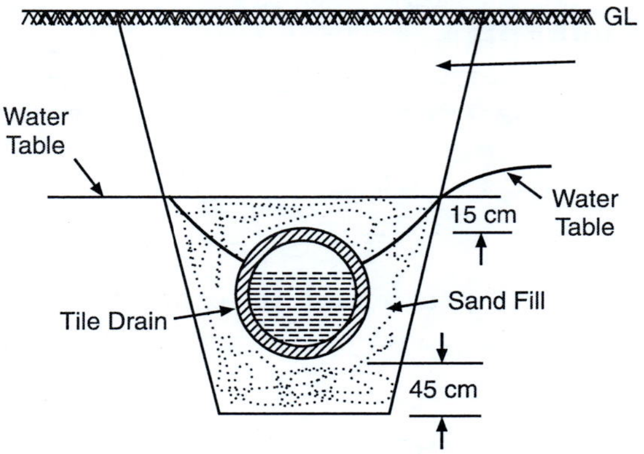

Fig. 11.2. Sub-surface drain

The tiles are laid with open joints. The tiles are closely laid to butt against each other. The open joints are covered with tarred cloth. This covering prevents the sand and silt from entering into the pipeline.

Layout: Since the tile drains are buried in ground their layout can be done in different ways to suit topographic features of the area to be drained.

Spacing of tile drains: Table 11.4 gives the minimum spacing of two sub-surface tile drains for various types of soils.

Table 11.4.

Soil type	***Spacing in m***
Gravelly soil ...	45 to 60
Alluvial soil ...	30
Granular soil ...	20 to 25
Clayey sand ...	18
Clay ...	10 to 12

11.13. DEFINITION AND OBJECTIVES OF CONJUNCTIVE USE OF SURFACE AND GROUNDWATERS

Conjunctive use of surface and groundwater (*S* and *G* waters) implies not only their joint use but their planned and coordinated use in such a way that the total water resources are used in an efficient and economical way to get maximum agricultural production through irrigation and drainage.

At the same time it does not disturb environment, on the contrary it makes regional climate healthy.

Two main objectives of the conjunctive use are:

(*i*) To achieve optimal utilisation of the water resources and maximise the agricultural production per unit of water, and

(*ii*) To reclaim waterlogged lands by gradually declining the ground water table.

Firstly, the main objective of conjunctive use of surface and groundwater is to achieve optimal utilisation of the water resources and maximise the agricultural production per unit of water. For best agricultural production it is essential that the crop is supplied with requisite quantity of water at the various critical stages of its growth. The total number of waterings (quantity-wise and time-wise) required cannot usually be fully met from either surface or groundwater individually.

Major and medium surface water schemes do not have sufficient flexibility and the roster of operating the channels cannot be adjusted to provide timely irrigation in their command areas for various crops with different base and critical periods. The needs can not also be economically met from groundwater alone on account of pumping effort required for lifting the water with the introduction of high yielding varieties with their more exacting water requirements, the situation has become all the more difficult. Howeverconjunctive use of the two sources helps to meet the requirements both with respect to time and quantity.

Secondly, surface water irrigation without proper drainage and/or adequate groundwater development has resulted in alarming rise of the water table in some parts of the country, creating problems of waterlogging. This waterlogging has affected crop growth adversely and has rendered large areas unproductive. The situation is likely to deteriorate further with increasing intensities of irrigation and the tendency on the part of the farmers to over apply irrigation from surface waters. Integrated use of water will lead to gradual decline of groundwater table and help in reclaiming the waterlogged lands.

11.14. PLANNING FOR CONJUNCTIVE USE OF 'S' & 'G' WATERS

In planning introduction of surface and groundwater use in integrated and conjunctive manner we have to take following steps:

(*a*) **Assessing availability of water resources:** The first necessity in the planning for integrated development and conjunctive use of surface and groundwaters is to assess their availability and distribution. It is also necessary to have a water budget for all sources of water for each area/sub-basin/basin/region.

In addition, knowledge of land resources is also required so as to properly evaluate the potentialities of integrated and conjunctive use.

As regards surface waters, the seasonal availability and areal distribution are generally known though not accurately.

In case of groundwaters, because they are hidden under the earth and their occurrence, availability and distribution are affected by complex factors like geological formations, void space, permeability etc., knowledge is comparatively scanty particularly for hard rock areas. The surface flows being visible can be quantified more easily than the hidden groundwaters which requires elaborate arrangements for their exploration and assessment.

(*b*) **Determination of quantity, quality, number and interval of water applications:** To increase crop yield, a number of agricultural inputs like better seeds, proper fertilizers, improved management and requisite irrigation water are necessary. Proper development and management of the water input, however, is of overriding importance since the success and efficiency of other inputs are dependent on the quantity, quality and timing of the water supply, the methods of its use and the adequacy of control over it.

A judicious integration of the surface and groundwaters helps in meeting the demands of the water input. In order to devise suitable methodology for such integration, a knowledge of the crops, their seasons, water requirements and critical stages for irrigation is essential. It is also necessary to know the less critical stages so that in case the available supplies are inadequate (whether in terms of quantity or timeliness), irrigation may be reduced or omitted at those stages with minimum loss in crop production.

The traditional cropping patterns in the country mostly depend on the rainfall season and availability of water for irrigation which meant utilising water as and when available. By integrated and conjunctive use, it is possible to provide water for irrigation when required for various stages of crop growth.

11.15. METHODS OF ACHIEVING CONJUNCTIVE USE

The conjunctive use of surface and groundwater can be achieved in number of ways. Various methods are described below.

1. **Storage of surface water for use during necessities:** The water can be stored when available and distributed during the times of need. Surface water is mostly available only during few monsoon months in a year. Only a part of this water can be stored in surface reservoirs and considerable portion flows waste to the sea. Part of this unused water can be used for charging underground storages. Surface water infiltrates into the underground aquifers by natural processes. However, artificial measures can also be undertaken to supplement the natural recharge. Thus extra water becomes available for use as groundwater component during needy periods.
2. **Removal of deficiency of one source of water by the other:** Cultivaters have keenness for using surface waters wherever and whenever available as in most cases it does not involve any lifting or pumping effort on their part. However, the availability of surface waters is subject to seasonal variations. These fluctuations can to some extent be regulated with help of groundwater.

   The integration of surface and groundwaters can be achieved in a variety of ways, as listed below:

   - Groundwater may be used for irrigation of pockets within or near a commanded perimeter which cannot be served by the surface irrigation canal system.
   - Groundwater may be used for supplementary irrigation in areas, where a surface supply is available, to give extra crop-growing flexibility.
   - Groundwater may be used to supplement surface waters and thus reduce peak demands, size of canals and other structures.
   - Groundwater may be used to guard against the insufficient surface water supplies in years of low surface runoff due to low rainfall. In the post-monsoon season of lean river flows and consequent low supplies to diversion canals, ground water is an important supplementary source. The groundwater can either irrigate directly or be pumped into the cannal system to augment its supplies. Surface and ground waters can be pooled at tails of channels which are subject to chronic shortages.
   - Groundwater may be used to meet the demands of a part of the existing command of a surface water scheme and the released surface water utilized in areas where there is a deficit supply of water.
   - Surface water may be used over as extensive an area of the command as possible with reduced number of waterings. Groundwater may be used conjunctively for providing intensive irrigation to the extent possible.
   - Saline/brackish groundwater may be mixed with good quality surface water to make the former suitable for irrigation.
3. **Earmarking time sequence of application of water by each source:** Conjunctive use can be achieved by sequencing application of irrigation water from two sources viz., surface and groundwater as mention below:

(*a*) In many regions of the country, just prior to the advent of monsoson season, the reservoirs are depleted and the river supplies are low and there is general lack of surface water supplies. Water is required in this period for raising of paddy nurseries. It is not economical to meet the small demands of the nurseries which are scattered all over through surface water sources as its conveyance entails considerable wastage. In such cases local groundwaters can be used to meet this demand.

(*b*) The demands of crops in an area in the monsoon season should be covered fully by surface water supplies through the canal system. During the dry months, irrigation coverage should be provided to the crops by groundwater to the maximum extent possible. Such a sequencing offers the advantage of quicker replenishment of the groundwater through greater recharge during the rainy season with the help of surface irrigation network.

(*c*) Time sequence permits a phased and economical development of water resources. Groundwater may be used in the initial stages of irrigation development before the availability of surface water, thus requiring smaller capital investment and allowing for training of farmers in irrigation methods.

11.16. ADVANTAGES OF CONJUNCTIVE USE

Apart from fulfilling two main objectives, the integrated and conjunctive use also has certain advantages to control some of the ill effects of irrigation.

Advantages of conjunctive use are as below:

(*i*) **The water requirements of crops is met fully to ensure optimum crop yield:** Timely supply of irrigation water in adequate quantities is essential for a good crop yields. It is all the more necessary in case of high yielding varities. Surface water schemes do not have sufficient flexibility and so, the roster of the operating channles cannot be adjusted to provide timely irrigation in the command area for various crops with different base and critical periods. The needs can not also be economically met from groundwater alone on account of pumping efforts required for lifting the water. Thus, conjunctive use can help meet the requirements both in respect of quantity and time.

(*ii*) **Controls waterlogging and salinisation:** Continuous and excessive use of surface water in canal command areas without proper surface drainage and/or adequate groundwater development results in alarming rise of waters table, creating problem of waterlogging and salinisation, affecting crop growth adversely and rendering large areas unproductive. With increasing intensities of irrigation and tendency on the part of the cultivators to over supply irrigation from surfacewater, the problem gets aggravated further. Simultaneous development of groundwater specially through dug-wells and shallow tube wells lowers water table, provides veritcal drainage and thus can prevent waterlogging and salinisation. Areas which are already water logged can also be reclaimed.

(*iii*) **Remedies problem of salinity ingress:** In case of coastal areas, the excessive pumpage of groundwater has been responsible for causing gradual movement of seawater into inland aquifers. This salt water intrusion makes fresh groundwater saline, rendering it unfit for many purposes. Such a condition can also occur in the inland areas due to lowering of water level as a result of excessive withdrawal of fresh water in the vicinity of a saline water zone. This situation can be controlled by increased application of surface waters by encouraging conjunctive use.

(*iv*) **Control on over-pumping of groundwater reservoirs is made possible:** Continous increased withdrawals from a groundwater reservoir in excess of replenishable recharge has resulted in regular lowering of waters table leading to mining of the groundwater. In such a situation a serious problem is created resulting in drying of shallow wells and increase in puming head for deeper wells and tube wells. The remedy lies in providing more surface water irrigation with the help of storage reservoirs, inter-basin transfer, etc. In some cases recourse will also have to be taken for providing artificial recharge to groundwater.

(*v*) **It makes use of saline water possible:** In certain areas the surface water is not able to meet full demand of irrigation water. At the same time, groundwater being saline, direct application is not possible. In such cases, conjunctive use can be made by construction of augmentation tube wells and mixing the saline water with canal water to the extent that the quality of mixed water remains within tolerable limits of corps.

(*vi*) **Helps in augmentation of water resources in project command:** In certain areas, ground water is available but there is no direct use in that area. In such cases, it can be transferred to other needy areas by construction of augmentation canals fed by battery of augmentation tube wells.

This has already been done in Haryana and Punjab. In U.P. it is proposed in Gandak and Sarda Sahayak Commands. In respect of augmentation tube wells (ATW), Haryana was the first State to install 160 ATWs for conjunctive use in the year 1972, in Western Yamuna canal tract to feed the 75 km long augmentation canal to the extent of 14 cumec during the lean season.

(*vii*) **Helps in recovering lost seepage water for irrigation use:** Leakage from unlined canals may be recovered in certain areas by pumping in preference to control of seepage through costly canal lining; this water may be pumped back into the canal or used directly for irrigation.

11.17. CONSTRAINTS IN CONJUNCTIVE USE OF 'S' & 'G' WATERS

There are certain constraints which inhibit integrated and conjunctive use of surface and groundwaters. Some of these are indicated below:

(*i*) **Economic and financial constraint:** The structure of water charges is not conducive to conjunctive use. Wherever surface water is available at a cheaper rate, groundwater is not utilized to the fullest extent possible even though its availability may be quite adequate. This also leads to over-application and wastage of surface water which contributes to the drainage problem.

Major portion of groundwater development has been by individual enterprise through backing of financial institutions. In practice, however, a large section of the farmers have very small (2 hectares and less) holdings and they are unable to afford or even obtain loans for exploitation of groundwater; this is an impediment to development of conjunctive use.

(*ii*) **Legal constraints:** Most of the existing laws are concerned with surface waters because they are framed to safeguard the interest of revenue from surface water schemes which are financed by the government. There is till now no legislation in various States to control development of groundwater by private enterprise. In many areas indiscriminate exploitation has adversely affected the discharges of existing shallow wells and tube wells in the area. The development of groundwater may also be at the expense of committed surface water uses due to reduction in effluent flows in the river streams on the downstream.

(*iii*) **Administrative and organisational constraint:** Surface and groundwaters are assigned to different departments in the States without proper coordination of their activities. Surface and groundwater schemes are often executed in isolation of each other. Such an arrangement is not conducive to development of integrated use. There is also lack of reliable hydrological data required in connection with integrated development and use.

(*iv*) **Monitoring and research constraint:** There is hardly any monitoring of integrated and conjunctive use and their resultant effects in an area. Monitoring is very essential for devising proper methodology for the use and the strategy for further extension.

There are many problems of assessment, development, operation and management which require to be solved in order to derive full benefits of integrated and conjunctive use. Considerable and coordinated research employing the latest techniques is required in order to develop suitable technology and achieve optimum results.

11.18. SCOPE OF DEVELOPMENT OF CONJUNCTIVE USE

Ground water has been utilised for irrigation in our country for many centuries. Similarly tanks also provided water for irrigation. Over the years tube wells, battery of tube wells, tank systems have been developed. Similarly from nineteenth contury onwards large scale river diversion and storage schemes for irrigating lands have come up. However these developments were made in an isolated manner without coordination with other source of water.

From the nineteen sixties onwards, on account of the rapidly increasing demand for food grains and rise in their prices, the attention of Central and State governments was focussed not only towards increased development of both the sources of water but also towards increased integrated and conjunctive use. The introduction by mid-sixties of high yielding varieties of crops with their exacting demands for irrigation supplies gave a big boost to integrated and conjunctive use. However, such development was mostly achieved by the enterprise of the individual farmers who utilised groundwater from their own wells and surface waters from the government owned canals.

In the twenty-first century the need for optimisation of water use both by surface and groundwater has increased many fold to feed ever incresing population of the country. It requires systematic and scientific study to develop conjunctive and integrated use of all sources of water.

The study of water balance in the various river basins has to be intensified in order to assess the quantum of utilizable surface and ground waters before taking up integrated and conjunctive development of water resources on a large scale. The detailed water balance studies are likely to take considerable time. As such in near future possibility of integrated and conjunctive use in the commands of major and medium irrigation projects appears to be achievable target in the first instance.

QUESTIONS FOR REVISION

1. (*a*) What do you understand by waterlogging of land?
   (*b*) Why soil is rendered unproductive and infertile when it is waterlogged?
2. (*a*) What are the causes of waterlogging?
   (*b*) What anti-waterlogging measure you suggest?
3. (*a*) What is object of providing drains?
   (*b*) What benefits are derived from a good drainage scheme?
4. (*a*) Define conjunctive use and mention various advantages.
   (*b*) What are constraints in introduction of systematic conjunctive use?

❑❑❑

12

Regulation and Control of Canal System

12.1. DEFINITION AND FUNCTIONS OF HEADWORKS

The river discharge is never constant and regular. Obviously when a permanent canal system is taken-off from the river some works are to be provided at the off-take to take care of irregularities in river flow condition. Since these works are constructed at the point of take-off, that is at the head of the canal system they are termed as canal **headworks**.

The main object of the canal headworks is to divert the water from the river into the canal. It is sometimes termed diversion headworks also.

To achieve the main object of water diversion form the river into the canal successfully and efficiently following essential steps are taken:

1. **An obstruction or a barrier** is constructed across the river. It serves the following purposes:
    - It obstructs and diverts the river flow into the canal for irrigation.
    - It raises the water level in the river locally at the canal take-off point and helps in maintaining gravity flow. Thus it reduces the excessive cutting of the canal in the head reach.
    - It reduces fluctuation in the level of the river.
    - As the flow of water is obstructed the velocity is reduced and water drops down its silt load before it passes in to the canal system. It creates a still pond on the upstream.
2. **A regulatory structure is constructed** at the canal entrance or the take-off point. It gives proper control over the supply taken into the canal for irrigation purposes. It performs following functions in the process:
    - It prevents the river flood from entering into the canal.
    - It allows only predetermined discharge to pass into the canal.
    - It prevents excessive silt entry into the canal.

In other words the above mentioned points are the main functions served by the canal headworks.

12.2. SELECTION OF SITE

The site for headworks should be selected after giving proper consideration to the following points:

1. **Economical considerations**
   - The construction materials should be available in large quantities in close proximity of the site.
   - At the site the river section should be preferably narrow, straight with high non-submersible banks. It reduces the cost of river training work.
   - The cost of connecting the site by rail and roadway should be less.
   - To reduce the cost of construction of the canal system the site should be close to the culturable area.
   - Number of cross drainage works required to be constructed should be minimum.
2. **Functional considerations**
   - Moderate height of the obstruction should give good command.
   - So far as possible the canal take-off should be at right angles to the river.
   - The river reach should have central flow without tendency of scouring of silting.
   - The canal should give maximum command.
3. **Structural considerations**
   - Good foundation should be available at moderate depth.

12.3. COMPONENTS OF CANAL HEADWORKS

Now it is clear that the canal headworks is an assemblage of various hydraulic structures. After knowing the functions to be served by the diversion headworks various hydraulic structures which constitute the headworks may be listed. They are:

- Weir or an anicut.
- Divide wall.
- Under sluices or scouring sluices.
- Fish ladder.
- Head regulator for a canal.
- Silt exclusion device.
- River training works.

The arrangement of these structures is as shown in Fig. 12.1. It gives typical layout of the diversion headworks. Canals are taken-off from both the banks.

12.4. WEIR OR ANICUT

It is already mentioned that a weir is an obstruction or a barrier constructed across the river. The obstruction is of smaller height in comparison with a dam. It raises the water level locally and supports the water against its face. Thus the diversion of the water from the river into the canal takes place. So far as possible weir is aligned at right angles to the direction of flow of water in the river.

There are various types of weirs in use. The classification of weirs, may be done in number of ways.

1. **Classification of weirs according to the functions served by them:** In this category we have:
   - **Storage weirs:** They are constructed purely for storing water. Hence they are termed as low dams also.
   - **Pick-up weirs:** In a storage project the reservoir water is discharged in the river through supply sluices. This released water is picked up by constructing a

weir across the river where canals take-off. A series of such pick-up weirs may be constructed to utilise the available water.

- **Diversion or intake weirs:** They are constructed as a part of headworks. They raise the water level by obstructing the river flow. The water is then diverted into the canal.
- **Waste weirs:** They are generally used as spillways for reservoirs. They are located on the periphery of the reservoir basin. During floods when water level in the reservoir rises above FRL waste weir comes into operation. Thus all excess flood water is discharged safely. It protects the reservoir and the main storage dam.

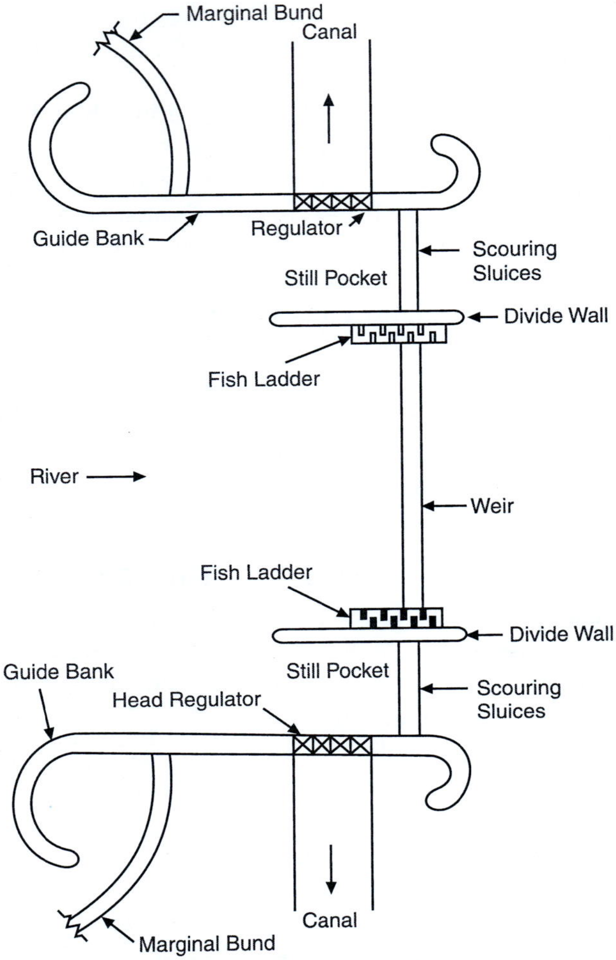

Fig. 12.1. Layout of headworks

2. **Classification according to the device provided on the crest for controlling flow:**
   - **Barrage:** It is a low obstructive barrier constructed across the river. Gates are provided on the crest of the barrier. The gates are housed in the grooves made in the piers and abutments. The piers are constructed on the crest. They support road or rail bridge. They also support a platform used for lifting and lowering the gates. Thus the flow is perfectly controlled by the gates.
   - **Weir:** All other small obstructions are simply termed as weirs. They may be designed with the crest at pond level or the crest may be fixed at a low level with some controlling device, for example, falling shutters or tainter gates or drum gates. Generally weirs with the crest at a lower level and with some controlling device is preferred to weirs with high crest level. The reason is that the latter causes excessive afflux when the river is in floods. Its another advantage is that when flood comes controlling device lowers and makes extra space available to discharge the flood quickly.
3. **Design aspect as a basis of classification of weirs:**
   - **Gravity weirs:** Gravity weirs are those which are stable due to the weight of the mass in the body of the weir. Self weight is the only restoring external force and counteracts all the disloding forces, *e.g.,* uplift, water pressure etc. All weirs except concrete weirs are gravity weirs. Even individual component parts of the weir are stable due to self-weight.

- **Non-gravity weirs:** In non-gravity weirs the structure as a whole is safe and stable against all forces. But individual component parts of the weir may not be stable since less material is used in costruction. Good example of this type is a concrete weir. Reinforcement is used to make the whole body one solid rigid mass. Then it is safe and stable.

4. **Classification of wires according to the materials of construction:**

- **Weirs constructed with masonry:** In this type of weir vertical drop is given as shown in Fig. 12.2. Shutters are generally provided to store the water upto the required level. The structure is constructed with dressed stones.

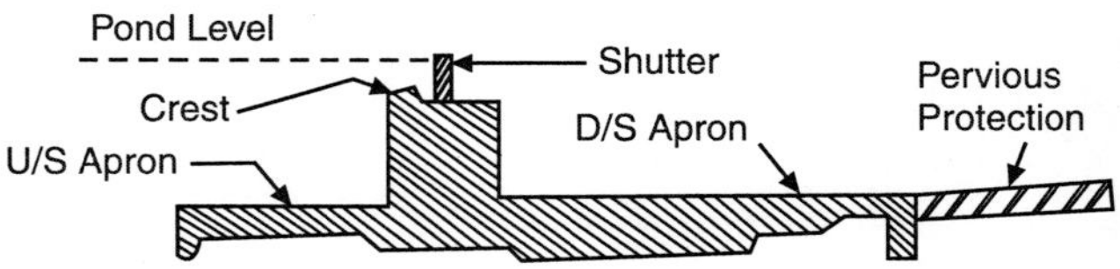

Fig. 12.2. Masonry weir

- **Weirs constructed with rock fill and masonry combined:** It is constructed with dry rockfill. It requires a very big section. Naturally for achieving economy material should be locally available at moderate cost. The crest of this weir is also controlled by the shutters. The top surface is plastered. In between masonry retaining walls are constructed to support the rockfill. (Fig. 12.3)

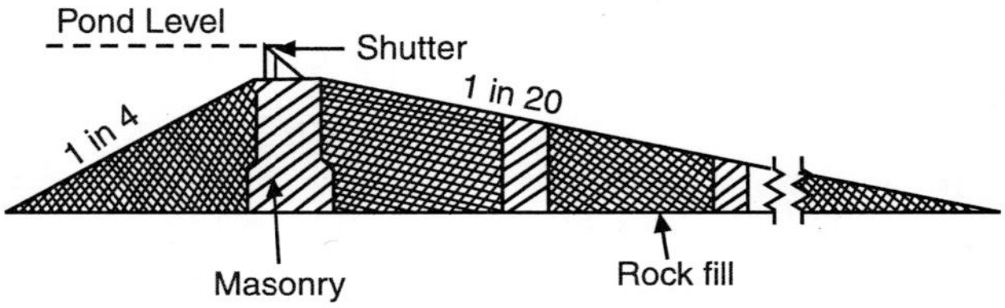

Fig. 12.3. Masonry and rockfill weir

- **Weirs constructed with concrete:** In this type a sloping glacis is provided on the downstream side. It helps in creating a hydraulic jump on the sloping face. (Fig. 12.4)

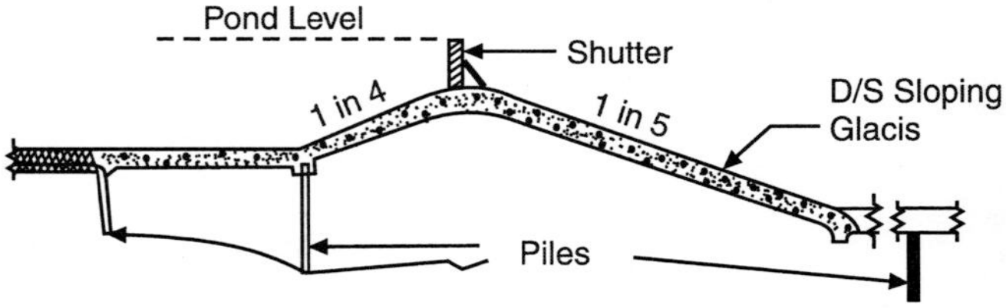

Fig. 12.4. Concrete weir

When the hydraulic jump is created the energy of flow is destroyed. Thus the bed is protected below the weir. The flow over the weir is controlled by providing shutters. This type weir is best suited on pervious foundation.

12.5. COMPONENTS OF WEIR AND BARRAGE

In a headworks the weir or barrage is the main structure. It is very essential to know various parts of a weir before actually designing the weir. Figure 12.5 give the common parts of the weir structure.

Generally weirs consists of the following parts:

- Upstream boulder pitching for bed protection.
- Upstream curtain wall.
- Upstream or fore apron.
- Shutter on the crest of the weir.
- Weir body wall.
- Downstream or rear apron.
- Downstream curtain wall.
- Downstream protection for channel bed.

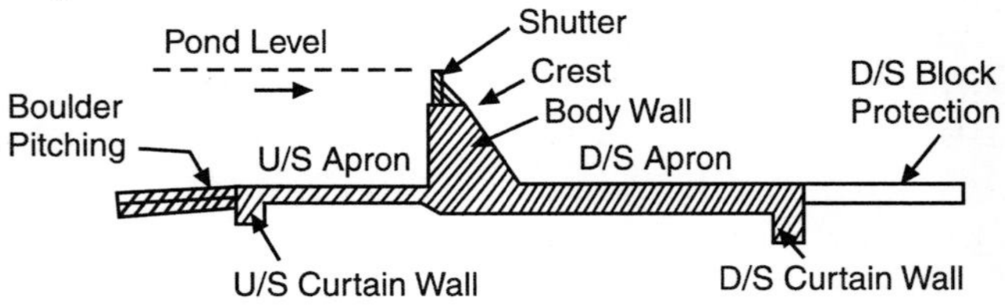

Fig. 12.5. Parts of weir

Figure 12.6 shows the common parts of the barrage structure. Generally barrages consist of the following parts:

- Upstream stone pitching for bed protection.
- Upstream sheet pile.
- Upstream apron.
- Upstream sloping glacis.
- Barrage crest.
- RCC pier with rail/road bridge on top.
- Gate on the crest between RCC piers.
- Downstream glacis.
- Downstream apron.

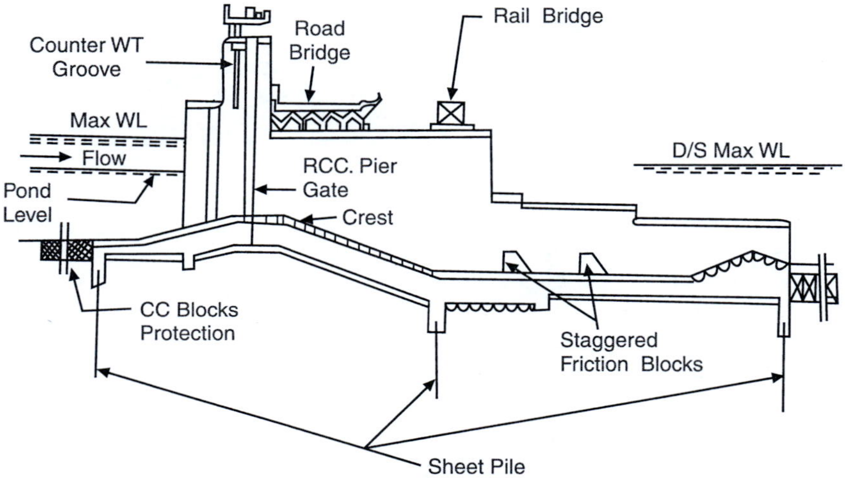

Fig. 12.6. Parts of barrage

- Energy dissipation devices like friction blocks, end sills etc.
- Downstream and intermediate sheet piles.
- Downstream stone pitching for bed protection.

12.6. DIVIDE WALL

As the name indicates it is a long solid wall or a groyne. It is contracted at right angles to the weir axis. It divides the river channel in two compartments. Thus in the smaller compartment, which is nearer to the head regulator, a still pond is created. The obvious reason is this smaller compartment is partially cut off from the main river channel where wide fluctuations in the water levels take place. The still pond is created in front of the head regulator.

Divide wall extends from the weir body to a little beyond the length of the head regulator.

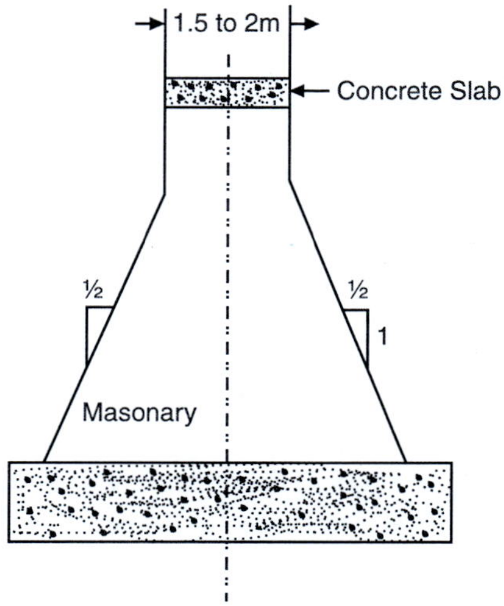

Fig. 12.7. Divide Wall

Generally divide wall is constructed with masonry (Fig. 12.7).

The top width of the wall varies from 1.5 to 2 m. The section of the divide wall is as shown in Fig. 12.7. The divide wall should always be founded on a strong foundation. Generally it is necessary to provide a well foundation for at least 30 m. It is clear that the divide wall retains the water on its both faces. Of course on one side the water is still in the pocket. Whereas on the other side there is river flow. The divide wall should be designed after taking this point in consideration.

12.7. UNDER SLUICES OR SCOURING SLUICES

They are openings provided in the body of a weir or anicut at low levels. They are located in the smaller compartment in still pond. These sluices are perfectly controlled by means of gates which are operated from top.

When a still pond is created in front of the head regulator silting takes place in the pocket. If this silting is allowed in the pocket for a long time the capacity of the pocket gets reduced. The sluices are used to remove or to scour this deposited silt. Naturally the sluices should be located quite below the crest of the head regulator (Generally 1.25 m below).

Thus the scouring sluices maintain the channel clear and defined in front of the head regulator.

The sluices can also be used during the peak flood period to lower the discharge over the crest of the weir.

The capacity of the sluices should be kept about two times the canal discharge. Then it ensures efficient scouring action. When the floods are to be passed through these sluices the capacity of the sluices should be properly designed to accomplish the purpose.

12.8. FISH LADDER

When a weir is constructed across a river with a view to check the flow of water the passage is fully closed. Naturally the fishes, which are always present in river are obstructed from moving freely. If some provision is not made for them the fish life may be perished. The structure provided for allowing free passage to the fishes is called a fish ladder. Figure 12.8 shows a plan and elevation of the fish ladder. The fish ladder is designed in such a way that the velocity of flow is between 3 to 3.7 m/sec. This velocity is such that the fish can travel in or against the direction of flow easily.

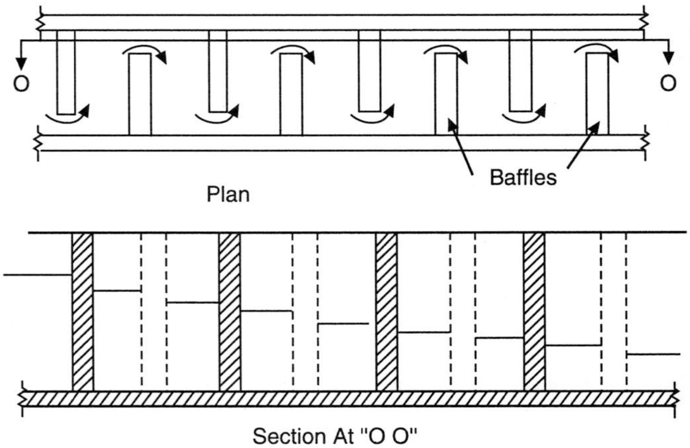

Fig. 12.8. Fish ladder

The fish ladder is generally provided on the other side of the divide wall in a bigger compartment. The baffles are arranged to stagger the flow of water.

12.9. SILT CONTROL AT HEADWORKS

The river carries appreciable amount of silt with the flow. Excessive silt entry into the canal has to be checked. If free passage is allowed to the silt it drops down and deposits on the bed and the banks. It reduces capacity of the canal. It is true that the silt entry in the canal cannot be checked cent percent. Very fine silt always remains in suspension in the canal water. Actually there should be some amount of silt in the irrigation water as it has manurial property. Silt entry in the canal can be controlled in number of ways.

1. **Still pond system:** It is adopted at every headworks. It is done by creating a pocket by constructing a divide wall. From Fig. 12.9 it is clear how a still pond is created in the pocket. Pocket is closed from three sides, head regulator being on one side, weir on the otherside and divide wall on the third side. Thus when the velocity of the water is destroyed in the pocket water drops down its silt load and only clear water enters the canal through opening in the head regulator.

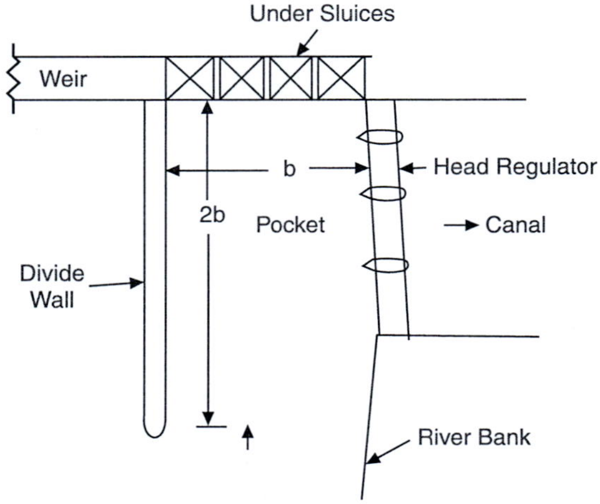

Fig. 12.9. Still pond system for silt control

2. **Regulator gates:** The silt is kept in suspension mostly in the bottom layers of the flow. Obvioulsy when only top layers are allowed to enter the canal the silt load is prevented from entering the canal. This can be achieved by providing shutters on the top of the sill or regulator crest. Most convenient method is to provide the shutters in 2 or 3 tiers. It reduces the extraordinary height required for operating the shutters. It also accomplishes the purpose of silt control very effectively. The shutters may operate in same or different grooves made in the piers and abutments. The bottom tier is usally kept closed to allow only surface water to enter into the canal. Bottom shutter goes behind the sill when it is opened. Whereas other shutters go up when ventway is to be opened.
3. **Silt excluder:** It is a structure which excludes the silt from irrigation water as the name implies. It separates the lower silt ladden portion of the water from the upper silt free portion. It consists of a series of parallel tunnels of low height. The tunnels are constructed in the pocket parallel to the flow of water in the river. The height of tunnels depend upon the silt distribution in the flow of water. Figure 12.10 shows a plan and section of a silt excluder.

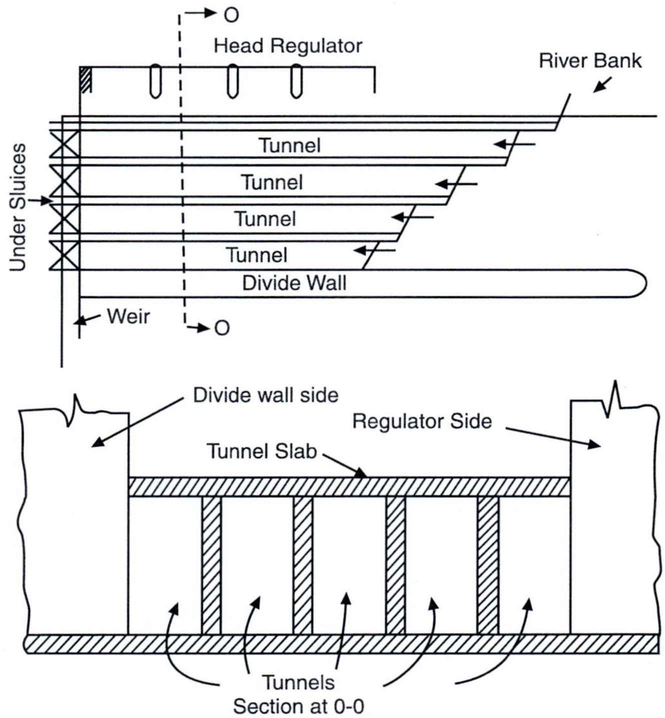

Fig. 12.10. Silt Excluder

The lower portion of the flow which contains heavy silt load enters the tunnels. The coarse silt load is then driven towards the scouring sluices. This water passes to the downstream side of the weir through the sluices. Thus only clear water is allowed to enter the canal.

12.10. TYPES OF REGULATORS

For equitable and efficient distribution of irrigation water it is very essential to regulate the supply. A hydraulic structure constructed to regulate the supply is called a regulator. The regulators are designed not only to regulate the irrigation water supply but also to control the silt entry into the irrigation canal.

To regulate the irrigation water supply it is essential to have various ventways with shutters to control the openings. The ventways are provided by constructing abutments and piers in the canal cross-section. Grooves are made in the piers and abutments in which the shutters operate.

To control silt entry the shutters are provided in tiers. The lower tier is usually kept closed. Thus the sill of the regulator opening is raised. As a result only top layers of silt free water enters the canal through the regulator.

For efficient and sucessful regulation it is essential that the regulator should serve the purpose of a meter also.

For economy a road or a rail bridge, if any, is combined with the regulator. As then the piers, abutments and the foundation work is common to both the structures.

Depending upon the location of a regulator following broad classification of regulators may be recognised:

- Canal head regulator.
- Canal cross regulator.
- Distributary head regulator.

12.11. FUNCTIONS OF HEAD REGULATORS

It is a strucutre constructed at the entrance (the head) of the canal where it takes off from the river (Fig. 12.11). The regulator serves the following purposes:

- It regulates the flow of irrigation water entering into the canal.
- It can be used as a meter for measuring the discharge.
- It regulates and prevents excessive silt entry into the canal.

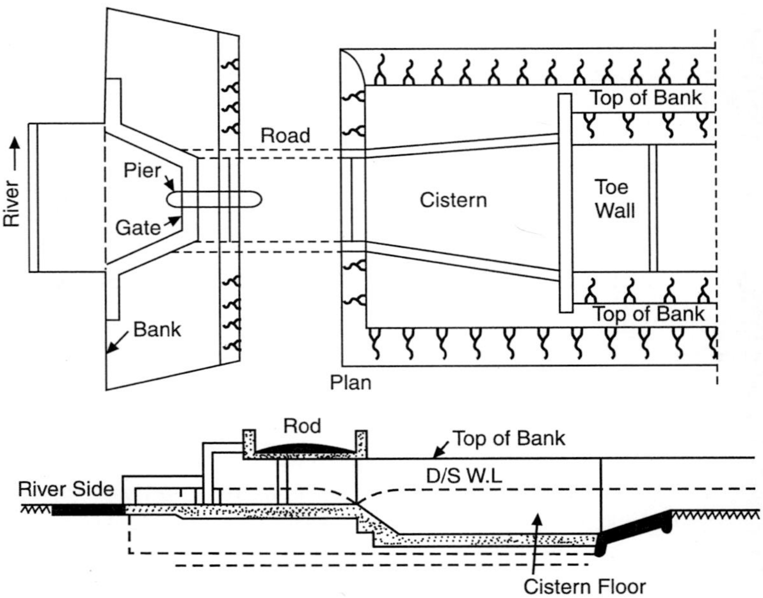

Fig. 12.11. Head regulator

It consists of a raised crest with abutments on both sides. The crest may be subdivided in various bays by providing piers on the crest. The piers support roadway and a platform for operating gates. The gates control the flow over the crest. They are housed and operated in grooves made in the abutments and piers. Sill of the regulator crest is raised to prevent silt entry. Sometimes the gates are provided in tiers. Then lower tiers may be kept closed to raise the sill of the regulator.

The head regulator is generally constructed with masonry. It should be founded on a good rock foundation. It should be safe against shear, sliding and overturning. It should be flanked with adequate wing walls. The head regulator should also be given proper protection by providing aprons on upstream and downstream side of the barrel. To prevent seepage cut-off is also essential. To take irrigation water at low velocities waterway of the head regulator should be sufficiently big.

12.12. FUNCTIONS OF CROSS REGULATOR

It is a hydraulic structure constructed across the canal to regulate the irrigation water supplies. It may be constructed across any type of canal main, branch or a distributary. Following considerations make it necessary to construct a regulator across the canal:

(*i*) When due to inadequate supply the water level is lowered the off-taking channels do not get their proper share. A cross regulator is provided to raise the water level.

(*ii*) Sometimes it becomes necessary to carry out some repair works on a canal. The cross regulator if existig above that reach of the canal, it can be closed and repairs can be done efficiently.

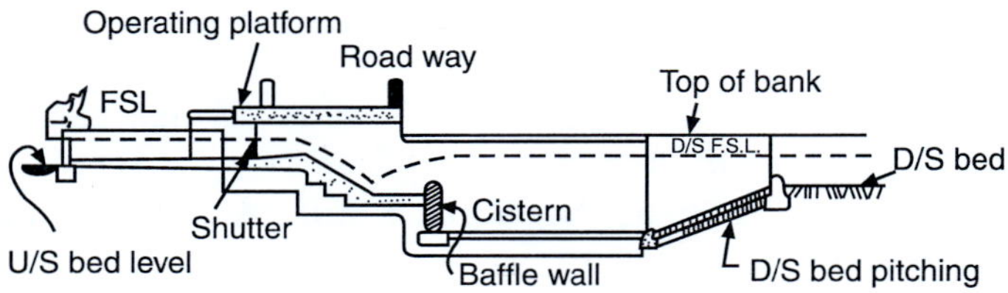

Fig. 12.12. Fall combined with a cross-regulator

(*iii*) Sometimes it is necessary to close the canal below a particular point. Say when there is no demand for irrigation water during a particular period.

(*iv*) When the costly headworks are not constructed in the initial stages, the cross-regulator helps in regulating the canal supplies.

(*v*) Cross regulators divide long canal reach into smaller ones and make it possible to maintain the reach successfully and efficiently. For efficient functioning they should be spaced 10 to 13 km apart on the main canal and 7 to 10 km on the branches.

A cross-regulator is often combined with rail or a road bridge. When a fall is available on the canal the cross regulator is constructed as a fall-regulator (Fig. 12.12). The cross-regulator may be flumed at the site. It is similar in construction to the head regulator.

12.13. FUNCTIONS OF DISTRIBUTARY HEAD REGULATOR

It is a hydraulic structure constructed at the head of a distributary. This regulator performs the same functions as that of a head regulator.

- It regulates the supply of the distributary.
- It can be used many times as a meter.
- It is also a silt selective structure.
- Distributary head regulator controls the flow in the distributary. By closing the gates distributary can be dried to carry out repairs or maintenance works.

The points to be considered in design are similar to those considered in the design of a head regulator. Only difference is that the distributary head regulator is much smaller in magnitude as compared to the head regulator. Figure 12.13. shows sectional end view of a distributary head regulator.

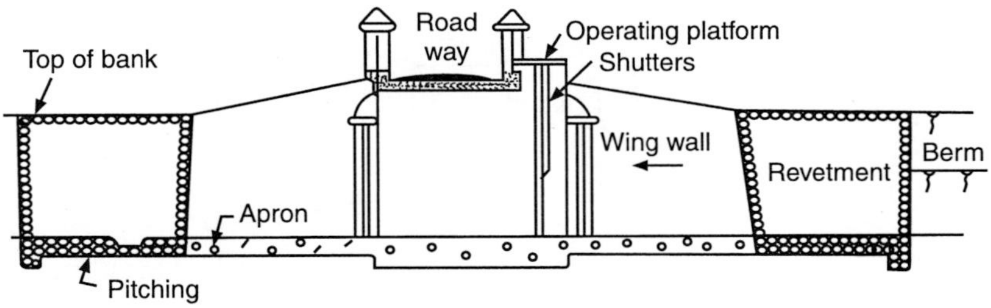

Fig. 12.13. Distributary head regulator

12.14. NECESSITY AND DEFINITION OF A FALL

Necessity: When an irrigation canal is constructed depending upon the design conditions a fixed slope is given to the bed of the canal. It is clear that natural surface never follows a uniform slope. As such when the natural surface falls very rapidly the canal bed goes very high and the canal then comes in filling. It increases the cost of construction appreciably. Obviously then it becomes necessary to bring the bed level of the canal down, from a higher level to such a lower level that the canal ultimately comes in balancing depth. Here lies the necessity of providing falls in the canal.

Definition: The structure constructed to achieve the object of lowering the bed level of a canal suitably is called a fall. When the change of elevation of water level takes place potential energy of the water is converted into kinetic energy. This excess energy increases the velocity of flow and damages downstream portion of the canal by sourcing action. Here lies necessity of some device for dissipating excess energy of flow. Thus a **complete fall** is a hydraulic structure and is a combination of a water lowering structure and an energy dissipation device (Fig. 12.14). The main parts of a fall are u/s approach, throat, d/s glacis, d/s expansion and an energy dissipator. Following requirements should be given due consideration while designing a fall:

- A fall should be able to admit of variations of water levels in the canal.
- Bed, banks and downstream portion of the structure should be safe against erosion and damage due to excess energy of flow.

12.14.1. Selection of Site for a Fall

Following points are given due consideration while selecting the site for a fall.

- So far as possible the fall should be combined with a rail or road bridge. The obvious reason is then abutments, wing walls and foundation are common to both the structures. Such combination makes a project economical.
- For achieving economy, a fall should be combined with a regulator which is required for bifurcating the canal.
- The site for a fall on the distributaries and minors should be selected in such a way that the command is not sacrificed in process of lowering the bed. Actually when the falls are provided the FSL below the fall goes below natural surface level for some length. Then it becomes necessary to provide outlet on the upstream side of the fall to command the area which ties immediately below the fall.

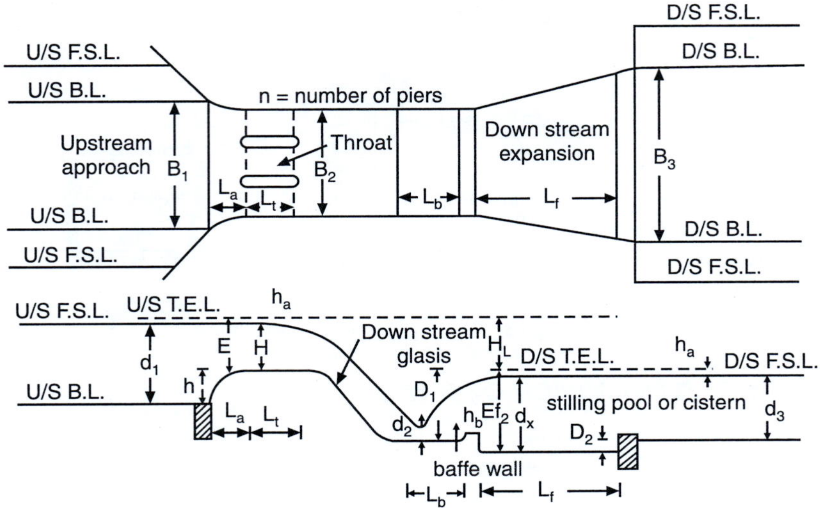

Fig. 12.14. Components of a fall structure

12.15. CROSS DRAINAGE WORKS

When a natural drain crosses or intercepts an irrigation canal it becomes necessary to construct some suitable structure to carry forward the canal safely. As these works are constructed for crossing the drainage, they are termed as cross drainage work. They are also called *CD* works. It is needles to mention that such additional works on the canal increase the cost of the project. Hence, so far as possible such works should be avoided. It may be achieved in two ways.

Firstly the alignment of the irrigation canal should be changed to avoid its crossing with the drain.

Secondly the drain itself may be diverted to the adjoining stream to avoid the crossing.

But in practice it may become impossible to avoid such a crossing. Then there is no other alternative but to construct a structure to carry the canal across the drain.

12.15.1. Main Types of Cross Drainage Works

An irrigation canal may cross a natural drain in three possible ways. They are:

- Irrigation canal bed is sufficiently above the bed of a drain.
- Irrigation canal bed is sufficiently below the bed of a drain.
- Irrigation canal and the drain cross each other at the same bed level.

Depending upon the above mentioned possibilities following four cross drainage works may be constructed respectively:

- **Adequate:** It carries an irrigation canal over a drain.
- **Super-passage:** It carries a drain over an irrigation canal.
- **Level crossing:** This structure makes it possible to dispose off drain water safely at the same level as that of a canal.
- **Inlet and outlet:** When possible drain water is taken in the canal to be discharged afterwards into a drain at suitable location.

12.16. SELECTION OF SITE AND TYPE OF CD WORK

The selection of a particular type of cross drainage work and the selection of its site depends upon various factors which can be summarized as follows:

- Nature of the foundation available at site.
- Existing condition of a natural drain.
- Bed levels of the irrigation canal and the drain.
- Relative water levels of the canal and the drain.
- Magnitude of the drain and the irrigation canal.
- Angle of crossing of the canal and the drain.
- Other available constructional facilities.

Usually the canal and the drain should cross each other at right angles. If such a crossing is not available the alignment of the canal may be altered to achieve normal crossing.

When the drainage is much bigger in size in comparison with the canal it is advisable to allow the drain to flow unaltered.

When the bed level of the canal is sufficiently higher than the High Flood Level (HFL) of the drain aqueduct should be constructed. (Fig. 12.15)

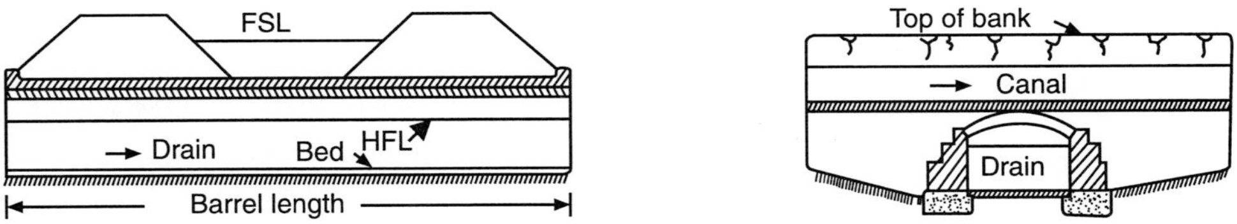

(a) Cross-section (b) L-section

Fig. 12.15. Aqueduct with unobstructed canal

Alternatively when bed level of the drain is sufficiently higher than the Full Supply Level (FSL) of the canal super-passage may be constructed. (Fig. 12.16)

When the level clearances are not sufficient for such crossings bed level of the conveyance channel passing at a lower level can be depressed at the crossing to lower the water level suitably. Such lowering is possible in case of aqueduct as well as super-passage. It is then called syphoning.

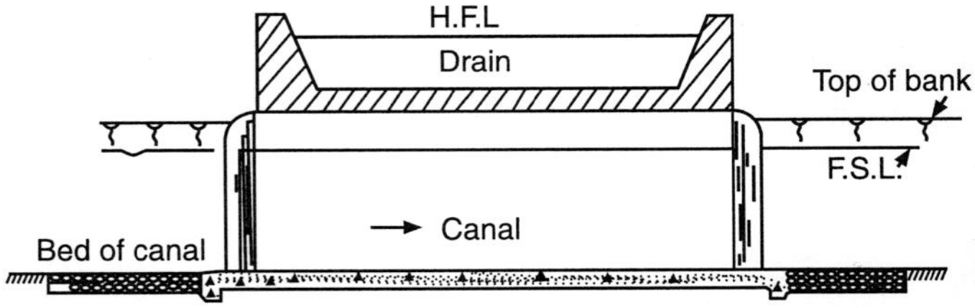

Fig. 12.16. Super-passage

To increase the clearances the alignment of the canal can be taken upstream or downstream of the drain as the situation demands.

When the drain is too big to be interfered with and levels so permit level crossing may be provided. In such a situation canal water is dropped in the drain/river from one bank and the canal is again taken-off from the other bank. This arrangement, however, requires an assemblage of hydraulic structures similar to and as big as any diversion headworks.

When the canal and the drain cross each other at nearly same level, crossing may be provided.

When a big canal crosses small drains, the drains may be let into the canal. Then the added flow of two or three drains can be discharged out of the canal at a suitable point to dispose off the additional water through a drain on the other bank. For admitting the water into the canal inlets may be constructed and for discharging the water outlets may be provided.

12.17. TYPES OF SILT CONTROL WORKS

Two steps are essentially required to gain control over the silt. They are:

- Prevention of silt entry into the canal, and
- Removal of silt which has entered into the canal system.

Prevention of silt entry into a canal is the first and an essential step because it affords better control on the silt problem. Various silt excluding devices are built in conjunction with the canal head regulator and form essential component of a canal head-works. The silt which has entered the canal and which is likely to disturb the regime condition may be removed in various ways by constructing silt ejection devices. Thus, the structures constructed outside the canal section are called **silt exclusion devices** and those constructed across the canal section are called **silt ejection devices**. Depending upon the nature of silt problem and site conditions silt ejection devices may be used singly or in combination or in conjunction with silt exclusion devices constructed at the head of the canal system.

12.18. SILT EJECTOR

It is a structure constructed across the canal. It extracts the bottom layers of silt laden water from the canal to be disposed off suitably. It consists of a series of tunnels constructed on the bed of the channel. The tunnels are parallel to the flow at entrance. The tunnels are then turned through 90 degrees to take out the water from the canal (Fig. 12.17). The tunnels are suitably subdivided into compartments to increase the velocity in the tunnels. To avoid sudden increase in velocity and consequent disturbance of flow condition a diaphragm which covers the piers to create tunnels is constructed as a cantilever slab and overhangs the tunnel entrances by sufficient margin.

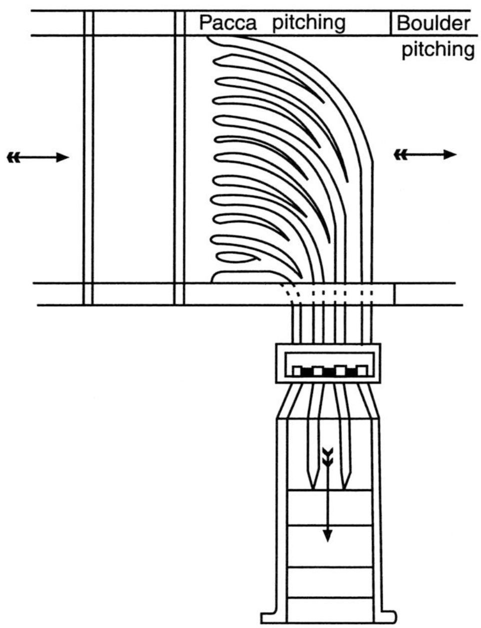

Fig. 12.17. Silt ejector

The tunnels are formed by constructing the piers in the bed of a canal and then covering the compartments so formed by a slab. To increase the efficiency of the ejector the canal should be widened just upstream of the ejector. The increased waterway helps in reducing the velocity of flow. Consequently the silt load then remains confined to the bottom layers. The bed of the tunnels may be depressed below the bed of the canal for achieving efficiency in extraction. The tunnel bed may be given suitable slope for maintaining self cleaning velocity. Once the silt laden water comes out of the canal it may be disposed off suitably in some natural drain. The proportion of the canal discharge from the bottom layers that can be used in driving silt out of the canal depends on the design of the structure and local conditions. However, if too less discharge is provided for flushing the silt it may be ineffective whereas too much provision for flushing the silt may create disturbances in the upper portion of flow thereby disturbing silt distribution in the canal flow. Also it may lead to loss of valuable quantity of irrigation water. Depending upon the design the ejector discharge may vary from 15 to 25 per cent of the canal discharge. This percentage is however a matter of judgement and depends on silt charge and silt grade.

QUESTIONS FOR REVISION

1. Mention the points which affect selection of a site for headworks.
2. (*a*) Draw a typical layout plan of headworks.
   (*b*) Mention the functions of each part in brief.
3. What is a weir ? Give classification for weirs.
4. Differentiate between the following:
   (*i*) Weir and dam
   (*ii*) Barrage and weir
   (*iii*) Divide wall and spur
   (*iv*) Barrage and dam.
5. Why it is necessary to control silt entry in the canal ? What methods are adopted for the purpose?
6. Why canal regulators are provided?
7. Write short notes on the following:
   (*a*) A fall
   (*b*) Silt ejector
   (*c*) Aqueduct
   (*d*) Super passage.

❑❑❑

13

Irrigation Outlets

13.1. REQUIREMENTS OF IRRIGATION OUTLETS

Outlets are the openings constructed in the banks of distributaries and minors. Field channel takes off from this point and gets irrigation water through outlet. It is also called **turnout** or **sluice** in some parts of India.

The field channels below the outlets are maintained by the cultivators. Once the water comes out of the outlet irrigation department has no control over it. Further, in a canal system there are numerous outlets. Hence a perfect outlet should fulfil following requirements:

- Cost of construction of an outlet should be low.
- Design of outlet should be simple and the construction should be speedy.
- Depth of flow in minors and distributaries is quite low. Hence the outlet should work under low head.
- The outlet should draw the silt in proportion to its discharge.
- It should be difficult for cultivators to tamper the outlet.
- There should not be moving or loose parts in the outlet. They are liable to be damaged.
- The outlet should discharge a known and constant quantity of water. It should be independent of water depths in distributary and the field channel.
- The outlet should work equally well when head is high and when head is low.
- The outlet should provide a measure of the rate of flow passing through it.

Expensive canal system with its accessories is planned and constructed for making irrigation water available to the cultivators. Hence, it should be remembered that in achieving economy efficient functioning of the outlet is not sacrificed because then the whole irrigation project becomes unsuccessful from the point of view of the cultivators.

13.2. CLASSIFICATION OF OUTLETS

Depending upon their principle of performance outlets may be classified as shown below:

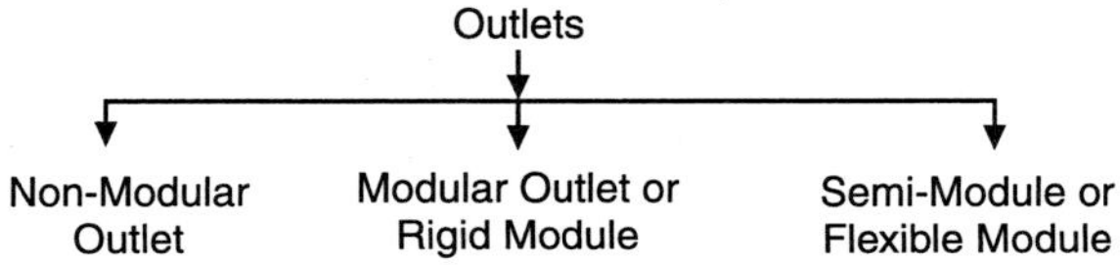

A **non-modular outlet** is an ordinary type of outlet in which discharge is directly dependent on the working head. That is, it depends on the difference of water level in the parent channel and in the field channel. Thus, when working head is less discharge is less. On the contrary when the head is more discharge is more. Submerged pipe outlets, masonry sluices and orifices come under this category.

A **modular outlet** is also termed as a rigid module. In this type of outlet the discharge is maintained constant and it is not affected by the water levels in the parent and the field channel. Gibb's rigid module comes under this category.

Sometimes a **semi-modular outlet** is also termed as a flexible module. This type is intermediate between the two types mentioned above. In this type of outlet discharge directly depends on the water level in the parent channel. But it is independent of the water level in the field channel. Crump's adjustable proportional module and Kennedy's semi-module come under this category.

From the above discussion it is clear that the modular outlets are better from the point of view of equitable distribution. But in modular outlets loss of head is more. Hence, non-modular outlets are also used in some areas (for example, in Uttar Pradesh). Though discharge varies with working head this type is simple and cheap. When the range of fluctuation of water levels is not large semi-modular type of outlet is best suited.

13.3. DEFINITIONS

Before proceeding with the subject it is useful to understand the following terms:

1. **Module:** It is a type of irrigation outlet which discharges constant quantity into the field channel irrespective of the available working head. Working head being the difference in water levels of the parent channel and that of the field channel.
2. **Semi-module:** It is that irrigation outlet in which discharge is dependent on the water level in the parent channel. However, it is independent of the fluctuation of water level in the field channel.
3. **Flexibility:** It provides a criteria for judging the behaviour of modules and semi-modules. It is generally denoted by letter 'F'. It is the ratio of the rate of change of outlet discharge to the rate of change of discharge of the parent channel.

   Thus, it can be put in the form of an equation as below:

$$F = \frac{dq/q}{dQ/Q}$$

where Q is the discharge of parent channel

and q is the discharge of outlet.

It is known that the discharge of the parent channel is given by equation

$$Q = C \cdot D^n$$

where D is depth of flow in canal

C is a constant, and

n is a canal index.

Differentiating the equation, we get

$$dQ = n \cdot C. D^{n-1} \cdot dD$$

$$\therefore \quad \frac{dQ}{Q} = \frac{n \cdot C \cdot D^{n-1}\, dD}{C \cdot D^n} = n\frac{dD}{D}$$

Similarly, let discharge of the field channel is given by

$$q = K \cdot H^m$$

where H ishead acting on the outlet

K is a constant, and

m is outlet index.

K and m depend on the type of outlet.

Differentiating the outlet discharge equation, we get

$$dq = m \cdot K \cdot H^{m-1}\, dH$$

$$\therefore \quad \frac{dq}{q} = \frac{m \cdot K \cdot H^{m-1} \cdot dH}{K . H^m} = m\frac{dH}{H}$$

Substituting values of $\frac{dq}{q}$ and $\frac{dQ}{Q}$ we get

$$F = \frac{dq/q}{dQ/Q} = \frac{m \cdot dH/H}{n \cdot dD/D} = \frac{m}{n} \cdot \frac{D}{H} \cdot \frac{dH}{dD}$$

As a change in the water depth of the parent canal results in an equal change in the head on the outlet.

$$dD = dH$$

$$\therefore \quad F = \frac{m}{n} \cdot \frac{D}{H}$$

4. **Proportionality and Proportional module:** Proportionality is said to have been achieved if flexibility 'F' for a module is equal to unity. The module is then said to be proportional module. It is a type of semi-module in which the rate of change of outlet discharge is equal to the rate of change of parent channel discharge.

   Thus, for the proportional outlet

$$F = \frac{m}{n} \cdot \frac{D}{H} = 1$$

or

$$\frac{H}{D} = \frac{m}{n}$$

5. **Setting:** It is the second criteria for judging the behaviour of modules and semi-modules. It is the ratio of the depth of sill of a module below the FSL of the parent channel to the FS depth of the parent channel.

   Obviously $\frac{H}{D}$ represents setting.

   The significance of the setting can be explained as follows:

   (*a*) In case of trapezoidal canal with $\frac{1}{2}$: 1 side slopes

   $Q \propto D^{5/3} \quad \therefore \quad n = 5/3$

   (*b*) In case of weir type outlet

   $q \propto H^{3/2} \quad \therefore \quad m = 3/2$

   (*c*) Similarly in case of orifice type outlet

   $q \propto H^{1/2} \quad \therefore \quad m = 1/2$

   We know that for proportionality

$$F = 1 = \frac{m}{n} \cdot \frac{D}{H} \quad \text{or} \quad \frac{H}{D} = \frac{m}{n}$$

Taking a case when weir type outlet is provided in the banks of a trapezoidal canal, for proportionality ($F = 1$)

$$\text{Setting} = \frac{H}{D} = \frac{m}{n} = \frac{3}{2} \div \frac{5}{3} = \frac{9}{10}$$

or $$H = 0.9 \text{ D}$$

In other words the weir type outlet will be proportional if the sill is set at 0.9 D (D is FS depth of the parent channel) below FSL of parent channel.

Similarly, when an orifice type outlet is constructed in the banks of trapezoidal canal it will be proportional $\left(i.e., \frac{H}{D} = \frac{m}{n} \right)$

when $$\frac{H}{D} = \frac{m}{n} = \frac{1/2}{5/3} = 0.3$$

or $$H = 0.3\ D$$

This means outlet has to be set at 0.3 D

(where D = FSD) below FSL of parent channel.

6. **Hyper-proportionality:** When flexibility is greater than one, the setting $\frac{H}{D}$ has to be less than $\frac{m}{n}$ because $F = \frac{m}{n} \cdot \frac{D}{H}$.

   When $F > 1$ the outlet is called hyper-proportional outlet.

7. **Sub-proportionality:** Similarly when $F < 1$ the setting $\frac{H}{D}$ has to be more than $\frac{m}{n}$.

   Thus, when $F < 1$ the outlet is called sub-proportional outlet.

8. **Sensitivity:** It is a third criteria for judging the behaviour of semi-module and modules. It is generally denoted by leter S. It is the ratio of the rate of change of outlet discharge to the rate of change of water level of the parent channel when referred to the normal depth of water in the parent channel. It can be stated by mathematical expression as follows:

$$S = \frac{dq/q}{dG/D}$$

   where G is the gauge reading so adjusted that when $q = 0$, $G = 0$. Also since $dG = dD$

$$S = \frac{dq/q}{dD/D} \qquad ...(1)$$

   It is already mentioned that

$$F = \frac{dq/q}{dQ/Q} = \frac{dq/q}{n\, dD/D} \qquad ...(2)$$

   From (1) and (2), $S = n \,.\, \boldsymbol{F}$

   Since the rigid module maintains a constant discharge its sensitivity is zero.

9. **Drowning ratio:** It is the ratio of the depth of water over the crest on the downstream side to the depth of water over the crest on the upstream side.

10. **Efficiency of outlet:** It is a measure of conservation of head by an outlet. Efficiency is given by the ratio of head retained by an outlet to the head acting on the outlet. In weir outlet the efficiency is same as the drowning ratio. An outlet which functions with minimum working head can be considered to be more efficient.
11. **Minimum modular head:** It is the minimum head required to keep an outlet operative.
12. **Modular limits or modular range:** The upper and lower limits of any one or more outlet design parameters beyond which an outlet becomes inoperative (*e.g.*, a module or semi-module specify the modular limits or modular range). Obviously a semi-module or a module has to be necessarily designed within the modular range to make it successful.
13. **Non-modular outlet:** It is that type of outlet in which discharge is directly dependent on the working head.

13.4. NON-MODULAR IRRIGATION OUTLETS

13.4.1. Pipe Outlet

It is provided in the form of a simple opening made in the canal banks which leads water from the parent channel to the field channel. The opening may be circular or rectangular in shape. In the former pipeline may be used. The rectangular tunnel or barrel may be constructed of masonry. Figure 13.1 shows the longitudinal section of a non-modular pipe outlet. The diameter of the pipe may range from 10 to 30 cm. The pipeline is laid on a light concrete foundation to prevent possibility of settlement.

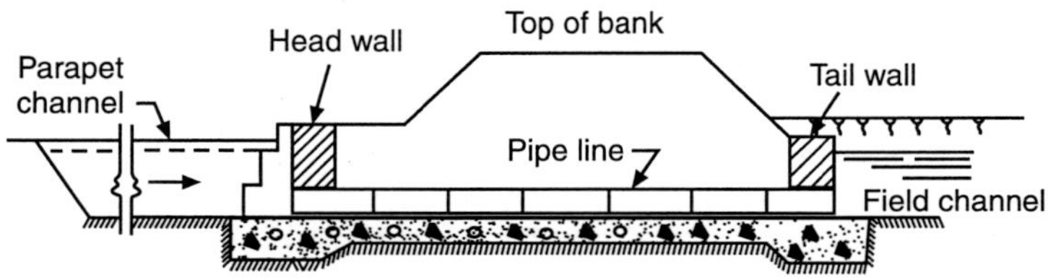

L- Section

Fig. 13.1. Pipe outlet (Non-modular type)

The opening is generally drowned and hence the outlet discharge depends on the difference of water level of parent and field channel. The loss of head through pipe is given by well known relation

$$H = \left(0.5\frac{v^2}{2g}\right) + \left(\frac{4flv^2}{2gd}\right) + \left(\frac{v^2}{2g}\right)$$

First term gives entry loss, second friction loss and third velocity at exit. Discharge is given by $q = KA\sqrt{H}$.

So, far as possible the pipe line or the rectangular tunnel is constructed at right angles to the parent channel. The pipeline or barrel is generally laid in horizontal position. When the outlet is feared to draw more silt share the pipe line may be laid in inversely inclined position with a rise of 1 in 12 (Vertical : Horizontal). Then the parent channel end of the pipe is depressed whereas an exit end is raised. The location of the entrance end depends on the type of parent channel. For channels in which the discharge variation is more sill of the opening is kept at the bed level of the channel. Whereas when there is no appreciable change in discharge conditions the opening may be kept slightly below FSL of the parent channel. To regulate the discharge through the outlet shutter may be fixed at entrance end with some type of locking arrangement.

13.5. MODULAR IRRIGATION OUTLETS

As the outlet discharge of this type is independent of the difference of water levels of the parent channel and field channel it is also called **rigid module**. Modular outlets may be constructed with movable parts. But then the movable parts are liable to be damaged or choked. Hence, this type is not used in practice. As a result modular outlets with immovable parts are evolved. They are Foote module, Spanish module, Khanna's module, Gibb's module, etc. Description of Gibb's module is given here.

13.5.1. Gibb's module

It is a modular outlet. Irrigation water is taken through an inlet pipe to a rising pipe. The rising pipe is in the form of a spiral. Generally, it is semi-circular. The water when flows through it is turned through 180°. During the movement in the rising pipe vortex motion is developed. As the flow is continuous angular velocity of the flow is same.

Angular velocity $\omega = v.r.$

where v is tangential velocity, and r is radius of flow.

Obviously, the tangential velocity of flow at inner radius of the rising pipe is greater than that at outer radius. Also there is centrifugal head impressed on the water. As a result depth of water at outer radius is more than that at inner radius of the rising pipe.

The rising spiral pipe is connected to an eddy chamber. Figure 13.2 shows the plan and longitudinal section of Gibb's module. It gives clear idea about the arrangement of the component parts.

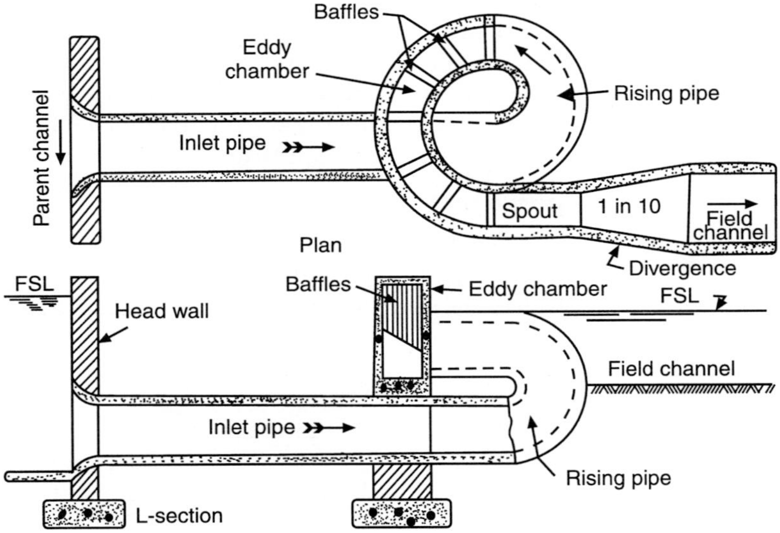

Fig. 13.2. Gibb's module

The eddy chamber is rectangular in section but semi-circular in plan with horizontal floor. It takes back the water in the original direction of flow. In the eddy chamber the baffles are provided at equal distance to dissipate excess energy of flow and to maintain a constant discharge. The baffles do not rest on the bottom of the eddy chamber but there is an opening left between the floor of the chamber and the lower end of the baffles. This bottom opening is not rectangular in-shape but the height of opening reduces towards the inner side of the chamber. Thus, the lower end of the baffle is not flat but it is kept sloping. This arrangement helps in maintaining constant discharge. When the energy of incoming water is more for perfect dissipation of energy the length of eddy chamber and in turn the number of baffles is increased. This is attained by giving one full turn to the eddy chamber in addition to the previous half turn. Thus, the eddy chamber is given one and a half turn.

After the excess energy of flow is destroyed and discharge is made constant the water from eddy chamber is taken into a spout. The spout is connected to a field channel by means of expansion walls. The walls are generally splayed with 1 in 10 (lateral : longitudinal) expansion.

The Gibb's module can be designed to give 0.03 cumec constant discharge for modular range of 0.3 m. The minimum working head required to maintain this discharge is 0.12 m.

At this stage it may be recognized that as modular outlets require complicated arrangement of parts it is quite expensive. Secondly, in alluvial tracts the silt trouble is more. Outlet gets choked up with silt. Hence, this type is not much in practice.

13.6. SEMI-MODULAR OUTLETS

It is already mentioned that in this category outlet discharge is independent of the water level in the field channel. Hence, this type may be correctly recognized as intermediate type to modular and non-modular outlets. It is designed to utilize the advantages of both the types in a limit. When the water level in the parent channel is high all outlets derive proportionately more discharge and protect the channel from being damaged. Also when the level in the parent channel is low all the outlets derive correspondingly smaller discharge to maintain equitable distribution even at the tail of the channel. Thus, this is most suitable type of irrigation outlet and hence widely used.

There are various types of semi-modules namely, Free discharging pipe outlet, Kennedy's gauge outlet, Scratcheley outlet, Harvey Stoddard module, Crump's open flume outlet, Crump's adjustable proportional module, etc. Out of all these types Crump's adjustable proportional module is widely used in Punjab. Crump's adjustable proportional module and Kennedy's semi-module, open-flume outlet, pipe outlet are described below.

13.6.1. Crump's Adjustable Proportional Module

Generally abbreviation APM is used for this type. It is also called **Adjustable Orifice Semi-Module** (A.O.S.M.). Figure 13.3 gives the plan and longitudinal section of APM.

In this type a cast iron roof block is provided by bolts in masonry at the entrance end. This block is given lemniscate curve at the lower end on the entrance side. It is given a tilt of 1 in 7. At the sill also a cast iron base is provided. A check plate 0.3 m wide is also provided. To facilitate smooth water entry upstream wingwall is made smaller is length. There is a throat of uniform width for about 0.60 m. Then the side walls diverge out with a radius of 7.625 m. The bed of the outlet is laid with a slope of 1 in 15 till it joins the bed of water-course. Whole outlet is constructed with masonry. Thus, this outlet is perfectly rigid once the roof block is fixed. But at the same time after dismantling the masonry slightly the opening can be adjusted by lowering or raising the roof block. Velocity of water in the outlet barrel is above critical. As a result hydraulic jump occurs on the sloping bed of the outlet downstream of the crest. This makes the outlet discharge independent of the flow conditions in the field channel.

The discharge through the outlet is given by formula

$$q = cd \cdot \sqrt{2g} \cdot B \cdot Y \sqrt{h}$$

where q = outlet discharge in cumec

B = width of outlet opening in m

cd = constant = 0.91
Y = height of outlet opening above crest in m
h = working head in m
= distance between canal FSL and lowest point of the roof block in m

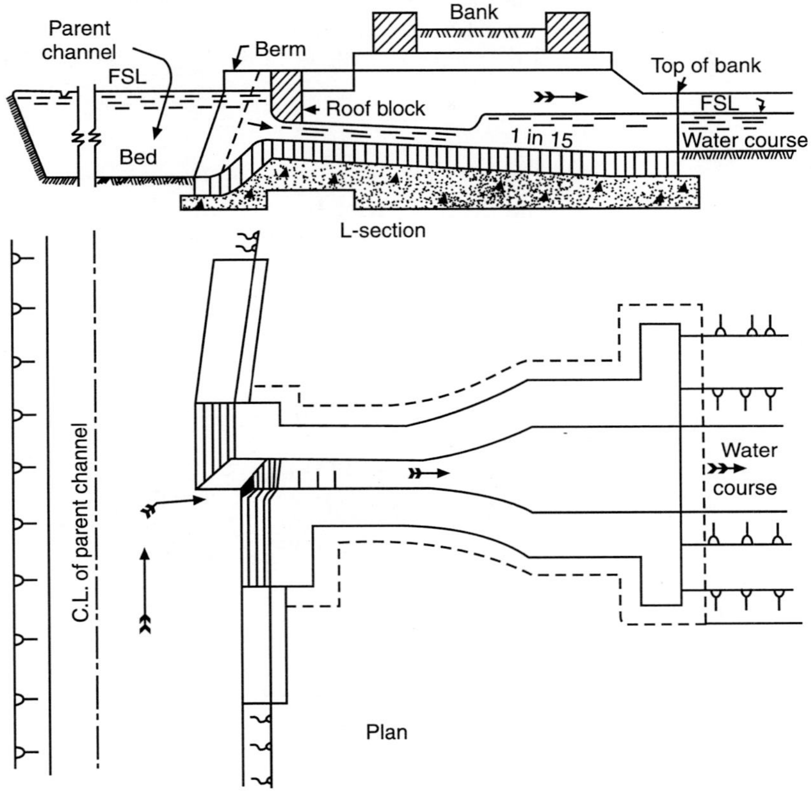

Fig. 13.3. Crump's Adjustable Proportional Module (APM)

Therefore, above equation reduces to

$$q = 4.04 B \cdot Y \cdot \sqrt{h}$$

For Minimum modular head

Mr. N.D. Gulati gave following relation

$$\text{MMH} = 0.82\ h - 0.5\ \text{B}$$

For APM $\qquad m = 1/2$ and $n = 5/3$

$\therefore$ For proportionality (*i.e.*, $F = 1$), setting should be

$$\frac{H}{D} = \frac{h}{D} = \frac{m}{n} = \frac{1}{2} \div \frac{5}{3} = \frac{3}{10}$$

or $\qquad h = 0.3\ D$

13.6.2. Crump's Open Flume Outlet

This type was first constructed on the Bari Doab canal in Punjab. Later on this type was modified slightly and a standardised **Punjab open flume outlet** was evolved and widely adopted. The main features of the two subtypes are described below.

1. Crump's open flume outlet: It is nothing but a weir with contracted throat followed by an expanding flume on the downstream (Fig. 13.4). The length of weir crest is 2.5 G, where G is head over weir crest in m.

Owing to fluming hypercritical velocity is generated on the d/s of the crest and hydraulic jump occurs. It is also, therefore, independent of the water level in the field channel as in Crump's APM. Upstream (u/s) wing wall is made smaller by the distance equal to width of opening of the outlet at its mouth. If it is equal to W, wing wall is set back by W and its value is given by

$$W = \frac{q}{Q}$$

The set back was provided to enable the outlet to take fair share of silt. The length of flume d/s of crest will naturally be equal to the horizontal length of the distributary bank. The slope of d/s glacis depends on the bed level in water course.

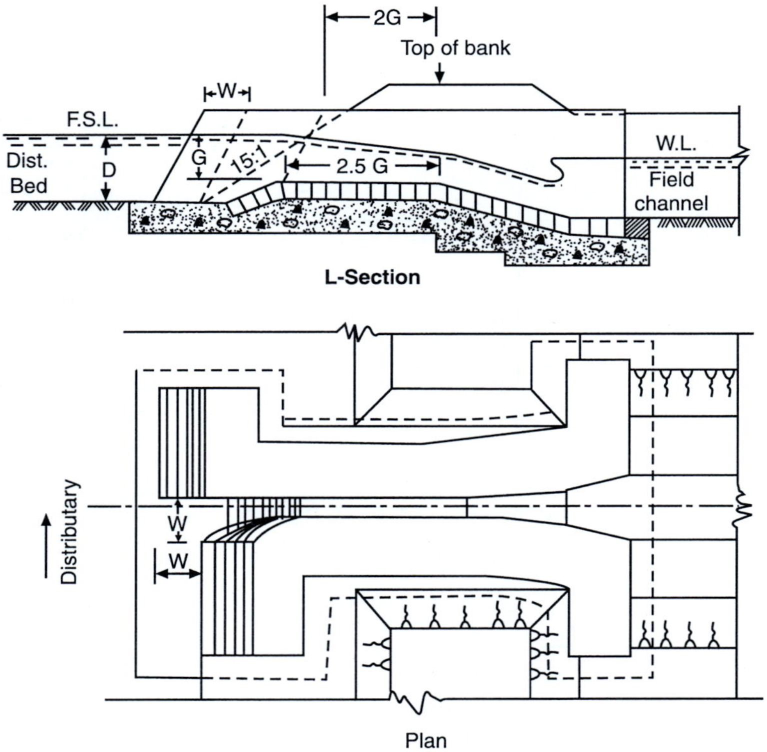

Fig. 13.4. Crump's open flume outlet

The discharge of outlet is given by equation

$$q = K \cdot B \cdot G^{3/2}$$

where G = Head over crest in m

and K = Coeff. of discharge with theoretical value of 1.71.

Owing to losses for different widths of throat, value of K differs and can be taken as follows:

B in m	K
0.06 to 0.09	1.60
0.09 to 0.12	1.64
over 0.12	1.66

The minimum modular head (MMH) is adopted to be 0.2G.

The setting $\left(\frac{H}{D}\right)$ for proportionality (*i.e.*, $F = 1$) is given as follows:

(for weir) $m = 3/2$ and $n = 5/3$ (for trapezoidal canal)

$$\therefore \quad \frac{H}{D} = \frac{m}{n} = \frac{3}{2} \div \frac{5}{3} = \frac{9}{10} \quad \text{or} \quad H = 0.9\text{ D}$$

2. Punjab Open Flume Outlet: Figure 13.5 shows the Punjab open flume outlet. Only differences are that the approaches have been modified to induce more silt into the outlet and the length of the throat is kept equal to 2G.

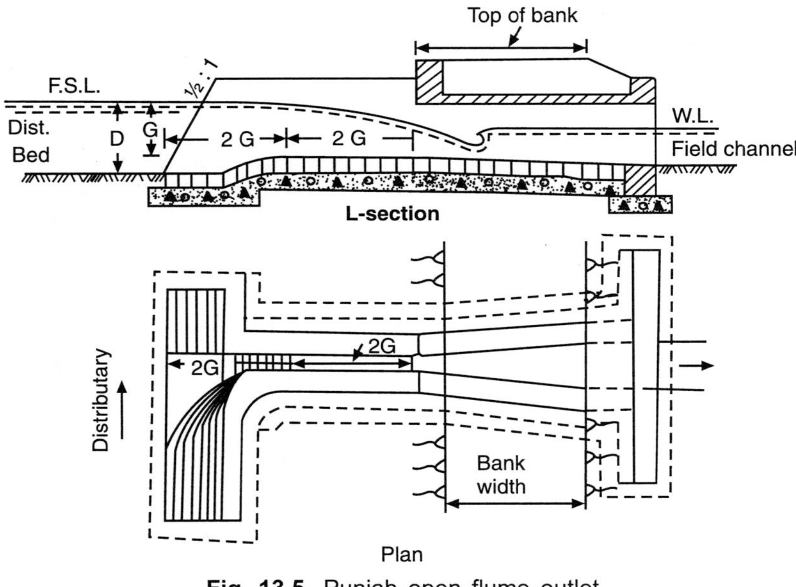

Fig. 13.5. Punjab open flume outlet

13.6.3. Pipe Outlet

When a pipe outlet discharges freely in the atmosphere the discharge through the outlet is not dependent in any way on the water level in the water course. In such cases pipe outlet can be said to be working as a semi-module.

For proportionality (*i.e.*, $F = 1$) setting should be equal to

$$\frac{H}{D} = \frac{m}{n}$$

We know that in this case $m = \frac{1}{2}$ and $n = 5/3$

$$\therefore \quad \frac{H}{D} = \frac{1}{2} \div \frac{5}{3} = \frac{3}{10}$$

or $$H = 0.3\ D$$

13.6.4. Kennedy's Semi-module

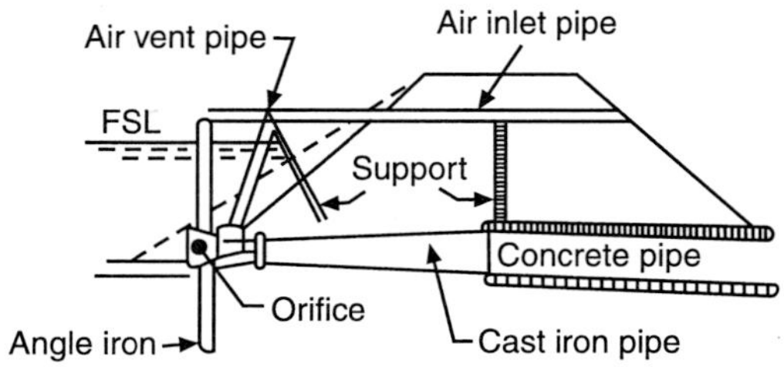

Fig. 13.6. Kennedy's semi-module

It consists of a bell mouth orifice. It is made of cast iron. The orifice abutts against a truncated cone which is slightly bigger in diameter than the orifice. An air vent pipe is fitted at the junction of the cone and the orifice. The air vent pipe is kept sloping and is protected by an angle iron on the outer side. An enamelled gauge is fixed on the angle iron (Fig. 13.6).

The air vent pipe is fitted to permit the orifice to discharge into free air at atmospheric pressure. The air vent pipe is connected to an air inlet pipe at the top. Air inlet pipe is horizontal perforated pipe laid on dry ballast. It makes the discharge independent of the water level in the field channel so long as the minimum modular head is available. Minimum modular head is 0.22 H, where H is the depth of water over the centre of the orifice. The water discharges at atmospheric pressure from bell mouth orifice into the truncated cone. The water is led further through cast iron expansion pipe to a concrete pipe and from there to the water course.

The outlet is cast in definite sizes for fixed value of discharges. The intermediate discharges may be obtained by raising or lowering the outlet orifice. This type of outlet is open to tampering easily by closing air vents. This causes the pressure drop in the chamber because the entering water jet sucks the air of the chamber. This increases the discharge of the outlet. This type of outlet is not therefore very much in use. The outlet discharge is given by the following formula:

$$q = A \cdot C \cdot \sqrt{2gH}$$

where A is cross sectional area of the pipe at throat
H is depth of water from centre of orifice to FSL
C is coefficient of discharge = 0.97

13.7. SELECTION OF TYPE OF OUTLET

The selection of a particular type of outlet should be done after considering the following points:

- Fluctuation of water level in the parent channel.
- Whether the site of an outlet is in a head reach or in a tail reach of a channel or it is above a cross regulator. In short, position of the parent channel should be taken into consideration.
- Provision for allowable loss of head.
- Condition of a parent channel, whether it has attained its regime or still in initial regime.
- Condition of a field channel.

13.8. TAIL CLUSTERS

A structure constructed at the tail of an irrigation channel is called a tail cluster. It is nothing but an assemblage of more than one similar outlets. The discharge of the canal reduces with increase in its length. When a distributary or a minor carries 0.14 m^3/sec or less water distribution is done by the tail cluster. It assures equitable distribution. The tail cluster consists of similar outlets. They are always arranged symmetrically in a group. The most suitable type of outlet in such situation is an open flume semi-modular outlet.

QUESTIONS FOR REVISION

1. (*a*) What are outlets, enumerate different types of outlets which are in common use.
   (*b*) Draw a figure of Crump's adjustable proportional module and explain its working.
2. What is meant by the following terms when they are used in connection with outlets:
   (*a*) Fixibility
   (*b*) Setting
   (*c*) Sensitivity
   (*d*) Proportionality
3. (*a*) State the requirements of an ideal outlet.
   (*b*) On what considerations a type of outlet is selected.
4. Write short notes on the following:
   (*a*) Tail cluster
   (*b*) Proportional module
   (*c*) Semi-modular pipe outlet
   (*d*) Drawning ratio.

14

River Training

14.1. IMPORTANCE OF RIVER TRAINING

It is already mentioned that when rainfall occurs or when the snow melts at high altitudes the consequent run-off takes place, to drain off the rainwater natural drainage channels are formed. With time these channels become more and more defined. The defined channels are nothing but rivers. Thus it is clear that the formation and development of a river takes years together. The draining water bring and carries along with it sand and silt load from the catchment area.

The flow in a river is never regular and constant. In rainy season floods occur due to incessant and intense rainfall. It is observed that the discharge of river during flood period may increase 50 to 100 times that of normal discharge of the river.

It may be stated here that due to varied flow conditions and due to construction of various works (bridge, weir, etc.) across the river channel following problems are met with in river engineering:

- Bad effects of closing or opening of certain channels in multiple channel river.
- Hydraulic losses at constructions, for example, bridge piers, coffer dams, etc.
- Probable scouring around the hydraulic structures which are constructed across the river channels.
- Erosion of banks which may endanger the populated and cultivated areas along the banks. In other words problems regarding shifting of a river channel.
- Flood control of rivers.
- Problems concerning navigation.
- Lastly tidal and estuary problems.

Any work constructed across the river channel interferes with the natural regime condition of the river. To ensure safety of the region through which the river flows and to make the river water available at all times for various purposes, including irrigation, it becomes essential either to control the river or to give proper training to the flow of water in the river channel. This can be achieved by understanding river hydraulics and devising suitable criteria for river control.

14.2. CLASSIFICATION OF RIVERS

Now it is clear that the rivers find their origin at high altitudes. The rives then follow natural slope of the ground and finally join sea. Of course it is not essentially true that finally each river joins the sea. It may join some big river or tank or may die or dry up in its course. Ordinarily each river should have same sequence of stages. But depending upon the regions through which the river flows following classification of rivers in a broad sense may be given for Indian territory:

- **Himalayan rivers:** The rivers which originate from the mountain Himalaya are termed Himalayan rivers. They derive their supply from melting of snow during the spring and summer season and also from rains during monsoon season. Thus the water contribution is continuous throughout the year. Hence, they are also termed as perennial rivers. The yield from such rivers is more or less of dependable nature.
- **Non-Himalayan rivers:** They originate from some other mountains in India. Obviously rivers of South and Central India come under this category. As they are not snow fed they are non-parennial mostly. The supply of water is mainly in rainy season. The rivers practically dry up in the summer season.
- **Flashy rivers:** In this type of rivers the flood occurs and recedes very suddenly. The flood may rise and recede in a short period of one or two days. The characteristic of this type is that the flood hydrograph is very steep.
- **Virgin rivers:** This type of river dries up before joining any river or sea unlike other rivers. The river dries up because of excessive evaporation and percolation losses. Such rivers are of common occurrence in desert areas. In India there are several virgin rivers in Rajasthan and Kutch area.

14.3. STAGES OF RIVER

From the above discussion it is clear that a river may be subdivided mainly in two main divisions namely:

- River in upper reach (in hilly region)
- River in lower reach (in alluvial plain or in flood plain).

Upper reaches of rivers can be further subdivided in various stages as follows:

14.3.1. Mountain Stage

It is a first stage of the river course. In this stage the river flows through steep valleys in the hills. The river bed in generally composed of rocks. The slope of the bed is also very steep and it may range from 1 in 100 to 1 in 500. As the region is hilly there exist many abrupt falls. The river water is very clear in this stage.

14.3.2. Sub-mountainous Stage

Here river leaves the mountain and enters sub-mountainous tract. The slope of the river is obviously flatter. It may range from 1 in 500 to 1 in 1000. This river stage is subdivided into two categories.

1. **Incised stage:** In this type mostly the river channel is formed by the process of degradation. The bed and the banks of the river channel are highly resistant to erosion. Naturally all the load which the river water carries is brought from the catchment area due to the process of erosion and denudation. Hence mostly there is no similarity in character between the load carried by the river and its bed.
2. **Boulder stage:** In this stage the bed is made up of a mixture of sand, shingle, gravel and boulders. As the bed material is very pervious there exists good amount of subsoil water flow. Rivers in this stage have straighter courses with wide bed and non-submersible banks.

   During flood period the coarse material comes in the river channel together with the high velocity flow and when the flood subsides this heavy and coarse material piles up in heaps. The water flows round the heaps and the river flow is bifurcated. Thus shifting braided and interlaced channels are formed in the river course.

When the river enters lower reach after traversing upper reach normally it attains following stages before it joins the sea finally:

14.3.3. Through Stage

This stage of river is also known as river in flood plain or alluvial river because in this region it forms new land by depositing its excess load during the flood season. The slope of the bed is still flatter and ranges from 1 in 1000 to 1 in 2000. During floods the river inundates very large areas causing considerable damage. The sediment load which the river carries is similar is character to that of the river bed and the banks. The rivers in this stage generally meander.

The rivers in this stage are generally subdivided as follows:

1. **Aggrading type:** When the river builds its bed it is called an aggrading type of river. The reasons for this building up of the bed may be following:
   - Heavy load is carried by the river water.
   - Obstruction like a barrage or a dam across the river.
   - Extension of delta at the river mouth.
   - Sudden intrusion of sediment load from a tributary in the main river.
2. **Degrading type:** If the river bed is getting scoured from year to year it is called degrading type of river. The river becomes degrading river generally below a dam. The reason being that the sediment load in the flow gets deposited in the reservoir and clear water discharges over or through the dam. This water picks up sediment from river bed.
3. **Stable type:** When there is no scouring or silting in a river bed, that is when the river carries down only that load which it receives, it is called a stable type of river. Thus there is no appreciable change in the alignment and slope of the channel from year to year. But it is always true that even this type of river migrates within its khadir.

14.3.4. Delta and Tidal Stage

When the river is about to end its course by joining the sea it may split up into number of channels or branches. This interwoven structure of the river branches is called a delta.

The slope of bed is very flat and may range from 1 in 5000 to 1 in 10,000. Because of flat slope and consequent low velocities the channel gets silted up and when the water level rises, the water may spill over the banks to form new channels.

The delta and tidal stage of the river comes almost simultaneously. When river joins sea it may receive water from the tidal waves of the ocean. The ocean water enters the river during flood tide and recedes during ebb tide irrespective of the main flow in the river channel. A river reach in which this periodic change in water level occurs due to the tides is called a tidal stage of river. The length of this stage of river depends on:

- slope of the river,
- the tidal range, and
- configuration of the river.

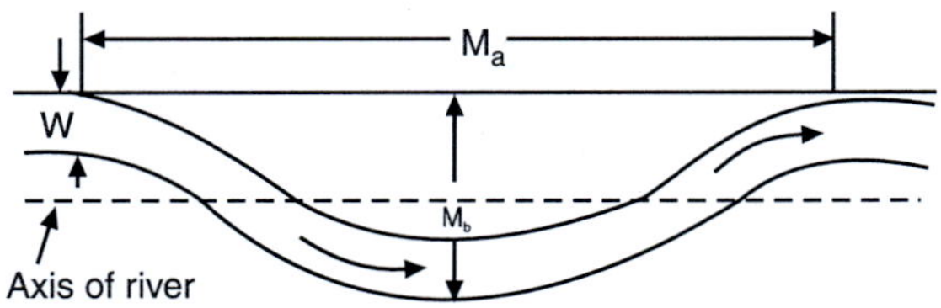

Fig. 14.1. Meandering of a river

14.4. MEANDERING OF RIVERS

It is already mentioned that the main characteristic of the trough stage of a river is to form meanders. When a river departs from its straight course and follows a sinuous winding path the river is said to form meanders. Figure 14.1 shows a full meander length.

From the figure it is clear that each meaner has two consecutive loops. In one loop the river water flows in clockwise direction, whereas in another loop it flows in anticlockwise direction. Meandering river approximately forms 's' type curve. With age the meanders become more and more pronounced and loops come closer to assume a shape of a spring coil if seen through bird's eye view. Various meander parameters are defined below.

In Fig. 14.1 M_a is the axial length of one meander. It is the tangential distance between the corresponding points of meander. It is called a **meander length**.

Tortuosity or sinuosity of a river is nothing but a measure of winding course adopted by the river. It is a ratio of actual length to meander length.

M_b is called the **meander belt**. It is the distance measured at right angles to the axis of the river. It is the distance between the outer edges of clockwise and anticlockwise loops.

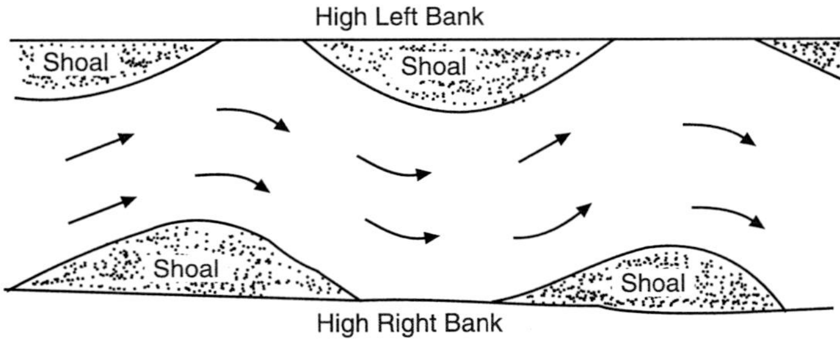

Fig. 14.2. Meander formation

The ratio $\frac{M_b}{M_a}$ is called the **meander ratio**.

A **crossover** is a portion of meandering river joining two meanders.

At this stage it is clear that the river meanders when the river flow departs from its uniform axial path due to some reason.

When the river flow turns away from its axis and flows along one bank, silting of other bank takes place. Consequently a shoal is created along the other bank as shown in Fig. 14.2.

The Fig. 14.2 clearly explains how the sinuous path is followed by the river. In years of time the river develops well defined meanders as progressive scouring of one bank and silting of the other bank takes place.

The main **causes of meandering** may be summarised as follows:

1. When there is heavy sediment load in the river water, which is more than that required for stability, river tends to build up a steeper slope by depositing the sediment load on the bed. If the banks are not resistant to erosion, increased velocity may erode the banks. Consequently the river flow may be deviated from its axis and meanders are formed.
2. If the equilibrium condition (stable condition) of the river is some how disturbed and if the bed is not degradable, the banks get eroded. Unequal erosion of the banks may deviate the flow from its axis to form meanders.

Thus it can be inferred that the process of meandering depends upon the following factors:

1. **Discharge of the river:** When the discharge of a river is more its capacity to carry silt load is also more. It is already mentioned that when the water carries excessive load meanders may be formed.

2. **Sediment load in the river water**: Composition of sediment load as well as its rate of movement affect the meandering process to a great extent.
3. **Bed slopes:** A change in the bed slope of the river also changes the meander pattern.
4. **Bed and bank resistance to erosion:** Meanders are formed after cutting up of the banks as already described. Obviously meandering process is affected by grain size, specific gravity, cohesion and roughness of the alluvium which is found in the bed and the banks. The reason for it is that the boundary resistance is characterised by these properties.

Inglis and others studied the mechanics of meandering by collecting data on various meandering rivers in Indian subcontinent. The analysis of the data enabled to arrive at following conclusions:

- In a fully developed meander formed in homogeneous valley material there exists a definite pattern of curvature, length and width at a constant river discharge.
- The width of river as well as meander length and belt roughly vary with the square root of discharge.
- Increase in sediment load increases river slope.
- Increase in discharge or slope makes meandering more pronounced. In other words the meandering pattern changes with discharges, slope and sediment load carried by the river.

The study of river meanders formed in flood plains and those formed when the river has cut its section below natural ground surface (incised rivers) and not able to overflow showed that the river width, meander length and meander belt varied with discharge as given in Table 14.1 The discharge is in cumec and other parameters are in metres.

Table 14.1.

Meanders in flood plain	*Meanders in incised river*
$M_a = 53\ Q^{1/2}$ $M_b = 153.4\ Q^{1/2}$ $W = 8.8\ Q^{1/2}$	$M_a = 46.0\ Q^{1/2}$ $M_b = 102.1\ Q^{1/2}$ $Q = 4.5\ Q^{1/2}$

From the above it can be seen that the meander ratio $\frac{M_b}{M_a}$ in flood plains has a value of about 2.85 whereas in case of incised rivers it nearly equals 2.2. The mander belt occupies a space nearly 17 and 27 times width of the river for the two situations respectively.

14.5. CUT OFFS

When the development of a meander reaches the extreme condition the two loops of a meander come closer to form a bottle neck as shown in Fig. 14.3.

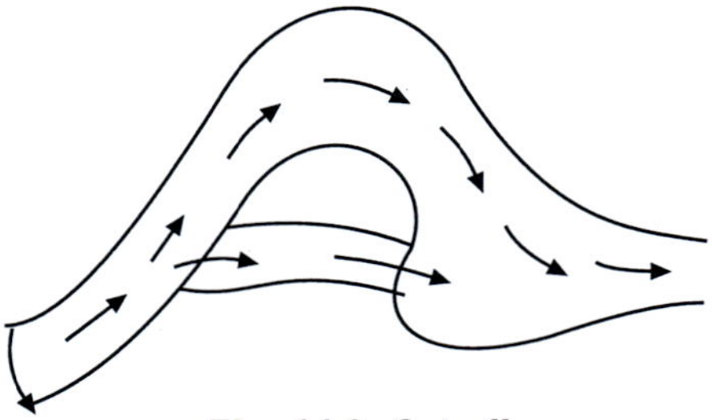

Fig. 14.3. Cut offs

During heavy flow period the loop may he straightened out. It is shown by dotted lines in Fig. 14.3. Thus, the curved length of the river course is cutoff and then the river flows straight. The cutoff is a natural process here.

Sometimes the cutoff may be made artificially to make the river flow straight. It helps in reducing the river length and also the land encircled by the loop is reclaimed. It is a simple river training work.

14.6. OBJECTIVE OF RIVER TRAINING

The three main objectives of river control may be put as:

- **To provide minimum required depth** for allowing uninterrupted navigation.
- **To prevent the river from changing its course:** When a river changes its course surrounding valuable property falls in the river course. Thus, the useful land is turned useless.
- **To prevent flooding of surrounding area:** It is needless to say that the floods damage crops, villages, towns, etc.

The river training works may be classified in various ways. First classification may be:

- Training works for protection.
- Training works to direct the river flow in desired direction.

Second type of classification may be:

- Temporary river training works.
- Permanent river training works.

Third type of classification may be:

- Training works for low water stage.
- Training works for high water stage.

At this stage it is clear that the first and foremost step in river control is to regulate the river discharge. It is already mentioned that in high floods the river discharge may vary from 50 to 100 times that of normal. Naturally absolute regulation seems to be and practically it is impossible. But various steps towards the fulfillment of the objective may be taken and are mentioned below.

- Lowering the intensity of floods to reduce peak flood discharge. Various methods adopted for the purpose are:
  - ➢ Afforestation.
  - ➢ Construction of check dams.
  - ➢ Construction of detention reservoirs.
- Floods can be carried at lower levels by widening and deepening the river. It is of course very costly process.
- Floods may be completely checked from submerging the surrounding area. It may be achieved by constructing earth embankment (usually upto 12 m in height).

  Now the list of various types of training works may be given which are dealt separately in subsequent articles.
  - Embankments or levees.
  - Guide banks.
  - Spurs or groynes.
  - Artificial cutoffs.
  - Bandalling and bed panels.
  - Bed pitching and bank revetments.
  - Widening and deepening of river by dredgers.

14.7. EMBANKMENTS

The floods may be prevented from submerging the country by constructing earth embankments. They are generally constructed upto a height of 12 m. They are designed and constructed in the same way as an earth dam.

The embankments are generally constructed parallel to the river channel. Depending upon the position of the embankments subdivisions made are:

- Marginal embankments or dykes or levees,
- Retired embankments.

The **marginal embankments** are constructed as close to the banks as possible to restrict the flood water from submerging the area behind them. Figure 14.4 shows the position of marginal embankments.

They are designed to hold up the water upto a maximum anti-cipated HFL without the possibi-lity of overtopping and with a view to withstand all external pressures. This condition is met with by providing sufficient free-board, bed width, top width and stone protection on adequate slopes. As the height of the embankment increases it becomes necessary to provide key treach, zoned section etc., to make the embankment stable.

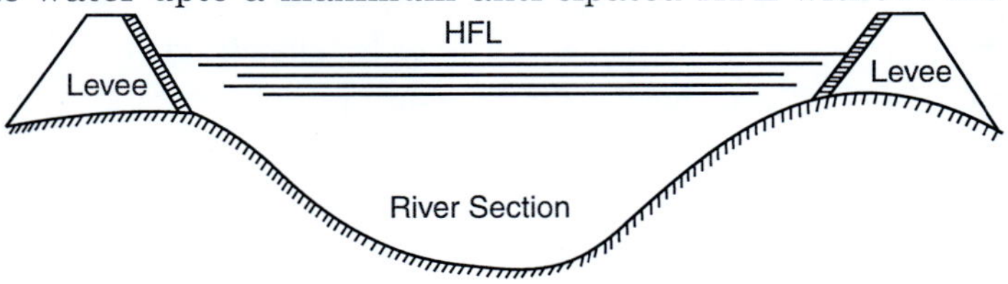

Fig. 14.4. Marginal embankments or levees

Like earth dams embankments are also likely to fail due to overtopping, piping, rat holes, seepage and caving in of river side sloping face. It is therefore necessary to adopt adequate sections for various heights.

The following sections are generally adopted for various heights. (Fig. 14.5 *a*, *b*, *c*)

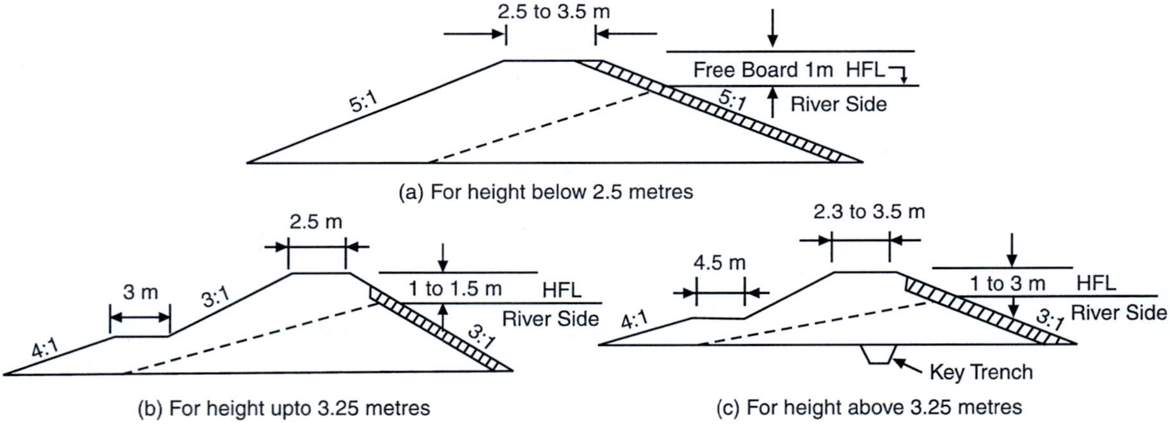

Fig. 14.5. Sections of marginal embankments

Advantages of embankments

(*i*) They are very widely used river training work.

(*ii*) It is cheaper and quick as well as simple in construction. They can be constructed with locally available material.

(*iii*) Maintenance of embankments is similar to canal bank maintenance and does not involve intricate methods.

(*iv*) Embankment can be constructed reach by reach to extend extent of protection.

(*v*) They protect large areas by comparatively small investment.

Disadvantages of embankments

(*i*) By restricting the waterway it raises the flood levels.

(*ii*) Unpredictable flood flows attack the embankment and hence chances of its failure is quite high.

(*iii*) During flood constant vigil is required on the embankments. It increases cost of maintenance.

(*iv*) They interfere in laying irrigation canal system and also reduce cultivable area.

Retired embankments are constructed at a distance from the river banks. Thus retired embankments is the intermediate type between the case of marginal embankments and river with no embankments. Retired embankments are generally constructed on a lower ground away from the banks.

Though they are costly due to increased height and risky, they have some mentionable advantages.

- They do not interfere in the process of raising of the ground by deposition of silt.
- They make it possible to store more water for longer period.
- They provide wider waterway in times of high floods.

14.8. GUIDE BANKS OR BELL'S BUNDS

Rivers in flood plains submerge very large areas during flood periods. Naturally when some structure is to be constructed across such a river (for example, bridge, weir, etc.), it is very expensive to construct the work spanning whole width of the river. To economise some training work may be constructed to confine the flow of water within a reasonable waterway. Guide banks are meant for guiding and confining the flow in a reasonable waterway at the site of the structure. The design of the guide banks is based on the theory developed by Mr. Bells. Hence, guide banks are also known as Bell's bunds.

This river training work has been devised from a study of the natural river channel in alluvial reach. The river has a tendency to meander over large width of low lying land thereby flooding it occasionally. But it was observed that the same stream passes through narrow and deep sections where high and stiff permanent banks are available on either side without appreciable afflux or abnormal velocity.

The guide banks guide the river flow past a bridge or any other hydraulic structure without casing damage to the work and its approaches. The guide banks are constructed parallel or approximately parallel to the direction of flow. They extend both upstream and downstream of the abutments of the hydraulic structure. The guide banks may be provided on either side of the hydraulic structure or on one side as required.

The guide banks consist of four parts mainly:

- upstream curved head or impregnable head,
- downstream curved head,
- shank or a straight portion which joins the two curved heads, and
- slope and bed protection, it includes apron.

Generally the core of the bund is built with sand. The sloping faces are protected with stones. An apron is also provided for protecting the bed against scouring. Sufficient freeboard and top width are also provided. The curved heads are laid with adequate curvature.

Guide banks mainly serve two purposes:

- They protect the approach embankment for the bridge from attack of the water. Approach embankments extend from the bank of the river to the guide banks generally in perpendicular direction to both.
- They control the river and induce it to flow through the bridge more or less axially.

14.8.1. Selection of Site and Section of Guide Banks

The site for guide banks should be selected in such a way that there is no side channel flowing parallel to the guide banks. The side channel if present may breach the approach embankment. The guide banks should be so designed that no swirls are produced.

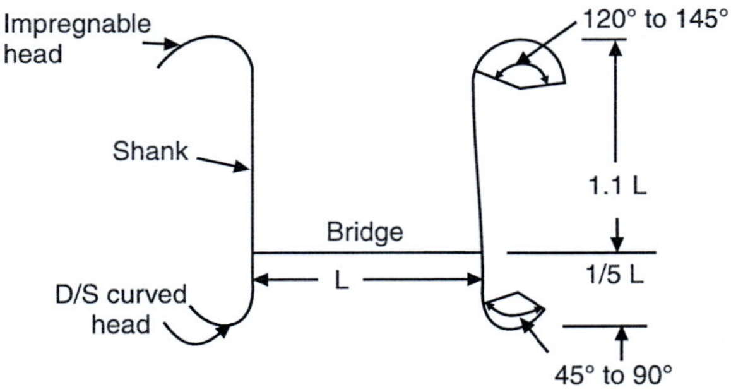

Fig. 14.6. Dimensions of guide banks

The top width of bank should not be less than 3 m. Side slopes should be 2:1 and free board 1.25 to 1.50 meters. While providing the free board due weightage should be given for heading up of the water and also for settlement of banks (generally 10 per cent of height). The inside slope should be protected with stone pitching and outside slope with good earth.

The waterway is given by Lacey's regime perimeter formula:

$$P_w = 4.825\ Q^{1/2}$$

where P_w is waterway in metres and

Q is discharge in cubic metres/sec.

The length of upstream part of the guide bank should be 10 per cent more than the length of a bridge or any other structure between the abutments. The length of downstream part of the guide bank should be 1/5 of the structure. (Fig. 14.6).

The radius of curvature of the upstream curved head should be such as not to cause intense eddies. The radius of downstream curved head may be kept half that of upstream curved head. The heads should be curved well round to the back of the guide bank. Upstream curved head generally subtends an angle from 120° to 145° to the centre and downstream head from 45° to 90°.

The upstream curved head is also called "impregnable head".

To protect the face of the guide bank at the river bed level a thick stone cover is laid on the bed. It is called an apron. When the scour undermines the river bed the apron comes down or launches to cover the face of the scour. Hence it is called Launching apron also. The quantity of stone in the apron should be adequate to insure complete protection of the scoured face. Figure 14.7 shows the details of a guide bank. After launching, the apron does not remain uniform in thickness. Generally apron thickness is kept 1.25 times thickness of the pitching. For rivers in which deep scour is likely to take place thickness of the apron may be increased to 1.5 times.

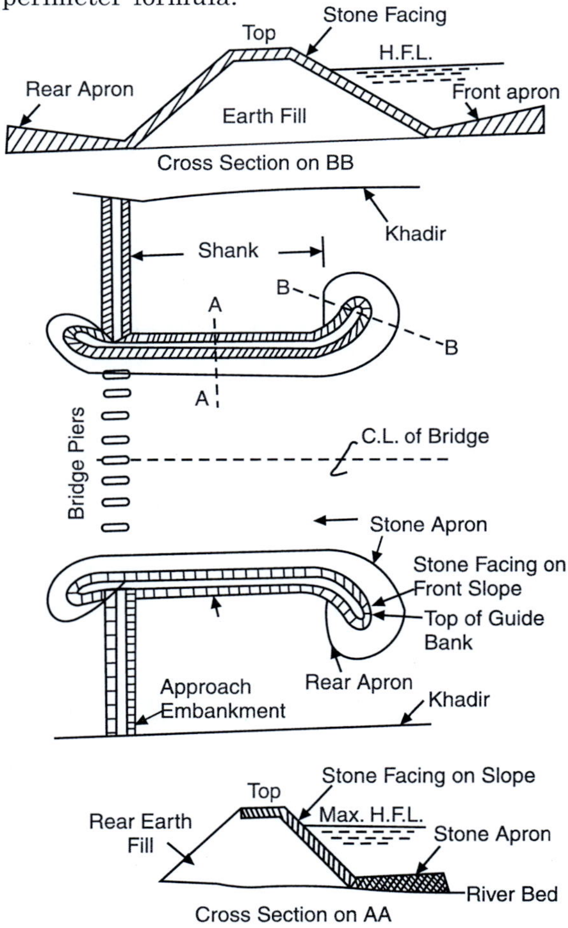

Fig. 14.7. Details of guide bank

14.9. SPURS OR GROYNES

They are the structures constructed transverse to the river flow. They extend from the bank into the river. Groynes serve following purposes:

1. They protect the river bank by keeping the flow away from it.
2. They create still pond along a particular bank with the aim of silting up the area in the vicinity.
3. They train the river to flow along a desired course by attracting, deflecting or repelling the flow.
4. They contract the wide river channel for improving the navigation depth.

Classification of Groynes

Various classification of spurs or groynes may be given as mentioned below:

1. Classification according to the method of construction.
   (*a*) Permeable, and
   (*b*) Impermeable.
2. Classification according to the height of the spur below high water.
   (*a*) Submersible, and
   (*b*) Non-submersible.
3. Classification according to the functions served.
   (*a*) Attracting type,
   (*b*) Deflecting type,
   (*c*) Repelling type, and
   (*d*) Sedimenting type.
4. **Special type:** For example, Denehy's 'T' headed groynes, Hockey spurs, etc.

When a river is to be confined to a definite channel impermeable type of groyne is most suitable. For excessively silt-laden rivers permeable groynes are suitable.

The groynes may be used singly or in series or in combination with other training work depending upon the problem in hand. When training or protection is to be given over a long and straight river reach groynes are used in series. Spacing of 2 to 2.5 times the length of groynes is a general practice. In a curved reach river can be trained by limited number of spurs. They can also be used in combination with other training measures.

14.10. IMPERMEABLE GROYNES

The groynes may be aligned either perpendicular to the bank or inclined, pointing upstream or downstream.

When a groyne points upstream then it is called a repelling groyne. The reason being, this type has a property of repelling the river flow away from the bank (Fig. 14.8). This is accomplished by creation of a still pond on the upstream. Obviously the river starts following beyond the still pond and in the process the river flow goes away from the bank.

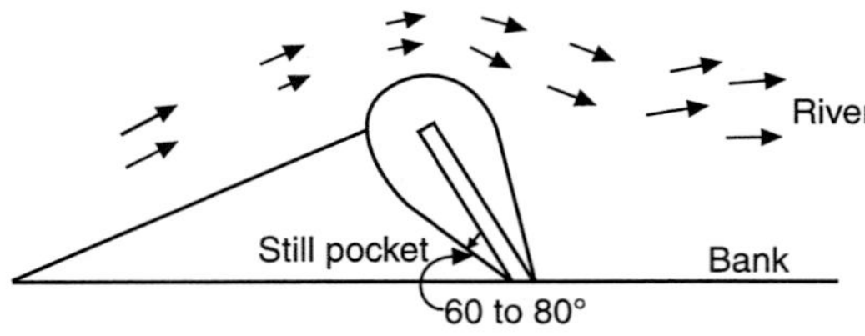

Fig. 14.8. Repelling groyne

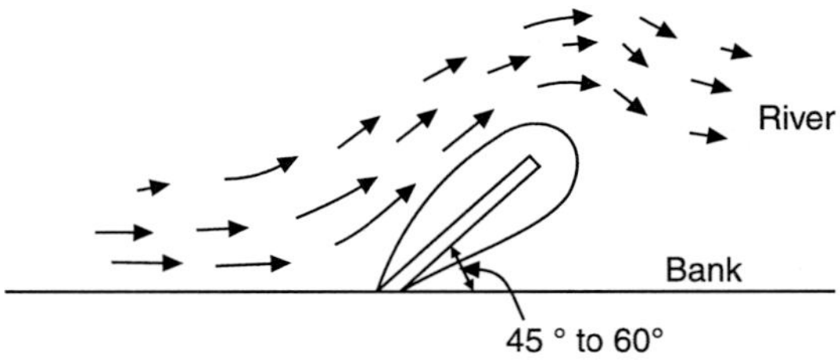

Fig. 14.9. Attracting groyne

On the contrary, when a groyne points downstream it is called an attracting groyne as it attracts the river flow towards the bank from which it takes off (Fig. 14.9). In this case the groyne actually provides a body against which the river current keeps hugging. The river flow thus remains along the bank permanently.

When a groyne of short length is taken perpendicular to the bank, it only deflects the flow locally. Hence, it is called deflecting groyne (Fig. 14.10).

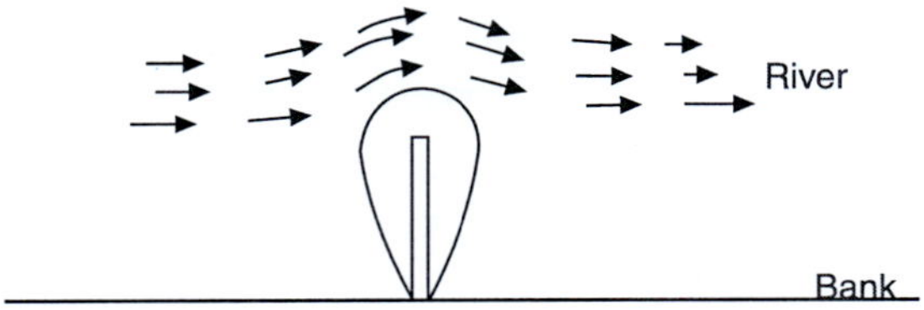

Fig. 14.10. Deflecting groyne

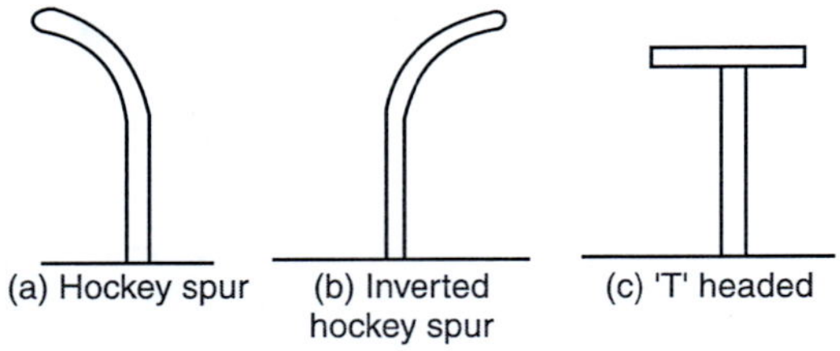

Fig. 14.11. Special types of spurs or groynes

After successfully conducting model experiments various designs for groyne heads have been evolved. A groyne with head normal to the groyne direction of called '*T*" headed groyne (Fig. 14.11).

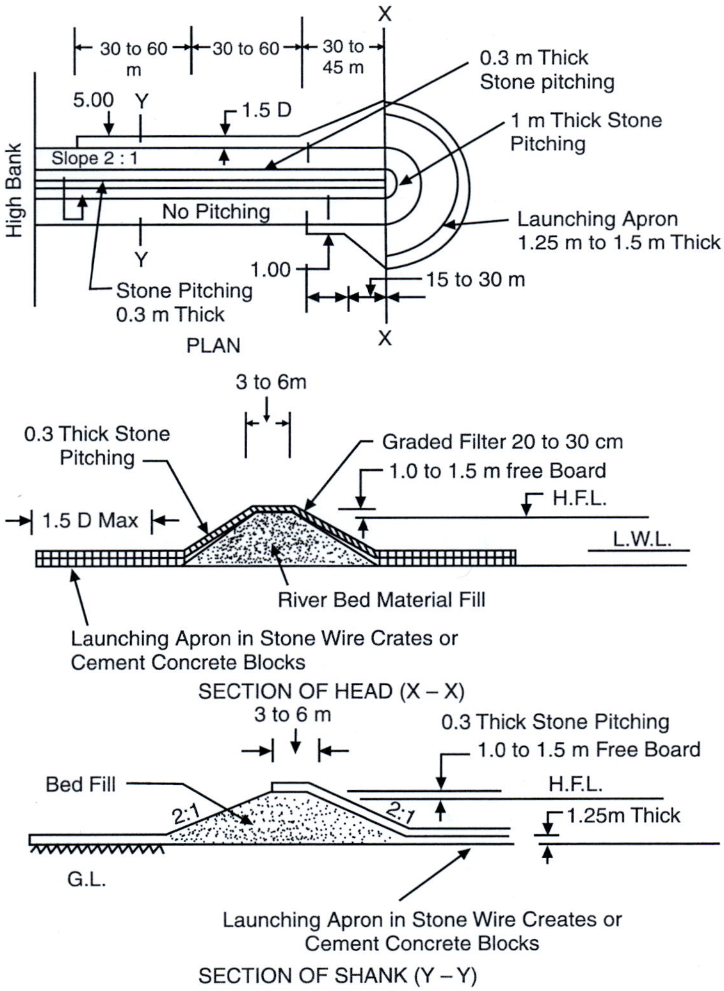

Fig. 14.12. Details of spur or groyne

From the above discussion it is clear that deflecting, repelling, attracting, T headed, hockey type, etc., all come under the impermeable type of groynes. The section of groyne is like a guide bund or an embankment (Fig. 14.12). It is protected on both sides by stone pitching or concrete blocks etc. At the river bed launching apron is also provided. Top of spur is generally kept 3 m wide. Side slopes of 2:1 is general practice. The spurs are built by sand, gravel and boulders.

14.11. PERMEABLE GROYNES

Common type of permeable groynes are tree groynes and pile groynes. They are temporary in nature and get washed away during floods. Therefore they are constructed every time before floods.

A tree groyne consists of a thick wire rope (2.5 cm dia) firmly anchored at one end to the bank and tied at the other to a heavy buoy. Sometimes this wire may be stretched across the river and anchored at its ends. It may be supported at intermediate points on tripods.

Entire leafy trees are taken and about 30 cm up the stem a hole is drilled through each tree. Then an iron ring is drawn through the whole and attached to the wire rope. Dimensions of trees may vary from 6 to 12 m in height and 0.50 to 1.2 m in girth.

A pile groyne consists of a series of piles driven 6 to 9 m into the bed 2.5 m to 3 m apart. There may be two or three rows. The rows are spaced 1 to 2 m apart. Each row is closely intertwined by brushwood branches. For stability upstream row is braced to the downstream row by transverse laterals and diagonals.

The permeable groynes lower the velocity of flow. As a result sedimentation occurs. Hence permeable groynes may be said to be of segmenting type according to the function served.

The cost of construction of this type is about 40 per cent that of impermeable type of same length. This type of groynes may be constructed even if there is flow in the river. Thus construction is easy and rapid.

To summarise, the factors which influence the choice and design of groynes are:

- Fall and velocity of flow in the river.
- Character of bed load carried by the river.
- Depth of waterway, maximum HFL and nature of flood hydrograph.
- Width of waterway, at high water, low water, and mean water.
- Availability of funds and construction materials.

14.12. BED PITCHING AND BANK REVETMENT

Sometimes to protect the bed and bank against action of water, protection is provided by laying a closely packed stone blocks or boulders or even concrete blocks. This permanent revetment and pitching counteracts the general tendency of the water to notch away the material from bed and banks.

14.13. DREDGING OF RIVER

To improve navigability of the river channel the river section may need to be excavated. This excavation is carried out to increase the depth of flow even when there is flow in the river. The process of under water excavation is termed dredging. The machinery used for the purpose is called a dredger. Various types of dredgers are in use for example, dipper dredger, grab dredger, bucket dredger, suction dredger etc. Figure 14.13 (*a*, *b*, *c*, *d*) shows various types of dredgers. General principle of working of each type of dredger is as given below:

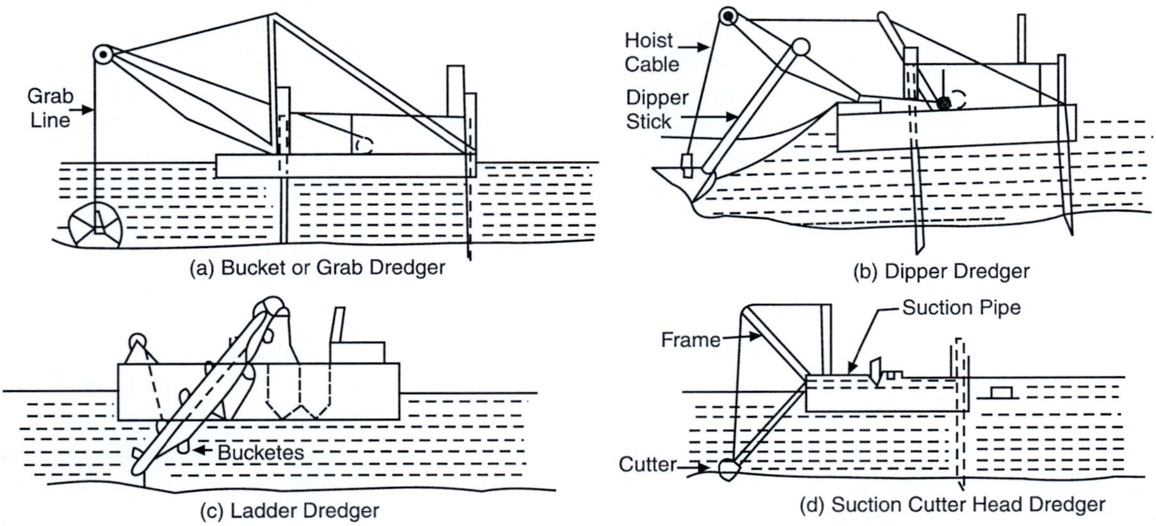

Fig. 14.13. Types of dredger

1. **Bucket or grab dredger:** It is essentially a stiff leg derrick, or crane fitted with bucket and these are mounted on a barge or a self propelled vessel. The buckets are of various weight for different kind of material to be dredged. These also have two types of cutting edges (plain and toothed). The material is either loaded in the hopper in the vessel (in case of large dredger) or dumped in hopper barges or flat barges for transporting to dumping places. The dredgers are kept in desired place with the help of anchors and three spuds provided in the dredger. The capacity of bucket varies from one cubic metre to eight cubic metres.
2. **Dipper dredger:** This consists of a floating power shovel and except for the dragging equipment, it is same as a Grab dredger. It is used when material is hard such as soft rock, boulders and require breaking.
3. **Ladder dredger:** This consist of a series of buckets attached to an endless chain operating around sheaves at each end of the frame. This type of dredger is suitable for dredging silt, mud and sand. When self propelled and provided with hopper the material is filled in hopper and carried away for dumping. Generally two units of hopper barges and tugs are used to carry the material when continuous operation is necessary and the dumping ground is far off. Generally these dredgers are not provided with hoppers so that draft could be reduced.
4. **Suction cutter head dredger:** It has a rotary cutter head which carves clay, breaks off chunks of soft rock such as coral and shale and stirs up gravel and sand so that the pipe carries material to its capacity. The cutter diameter ranges from 1 to 3 m and the cutter speed is 25 to 30 rpm. Impeller diameter of the pump range from 75 to 240 cm and pump power 100 HP to 5000 HP with pump speed varying 600 rpm to 140 rpm. The dredger is provided with large spuds round or square for anchoring.

The process of dredging is very costly and should be adopted only when the situation demands it seriously.

QUESTIONS FOR REVISION

1. (*a*) What is the necessity of river training works ?
   (*b*) Explain meandering of rivers and factors upon which it depends.
   (*c*) Describe in short various types of river training works.
2. Distinguish between the following:
   (*i*) Meander and cutoff,
   (*ii*) Embankment and a groyne.
   (*iii*) Repelling and attracting groynes.
   (*iv*) Marginal and retired embankments.
3. (*a*) Describe briefly how embankments are constructed to protect areas from flooding.
   (*b*) What is a guide bank or Bell's bund ? Draw a good sketch of a guide bank and explain its different parts.
4. What are groynes ?
   Explain various impermeable and permeable groynes you know.

❐❐❐

15

Measurement of River Discharge

15.1. INTRODUCTION

Several methods may be used to measure a discharge of regular irrigation channels and irregular channels like rivers, streams etc. The discharge may be measured either by passing the whole flow into or through some suitable measuring device or by measuring the velocity at number of points in the wetted cross-section of the channel. Methods of discharge measurement widely differ in accuracy of the result obtained from them, the methods of application and the cost. The suitable method to be adopted in any case depends largely on the degree of accuracy required and the size of the channel whether regular or irregular. Principal methods commonly used are mentioned below in Table 15.1.

Table 15.1.

For regular channels	*For irregular channels*
1. Area-velocity method	1. Area-velocity method
2. Chemical method	2. Chemical method
3. Weir method	3. Weir method
4. Meter flume	4. From measurement of catchment area
5. Stage discharge curve	5. Stage-discharge curve
	6. Flood discharge by empirical formulae
	7. Slope-area method

For small channels natural or artificial weir method is good.

For small artificial channel metre flume is suitable.

For large channels and rivers only satisfactory method is *area velocity method*. It is of course true that this method may be adopted in any case with suitable instruments and procedure for measuring area and velocity. As the name implies this method consists in determining the cross-sectional area of flow and the mean velocity of flow separately. Then the product of the two gives discharge.

15.2. VELOCITY DISTRIBUTION IN A CHANNEL

It is clear that the velocity of flow utilized for obtaining the discharge should be uniform throughout the section. But in practice it is not possible. The velocity is never uniform throughout the section. The obvious reason is, that frictional resistance offered by the sides and the bed of the channel reduces the velocity. Also the velocity of the top layer of water is reduced due to friction between the atmosphere and the water surface or due to any other reason. Thus as we go inwards from the channel boundaries the velocity of flow goes on increasing. The reduction in the velocity of flow at boundaries depends on the resistance offered by the surface in contact.

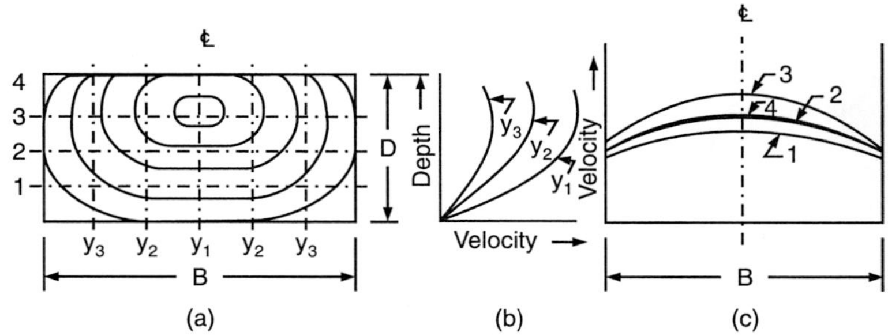

Fig. 15.1. Velocity distribution in a channel

In the channels the flow is mostly turbulent and hence it becomes slightly difficult to measure an average velocity at a particular point. But three to four observations at the same point give fairly average result.

For any type of flow, laminar or turbulent, the velocity distribution over the channel cross-section has some general characteristics. Hence to arrive at mean velocity of flow it is very essential to understand the nature of velocity distribution diagram. The velocity distribution over the cross-section of a rectangular channel is shown in Fig. 15.1, (*a*), (*b*) and (*c*). They give:

- Isovels in a section
- Velocity profile in vertical sections
- Velocity profile in horizontal sections.

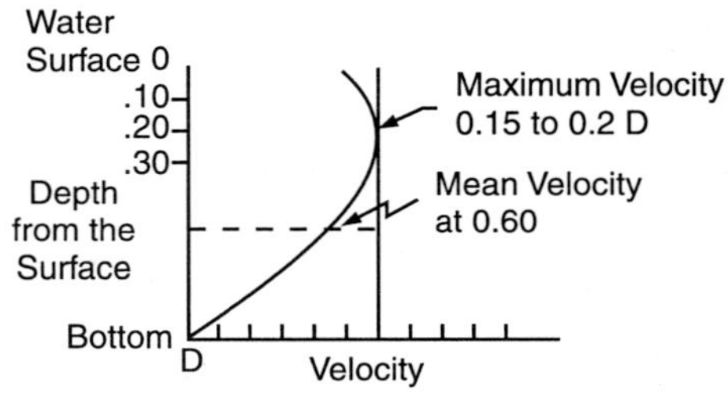

Fig. 15.2. Vertical velocity distribution

From Fig. 15.1 the nature of velocity distribution is very clear. The method of plotting the velocity distribution diagram consists in measuring the velocities at centres of the small squares formed by imaginary vertical and horizontal lines. Then the points of equal velocities may be joined with smooth curve to obtain various isovels. Thus isovel is a curve obtained by joining the points of equal velocities. The inner isovels are for higher velocities.

It has been experienced that the maximum velocity occurs at 0.15. D to 0.20 D below the water surface on the centre line of the channel. The mean velocity on the vertical section occurs at 0.6 D below the water surface. It has also been proved that the ratio of maximum velocity to mean velocity is usually 1.2. Figure 15.2 shows vertical velocity distribution at any section.

The velocity of flow may be measured by various methods. Some important methods are stated below:

15.3. MEASUREMENT OF VELOCITY BY FLOATS

This method is very simple and gives fairly accurate results. This method consists in noting the time taken by the floats to traverse a measured distance. Then

$$\frac{\text{Distance travelled by a float in m}}{\text{Time taken in sec.}} = \text{Velocity in m/sec.}$$

The floats may be classified into following categories:

- Surface floats
- Subsurface floats or Double floats
- Twin floats
- Velocity rods or rod floats

15.3.1. Surface Floats

They are made of light material, for example, cork. To make them easily distinguishable they are painted on top. The light floats are weighted to keep them floating just above the water surface. The floats should be of sufficiently small size so that they may be carried further by the velocity of the surface filament. Generally diameter of float varies from 8 to 16 cm. It is course true that the floats should not be too light as then the wind may affect the movement of floats. The measurement of velocity is done as mentioned. below (Refer Fig. 15.3).

Two wire ropes or cables are stretched across the channel at right angles to the flow. The ropes are marked to divide the channel width in various paths or tracks for floats. The two ropes are separated by a known distance. The distance between the ropes is called a reach or a run. This distance should be sufficiently long. For rivers and large irrigation channels it should be more than 50 metres. Usually 75 metres length is found to be sufficient. For lined irrigation channels and for small unlined channels minimum length of run should be 15 metres. A third marked rope is stretched across the channel upstream of the reach by about 15 to 20 metres. The floats are released at this point. By the time floats enter the run they acquire uniform velocity. The floats which follow their respective tracks are only considered. The floats may be observed through theodolites to note accurately the time taken by the floats to cross the run. As the velocity of the floats is affected by wind, a calm day should be chosen for taking observations. By this method surface velocity is obtained which is not the mean velocity of flow. Hence, this value should be multiplied by a suitable constant to arrive at a mean velocity of flow.

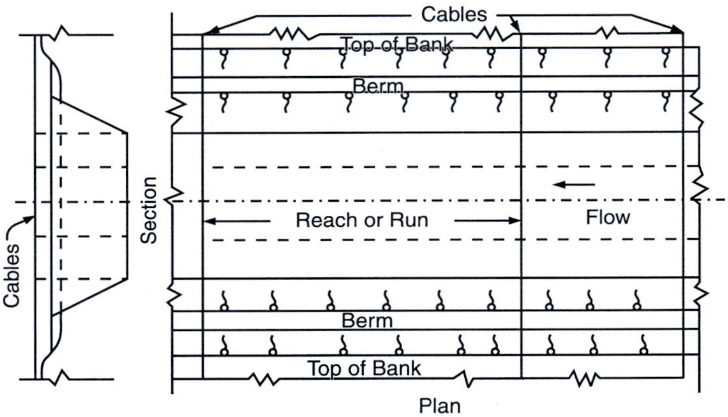

Fig. 15.3. Measurement of velocity by floats

15.3.2. Subsurface Floats or Double Float

It consists of a lighter surface float attached by a cord to a hollow metal sphere or a cylinder which is another subsurface float (Fig. 15.4).

Sub-surface float is slightly heavier than water. The depth to which sub-surface float is to be confined may be increased or decreased by adjusting the length of the cord. The surface float in this type may be taken of much smaller size to reduce the resistance to motion at surface. When the subsurface float is adjusted accurately (about 0.2 D above bottom) the velocity obtained by timing the surface float is nearly equal to the mean velocity. This method is superior than surface float. However, disadvantage of this method is, it is impossible to determine the exact position of the subsurface float.

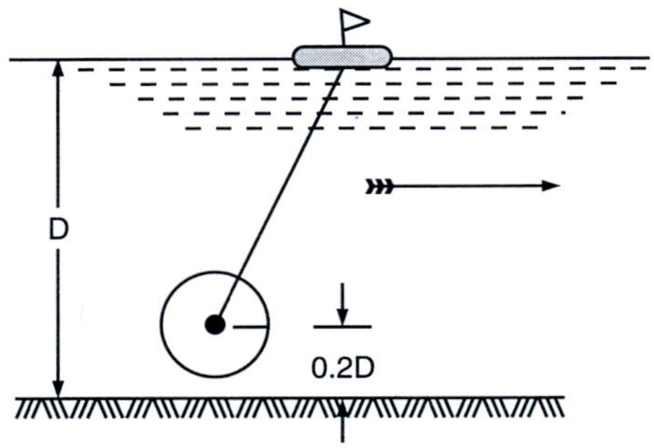

Fig. 15.4. Subsurface float.

15.3.3. Twin Float

It consists of two hollow metal spheres of equal size attached by a cord. One of them is so weighted that it remains vertically below the one which floats at the surface. The velocity obtained in this method is the mean of the velocities at the surface and at the depth of lower float. If the lower float is adjusted in such a way that it just clears the bottom, then the velocity obtained will be approximately the mean velocity for that vertical section.

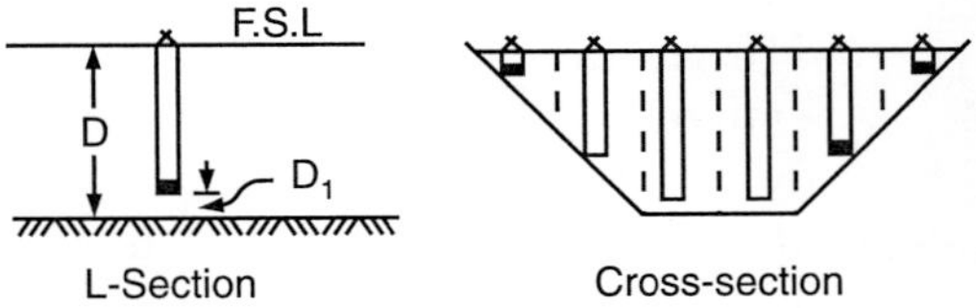

Fig. 15.5. Velocity rods or rod floats

15.3.4. Velocity Rod or Rod Float

A velocity rod or a rod float consists of a hollow metal tube or a wooden rod. Diameter of the float rod is about 3 to 5 cm. The rod is weighted at the bottom to keep the rod floating vertically with its top just above the water surface. The rod should be of sufficient length to cover entire depth. To suit different depths the rod is made in adjustable lengths that is, it is made telescopic. For large rivers the rod floats may be made of wooden logs of about 30 cm diameter weighted at bottom to keep them vertical. As the weeds interfere with the use of rods, the portion of channel selected as a run should be free from weeds. Figure 15.5 shows rod floats.

The velocity with which the rod travels in the channel is nearly equal to the mean velocity over its depth. If the length of the rod float is just equal to the depth of flow then the lower end of the float may drag over the bed. Then the velocity of flow will be inaccurate. To avoid this trouble the lower end of the rod is generally kept slightly above the bed of the channel, say by D_1 metres. Then the mean velocity of the section may be calculated from the formula

$$V_m = V\left\{1.012 - 0.116\sqrt{\frac{D_1}{D}}\right\}$$

where V is the observed velocity of a rod float.

D is depth of flow in a channel.

D_1 is clearance between lower end of the rod and the bed of channel.

The rod float is suitable for large channels and is better than any other type of float.

It may be mentioned here that the procedure of measuring velocity by floats in all the cases is similar to one described in the surface float method.

15.4. MEASUREMENT OF VELOCITY BY CURRENT METER

The velocity of flow at any point in the open channel can be most accurately and conveniently determined by means of a mechanical device named current-meter. Current-meter of common use may be classified in two categories according to the type of revolving part used.

Current-meter

(a) Current-metres in which conical cups are used
(i) Anemometer
(ii) Price current-meter

(b) Current-metres in which helicoidal vanes are used
(i) Haskell current-meter
(ii) Amsler current-meter

Figures 15.6 and 15.7 give clear idea about two types of current-meters.

In Fig. 15.6 Price current-meter is shown in which a wheel of conical cups (6 to 8) is mounted on arms of a vertical axis.

In Fig. 15.7 Amsler current-meter is shown in which a series of helicoidal vanes is mounted on a horizontal axis.

The first type of current-meter is better in the following respects:

- Frictional resistance is usually less.

- It starts working even in a current of small velocity.
- It does not revolve too rapidly under high velocity flow.
- It tends to measure full velocity of flow even if it is not oriented very accurately.
- In turbulent flow it over-registers the velocity. On the other hand propeller type current meter under registers velocities by about same amount.

Only disadvantages of conical type current-meter is that while lowering the cup wheel starts rotating even if the water is still.

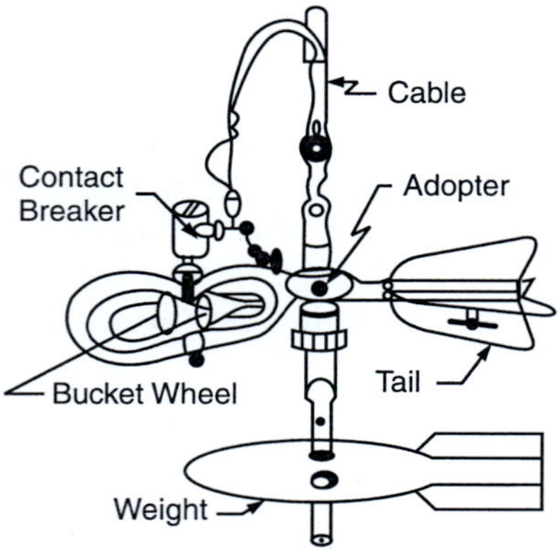

Fig. 15.6. Price current-meter

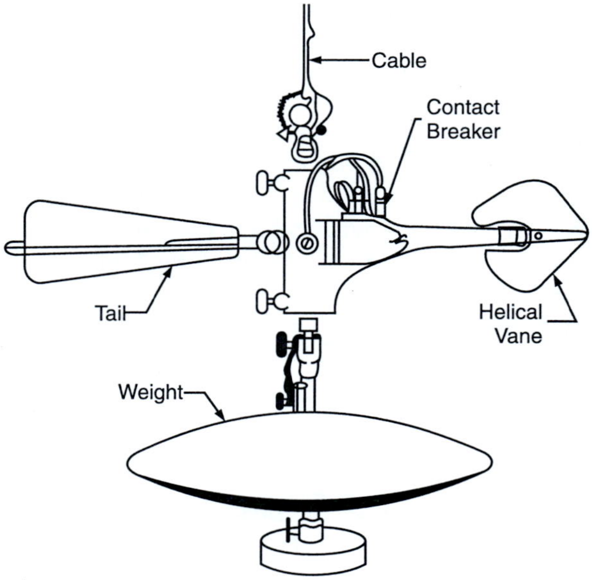

Fig. 15.7. Amsler current-meter

The current-meter with conical cups is more popular because of its accuracy and efficient working.

In principle the working of the two types is based on the same theory and can be outlined as mentioned below.

15.4.1. Working of Current-meters

When the meter is lowered in water and when it faces the current of water in the channel the wheel rotates. To keep the meter facing the direction of flow a tail is attached. This tail aligns the meter in the direction of flow. The meter is also fitted with a streamlined weight (fish weight) which keeps the meter in a vertical position. The rate of rotation of the wheel depends on the velocity of flow. A dry battery is kept on the shore or in a boat and an electric current is passed to the wheel from it. A commutator is fixed to the shaft of the revolving wheel. It makes and breaks the contact in an electrical circuit at each revolution. An automatic revolution counter is kept in the boat or on the shore with the battery which registers the revolutions. When an electric circuit is broken an electric bell in the boat rings or a head phone in the boat buzzes. Then the time taken for a required number of revolutions may be noted. The velocity of flow can be read from a rating table. The rating table is always provided with the meter.

15.4.2. Rating of Current-meter

The speed of rotation of the cup wheel is a function of the velocity of flow. The cup wheel rotates due to relative velocity between water and the cup wheel. This fact is utilized in calibration of the current-meter. Current-meter is rated in a long tank (say 150 m long) with sectional area 4 m × 2.5 m. On both sides of the tank rails are provided. A trolley runs over the rails with known velocities. Generally equipment is provided on the trolley to automatically record the current-meter revolutions, time and distance travelled. A current-meter is lowered in the still water of the tank from the trolley by a suspension rod. The trolley is run over the tank at different known velocities for number of times. The number of rotations of the current-meter cup wheel for various velocities are noted. From the readings a rating curve is prepared. It comes out to be a straight line and the equation is of the form $V = M R + C$ where V is velocity, R is revolutions of cup wheel per second and M and C are constants. Every current-meter whether new or old should be rated before putting it into use. In addition it is essential to calibrate every current-meter periodically.

15.4.3. Methods of Supporting Current-meters

So far as possible the current-meter should be suspended in the similar way as it is done for rating. When the depth of water in the channel is more, the current-meter is suspended by means of a fine cable or a wire rope and lowered to the required depth.

Here it becomes necessary to use a boat for suspending the current-meter. The boat should be equipped with the arrangement required for raising and lowering the meter (Fig. 15.8).

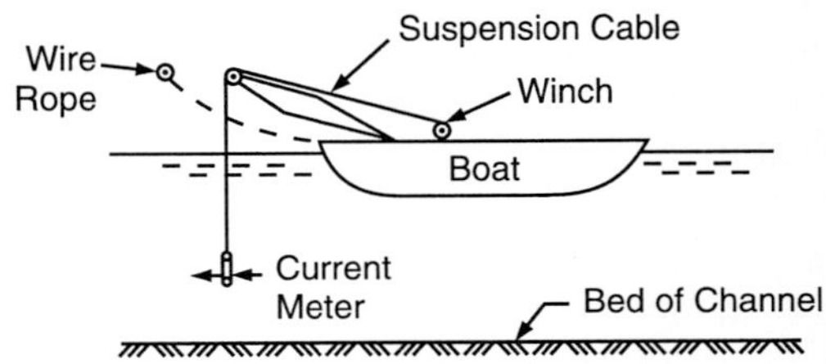

Fig. 15.8. Current-meter measurement in rivers

The boat is moored to the wire rope which is stretched across the river, at the determined point in the river.

When the channel carries water at shallow depths the suspension cable and the weight is removed and the current-meter is mounted on a wading rod. The rod is made to rest on the channel bed at required position for taking measurements. In this method as the water is shallow boat is not required. The current-meter may be fixed at any depth on the wading rod for observing velocity.

15.4.4. Methods of Measuring Velocity

When the velocity is measured by the current-meter at 0.5 D then to get mean velocity of the section the observed velocity should be multiplied with a coefficient 0.96.

Alternatively, to obtain mean velocity of flow at a particular section, velocities may be observed at 0.2 and 0.8 of depth from water surface. The mean of these two readings gives mean velocity of flow at that section. Generally mean velocity of flow at a section is obtained by keeping the meter at 0.6 D. In the observed velocity corrections are made for any obliquity of the current with the cross-section line and also for drift. It is necessary to conduct vertical velocity distribution experiment at each site to decide point of average velocity.

15.5. MEASUREMENT OF AREA OF FLOW

At this stage it is clear that the velocity distribution over the entire cross-section of the channel is so varied that it is practically impossible to arrive at one value of the velocity which represents mean velocity of flow. Hence, for measuring discharge the cross-sectional area is generally divided into small vertical strips, Fig. 15.9.

Width of the strips is selected in such a way that each strip may be taken as a rectangle for practical purpose. Then the velocity of each compartment can be calculated with fair accuracy by any one of the methods mentioned above. The product of area of strip and mean velocity of the strip gives the discharge of the strip. Discharge of the channel can be obtained by summing the discharges of all individual strips.

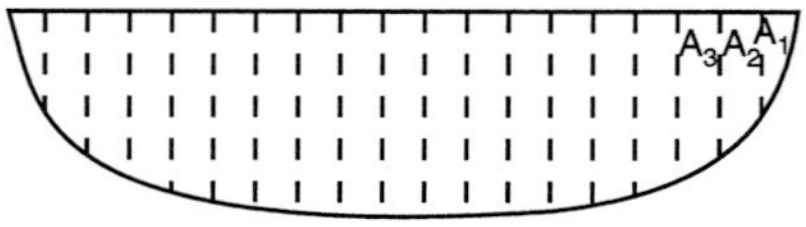

Fig. 15.9. Measurement of area

Thus if A_1, A_2, A_3 etc., are the areas of the strips and if V_1, V_2, V_3 are the corresponding mean velocities then the channel discharge is given by

$$Q = A_1V_1 + A_2V_2 + A_3V_3 \text{ etc.} = \Sigma A_1V_1$$

It is clear that measurement of area of flow involves the measurement of width and depth of the strip.

15.5.1. Measurement of Width

For measuring the width of each strip a steel wire rope is stretched across the channel. To recognize the width of the strip the rope is marked with pendants. It should be noted here that the stretched rope sags under its own weight. Hence it is essential to make correction in the width of the strip accordingly. For achieving accuracy the width of the strip should be measured by a theodolite. This method is useful upto a total width of 150 metres of a channel.

When the width of the channel is more than 150 metres (for example rivers) pivot point method is used. This method is based on the principle of similar triangles. The method is described below, for proper understanding (Refer Fig. 15.10).

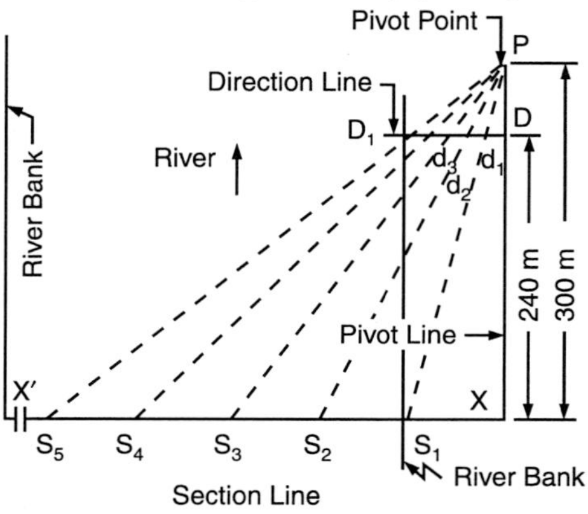

Fig. 15.10. Pivot point method

X and X' are two points fixed on the cross-section line on the bank of the channel. From point X a pivot line is erected at right angles to cross-section XX'. The length of the pivot line should be about half the width of the channel or 300 m whichever is more. Other extremity of the pivot line is called a pivot point, marked P in Fig. 15.10.

At 1/5 length of the pivot line from the pivot point a direction line DD_1 is drawn at right angles to pivot line. The direction line DD_1 is divided in suitable parts of equal length by points d_1, d_2, d_3 etc. Let the length of each part be 3 m. Then from similar triangles Pd_1d_2 and Ps_1s_2 the length of $s_1\,s_2$ is 5 times the length of d_1, d_2 (since PX is 5 times PD). The length of each strip s_1s_2, $s_2\,s_3$ etc., is 15 m. The points s_1, s_2 etc., can be located on the cross-section line very accurately with the help of a theodolite placed at P.

When the gauge site is to be made permanent the points D, d_1, d_2 etc., on the direction line, points X and X_1 on the cross-section line and the pivot point should be constructed with masonry in the form of blocks with a hole in the centre to fix a flag exactly on the point.

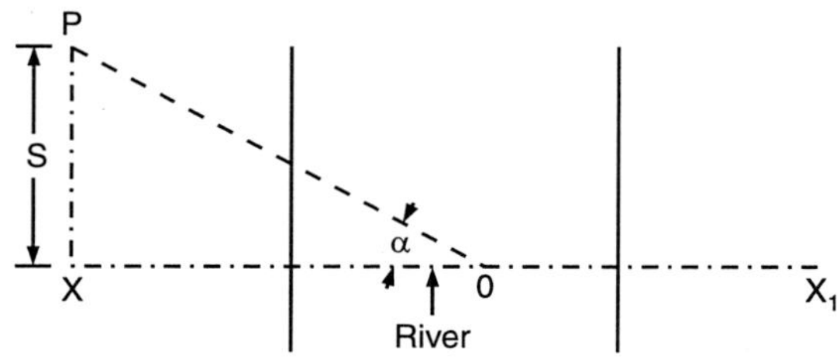

Fig. 15.11. Sextant method

When a channel is wider than 600 m the pivot lines should be erected on both the banks. The sounding points may be located accurately by adopting the same procedure. Each set up of pivot covers half width of the river.

Another method of measuring width is by using sextant (Fig. 15.11).

A boat is towed along the cross-section line XX_1 and made stationary at the observation point

O. From the point *O* angle *XOP* is measured by a sextant. The distance *S* is known. The distance of observation point *O* from the target *X* is given by *S*. cot α. For regular observation a chart can be prepared to give widths for various measured angles. This method gives fairly correct position of the boat. Only disadvantage is that at every point angle is to be measured by the sextant and the process becomes rather tedious than the pivot point method.

15.5.2. Measurement of Depth

Depth is measured just below each pendent on the rope which is stretched across the channel. The depth can be measured by observing the readings directly, for example, by sounding devices. Alternatively some instrument may be employed to take the reading and then from tables or graphs depth can be determined. Some common methods of measuring depth are mentioned below.

1. **Sounding rod:** It is a wooden rod 5 to 8 cm in diameter with markings on it. A bamboo pole may also be used as a sounding rod. Many times flat iron of 5 cm × 0.6 cm size may also be used for the purpose. The graduations are generally in tenth of a metre. To prevent sinking of a rod and to achieve accuracy in measurement the rod is provided with a flat base plate. The base plate is in the form of an iron disc 10 to 15 cm in diameter attached to the lower end of the rod.
2. **Lead line:** When velocity of flow is high and depth is more sounding rod is not preferable. Then lead line is used. It consists of a rope generally made of copper and covered with hemp. A weight or a lead is attached to one end of the rope. The shape of the lead is like frustum of a cone. It weighs from 5 to 30 kg depending upon the velocity of flow in the channel.

15.6. ECHO-SOUNDER FOR DEPTH MEASUREMENT

This method is based on an electrical principle. It consists in transmitting a sound impulse (by a transmitter) from the surface of water level to the bed of the river. When the sound waves reflect back in the form of an echo they are arrested by a receiver. There is an automatic arrangement for plotting the time of transmission and the time of reception. The velocity of sound in water is known. It is nearly 1470 m/sec. From this fact the depth of water is automatically computed. This method is of very common use on ships for observing depths. It can also be used successfully on rivers.

The construction features and the working of the echo-sounder is given in short in the following paragraphs. Figure 15.12 shows a schematic line diagram of the sounding machine. An echo-sounder mainly consists of four units. They are : (*i*) Driving unit (*ii*) Transmitting unit (*iii*) Receiving unit and (*iv*) Recording unit.

Driving unit consists of an electric motor with a governor 'G'. A 12 volts battery supplies current to the motor. Governor helps in maintaining constant speed of the motor under normal voltage. The motor is coupled to a gear box. The gear box has arrangement for maintaining two speeds. A drum is mounted on a shaft. The shaft is rotated by the motor through the gear box. The drum is made up of insulated discs. A stylus is mounted on the first disc of the drum as shown in the Fig. 15.12. The first disc is continuous and it receives the echo impulse in the electrical form from amplifier through the brush and is passed on to the stylus. Remaining two discs of the drum make the contact once in each revolution and as a result an electrical impulse is sent to the transducer.

Recording unit consists of a box. (It is also called a tank) in which a recording paper roll is wound over the rollers. One of the rollers is driven by a shaft from the motor. Thus the recording paper is drawn down at a regular speed (say 2.5 to 5 cm/minute). The paper moves over the face of the box or the tank. The paper is chemically treated with a solution of potassium iodide and starch. The stylus moves in anti-clockwise direction over the paper. When a small electrical impulse passes through the stylus it produces a brownish-purple stain at the point of contact.

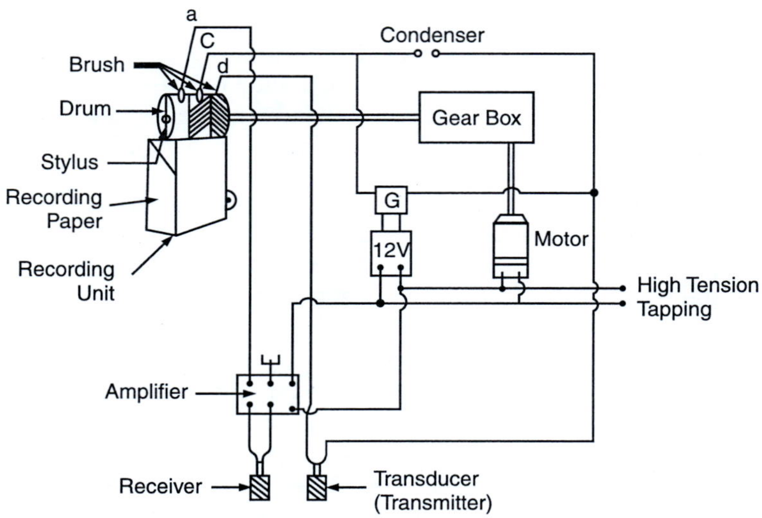

Fig. 15.12. Echo-sounder line diagram

Transmitting unit consists of a transmitter or a transducer. The 12 volt battery supplies current to a rotary converter C. It transforms the battery voltage of 12 volts to 400 volts DC. This voltage is used in operating the transmitter. In the transmitter a sound impulse is produced by a blow from a spring driven electromagnetically controlled hammer on a steel diaphragm. The diaphragm is in contact with the water.

Receiving unit consists of a receiver which arrests the reflected sound echo. A corresponding electrical impulse goes to the stylus via an amplifier. Current for operating the valve filaments of the amplifier is also obtained from the same battery. High tension supply to the amplifier is given from the tappings of the rotary converter.

Working of echo-sounder: When the motor is switched on the drum starts rotating in anti-clockwise direction. The stylus also rotates with the drum. During each rotation the contacts *c* and *d* made. As soon as the contact is established an oscillatory current passes to the transducer via condenser. The electrical impulse is converted into sound impulse in the transducer. The sound impulse is sent out into the water. After reflection from the bed the sound impulse is arrested by the receiver. A corresponding electrical impulse goes to stylus via amplifier through the brush '*a*'. During this period the stylus has rotated through certain angle corresponding to the echo interval. From the stylus the current passes to earth via paper and the recording unit. Thus the circuit is completed. The position of the stylus at the moment the impulse arrived is shown by a brownish-purple stain on the paper. The process is repeated for every revolution of the drum. The speed of motor is in correct ratio to the velocity of sound in the water. In the recording unit a scale is provided to give the depth of water directly.

SOLVED PROBLEM

Problem 15.1. Discharge measurement was done at a particular site on a small stream using current meter and velocity area method and following readings were taken:

RD on section (m)	*Depth of water (m)*	*Velocity m/sec at*		*Remarks*
		0.2 D	*0.8 D*	
0	—	—	—	L/B target
40	0.00	—	—	Water edge
50	0.80	0.450	0.400	
60	2.00	0.620	0.580	
70	1.40	0.540	0.460	
80	0.90	0.450	0.410	
90	0.70	0.420	0.380	
100	0.00	—	—	Water edge
120	—	—	—	R/B target

(a) Draw the cross-section of the river and measure total area of flow assuming that depth varies uniformly between two points.

(b) Calculate discharge of the river.

Solution: The cross-section of the river can be plotted as shown in Fig. 15.13.

The velocity is observed at two points viz., at 0.2 D and 0.8 D so mean velocity will be average of two. The data can be recorded and calculations done in tabular form as below.

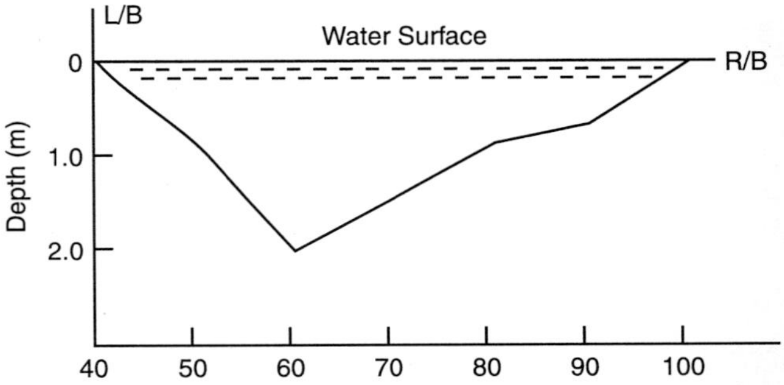

Fig. 15.13. Problem 1

RD on section (m)	*Depth of water (m)*	*Velocity m/sec at*		*Mean Velocity m/sec*	*D × V*	*Remarks*
		0.2 D	*0.8 D*			
0.0	—	—	—	—	—	L/B Target
40.0	0	—	—	—	—	Water edge L/B
50.0	0.80	0.450	0.400	0.425	0.34	
60.0	2.00	0.620	0.580	0.600	1.20	
70.0	1.40	0.540	0.460	0.500	0.70	
80.0	0.90	0.450	0.410	0.430	0.39	
90.0	0.70	0.420	0.380	0.400	0.28	
100.0	0	—	—	—	—	Water edge R/B
120.0	—	—	—	—	—	R/B Target
Total	5.80	—	—	—	2.91	

Area of flow $= \Sigma$ Depth $\times$ common width of segments $= 5.8 \times 10 = 58\ m^2$

Note: If the segments on two ends are not equal then correction has to be applied.

$$\text{Correction} = \frac{\text{Common segment width} - \text{Unequal segment length}}{2} \times \text{depth}$$

This correction is applicable for both sides.

In the problem since segments on L/B and R/B are equal no correction is called for.

$$\text{Discharge of river} = (\Sigma D \times V) \times \text{Common width of segments}$$
$$= 2.91 \times 10 = 29.1 \text{ m}^3/\text{sec.}$$

Since there is no area correction no discharge correction is called for.

15.7. RIVER GAUGING OR STAGE MEASUREMENT

The term stream gauging is used to mean measurement of discharge as well as river water level or stage in a general way. However, when a term gauge record is used it only means a record of river stages or water levels. Unlike the irrigation canals the depth of water varies along the cross-section of the river from one bank to the other. Hence, it is neither convenient nor of any practical significance to attempt to project average depth of flow as one of the flow parameters. On the other hand river stage or water level is uniform over the section and is therefore taken as one of the important river flow parameters. The procedure of measuring the river stage or the water level is commonly called river gauging. The uses of the river gauge record are the following:

- From the gauge record river levels can be obtained.
- Depth of flow at various points along the section can be assessed with fair accuracy to plan navigation, diversion and other water utilization proposals.
- The record of river stages can be used to determine the river discharge from the stage-discharge relationship.

The river water levels are generally measured with reference to a fixed benchmark which is always connected to GTS benchmark for maintaining uniformity and interrelation with other locations on the river. Depending upon the necessity the river stage or water levels may be observed once, twice or hourly everyday. Even continuous record of the river water levels can be obtained by installing self-recording water level recorders. The two commonly adopted methods of recording water levels are:

- Staff gauges; and
- Automatic water level recorders.

15.7.1 Staff Gauges

It is a simple device used to record the water level. The staff gauges are the wooden posts on which enamelled metal plates graduated to 0.01 m are fixed. Each staff gauge is generally 2 m in length. They are constructed along the slope of the bank as shown in Fig. 15.14.

They are so constructed that the top of the lower gauge continues the reading from the bottom of the next gauge. The gauges are generally constructed on both the banks on one section line with same levels. To secure the staff gauges they are fixed in the concrete blocks constructed along the banks. The staff gauges are read by an observer or a gauge reader. It is normally read twice a day and on hourly basis during the flood season. Sometimes the enamelled gauge plates are fixed or painted on the bridge piers to record water levels.

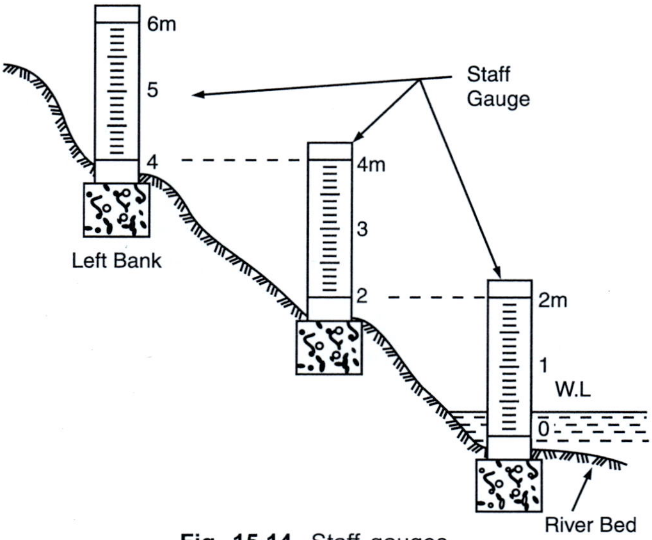

Fig. 15.14. Staff gauges

15.7.2. Automatic Water Level Recorders

Like the recording rain gauge the automatic water level recorder also works on the principle of a float. The most common type of automatic water level recorder uses a float line with a metal float at one end and small counter weight at the other end. The float line passes over a pulley and transfers the changes of water level to it. A recording stylus is attached to the pulley. It moves laterally and traces the water level fluctuations on a recorder chart. The recorder chart is a tracing quality strip paper wound over rollers or a drum. The recorder chart is connected to a clockwork mechanism which moves it at pre-determined speed continuously. The recorder chart thus gets wound slowly at a uniform rate and the recording stylus traces the water level fluctuations laterally on it to provide a continuous record of water levels. Depending upon the type of clockwork mechanism, available fall for clock counter weight and length of the strip chart the continuous record may be obtained from 1 to 6 months. Figures 15.15 and 15.16 show the automatic water level recorder and the strip chart with recorded water levels respectively.

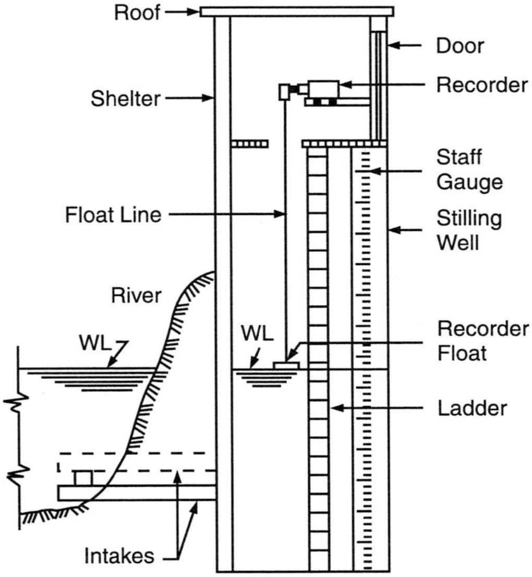

Fig. 15.15. Automatic water level recorder

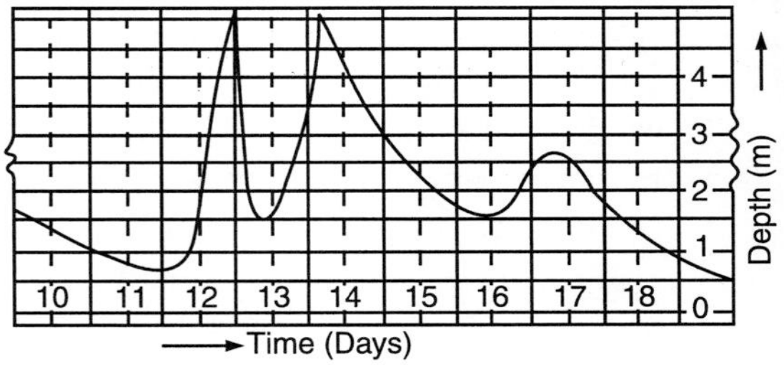

Fig. 15.16. Recorded strip chart

The water level recorder is a compact unit which is installed on a platform constructed over a stilling well. The stilling well may be constructed with corrugated iron sheet, concrete or masonry. On the top a roofed shelter with locking arrangement is constructed to house and secure the recorder. The stilling well is connected to the river by means of intakes. A ladder is provided in the stilling well for inspection and cleaning purposes. Figure 15.15 shows the general arrangement of the automatic water level recorder. An enamelled gauge plates are also fixed on the stilling well from outside to record the water level by visual observations during inspection visit. During each inspection visit the recorder is inspected, the datum of the recorded water levels compared and suitably adjusted with date and signature of the inspecting officer on the chart. It helps in computation of water levels in the office subsequently.

15.8. STAGE DISCHARGE CURVES

With the exception of few occasions of severe floods the river normally maintains its section and therefore the stage or water level of the river bears a definite relation with the river discharge. In practice it is observed that the relation of stage to discharge is usually controlled by a section or reach of channel below the gauge. It is called station control. It eliminates the effect of all other downstream conditions on the velocity of flow at the gauge. The section control may be either natural or constructed. It may consist of a ledge of rock across the channel, a boulder covered riffle, an overflow weir or any other physical feature capable of maintaining a fairly stable relation between stage and discharge. Section controls are generally effective only at low discharges. At medium and high discharges channel control becomes effective. The channel control consists of all physical features of the channel. These features include the size, shape, slope, alignment, expansions, constrictions and roughness of the channel.

Making use of the above fact a continuous record of the river discharge at a river gauging station can be computed from the records of the stage and the stage-discharge rating for the channel. The stage-discharge ratings at gauging stations are usually determined experimentally from measurements of discharge and stage. The discharge measurements are usually made accurately by the current meter method at the gauging stations covering various stages of the river. It should cover the full range of discharges to obtain a good relation. Then using this data a curve is plotted between the stage of the river and the corresponding discharge. It is called a rating curve. The stage-discharge relations or rating curves are developed from a graphical analysis of the data plotted on either rectangular coordinate (Fig. 15.17) or logarithmic plotting paper.

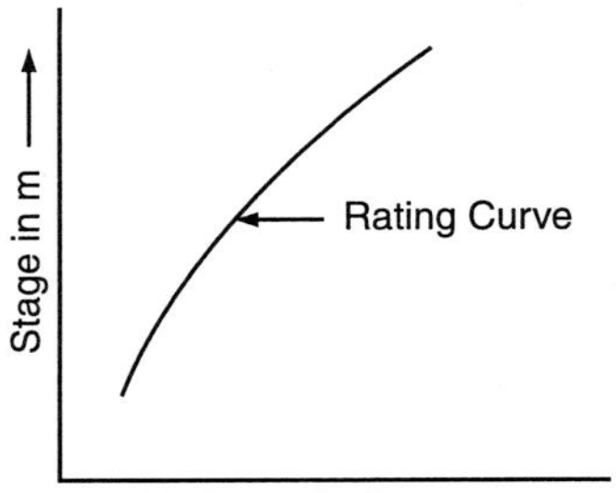

Fig. 15.17. Rating curve on rectangular coordinates

There are three advantages in plotting the data on logarithmic scale, namely:

- the rating curve nearly assumes a straight line in most of the cases;
- it is easy to identify the range in stage for which different controls are effective ; and
- parallel lines represent constant ratios and it is easy to spot measurements that deviate greatly from the established rating.

Figure 15.18 shows a rating on the log-log paper.

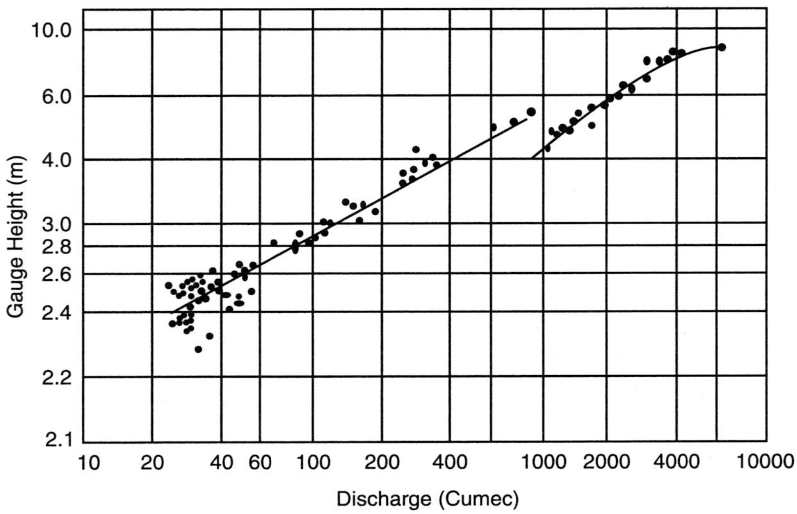

Fig. 15.18. Stage-discharge rating on the log paper

Once the rating curves are plotted it can be conveniently used at a later date to predict the discharge simply by noting the stage of the river. From the rating curve rating tables are generally prepared to facilitate quick calculations.

The stage-discharge relation though simple they are rarely permanent because of the changes in the river channel such as scour and fill, aquatic growth, change in bed roughness etc. Therefore frequent discharge measurements with current-meter method are necessary to define the position of the stage-discharge relation at any time. Secondly, an important point to be remembered in this method is that the gauge site should not be located just upstream of any hydraulic structure constructed across the channel. The obvious reason is that the back water effect affects the accuracy of the rating.

Thirdly, knowledge of the channel features which control the stage-discharge relation is very essential. This is so because the development of stage-discharge curves where more than one control is effective and the number of measurements is limited usually requires sound judgement in interpolating between the measurements and in extrapolating beyond the highest measurement.

15.9. SLOPE-AREA METHOD

When it is not possible to make direct measurement especially during floods, slope-area method is a most commonly used indirect method of measurement. In this method discharge is computed on the basis of a uniform flow equation involving channel characteristics, water surface profile and a roughness coefficient. The drop in water surface profile for a uniform reach of channel represents losses caused by bed roughness. The well known Manning's equation is used in this method. The discharge is given by equation.

$$Q = \frac{1}{n} A \cdot R^{2/3} S^{1/2}$$

where Q is discharge in cumec

A is cross-sectional area in m^2

R is hydraulic mean radius

S is friction slope, and

n is roughness coefficient

Although Manning's equation was developed for uniform flow it is assumed to be valid for non-uniform reaches that are encountered in case of natural channel.

The friction slope S to be used in Manning's equation is given by following relation

$$S = \frac{h_f}{L} = \frac{\Delta h + \Delta h_v - k(\Delta h_v)}{L}$$

where Δh is the difference in water surface elevation at the two sections

L is the length of the reach

Δh_v is u/s velocity head minus d/s velocity head

and $k(\Delta hv)$ is energy loss due to acceleration or deceleration in a contracting or expanding reach (k being coefficient)

It is difficult to correctly assess the value of the coefficient k. So value of S is calculated as follows:

$$S = \frac{\Delta h + (\Delta h_v / 2)}{L} \quad \text{when } \Delta h_v \text{ is positive.}$$

and $$S = \frac{\Delta h + \Delta h_v}{L} \quad \text{when } \Delta h_v \text{ is negative.}$$

In using Manning's equation the quantity $\left(\frac{1}{n} A \cdot R^{2/3}\right)$ is called conveyance and is denoted by a capital letter K. The conveyance for each section of the river will be different. The mean value of the conveyance has therefore been taken as a geometric mean of the conveyance values of the two sections at both ends of the selected reach L. The discharge is thus given by the equation.

$$Q = \sqrt{K_1 K_2 S}$$

15.9.1. Selection of Reach for Slope-Area Method

It is very necessary to select a suitable reach for slope-area measurement. Following points need due consideration in selecting the reach:

- There should be good high water marks available if staff gauges are not available to give gauge record.
- The reaches should be as uniform as possible because it is difficult to correctly calculate coefficient k to assess velocity head loss or gain.
- It is desirable to have a reach in which flow is confined to a single trapezoidal channel so that single value of n can be used for the whole section.
- Over the whole reach the cross-section should be as uniform as is practically possible. Sudden deepening of bed, constrictions, expansions etc., should not be present.
- The length of reach should not be less than 75 times mean depth in the channel.
- The fall of water profile in the reach should be equal to or greater than the velocity head.
- The fall in the reach should not normally be less than 15 cm.

Problem 2. During a severe flood it was not possible to actually measure the discharge of a river. The discharge site was visited at a later date and the following section parameters were collected:

(*i*) The length of reach between sections 1-1 and 2-2 was 120 m.
(*ii*) The difference of high water marks was 0.97 m.
(*iii*) Using slope-area method calculate the highest discharge that passed through the river.

	Area of section (m^2)	*Wetted perimeter (m)*	*Rugosity coefficient (n)*
Section 1-1	**208**	**67.3**	**0.045**
Section 2-2	**203**	**58.5**	**0.045**

Solution: Step 1: The discharge is given by the manning formula

$$Q = \frac{1}{n} A \,.\, R^{2/3} S^{1/2}$$

This formula is further modified to take average conveyance into account as follows:

$$Q = \sqrt{K_1 \cdot K_2 \cdot S}$$

where $$K = \frac{1}{n} A \cdot R^{2/3} \text{ for each section}$$

and $$S = \frac{h_f}{L}$$

Step 2: Conveyance for section 1.1

$$K_1 = \frac{1}{0.045}\left(\frac{A}{P}\right)^{2/3} \cdot A = \frac{1}{0.045}\left(\frac{208}{67.3}\right)^{2/3} \times 208$$

$$= \frac{1}{0.045} \times (3.09)^{2/3} \times 208 = \frac{2.12 \times 208}{0.045} = 9800$$

Step 3: Conveyance for section 2.2

$$K_2 = \frac{1}{0.045}\left(\frac{203}{58.5}\right)^{2/3} \times 203 = \frac{2.29 \times 203}{0.045} = 10340.$$

Step 4: As a first approximation

$$S = \frac{h_f}{L} = \frac{\text{Fall in reach}}{\text{Length of fall}} = \frac{0.97}{120} = 0.00808$$

Step 5: Peak discharge $Q = \sqrt{9800 \times 10340 \times 0.00808} = 906$ cumec.

Step 6: With this discharge average velocity of flow at both the sections can be calculated to calculate energy loss due to constriction/expansion

$$V_1 = \frac{906}{203} = 4.36 \text{ m/sec.}$$

$$V_2 = \frac{906}{203} = 4.46 \text{ m/sec.}$$

$$\Delta h_v = \frac{V_1^2}{2g} - \frac{V_2^2}{2g} = \frac{(4.36)^2}{2 \times 9.81} - \frac{(4.46)^2}{2 \times 9.81} = -0.045$$

Step 7: Since Δh_v is –ve

$$S = \frac{\Delta h + \Delta h_v}{L} = \frac{0.97 + 0.045}{120} = 0.0085$$

Step 8: With the value of S the discharge can be recalculated.

$$\therefore \quad Q = \sqrt{9800 \times 10340 \times 0.0085} = 928 \text{ cumec.}$$

QUESTIONS FOR REVISION

1. What is a current-meter ? Describe any type with a sketch. How it is used for measurement of velocity ?
2. (*a*) What is a pivot point method ?
   (*b*) Explain the working of an echo-sounder.
3. (*a*) Explain briefly how discharge is measured by Slope-area method.
   (*b*) Describe the method of floats for measuring velocity of flow.

16

Fundamentals of Groundwater

16.1. BASIC CONCEPTS OF GROUNDWATER

In a very broad sense, all the water which exists below the surface of the earth may be called groundwater. During and after rains large portion of the rainwater is absorbed by the surface soil depending upon texture and structure of the underlying soil mass. This absorbed water then travels downwards. While moving downwards it replenishes the soil moisture deficiency and then travels further down and at the end becomes static. As such soil cover can be divided into two main zones.

- Unsaturated zone of soil, and
- Saturated zone of soil.

The water which exists in unsaturated zone is called soil-water and is held in the pores of soil particles by some force or the other. This water is readily used by plant roots. It has been dealt quite in good length previously.

In saturated zone of soil the water particles surround the soil particles to such an extent that all soil particles are completely submerged in the water. Whenever some extra amount of water comes it in turn submerges soil particles of the layers of the soil or water bearing strata just above the saturated zone. There is no scope for this trickling water to go further down. Thus the extent of zone of saturation increases. Saturated zone of soil can be sub-divided into two categories:

(*a*) Zone of Intermittent Saturation: In rainy season infiltration is more so the amount of water coming down to the saturation zone is much more. The extent of zone of saturation due to addition of this water obviously increases. This increase in saturated storage is temporary because during dry period the water which has come as infiltration water gets mostly used up. So the extent to which this temporary zone of saturation goes is called a zone of intermittent saturation. This zone remains saturated during wet season and becomes slowly unsaturated in dry season. This zone thus every year gets recharged in wet season and is depleted completely during dry season.

(*b*) Zone of Permanent Saturation: It is the zone of soil mass which remains saturated for the whole year. The lower limit of this zone is indicated by some type of impervious rock underlying the soil mass. The upper limit of the zone depends on the various factors including rainfall, temperature, underground water usage, etc.

The water of unsaturated zone is called *soil water*, and can be subdivided as hygroscopic water, capillary water and gravitational water.

The upper most surface or top level of the zone of saturation is called the *water-table*. It can also be defined as the level at and below which soil particles are completely saturated. Water table generally follows the relief of the ground. Thus water table below hills is arched up and depressed in valleys. Figure 16.1. gives clear visualization.

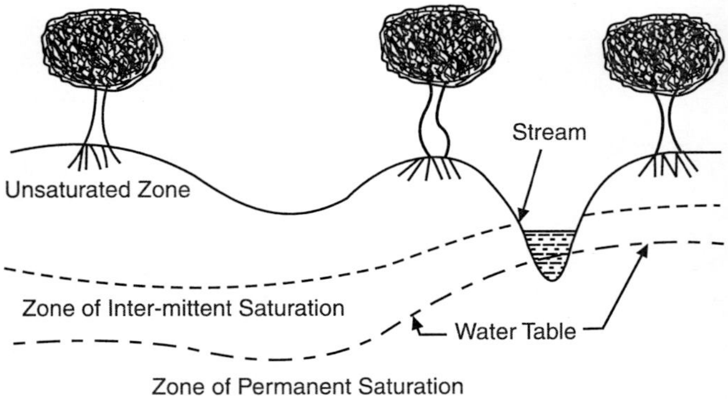

Fig. 16.1. Groundwater and water table

If at some place natural ground is depressed locally to such an extent that it falls below the water-table then the water from saturated zone of soil under high hydrostatic pressure, finds its way into the depression to form a stream.

Some times because of excessive rainfall and subsequent infiltration water table rises so fast that it coincides the ground surface. This completely fills the root-zone-depth. As no air is then allowed to circulate in the root-zone depth soil becomes useless for crop production. This phenomenon is termed as waterlogging.

When at some point water table meets the ground surface, springs, streams, swamps, ponds and lakes may be formed.

Further, during the long periods of consecutive dry spells the water table lowers considerably. This is the reason for drying out of wells, springs etc., which are dependent on upper portion of the groundwater.

16.2. GROUNDWATER IN HYDROLOGIC CYCLE

Groundwater is one component of the complete hydrologic cycle of the earth. For all practical purposes groundwater originates from surface water source. Water infiltrates into ground by natural recharge process. Rainfall, water from stream flow, reservoirs and lakes are the sources for infiltration. Also surface irrigation contributes artificial recharge.

This infiltrated water then moves further into groundunder the force of gravity. When a zone of saturation is reached the water starts flowing laterally. The direction of flow is controlled by the hydraulic boundary conditions.

Discharge of groundwater represents return of water above ground surface mostly in the water bodies, in the form of springs and seepage into streams. Evaporation and transpiration are other modes of discharge. Artificial discharge is in the form of pumping of groundwater. Hydrologic cycle shown in Fig. 16.2 makes the contribution of groundwater clear.

Essentially all groundwater is in motion. Groundwater moves in response to a hydraulic gradient in the same manner as water flowing in an open channel or a pipe. However, the flow of groundwater is appreciably restricted by friction with the porous medium through which it flows. This results in low velocities and high head losses. Velocities may vary from few centimetres per year to few metres per day. As can be seen from Fig. 16.2 permeable sub-soil strata serves as a conduit for transmission of water. It also acts as groundwater reservoir by storing water for various periods of time. Thus hydrologic cycle is completed. When sufficient water vapour gathers in the atmosphere the cycle again repeats.

The components of groundwater in hydrologic cycle for any area can be represented by an equation:

$$\Delta s = \text{recharge} - \text{discharge} \quad \text{... (1)}$$

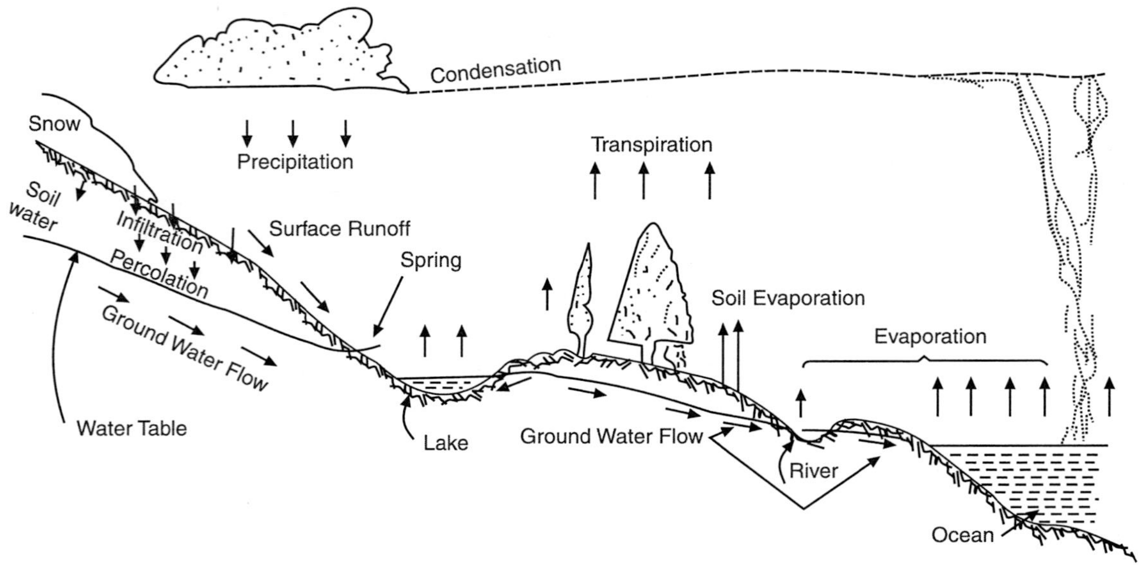

Fig. 16.2. Main features of the hydrological cycle and ground water

In this equation Δs represents the change in ground water storage during the period under study. Academically, over a long period of time and under natural conditions Δs will be zero as recharge equals discharge. However, due to human activities this rarely happens, which add artificial recharge as well as cause discharge to make the equation complex. If ground water storage in an area is less at the end of the selected period of time, than at the beginning, discharge is indicated as having exceeded recharge. On the contrary if ground water storage at the end of period exceeds that at the beginning, recharge is indicated as having exceeded discharge.

Terms on the right hand side of the equation (1) above namely, recharge and discharge are constituted as described below:

16.2.1. Recharge to Aquifers

Recharge from natural sources include the following:

(*i*) **Deep percolation from precipitation:** Deep percolation of precipitation is one of the most important sources of groundwater recharge. The amount of recharge in a particular area is influenced by vegetative cover, topography, and nature of soils; as well as the type, intensity and frequency of precipitation.

(*ii*) **Seepage from streams and lakes:** Seepage from streams, lakes, and other water bodies is another important source of recharge. In humid season, when groundwater levels may be high, the influence of seepage may be limited in extent. However, in arid regions where the entire flow of streams may be lost to an aquifer, seepage may be of major significance.

(*iii*) **Underflow from another aquifer:** An aquifer may be recharged by underflow from a nearby, hydraulically connected aquifer.

(*iv*) **Artificial recharge:** Artificial recharge to the groundwater may be achieved through planned systems or may be unforeseen or unintentional. Planned major contributions to the groundwater reservoir may be through spreading grounds, infiltration ponds, and recharge wells. Irrigation applications, and other sanitary and domestic activities have a similar, but usually unintentional effect. Seepage from reservoirs, canals, drainage ditches, ponds, and similar water impounding and conveyance structures may be local sources of major ground water recharge. Recharge from such sources can completely change the ground water regimen over a considerable area.

Groundwater Discharge: Losses from the groundwater reservoir occur in the following four ways:

(*i*) **Seepage to streams:** In certain reaches of streams and in certain seasons of the year, ground water may discharge into streams and maintain their base flows. This condition is more prevalent in humid areas than in arid or semiarid areas.

(*ii*) **Flow from springs:** Springs exist where the water table intersects the land surface or a confined aquifer outlets to the surface.

(*iii*) **Evaporation and transpiration:** Groundwater may be lost by evaporation if the water table is near enough to the land surface to maintain flow by capillary rise. Also, plants may transpire groundwater from the capillary fringe or the saturated zone.

(*iv*) **Artificial discharge:** Wells and drains are imposed artificial withdrawals on groundwater storage and in some areas are responsible for the major depletion.

16.3. AQUIFERS AND THEIR TYPES

A geologic formation which is water bearing and is saturated with water and in addition is capable of yielding sufficient quantity of water for economic exploitation is called an *aquifer*. Such aquifers form huge groundwater reservoirs. The lateral continuity and vertical boundaries of such reservoirs are generally not well defined owing to complexities of subsurface formations. The size of aquifers may vary from few km^2 to several hundred km^2. Depending upon the hydraulic characteristics the aquifers can be classified into following three types:

- Unconfined aquifer,
- Perched aquifer,
- Confined aquifer, and
- Leaky aquifer or semi-confined aquifer.

Unconfined aquifer: An aquifer which is not overlain by any confining layer but has a confining layer at its bottom is called unconfined aquifer. It is normally exposed to the atmosphere and its upper portion is partly saturated with water. The upper surface of saturation is called water table which is under atmospheric pressure therefore this aquifer is also called phreatic aquifer.

Perched aquifer: It is a special case of an unconfined aquifer. This type of aquifer occurs when an impervious or relatively impervious layer of limited area in the form of a lens is located in the water bearing unconfined aquifer. As shown in Fig. 16.3 the water storage created above the lens is perched aquifer and its top layer is called *perched water table.*

Confined aquifer: It is also called *artesian aquifer*. It is a type of aquifer overlain as well as underlain by confining layers. The water within the aquifer is therefore held under pressure. It is sometimes called pressure aquifer also. If the aquifer has high outcrop laterally than the ground surface there will be positive hydrostatic pressure to create conditions for a flowing well. Water from such well comes to the surface without pumping. The imaginary level upto which the water will rise is called piezometric surface.

Leaky aquifer: In nature, truly confined aquifers are rare because the confining layers are not hundred per cent impervious. An aquifer which is overlain or underlain by a semi-pervious layer (*aquitard*) through which vertical leakage takes place due to head difference is called leaky aquifer or semi-confined aquifer. The permeability of the semi-confining layer is usually very small as compared to the permeability of the main aquifer. Thus the water which seeps vertically through the semi-confining layer is diverted internally to proceed horizontally in the main aquifer.

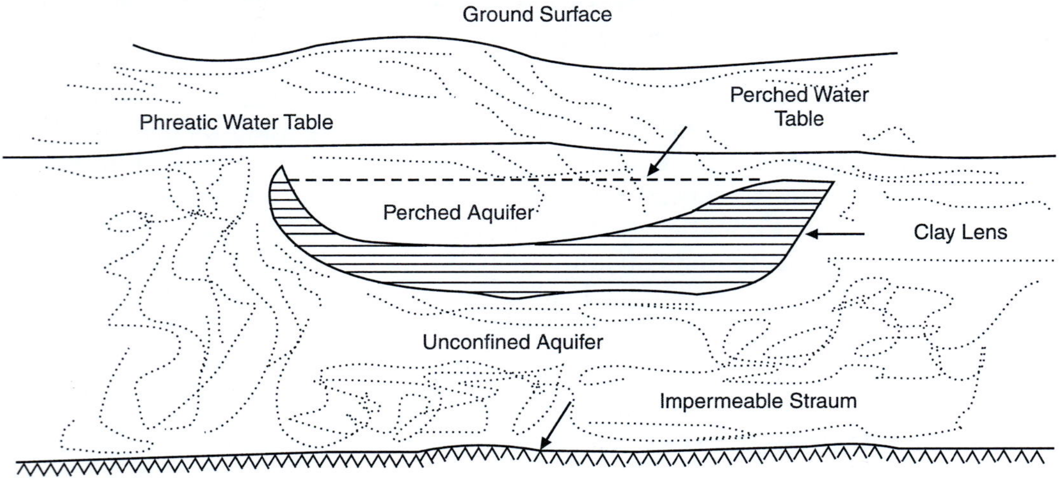

Fig. 16.3. Unconfined and perched aquifers

Figure 16.4 shows various typical groundwater structures and terminology which have been defined in this chapter.

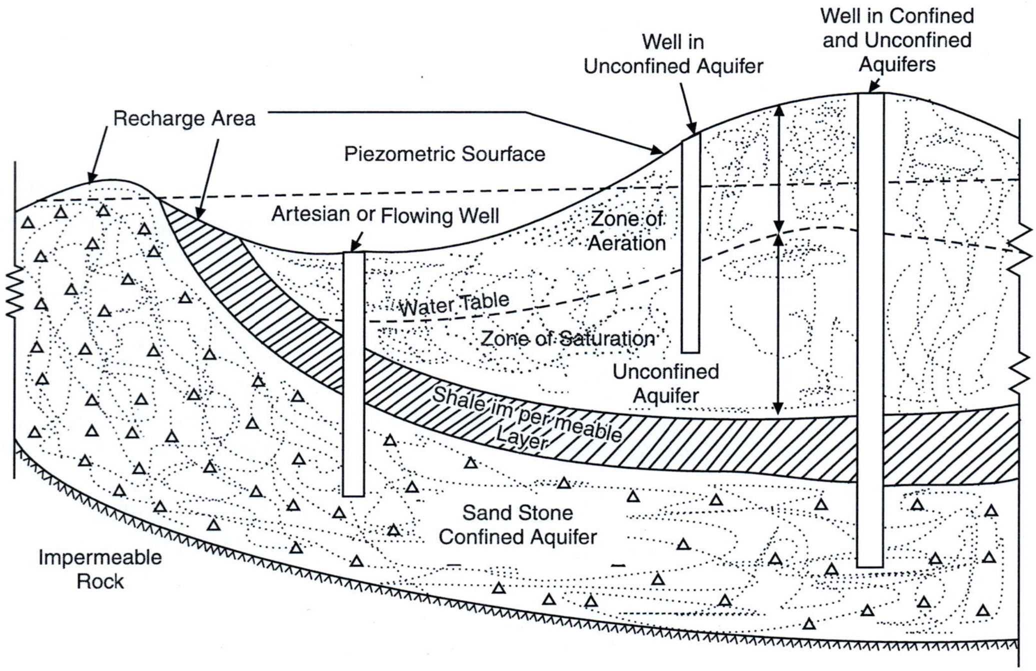

Fig. 16.4. Groundwater terminology model

16.4. IMPORTANT TERMS

16.4.1. Safe Yield

Safe yield of an aquifer is the amount of groundwater that can be withdrawn from the aquifer without resorting to mining of water.

Consider that a stage of broad equilibrium has reached in the groundwater hydrologic cycle. Therefore, only average annual replenishable amount of groundwater gets depleted annually. Obviously the limits of zone of intermittent saturation are maintained. Any depletion of groundwater from the zone of permanent saturation could be termed as overdraft or mining of groundwater because such excess withdrawal is not likely to be replenished in the following average rainfall year. If such withdrawals continue year after year the zone of intermittent saturation will grow in size and following adverse effects can be noticed:

- Permanent lowering of water table.
- Additional cost of pumping due to increased lift.
- Reduction in yield.
- Deterioration of water quality.
- Lowering of water table in connected adjacent groundwater reservoirs.
- Intrusion of saltish water if groundwater reservoir is located close to sea.

16.4.2. Optimal Yield

From the definition of safe yield it may be seen that main criteria to be observed is not to deplete groundwater to such an extent as to enter zone of permanent saturation by resorting to overdraft. Obviously since time for withdrawal is not specified safe yield at different times will be different. For example, at the end of monsoon safe yield is maximum because maximum amount of water is available in the intermittent zone of saturation whereas at the end of dry season or just before the beginning of monsoon safe yield will be minimum because minimum amount of water is available then in the intermittent zone of saturation.

On the other hand *optimal yield* indicates maximum amount of groundwater that can be withdrawn from the aquifer without resorting to mining of the groundwater. Thus maximum utilisation of annual gross recharge could be called optimal yield. It calls for early withdrawal of infiltrating groundwater to restrict unwanted effluent discharge.

16.4.3. Specific Yield

It is already mentioned that the water holding capacity of a soil mass depends on the grain size, shape and distribution of pores. The soils with small pore spaces retain more water. The retention varies inversely as the size of the pores. Since temperature and salt content affects surface tension, viscosity and specific gravity of water they also affect the water retention property of the soil mass. It is also mentioned that under the force of gravity only part of the water retained in the pore spaces drain freely if suitable drainage facilities are made available. Some part of water is also retained by the soil mass owing to molecular and capillary forces. In the light of these properties following three definitions can be given:

$$\text{Porosity } (n) \text{ of aquifer} = \frac{\text{Volume of voids}}{\text{Total volume of aquifer mass}} \qquad \text{... (1)}$$

$$\text{Specific yield } (S_y) \text{ of aquifer} = \frac{W_y}{V} \qquad \text{... (2)}$$

where W_y = is volume of water released from soil or rock mass by the force of gravity.

and V = is total volume of aquifer mass.

Thus *Specific yield* can be defined as the ratio of the volume of water drained to the total volume of the aquifer mass. It is sometimes also termed as effective porosity. The range of specific yields of different types of aquifer materials is given in Table 16.2.

Table 16.1. Ranges of Specific yield of aquifer

Type	*Specific yield (%)*
Clay	1 – 10
Mixed Sand	10 – 30
Gravel	15 – 30
Gravel and Sand	15 – 25
Sand Stone	5 – 15
Shale and Lime Stone	1/2 – 5

$$\text{Specific retention } (S_r) = \frac{W_r}{V} \qquad \dots (3)$$

where W_r = is volume of water retained due to molecular and capillary forces.

From (1), (2) and (3)

$$n = S_y + S_r$$

In an unconfined aquifer when the water is withdrawn from the aquifer by means of pumping out the well water the water level in the well drops down very fast. The depleted water is replenished from aquifer storage by three processes, namely,

- gravity drainage,
- compaction of aquifer, and
- expansion of water itself as pressure on the groundwater reduces.

It is a matter of common experience that the gravity drainage from the aquifer into the well is not immediate. This is invariably true in case of fine grained sediments. The delay in (yield) drainage from the aquifer depends on its drainage property.

The gravity yield can be defined as the percentage of water draining out under the action of gravity during a given period of time to total volume of the rock. The gravity yield equals specific yield when drainage of water from the aquifer continues for a long period.

16.4.4. Storage Coefficient (*S*) or Storativity

It is the volume of water released from or taken into aquifer storage per unit surface of the aquifer per unit decline or rise of head. It is also called storativity of an aquifer. In case of unconfined aquifer, the storage coefficient (*S*) for unit thickness of aquifer is equal to the specific yield provided gravity drainage is complete.

For unconfined aquifers (*S*) ranges from 0.02 to 0.30.

The case of confined aquifers is different than unconfined aquifers. In case of confined aquifers the storage coefficient depends on the volume of water forced out of the pores due to elastic compression of the aquifer resulting from change in hydrostatic pressure caused due to pumping of water from the well.

The storage coefficient can be mathematically calculated from the following equation:

$$S = n \cdot \gamma_w \cdot b \left(\beta + \frac{\alpha}{n}\right)$$

where γ_w is unit weight of water

b is thickness of the aquifer

β is the compressibility or the reciprocal of the bulk modulus of elasticity of water ($= 4.7 \times 10^{-6}$ cm^2/kg)

and α is vertical compressibility of aquifer material
(= 2.8×10^{-5} cm^2/kg approx.)

Typical values of 'S' range from 0.00005 to 0.005. This indicates that changes in hydrostatic pressure cause very small yields of water from confined aquifers.

Problem 16.1. An aquifer is 80 ha in area. The water table is lowered by 10 m over whole area. If the porosity of the aquifer is 30% and the specific retention is 5%. Calculate the amount of water actually drained from the aquifer.

Solution: Porosity $(n) = S_y + S_r$ (Specific retention)

$\therefore$ Specific yield $(S_y) = n - S_r = 30 - 5 = 25\%$

Also Specific yield $= \dfrac{\text{Volume of water drained}}{\text{Volume of aquifer mass}}$

or

Volume of water drained = Specific yield × Volume of aquifer
$= 0.25 \times 80 \times 10 = 200$ ha.m.

Problem 16.2. A confined aquifer is 40 m thick and 100 ha in area. The piezometric head is lowered by 5 metres. Determine the volume of water released from the aquifer storage. Take storage coefficient 'S' = 2×10^{-4}.

Solution: Amount of water released = (S) × (Area of aquifer) × (Decline of piezometric head)
$= 2 \times 10^{-4} \times 100 \times 5 = 1000 \times 10^{-4}$ ha.m
$= 1000$ m^3 or 0.1 ha.m.

16.4.5. Transmissibility or Transmissivity of Aquifers

This parameter gives the capacity of an aquifer to transmit water through its entire thickness. It is defined as a product of coefficient of permeability of aquifer material and the saturated thickness of the aquifer. Thus

Coefficient of Transmissibility $(T) = K \cdot b$

where K is coefficient of permeability

and b is saturated thickness of aquifer

Thus it may be seen that specific yield S_y *or* storage coefficient S and the coefficient of transmissibility together fully describe the characteristics of the aquifer.

16.5. DARCY'S LAW OR FLOW OF WATER IN SATURATED SOILS

Now it is clear that groundwater is largely hidden and cannot be seen and understood as surface water. However, groundwater storage in saturated zone or groundwater reservoir forms a significant part of the total water resources.

In some respects, groundwater reservoirs are similar to surface water reservoirs. They also have a particular capacity and must have discharge and recharge.

The flow of water in saturated zone of soil is primarily governed by Darcy's law which states that the velocity of flow through a porous media is proportional to the first power of hydraulic gradient. It can be expressed as:

$$V = K\,i$$

or

$$Q = K\,A\,i$$

where Q is the quantity of water flowing in the soil.

K is a constant, it depends upon porosity and permeability of the soil and is called coefficient of permeability or hydraulic conductivity of the material.

i is the loss of head per unit length or hydraulic gradient,

Mathematically $$i = \frac{h_1 - h_2}{L}$$

V is velocity of water in soil

and A is cross-sectional area through which the water flows.

In deriving this equation Darcy assumed that there are number of capillary tubes in soils through which water passes in the similar way as it passes through pipes. Since the velocity in the laminar flow is also proportional to first power of hydraulic gradient, the flow obeying Darcy's law is essentially laminar. Applying same analogy as that of flow through pipes the Reynold's number can be used as an index to determine the range of applicability of Darcy's law.

The Reynold's number is expressed as

$$R = \frac{Vd}{\gamma}$$

where V is velocity of flow

d is average grain diameter

and γ is Kinematic viscosity of the fluid

$$\gamma = \frac{\mu}{\rho} = \frac{\text{Dynamic viscosity of fluid}}{\text{Density of fluid}}$$

It has been seen that in natural and artificial sandy medium of uniform grain size deviation of flow from laminar state to turbulent state starts for values of R between 1 and 10 depending on the shape and range of grain size. Almost in all cases the groundwater flow can be seen to be laminar and therefore obey Darcy's law. In tube wells also where pumping rates are quite high ordinarily the flow obeys Darcy's law upto the well face. But at the well face it may tend to become turbulent owing to high rates of pumping.

The observation of the waters table can be very easily and quite accurately done by noting the levels in the rivers, streams and existing wells.

16.6. FLOW OF WATER IN UNSATURATED SOILS

Darcy's law for flow of water through saturated soils also holds good for unsaturated soils. Water always flows in the direction of decreasing energy. Thus the equation $V = k\ i$ is also true for unsaturated soils. Unlike in saturated soil, cross-sectional area of flow is decreased since the moisture content is less. It also lowers permeability rates. Another complications in the unsaturated soil is about measuring negative pressure. For saturated soils energy can be very conveniently measured by inserting piezometer tubes in the soil. In unsaturated soil mass in addition to water flow there exists vapour flow also. It is very difficult to measure vapour flow through the soil. The flow of vapour increases with decrease in moisture content of the soil. Hence, no perfect analytical treatment of flow of water through the unsaturated zone of soil is still feasible.

16.7. WELL HYDRAULICS–SOME DEFINITIONS

The most commonly used structure for harnessing groundwater is a tube well from which the water is pumped out to utilise at surface. The amount of water pumped out is called groundwater draft.

Judicious and efficient extraction of groundwater is possible only when the tube wells are properly designed and efficiently and economically constructed. It calls for proper understanding of various laws governing the flow of groundwater through permeable formations and practical formulae for flow through aquifers. The whole subject matter comes under the heading of hydraulics of wells.

It is useful to understand following terms:

Drawdown: When the water is pumped out from a tube well, the water level in the well as well as in the vicinity of the pumped well is lowered. The lowering of the water level at any point as a result of groundwater pumping is called drawdown at that point. The drawdown is maximum at the well and goes on reducing away from the well till at some distance the drawdown is zero. Obvisously at and after this point water level remains uneffected.

Hydraulic gradient: Owing to the differential lowering of water level a difference of head is created between the water level at the well and original groundwater table. The slope or head difference per unit length is called hydraulic gradient.

Cone of depression: The existence of hydraulic gradient induces flow of water into the well from the surrounding aquifer. Since the water moves towards the well from all directions it forms successive cylindrical sections. As the water approaches the well the surface area of each cylindrical section goes on reducing. Although the surface area of cylindrical rings reduces the rate of flow is same through all cylindrical sections. As a result the velocity of flow is smaller at outer most ring which gradually increases towards the well. With increasing velocity the hydraulic gradient also increases as the flow approaches the well. This situation continuces to exist so long as head difference is maintained due to pumping. The form of surface created by increasing hydraulic gradient is cone shaped. It is therefore called cone of drepression and it represents drawdown at various points in the aquifer. Figure 16.5 shows various terms used in well hydraulics.

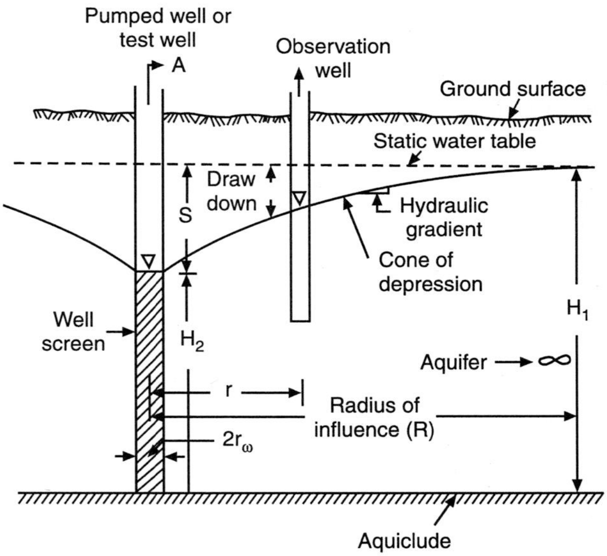

Fig. 16.5. Terminology used in well hydraulics

Radius of influence: It is distance from the centre of the well to the point at which the drawdown is zero. The point at which drawdown is zero indicates the outer limit of the cone of depression. The cones of depression are larger for wells sunk in the confined aquifer. Therefore, radius of influence is also greater in confined aquifer than in the unconfined aquifer.

It may be noted that drawdown, cone of depression and radius of influence are the three parameters of the same phenomenon, are closely inter-related and are characteristic features of every pumped well. They depend on the following:

- Aquifer characteristics;
- Well discharge, *i.e.,* pumping rate;
- Duration of pumping; and
- Slope of water table recharge within the influenced zone.

Steady flow: In a pumped well when an equilibrium is reached between the discharge of the pumped well and the recharge of the aquifer by an outside source this condition represents a steady or equilibrium flow. In other words steady state is reached when in the piezometers the changes in the drawdown with time have become negligible or when the hydraulic gradient has become constant.

For a semi-confined aquifer overlain by a semi-pervious layer the outside source could be an unconfined aquifer with constant water table which is situated over the semi-pervious layer. This constant water table aquifer can contribute required recharge to the semi-confined aquifer through the semi-pervious layer.

Unsteady flow: The unsteady flow indicates non-equilibrium condition. This type of flow exists from the moment groundwater pumping starts from the well till the steady state is reached. In a completely confined horizontal aquifer of constant thickness and infinite size which is pumped at a constant rate there will be unsteady flow for all the time in the absence of a source of recharge in the from of vertical leakage or a recharge boundary.

Conversely to the steady-state flow, when the water level in piezometers or hydraulic gradient changes with time during pumping unsteady flow condition exists.

Partially penetrating well: When the tube well does not extend to full depth of the aquifer but draws water from part depth of the aquifer the well is called partially penetrating well.

Specific capacity of a well: It is a measure of the productivity of a well. It is a ratio of the pumping rate and drawdown in the well. (Also refer chapter 18)

$$\therefore \quad \text{Specific capacity} = \frac{Q}{H_1 - H_2} = \frac{Q}{s}$$

where Q is well discharge
and $(H_1 - H_2)$ or s is drawdown.

16.8. COMPUTATION OF COEFFICIENT OF PERMEABILITY AND DISCHARGE CAPACITY OF AQUIFER

Darcy, Dupit, Theim, Theis and Jacob have analysed groundwater flow theoretically. While Dupit and Theim analysed steady flow in confined and unconfined aquifers, Theis and Jacob analysed unsteady flow.

The foundation of groundwater hydraulics is Daracy's law. It states that "the flow rate through a porous medium is proportional to the head loss and inversely proportional to the length of flow path". The law is applicable when the flow is laminar and without turbulance. The most common forms of the formulae are:

$$V = Ki$$

$$Q = KiA$$

or

$$Q = K.A\,\frac{h_1 - h_2}{L}$$

where V = velocity

K = permeability or hydraulic conductivity of porous medium (commonly called coefficient of permeability)

i = hydraulic gradient $= \dfrac{h_1 - h_2}{L}$

A = area normal to the direction of flow

Q = rate of flow

h_1 and h_2 = water levels or potential at two points on a line parallel to the direction of flow

L = length of flow path between h_1 and h_2

16.9. DUPIT'S FORMULA FOR STEADY ONE DIRECTIONAL FLOW

It is already mentioned above that steady flow of groundwater in a confined aquifer of uniform thickness behaves in accordance with Darcy's law; that is, the head decreases linearly in the direction of flow. However, in a free aquifer the water table is also a flow line. The shape of the water table determines the flow distribution, conversely, the flow distribution determines the shape of the water table. Therefore, general analytical solution of the flow is not possible. However, Dupit in an attempt to simplify analysis of undirectional flow, made the following assumptions:

1. The velocity of flow is proportional to the tangent of the hydraulic gradient rather than the sine as determined by Darcy; and
2. The flow is horizontal and uniform everywhere in a vertical section.

Dupit derived the following equation:

$$Q = \frac{K\left(h_2^2 - h_1^2\right)}{2L}$$

where, Q = flow of water per unit time, per unit width normal to the direction of flow.

h_1 = head of water in the aquifer at point 1 in the line of flow.

h_2 = head of water in the aquifer at point 2 in the line of flow from h_1, and

L = distance between the points 1 and 2 parallel to the direction of flow.

Despite the simplifying assumptions, the equation closely approximates the water table position where the sine and tangent of the slope of the water table are approximately equal.

16.10. DUPIT'S FORMULAS FOR STEADY OR EQUILIBRIUM RADIAL FLOW

In 1863, Dupit analysed steady or equilibrium flow in gravity wells as well as pressure wells sunk in unconfined and confined aquifers. He studied a single well in which steadystate or equilibrium condition was established by pumping the water from the well for a sufficiently long time till the rate of pumping became equal to rate of recharge. He presented two formulas one for steady radial flow in unconfined aquifer and other for steady radial flow in a confined aquifer. In doing so he made quite a few assumptions to simplify the matter. The assumptions are:

(*i*) The velocity of flow is proportional to the tangent of the hydraulic gradient rather than the sine as determined by Darcy.

(*ii*) The flow is horizontal and uniform everywhere in a vertical section.

(*iii*) The aquifer is homogeneous, isotropic and of infinite areal extent so that coefficient of permeability is constant everywhere.

(*iv*) The well penetrates fully and it receives water from the entire thickness of the aquifer.

(*v*) Pumping is so adjusted that equilibrium stage exists.

(*vi*) The flow is laminar and therefore it follows Darcy's law.

Dupits formula for confined aquifer is

$$k = \frac{2.303 Q \log_{10} \dfrac{R}{r_w}}{2\pi m (H_1 - H_2)}$$

or

$$Q = \frac{2\pi k.m (H_1 - H_2)}{2.303 \log_{10} \dfrac{R}{r_w}}$$

Also, Dupits formula for unconfined aquifer is

$$k = \frac{2.303 Q \log_{10} \dfrac{R}{r_w}}{\pi (H_1^2 - H_2^2)}$$

or

$$Q = \frac{\pi k (H_1^2 - H_2^2)}{2.303 \log_{10} \dfrac{R}{r_w}}$$

where,
k = coefficient of permeability
Q = discharge
m = thickness of confined aquifer
R = radius of circle of influence
r_w = radius of well
H_1 = height of static water table from well bottom
H_2 = height of water in the well from well bottom

16.11. THEIM'S FORMULAS FOR STEADY OR EQUILIBRIUM RADIAL FLOW

In 1906 C. Theim and P. Forchheimer independently derived equations for steady radial flow to a fully penetrating well with 100 per cent penetration and open hole using Darcy's law and Dupit's assumptions. The equations, known today as the steady state, Theim, Dupit-Forchheimer, or Theim-Forchheimer equations, can be used to determine the coefficient of permeability of an aquifer from measurements made during a pumping test using a fully penetrating well with 100 per cent open hole and two or more observation wells.

Like Dupit's formula, Theim-Forchheimer or equilibrium equations are also based on the following assumptions:

(*i*) Aquifer is homogeneous, isotropic, and of uniform thickness.

(*ii*) The discharging well penetrates and receives water through the entire thickness of the aquifer.

(*iii*) Coefficient of transmissivity or permeability (hydraulic conductivity) is constant at all times and at all locations.

(*iv*) Discharging has continued for a sufficient duration for the hydraulic system to reach a steady state.

(*v*) Flow to the well is horizontal, radial, and laminar, and originates from a circular open water source with a fixed radius and elevation which surrounds the well.

(*vi*) Rate of discharge from the well is constant.

The equilibrium equations, which yield values of permeability and transmissivity, are used for the analysis of discharging well tests. The general test procedure is to simultaneously pump from a test well at a constant, known rate and to periodically measure the drawdown in two or more nearby observation wells.

Theim and Forchheimer formula for confined or artesian aquifer is

$$K = \frac{2.303 Q \log_{10}\left(\frac{r_2}{r_1}\right)}{2\pi m (s_1 - s_2)}$$

or

$$T = \frac{2.303 Q \log_{10}\left(\frac{r_2}{r_1}\right)}{2\pi (s_1 - s_2)}$$

For an unconfined or free aquifer the formula is

$$K = \frac{2.303 Q \log_{10}\left(\frac{r_2}{r_1}\right)}{\pi (h_2^2 - h_1^2)}$$

or

$$T = \frac{2.303 Q m \log_{10}\left(\frac{r_2}{r_1}\right)}{\pi (h_2^2 - h_1^2)}$$

where K = permeability or hydraulic conductivity,

Q = discharge of the test well,

m = saturated thickness of the aquifer,

$T = k.m$ = transmissivity of the aquifer, or transmissibility of the aquifer,

r_1, r_2 = horizontal distances from centreline of the test well to centreline of observation wells 1, 2,

h_1, h_2 = saturated thickness or piezometric heads of the aquifer at distances r_1, r_2, from the test well,

s_1, s_2 = drawdown in observation wells at distances r_1, r_2 from the test well.

It can be seen from Articles 16.10 and 16.11 there is basically no difference between Dupit's formulas and Theim's formulas. The difference lies in the fact that in Theim's formula two or more observation wells have been considered within the distance of radius of influence (R) of the test well. Theim's equation gives a generalized equation. It is to be noted that although none of the assumptions really exist in practice, nevertheless these formulas are extensively used in solving the groundwater problems because they offer the best possible solution.

16.12. WELL LOSSES

When the water is pumped out from the well the groundwater starts moving from the surrounding area of the aquifer into the well.

It causes head loss and can be called aquifer loss. The loss of head or drawndown is maximum at the well face. But this loss of head is not all that occurs. In addition some loss of head (something like loss of head at entrance) takes place as the water flows through the screen. Also as the water inside the well enters the pump intake there also some loss of head takes place. Owing to these two causes additional drawndown takes place. This additional drawdown is called well loss. (Fig. 16.6)

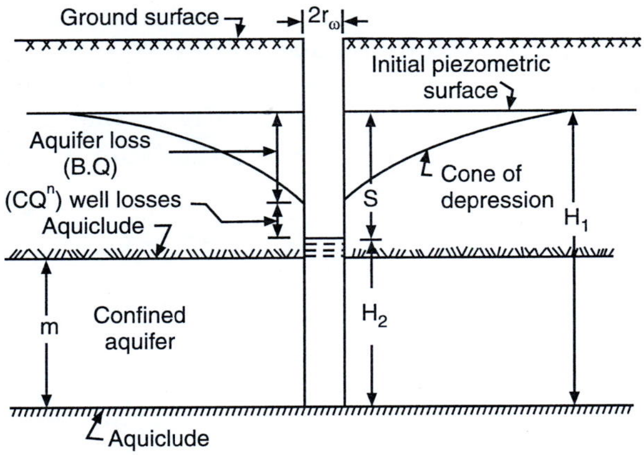

Fig. 16.6. Well losses

As the flow is generally turbulent at the well face this loss of head is proportional to n^{th} power of the discharge. Thus

$$\text{Well loss} \propto Q^n$$

or $$\text{Well loss} = C \cdot Q^n$$

Exact value of n cannot be stated due to large variations among individual wells. It can be stated that

$$1 < n < 2$$

C is a constant. It depends on the radius of well, manner of construction and condition of the well.

Total drawdown at the well face

$$= H_1 - H_2 = s$$

$$\therefore \quad s = \frac{2.303Q}{2\pi \cdot Km} \log_{10} \frac{R}{r_w} + C \cdot Q^n$$

For a particular well taking

$$\frac{2.303}{2\pi \cdot Km} \log_{10} \frac{R}{r_w} = B$$

$$s = B \cdot Q + CQ^n$$

The term BQ represents aquifer loss and CQ^n gives well loss.

The well loss can be minimised by adopting following measures:

- By keeping screen openings equal to the aquifer pores;
- By maintaining the well in good condition;
- By employing low pumping rates; and
- By building large diameter wells.

Effect of well loss on specific capacity:

It is already stated that

$$\text{Specific capacity} = \frac{\text{Discharge}}{\text{Drawdown}} = \frac{Q}{s}$$

or $$\text{Specific capacity} = \frac{Q}{B \cdot Q + C \cdot Q^n} = \frac{1}{B + C \cdot Q^{n-1}}$$

The above expression shows that specific capacity of the well is not constant but it decreases with increase in well discharge.

16.13. INTERFERENCE OF WELLS

Many times two or more wells are located in the same aquifer and are close to each other. Then it is possible that their cones of depression may intersect each other. When such a situation exists the wells are said to interfere with each other because the zone of influence of one well then overlaps the zone of influence of the other well.

Thus when closely spaced multiple well system exits in an aquifer causing interference following effects can be noticed:

(*i*) The total groundwater output is less than the sum of the discharging capacity of individual wells.

(*ii*) The efficiency of each well is decreased.

(*iii*) The drawdown is increased and as a result pumping lift has become higher.

(*iv*) The pumping cost has increased due decrease in efficiency and increase in pumping lift.

In view of the above effects it is necessary to properly demarcate the well field to provide most economical pumping system.

It is mentioned above that due to interference of wells the total pumping rate is not sum total discharging capacity of the wells taken individually. The reduction in the pumping rate from the wells is a function of the characteristics of the aquifer, distance between the wells and the pumping time. The magnitude of interference in any well is represented by the ratio of diminution in its discharge without interference.

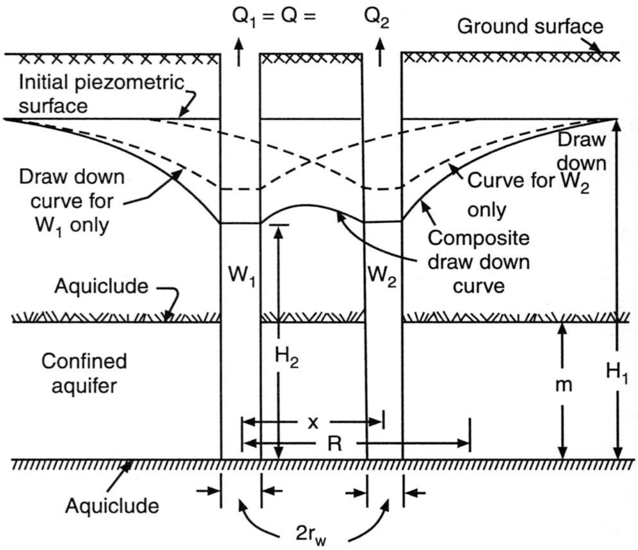

Fig. 16.7. Interference of wells

16.13.1. Confined Aquifer Case

When two wells are located at a distance '*x*' apart are discharging simultaneously and have same diameter and drawdown (Fig. 16.7) discharge through each well is given by following equation:

$$Q_1 = Q_2 = \frac{2\pi K m(H_1 - H_2)}{2.303\log_{10}\dfrac{R^2}{r_w.x}} \qquad ...(a)$$

Since the wells are interfering the radius of influence, $R > x$.

For a single well pumping individually discharge is given by Thiem or equilibrium equation

$$Q = \frac{2\pi K m(H_1 - H_2)}{2.303\log_{10}\dfrac{R}{r_w}} \qquad ...(b)$$

When wells interefere $R > x$

$\therefore$ $\frac{R^2}{r_w . x}$ term of equation (*a*) is greater than $\frac{R}{r_w}$ term of equation (*b*).

Obviously $Q > Q_1$.

$\therefore$ Magnitude of interference $= \frac{Q - Q_1}{Q}$

16.13.2. Unconfined Aquifer Case

When two wells are located at distance '*x*' apart in unconfined aquifer and are identical the discharge through each well is given by following equation:

$$Q_1 = Q_2 = \frac{\pi K \left(H_1^2 - H_2^2\right)}{2.303 \log_{10} \frac{R^2}{r_w . x}} \quad \ldots(c)$$

For a single well pumping individually discharge is given by Theim's equation

$$Q = \frac{\pi K \left(H_1^2 - H_2^2\right)}{2.303 \log_{10} \frac{R}{r_w}} \quad \ldots(d)$$

Since $R \geq x$ in $\frac{R^2}{r_w . x}$ term of equation (*c*) is greater than $\frac{R}{r_w}$ term of equation (*d*).

Obviously $Q > Q_1$

$\therefore$ Magnitude of interference in this case

$$\text{also} = \frac{Q - Q_1}{Q}$$

Problem 16.3. A tube well penetrates into an unconfined aquifer having a saturated depth of 80 m. When the drawdown was 12 m discharge from the well was 180 litres per minute. Assume steady state condition and radius of circle of influence same for both cases. Calculate the discharge at 18 m drawdown.

Solution: Step 1. Following equation can be used

$$K = \frac{2.303 Q}{\pi \left(H_1^2 - H_2^2\right)} \log_{10} \frac{R}{r_w}$$

It may be see that *K*, π, *R* and r_w have same values for different drawdown conditions. So taking

$$\frac{2.303}{\pi} \log 10 \frac{R}{r_w} = A$$

$$K = \frac{A \cdot Q}{\left(H_1^2 - H_2^2\right)} \quad \text{or} = \frac{K}{A} = \frac{Q}{\left(H_1^2 - H_2^2\right)}$$

Step 2. It is given that Q = 180 litres/minute and

$$H_1 = 80 \text{ m}, H_2 = 80 - 12 = 68 \text{ m}$$

$$\therefore \quad \frac{K}{A} = \frac{180}{6400-4624} = \frac{180}{1776} = \text{constant}$$

Step 3. Taking second drawdown condition

Q is unknown, H_1 = 80 m and H_2 = 80 – 18 = 62 m

$$\therefore \quad \frac{K}{A} = \frac{180}{1776} = \frac{Q}{6400-3844} = \frac{Q}{2556}$$

$$\therefore \quad Q = \frac{180 \times 2556}{1776} = 259 \text{ litres/sec}$$

Problem 16.4.

(*a*) In a confined aquifer 18 cm diameter well penetrates fully. If the length of the strainer is 20 m, what will be the yield of a tube well under 5 m drawdown. Assume steady state condition. Take K = 30 m/day and radius of influence R = 300 m.

(*b*) Will the discharge of the tube well be doubled by doubling the diameter of the well? Assume all other conditions to remain same.

Solution: (*a*) Formula to be used is

$$Q = \frac{2\pi K m (H_1 - H_2)}{2.303 \log_{10} \frac{R}{r_w}}$$

It is given that K = 30 m/day, m = 20 m; $(H_1 - H_2)$ = 5 m

R = 300 m; r_w = 0·09 m

$$\therefore \quad Q = \frac{2 \times \frac{22}{7} \times 30 \times 20 \times 5}{2.303 \log_{10} \frac{300}{0.9}} = \frac{2 \times 3.143 \times 20 \times 5}{2.303 \times 3.523}$$

= 2324·3 m^3/day or 27 litres/sec.

(*b*) Diameter of the well if doubled, r_w = 18 cm = 0.18 m

For this size well, other conditions remaining same

$$Q = \frac{2 \times \frac{22}{7} \times 30 \times 20 \times 5}{2.303 \log_{10} \frac{300}{0.18}} = \frac{2 \times 3.143 \times 30 \times 20 \times 5}{2.303 \times 3.222}$$

= 2541·4 m^3/day or 29 litres/sec.

Thus by doubling the diameter well discharge increases only from 27 l/sec to 29 l/sec.

Problem 16.5.

(*a*) Under a steady state condition a well is discharging 40 l/sec from a confined aquifer. Two observation wells are located at 10 m and 50 m from the pumped well. The difference in piezometric surface of the two wells was measured to be 1·5 m. Calculate transmissibility of the aquifer.

(b) If the diameter of the pumped well is 20 cm and well loss is 15 cm find out the difference in piezometric surface between pumped well and the observation well at 50 m.

Solution: (*a*) Thiem's equilibrium equation may be used to calculate trasnsmissibility which is expressed as

$$K \cdot m = T = \frac{2.303}{2\pi(h_2 - h_1)} \cdot \log_{10} \frac{r_2}{r_1}$$

It is given that $Q = 0{\cdot}040$ m³/sec; $h_2 - h_1 = 1{\cdot}5$ m;

$r_2 = 50$ m and $r_1 = 10$ m

$$\therefore \qquad T = \frac{2.303 \times 0.04}{2 \times \frac{22}{7} \times 1.5} \log_{10} \frac{50}{10} = \frac{2.303 \times 0.04 \times 0.699}{2 \times 3.143 \times 1.5} = 0{\cdot}007 \text{ m}^2\text{/sec.}$$

∴ Transmissibility of the aquifer is 0·007 m²/sec.

(*b*) Total drawdown in well = Difference in piezometric surface

∴ Difference in piezometric surface = Aquifer loss + Well loss

$$\text{or} \qquad h - H_2 = S = \frac{2.303Q}{2\pi T} \log_{10} \frac{r}{r_w} + \text{Well loss}$$

For part (*b*) it is given that $r_w = \frac{20}{2} = 10$ cm $= 0.1$ m and well loss = 15 cm

Using all other values from part (*a*) of the problem

$$h - H_2 = S = \left(\frac{2.303 \times 0.04}{2 \times \frac{22}{7} \times 0.007} \times \log_{10} \frac{50}{0.1} \right) + (0{\cdot}15)$$

$$= \left(\frac{2.303 \times 0.04 \times 2.699}{2 \times 3.143 \times 0.007} \right) + (0.15) = 5{\cdot}65 + 0{\cdot}15 = 5{\cdot}80 \text{ m}$$

∴ Difference between piezometric surface between pumped well and observation well at 50 m is 5.80 m.

Problem 16.6. An unconfined aquifer made up of fine sand and gravel of 20 m thickness is underlain by a clay bed. The groundwater level was observed 2.5 m below the surface. A pumping test was conducted in a fully penetrating well at steady state condition. The discharge obtained was 4 litres per second. Two observation wells were monitored during the test. The wells were located at 25 m and 50 m from the pumped well and the drawdown observed was 1.5 m and 0.5 m respectively. Calculate the coefficient of permeability of the aquifer.

Solution: Dupit's equation can be utilised to calculate the permeability of the unconfined aquifer, which can be expressed as

$$K = \frac{2.303 \times Q}{\pi(h_2^2 - h_1^2)} \log_{10} \frac{r_2}{r_1}$$

It is given that total saturated depth is 20 – 2.5 = 17.5 m

$Q = 0{\cdot}004$ m³/sec; For $r_1 = 25$ m, $h_1 = 17{\cdot}5 - 1.5 = 16$ m

and for $r_2 = 50$ m, $h_2 = 17{\cdot}5 - 0.5 = 17$ m

Substituting the values in above equation

$$K = \frac{2.303\times 0.004}{3.143\left(17^2-16^2\right)}\log_{10}\frac{50}{25} = \frac{2.303\times 0.004\times 0.301}{3.143\times 33}$$

$$= \frac{0.0028}{103.72} = 0{\cdot}0000269 \text{ m/sec} = 96{\cdot}84 \text{ mm/hour}$$

Problem 16.7. In a confined aquifer of 30 m thickness two wells of 20 cm dia, each are sunk fully. The two wells are 150 m apart. Adopting the following data calculate:

(*a*) Discharge of the well when only one well is pumped.

(*b*) Discharge of both the wells when both are pumped simultaneously.

Data given: Assume steady state condition

Depression head 5 m

Coeff. of permeability K = 30 m/day

Radius of circle of influence = 300 m

Solution: (*a*) When only one well is being pumped discharge is given by Thiem's equation

$$Q = \frac{2\pi K m\left(H_1 - H_2\right)}{2.303\log_{10}\dfrac{R}{r_w}}$$

Substituting the values

$$Q = \frac{2\times 3.143\times 30\times 30\times 5}{2.303\log_{10}\dfrac{300}{0.1}} = \frac{2\times 3.143\times 4500}{2.303\times 3.477} = 3532 \text{ m}^3\text{/day}$$

$$= 0{\cdot}041 \text{ m}^3\text{/sec or } 41 \text{ l/sec}$$

(*b*) When both the wells are pumped simultaneously their discharge are equal since they are identical and is given by following equation

$$Q_1 = Q_2 = \frac{2\pi K m\left(H_1 - H_2\right)}{2.303\log_{10}\dfrac{R^2}{r_w . x}}$$

Substituting the values $= \dfrac{2\times 3.143\times 30\times 30\times 5}{2.303\times\log_{10}\dfrac{300\times 300}{0.1\times 150}} = 3251 \text{ m}^3\text{/day}$

or $= 0{\cdot}038 \text{ m}^3\text{/sec or } 38 \text{ l/sec.}$

QUESTIONS FOR REVISION

1. Explain the following terms:
   (*a*) Confined aquifer
   (*b*) Unconfined aquifer
   (*c*) Perched aquifer
   (*d*) Specific yield of aquifer
   (*e*) Storage coefficient.
2. Define and clearly distinguish between safe yield and optimal yield of aquifers.

3. (*a*) State Darcy's law.
   (*b*) What is the limiting value of the Reynold's number for a laminar flow in sand aquifer?
   (*c*) What are limitations of Darcy's law ?
   (*d*) Is it applicable for unsaturated zone of soil?
4. Define the following:
   (*a*) Steady state flow
   (*b*) Specific capacity of a well
   (*c*) Radius of circle of influence
   (*d*) Drawdown
   (*e*) Transmissibility.
5. (*a*) What do you understand by well losses ? How can they be minimised ?
   (*b*) What is interference of wells ? What are the effects of interference ?
6. What is Dupit's theory ? What assumptions we make to simplify the theory ?

17

Open Well Irrigation

17.1. TYPES OF OPEN WELL

Open well: To tap the groundwater storage vertical hole of bigger diameter (2 to 10 metres generally) is some time dug or sunk in the rock or soil mass. The hole is sunk till it penetrates saturated underground material. The hole is further taken down to reach a depth quite below ground water-table. The depth of well or hole below the ground water-table should be such that even during a dry year there will be sufficient depth (3 to 4 metres) of water to cope up with the requirements then. The water surface in the well and free surface of water in the soil are at atmospheric pressure. The hole may be lined by some protecting materials or it may be left unprotected depending upon the nature of duties the well is supposed to perform. The shape of well is mostly circular. Open wells are suitable for rocky or hard soil formation. Average discharge of the well may be about 0.004 m^3/sec.

Depending upon the type of protection provided to the sides of the wells, first classification of open wells may be:

- Wells with pervious lining and
- Wells with water-tight lining.

17.1.1. Wells with Pervious Lining

In this type, the wells are lined with dry bricks or stone masonary. Water contribution to the well takes place through the sides.

This type is very suitable when subsoil is formed of gravel or coarse sand deposits. When the subsoil consists of sand layers, along with the seeping water sand also comes in the well through pervious lining. As a result hollow space is formed behind the wall lining. To check development of such hollow formation, brick ballast of about 20 mm size is packed behind the lining at least upto the ground water-table from the bottom. Figure 17.1 shows the typical cross-section of a well with pervious lining. To make the construction stable following procedure can be adopted. Construct 1.25 m of lining just above the curb in mortar and give about 1 m thick concrete plug in the bottom. Linning upto the gound water-table is laid dry but with

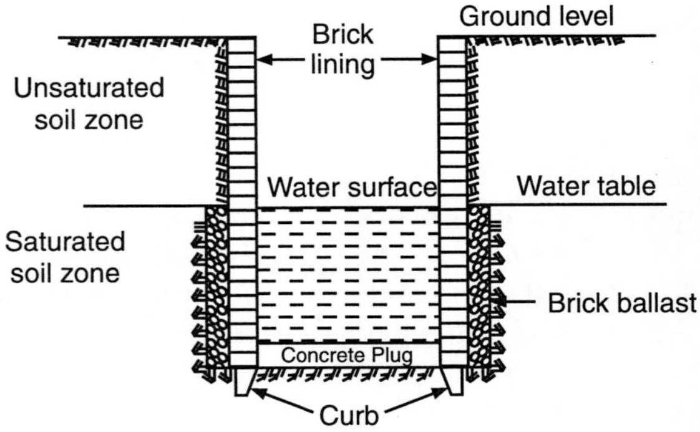

Fig. 17.1. Well with pervious lining

0.3 m strip of lining in mortar after every 1.25 m of dry lining. Above the ground water-table construct the lining in mortar upto the top. When the rate of water withdrwal is not more, this type is very economical and lasting.

17.1.2. Wells with Impervious or Pucca (Permanent) Lining

In regions with alluvial soil formation this type is mostly adopted. Once constructed, it forms a permanent source of water as long as ground water conditions remain favourable. This type is very suitable for irrigation purposes. The wells with impervious lining are deeper than the type mentioned above. In this type the depth should not exceed 30 m for efficient working. (Fig. 17.2)

If the well is dug deeper than 30 m then lifting of water becomes uneconomical. The well lining is set in lime mortar or cement mortar. Sometimes the well may meet some clay layer of small thickness. Then the well may be founded on it and a bore may be drilled to tap underlying water bearing stratum. Such an arrangement helps in keeping the depth of wells within specified limits.

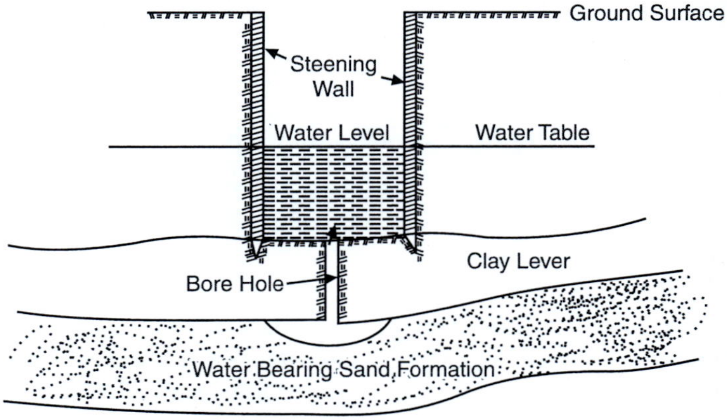

Fig. 17.2. Well with impervious lining

The thickness of steening wall varies from 0.3 to 0.7 m.

Table 17.1 shows suitable thickness for various depths.

Table 17.1.

Depth of well in metres	*Thickness in cm*	
	Bricks	***Masonry***
3	33	30
3–10	45	38
10 – 15	54	45
15 – 20	61	54
20 – 30	68	61

Another significant classification of open wells is acoording to the mode of water contribution to the well from the subsoil formation. Deep and shallow wells form two distinct sub-heads in this classification. Before actually defining deep or shallow wells it is essential to understand clearly the possibilities of meeting various types of subsoil formation.

Now it is clear that well is sunk or dug for derving water from pervious subsoil formation at any site. Whenever unlimited prvious soil or rock formation exists below the ground level, well can be taken down the waters table by about 4 to 5 metres to store sufficient quantity of water. Under the situation mentioned above water contribution to the well takes place through the sides and the bottom of the well depending upon the nature of lining.

Another possibility is, there may be a local patch of impervious stratum Fig. 17.3 (*a*) or full impervious stratum with good high out-crop, Fig. 17.3 (*b*), entrapped between the pervious soil or rock formations.

Alternatively there may be alternate layers of pervious and impervious nature, Fig. 17.3 (*c*).

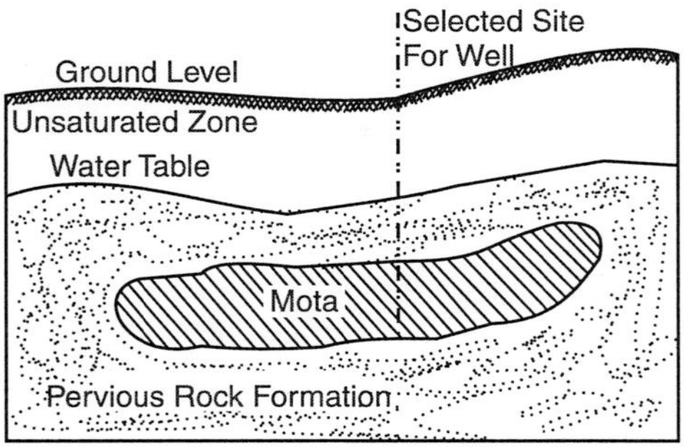

Fig. 17.3. (a) Section showing localised mota formation

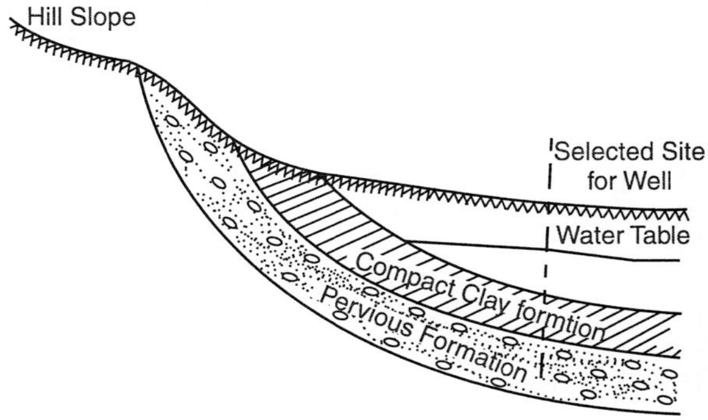

Fig. 17.3. (b) Section showing subsoil formation under hill slope

The occurrence of the impervious stratum in all the three above mentioned cases is due to local deposition of materials, in alluvial formation. The impervious stratum may be formed of clay, compacted and cemented sand and other hard materials. When such a layer occurs below the water-table and when it is entrapped between pervious layers at the top and bottom it is called a Mota layer. This type of subsoil formation is very common in the Indo-Gangetic plains and local names for mota are also **Mat barwa** and **Nagasan**.

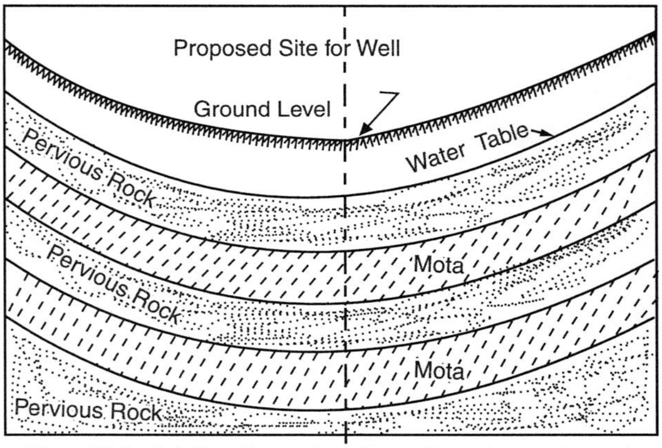

Fig. 17.3. (c) Section showing alternate mota and pervious layer formation

When the impervious layer occurs above the waters table it is not termed as mota. Essentially it should lie below waters table. Now at this stage definitions of shallow and deep wells can be given.

A **shallow well** is that well which goes below the waters table without meeting mota formation. Shallow well derives water from the underground pervious rock or soil formation through the bottom or sides of the well depending upon the nature of construction.

A **deep well** is that well which goes below the waters table and finally finds its foundation in the mota formation. It is clear that mota formation is impervious. Hence, to derive the water from the soil a bore (15 to 22 cm in diameter) has to be drilled in mota formation so as to tap the water of underlying pervious rock or soil layer (Fig. 17.4).

The water which enters the well through the bore brings fine sand along with it in initial stages because of high velocity and pressure of the incoming

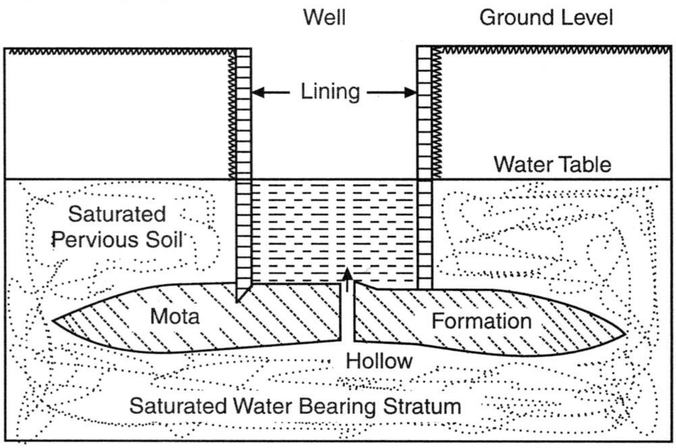

Fig. 17.4. Deep well on localised mota formation

water. Progressively a hollow space is formed below the bore and when the rate of discharge from the well equals the rate of entry of the water in the well only clear water enters the well.

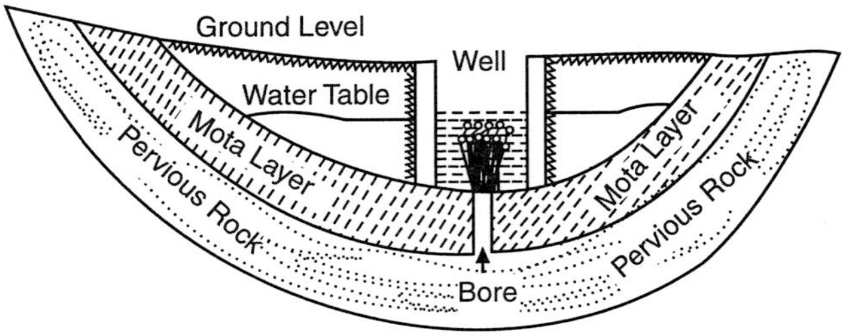

Fig. 17.5. Deep well on mota layer with high out-crop

If a pefect mota layer with a good high out-crop is present below the waters table then the pressure at the lowest point may be so high as to create artesian conditions. In such ceses water jet enters through the bore in the open well (Fig. 17.5) in the form of a fountain.

Some times more than one bore holes are drilled to derive more water from the underlying pervious rock.

Advantage of mota is that, it gives good foundation to the wells with heavy masonry lining.

After the above discussion it is needless to say that the shallow well does not mean that it is of leaser depth and the deep well is of greater depth. The names shallow and deep wells are given from the consideration of presence of mota foundation.

17.2. LOCATION OF WELLS

While selecting the site for a well it is very essential to give proper weightage to the following points:

1. A well should be sunk after ascertaining groundwater availability.
2. The well site should be such that the well after completion will be in a position to command the surrounding area to the fullest extent. The well, if situated in the centre of an area, gives good command. At the same time care should be taken to see that the well site is on the highest patch of the area so that water can flow by gravity.
3. Location of shallow wells create no problem when the capacity of the well is moderate and the underground conditions are favourable, that is when the water-table is sufficiently high and subsoil formation is pervious.
4. Wells should be located in a region where drainage water does not bring in water of bad quality. It is harmful to crop production. While locating all the wells, more particularly shallow wells, care should be taken to keep the selected site always away from the sources of pollution.
5. The wells should be located in such a way that their circles of influence under maximum drawdown conditions do not enter the neighbouring property lines. If the withdrawal of water is from beneath some property, it may damage the property itself or the structures above and well owner might be asked to pay for damages.
6. Two wells of any area should be spaced in such a way that the cones of depression or circles of influence of the wells do not cross each other. Mutual interference reduces the dependable groundwater storage of each well as shown in Fig. 17.6. In turn it reduces the discharge of the wells.

   From Fig. 17.6 it is clear that as drawdown goes on increasing, the area of mutual interference also increases. The spacing of wells is generally keep between 75 m – 175 m to avoid interference.
7. A well should irrigate the land which is not receiving irrigation from flow irrigation system.

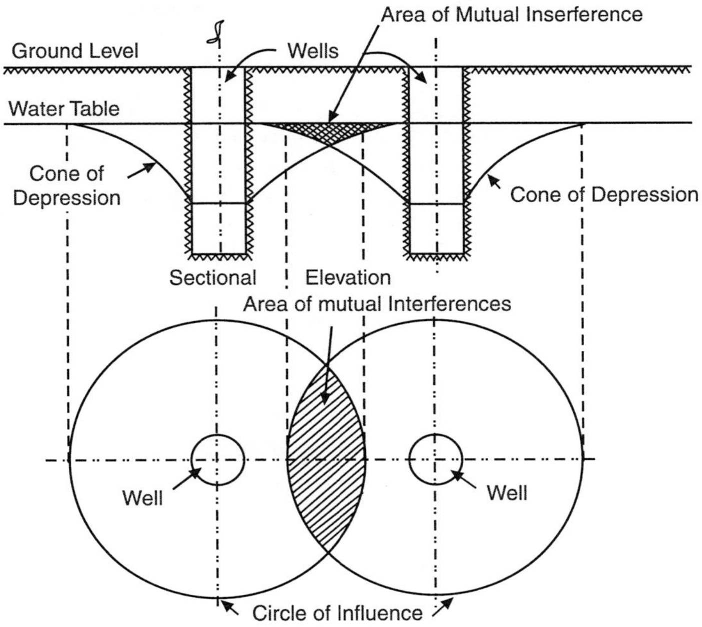

Fig. 17.6. Spacing of wells

8. A well if located in the command area of flow irrigation system should irrigate lands in such a way as to remove deficiencies of flow irrigation system. For example, either it should lower rising water table to keep a check on waterlogging or it should irrigate high patches of land left unirrigated by flow irrigation system.

17.3. OPEN WELL CONSTRUCTION

Well construction can be done in two ways. The two are, dug wells and sunk wells. The selection of a particular procedure depends on the nature of the subsoil formation.

Dug wells: When a subsurface is formed of rock or hard soil formations the construction involves excavation of subsurface by means of blasting or with the help of rock breakers. The hole is dug till it reaches an aquifer or water bearing stratum and goes sufficiently below the groundwaters table to hold sufficient quantity of water. The hole dug is kept bigger by ½ to 1 m in diameter than the proposed diameter of the finished well. After digging the hole to the required depth the well may be lined with brick or masonry laid in mortar in successive lifts of 1 to 1.5 m. If necessary the lining of the well may be raised above the ground level to provide the well with a parapet to facilitate lifting of water from the well. The hollow between the rocky surface and the lining may be filled with brick bats in mortar or simply with clay earth. This filter material is given good compaction by ramming.

Sunk wells: They are constructed in soft formation. The subsoil formation may be of sandy to clayey soils. This type is called sunk well because the lining is first constructed above the ground level and then it is sunk in the subsoil formation by putting load on the lining. This procedure of constructing the lining and loading is done in small lifts successively. The actual procedure may be outlined as mentioned below.

A hole 1 to 1.5 metre bigger in diameter than the finished well is dug in the ground. The depth of this digging is such that the natural sides stand vertical without a support. At this depth a curb* is placed on the bottom. The curb is of the same thickness as the thickness of proposed lining.

The inner diameter of the curb is same as that of finished well. Then the masonry is constructed on the curb in the form of a cylinder till it emerges 1 to 2 metre above the ground surface. A temporary circular ring platform is now constructed on the top of the lining cylinder and the platform is loaded with locally available material for example, rock blocks or sand filled in gunny bags. The diameter of the open space of the ring platform is about 2 meters. This opening is necessary for removing the earth from the bottom of the well while sinking is in progress. The platform extends beyond the side of the steening cylinder also to provide more area for loading. Under the load the cutting edge penetrates into the earth, the material below the curb and at the bottom of well can now be removed by some suitable device. Most common method is by bamboo baskets running over pulley. Depending upon the size of the well number of pulleys may be fixed at various points along the circumference of the well.

Thus the sinking of the well cylinder proceeds. When the sinking has taken place to such an extent that the platform touches the ground surface it is removed and the masonry cylinder is built again on the top of the old masonry. It is taken to a sufficient height above the ground surface and the platform is again mounted on it and load is applied. When the well reaches the depth of the waters table, water starts coming in, which can be removed together with excavated material from bottom by means of pumps etc. To see that sinking of well is done vertically four plumb bobs may be suspended from four sides of the well. After the well is sunk to the required depth in successive lifts platform is removed. The space between the lining and the natural face of the wall is then filled with clayey earth and compacted. Finally the well is finished to suit the requirements.

17.4. WATER LIFTING ARRANGEMENTS

The water lifting arrangements can be summarised in the following three categories:

- Lifting arrangement where man power is used.
- Lifting arrangement where animal power is used.
- Lifting of water by mechanical power.

Lifting arrangements making use of manpower include the following:

17.4.1. Doon

The construction of doon is based on the principle of a lever. From Fig. 17.7 system is very clear. It consists of a wooden trough open at one end and semicircular at the closed end.

* It is a circular cutting edge approximately triangular in section provided at the bottom of lining for sinking purposes. The width and height of curb is nearly the same. The curb may be made of wood, reinforced cement concrete or iron. The section of each is shown in following sketches:

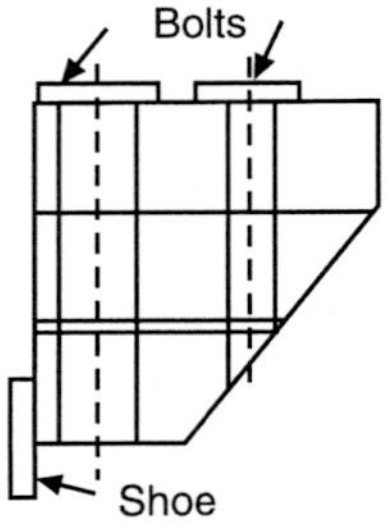

Wooden Curb

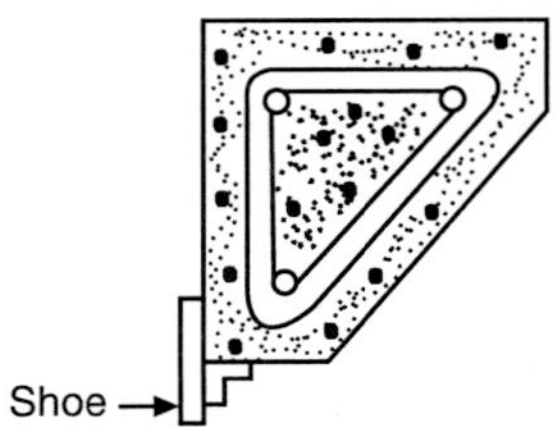

Concrete Curb

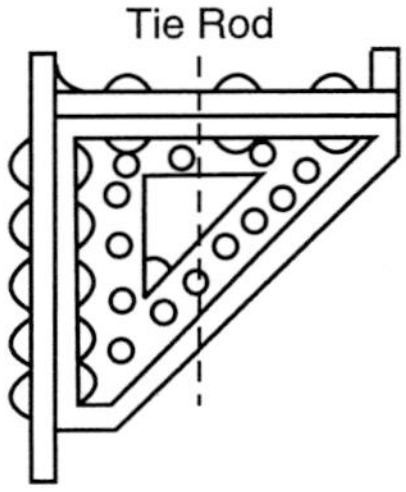

Iron Curb

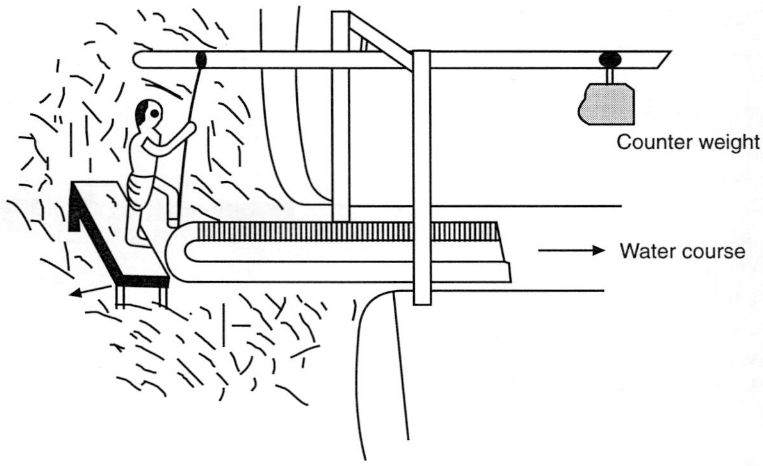

Fig. 17.7. Doon

The trough rests on a fulcrum about which it swings. A rope is attanched to the trough at closed end. Other end of the rope is attached to a lever rod. The lever rod has a counter weight fastened at the other end. The lever rod rests in a ring fulcrum. When water is to be lifted from an irrigation channel to a field channel or a water course a man standing on a stool can bring the trough down by applying pressure with hands and a leg. The trough is lowered till its closed end goes below the water level in the channel. When the pressure is releases, due to the counter-weight trough swings back and the water in the trough discharges in the water course. The procedure is repeated number of times.

17.4.2. Archimedian Screw

It consists of a helical coil fitted in a cylinder made of wood or metal. The cylinder is fixed in an inclined position in such a way that the centre of lower end of the cylinder remains in level with low water level in the stream or the channel. The lower end of the coil is attached to a shaft which rests on a bearing. The upper end of the coil also ends in a shaft and rests on a bearing. This end of coil is connected to a handle. When the coil is rotated with handle water is lifted up and discharged in the water course. Figure 17.8 shows the water lifting arrangement by means of Archimedian screw.

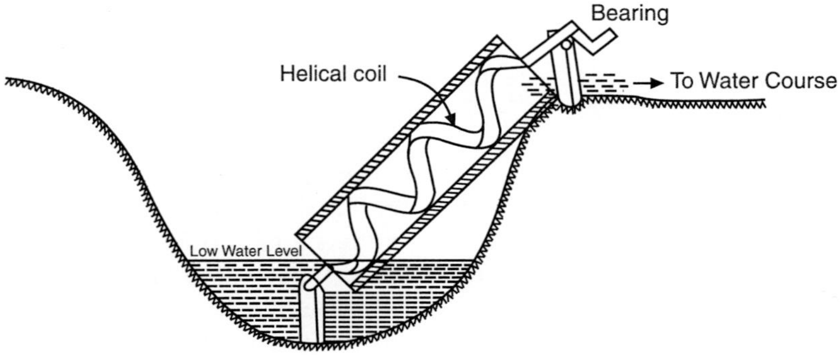

Fig. 17.8. Archimedian screw

17.4.3. Swinging Basket

In this method water is lifted by means of a basket from the low channels or ponds which are by the side of the field. The basket is made of any cheap material for example leather, tin, etc. The basket is slung by ropes on both sides as shown in Fig. 17.9.

Two persons hold the ropes at ends and dip the basket in the channel water. Then swing is given to the basket so that the basket full of water tilts over the water course and discharges water in the water course. This method is useful upto a height of 75 cm. If the lift is more the same method may be used in stages. In such cases in between the water course and the source of water temporary ponds may be constructed. The water from the source is transferred to the pond just above and from there onwards till water reaches the field channel. Such a case will be of course very costly, but in hilly tracts one has to adopt this method.

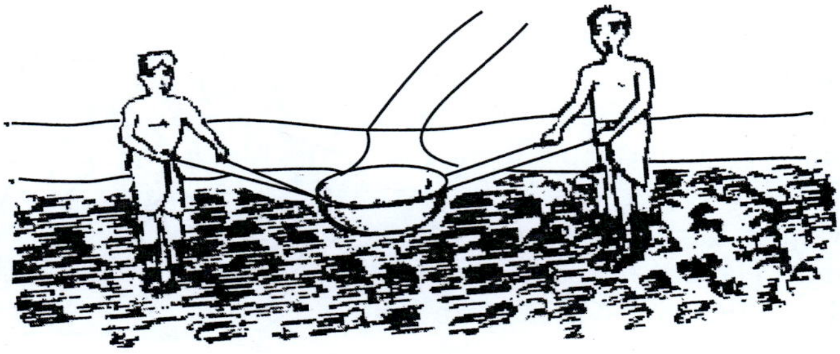

Fig. 17.9. Swinging basket

Methods of lifting water making use of animal power are the following:

17.4.4. Mote or Churus or Pur

This arrangement of lifting water is suitable for lined wells and it can be conveniently used for deep wells upto a maximum depth of 30 m. The system consists of a leather container whose shape is like a funnel. There is a spout on one side and shape of upper portion is that of a conical cylinder. It is open at both ends. The capacity of the leather container or mote is from 0.08 to 0.15 cubic metres. Two ropes are tied to two ends of the mote as shown in Fig. 17.10. The ropes pass over two pulleys. Upper pulley is a circular fixed pulley of bigger diameter whereas a lower pulley for the rope tied to the spout end is cylindrical in shape and is of smaller diameter. The pulleys are fixed in the wooden framework as shown in the figure. Other ends of the ropes are connected to a specially prepared yoke. A pair of bullocks is employed to haul the mote. The bullocks move forward and backward on a sloping bullock run or a ramp. When the bullocks reach other (lower) end of the ramp the spout rests on the lower pulley and the conical portion remains above the lower pulley in suspension from the upper pulley. Thus the water discharges through the spout. The discharging water may be collected temporarily in a trough constructed in front of the cylindrical pulley from where water is taken to the water course.

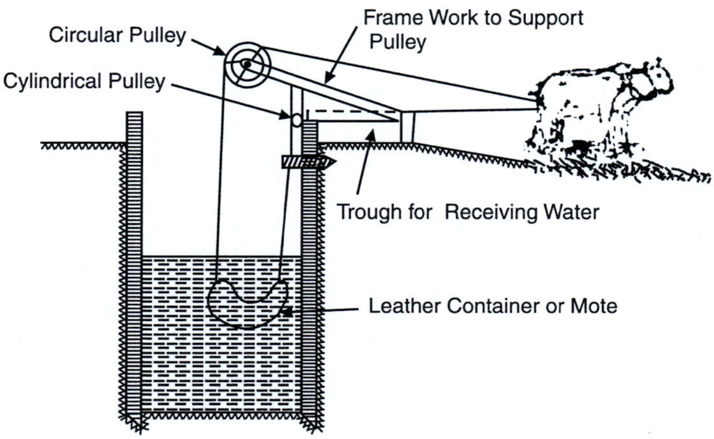

Fig. 17.10. Mote or churus

When the pair of bullocks come back to the well end of the ramp the mote goes down in the well and dips in the water and gets filled. When the pair moves forward because of the tension the spout end comes in level with the other end. Thus the water is not allowed to spill over or to discharge out of the mote.

Water can be lifted in two ways by the mote, namely, kili method and lagor method.

Logor method, is the one already described above.

In kili system, in addition to the normal ramp a secondary parallel ramp is provided. When the pair of bullocks and the driver of the mote reach lower end of the ramp the kili or the fastening of mote ropes to yoke is loosened and the mote goes down in the well under its own weight and the driver holding the rope ends goes up to the well end of the ramp due to a pull exerted by the descending mote. The pair of bullocks comes up freely by the secondary parallel rampway without any load on their neck. At the top of the ramp the rope is again fastened to the yoke and the pair of bullocks agan haul the mote. This process is repeated number of times. This method of lifting water by mote seems to be better as the pair of bullocks is relieved of their load while the pair comes back. The mote is generally used for gardening purposes and also on small fields.

17.4.5. Persian Wheel

This is another type in the same category. It is also known as Rahat in some parts of India. This arrangement is very suitable where the depth of well is from 12 to 20 metres. Once this method was very commonly used in Uttar Pradesh and Punjab. This method is very suitable for irrigating crops when the wells and persian wheels are in abundance and area supported by each is less. It is also very much sutiable for gardening. It consists of an endless chain of buckets made of G.I. sheets. The chain is mounted on a drum. The endless chain is suspended in the well in such a way that the chain dips in the water to sufficient extent. The drum is connected by means of a horizontal axle to a toothed wheel which is also held in a vertical plane as shown in Fig. 17.11. This vertical wheel is geared with a horizontal wheel. The vertical axle of the hroizontal wheel is connected to a horizontal shaft. A pair of bullocks or buffaloes or a camel is yoked to the shaft by means of ropes and a yoke. The horizontal shaft is rotated with the help of animal power about the vertial axle. When the animal moves round the vertical axle of the system rotates. As a result bucket chain also moves. The buckets filled with water move up and when they turn down at the top of the drum the water is discharged. This water is collected in a trough from where it goes to the water course.

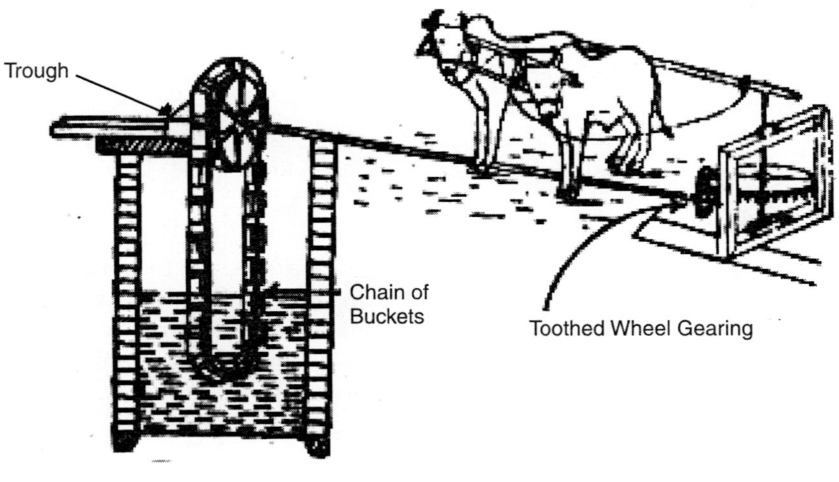

Fig. 17.11. Persian wheel

The empty buckets then move onwards and go down in the well finally dip in the water of the well in inverted position and when they emerge out they are upright and completely filled with water. Thus the buckets when ascend are full of water and when descend they are empty. In this method water is lifted continuously as long as system is kept in rotation. This method is very efficient.

Table 17.2 gives an idea about the capacity of various methods to lift the water in one lift.

Table 17.2.

Type	*Capacity in cubic metre*
Swinging basket	0.01
Archimedian screw	0.01
Mote	0.08 to 0.15
Persian wheel bucket	0.003

17.4.6. Pumps

Pumps of various types say displacement or centrifugal, operated by mechanical or electrical power are now very commonly used for lifting water to irrigate the fields. The water may be lifted up from streams, rivers, wells etc., with suitable accessories.

Most commonly used pump is a centrifugal pump. Centrifugal pump with a horizontal spindle is installed in a pump-sump suitably. It consists of blades or curved vanes mounted on a shaft. The assembly is placed inside a casing called volute chamber Fig. 17.12.

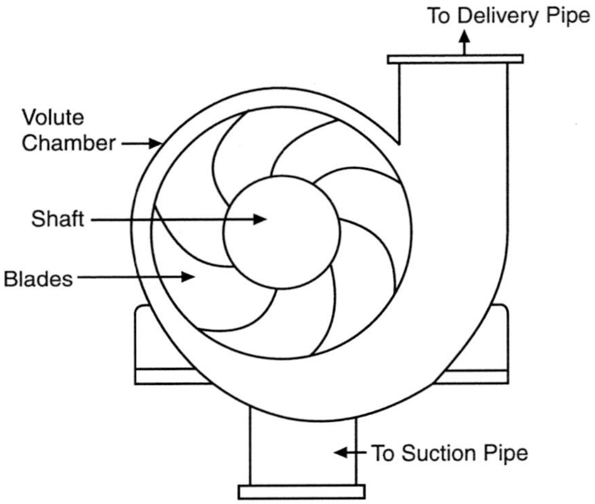

Fig. 17.12. Centrifugal pump

When the shaft is rotated by mechanical or electrical power with a high speed the centrifugal force is created in the chamber. The vanes then force the water out through a delivery pipe. The delivery pipe is attached tangentially to the chamber or the casing. The pump cannot work unless intially filled with water. For this reason a foot-valve is generally provided at the lower end of the suction pipe. The cost of maintenance of the centrifugal pump is very less. It is very compact unit and requires less floor area. The cost of installation is also low.

17.5. YIELD OF WELLS

It is well known that under the favourable condition water tries to maintain its own level. Hence, it is obvious that the level of water in a well approximately indicates the level of waters table under normal conditions of no withdrawal. As the water is pumped out or withdrawn from the well the level of water in the well falls more quickly than the ground water level and consequently if forms a cone of depression. The difference of level of waters table and the water level in the well now is called a head of depression. Actually under this head water percolates into the well through soil pores. Naturally when depression head is more the rate of water contribution to the well will also be more. If the depression head goes on increasing, due to continued water withdrawal form the well then a time may come when the increased velocity will dislodge the soil particles. At this stage the percolating water brings soil particles with it in the well. Naturally this stage is critical and hence various terms, for example, depression head, rate of percolation, (it is also termed yield of well and is expressed in cubic metres per hour or in litres per minute) velocity of percolation are prefixed with the term critical. It is very essential that the critical depression head should not be allowed to reach or be allowed to exceed for a particular well withdrawal as after that it may create unstable conditions for the well structure. Sufficient margin of safety or a factor of safety should be provided (Generally factor of safety is 3 to 4). It should be noted now that whenever "yield of well" is referred to, its means maximum safe yield unless otherwise stated. Yield of well is the rate at which water percolates into the well under the safe maximum working head or critical depression head. It is expressed in m^3/hr or lt/min.

The yield of open well can be determined by any one of the two methods, namely, pumping test and recuperation test.

Pumping test: In this method water is withdrawn from the well freely till a critical depression head or a safe maximum head is created. Once this stage is reached the rate of pumping is so adjusted as to maintain the constant water level in the well. Thus the depression head remains constant. Naturally at his stage the rate at which water is pumped out of the well will be equal to the rate at which water percolates into the well. This rate is expressed in m^3/hr or lt/min and will be obviously the yield of the well.

Recuperation test: In this method water level in the well is depressed by pumping to any level below the normal level. Then the pumping is stopped and time taken by the percolating water to fill the well to any particular level is noted. Total quantity of water percolated into the well is calculated by knowing cross-sectional area and rise in the water level after stoppage of pumping. The rate of percolation or the yield of well can be arrived at by dividing the quantity of water by the time. This test is carred out generally in a driest period to take worst condition into account.

Now it can be inferred that the actual pumping test of determining available yield is most reliable but it is difficult to conduct the test accurately. Whereas recuperation test is very simple to perform but it does not give the maximum safe yield. The reason is as the water level in the well rises the safe maximum working head is not maintained throughout the period of observation.

17.6. SPECIFIC YIELD OF WELLS

Rate of water percolation in the well or yield of a well in m^3/hr under a head of one metre is called the specific yield of the well.

From the above definition it is clear that the specific yield depends on: (*i*) position of the water-table, (*ii*) permeability and porosity of the soil formation, (*iii*) the rate of water withdrawal form the well, and (*iv*) quantity of water storage in the well.

Specific yield of the well is also called specific capacity of the well. It can be calculated from the followig formula:

$$K = 2.303\ \frac{A}{T} \log \frac{H_1}{H_2}$$

where K is specific yield of a well in m^3/hr under depression head of one metre.
A is area of well in plan in m^2.
T is total time of recuperation to bring water level from depth H_1 to H_2
H_1 is difference of water level in the well just after stoppage of pumping and the normal water level of the well
H_2 is difference of water level in the well after time T and normal water level of the well.

Above mentioned formula can be derived as follows with the help of Fig. 17.13.

Discharge of the well is proportional to depression head.

i.e., $Q \propto H$

or $Q = K \cdot H$ ... (1)

where K is a constant or specific yield of the well.

$O - O$ represents normal or orginal water level in the well.

$m - m$ represents the water level in the well after pumping.

$n - n$ represents the water level in the well at time T after stoppage of pumping.

$q - q$ represents the water level in the well at any time t after stoppage of pumping.

$p - p$ represents the water level in well at time $t + dt$ after stoppage of pumping.

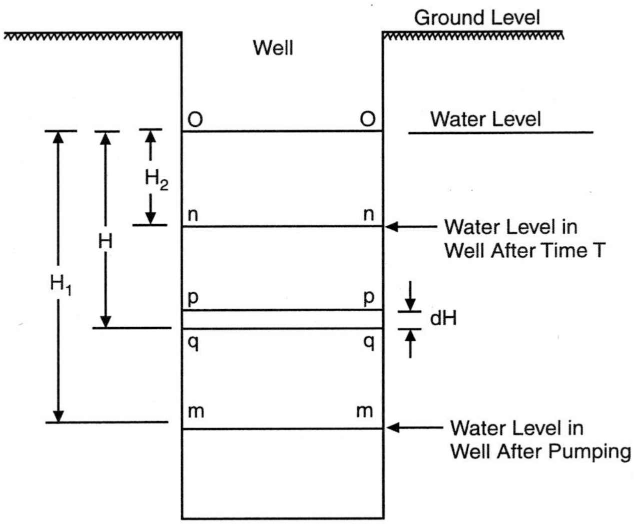

Fig. 17.13.

The amount of water percolated into the well in short time δt is given by

$$q = A \times dH$$

where dH is the rise in water level in the short time δt.

Also $q = Q \times \delta t$

$\therefore$ $Q \times dt = A \times dH$

Substituting for Q from (1)

$$K \times H \times \delta t = A \times dH$$

Separating the variables

$$\frac{K}{A}\delta t = \frac{dH}{H} \quad \text{... (2)}$$

Now integrating the equation (2) between limits

$$t = 0 \quad \text{and} \quad t = T$$

and $H = H_1$ and $H = H_2$

As the time increases depression head decreases hence negative sign will be taken for limits of H.

$$-\log_e H \Big|_{H_1}^{H_2} = \frac{K}{A} t \Big|_0^T$$

Taking limits

$$\log_e \frac{H_1}{H_2} = \frac{K}{A}.T$$

Converting into common logarithm

$$2.303 \log_{10} \frac{H_1}{H_2} = \frac{K}{A}.T$$

or $$K = 2.303 \frac{A}{T} \log_{10} \frac{H_1}{H_2} \quad \text{... (3)}$$

The equation (3) can also be written in the from

$$\frac{K}{A} = 2.303 \frac{1}{T} \log_{10} \frac{H_1}{H_2}$$

K/A is the specific yield of well per unit area of the well and expressed in $m^3/hr/m^2$.

When the limit of critical velocity is not exceeded it may be rightly assumed that *K/A* is constant for a well. For several types of subsoil formations Marriot has given the following Table 17.3.

Table 17.3. Marriot's Table of Specification

Type of subsoil	*K/A*
Clay	0.25
Fine sand	0.50
Coarse sand	1.0

Problem 17.1. A well 3 metres in diameter has its normal water level 3 metres below the ground level. By pumping, water level in well is depressed to 10 metres below ground level. In a time of 4 hours the water rises by 5 metres. Calculate specific yield of the well.

Solution: Using formula, $K = 2.303 \frac{A}{T}. \log_{10} \frac{H_1}{H_2}$

$$A = \frac{\pi}{4} \times d^2 = 0.785 \times 9 = 7.065$$

$$T = 4 \text{ hours},$$

$$H_1 = 7 \text{ m} \quad \text{and} \quad H_2 = 2 \text{ m}$$

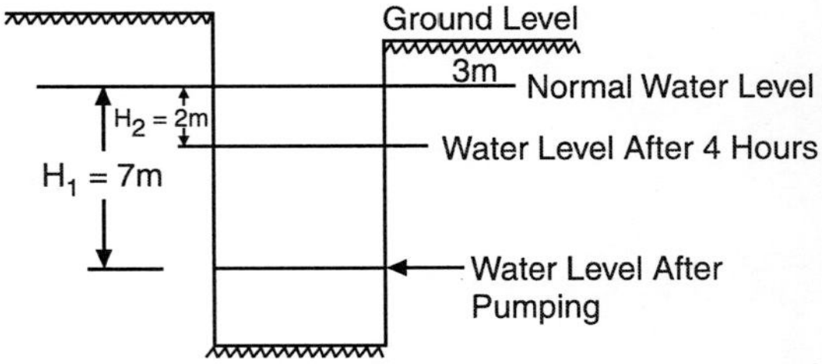

Fig. 17.14.

$$\therefore \quad K = 2.303 \times 7.065 \times \frac{1}{4} \log \frac{7}{2}$$

$$= 2.303 \times 7.065 \times \frac{1}{4} \times 0.5441$$

$= 2.213$ m^3/hr under the head of one metre.

$\therefore$ Specific yield of well is 2.21 m^3/hr under the head of one metre.

Problem 17.2. A well is to be constructed in a fine sandy subsoil formation. The discharge of the well is anticipated to be 0.004 m³/sec under the depression head of 4 metres. Find out the diameter of the well.

Solution: Yield of well $0.004 \times 60 \times 60 = 14.4$ m³/hour

Q in m³/hr or yield $= K \times h$

$\therefore \quad 14.4 = K \times 4$

$$\therefore \quad K = \frac{14.4}{4} = 3.6 \text{ m}^3\text{/hr under head of 1 metre.}$$

From Marriot's table value of K/A for given subsoil formation is 0.5

$\therefore \quad 3.6/A = 0.5$ or $A = 7.2$ m².

$$\therefore \quad \text{Diameter of well} = \sqrt{\frac{A \times 4}{\pi}} = 3.07 \text{ m}$$

$\therefore$ Diameter of well is 3.07 metres.

Problem 17.3. As open well was subjected to recuperation test and following data was observed:

(*i*) Depression of water level after puming = 3 m

(*ii*) Time of recuperation = 1.5 hours

(*iii*) Depth of recuperation = 2 m

Calculate: (*a*) Yield of a 4 m diameter well under a depression head of 4 m.

(*b*) Diameter of a well under the depression of 4 m if yield desired is 15 litres/sec.

Solution: Step 1. From the observed data specific yield of well per unit area can be calculated (K/A) which can be treated as a constant for the given soil formation. Using formula,

$$\frac{K}{A} = \frac{2.303}{T} . \log_{10}\left(\frac{H_1}{H_2}\right)$$

where in this problem $T = 1.5$ hr.

$H_1 = 3$ m

$H_2 = (3 - 2) = 1$ m

$$\frac{K}{A} = \frac{2.303}{1.5} . \log_{10}\frac{3}{1} = \frac{2.303}{1.5} \times 0.4.77 = 0.733 \text{ m}^3\text{/hr/m}^2 \quad \text{... (1)}$$

Step 2. (a) Yield of 4 m diameter will under 4 m depression head will be given by formula

$$Q = \left(\frac{K}{A}\right) . AH$$

$$A \quad \frac{\pi d^2}{4} = \frac{\pi}{4} . \; 4 \times 4 = 12.57 \text{ m}^2$$

$$H = 4 \text{ m} \quad \text{and} \quad \frac{K}{A} = 0.733$$

$\therefore \quad Q = 0.733 \times 12.57 \times 4 = 36.86$ m³/hour, or $= 10.2$ litres/sec.

$\therefore$ yield of 4 m dia well under 4 m depression head = 10.2 litres/sec

Step 3. (*b*) 1 litre/sec = 3.6 m³/hr

$\therefore$ 15 litres/sec = $15 \times 3.6 = 54$ m³/hr

So desired $Q = 54$ m³/sec

for $$\frac{K}{A} = 0.733 \text{ and } H = 4 \text{ m}$$

$$Q = \left(\frac{K}{A}\right) . A . H$$

or $$54 = 0.733 \times A \times 4$$

$\therefore$ $$A = \frac{54}{4 \times 0.733} = 18.42 \text{ m}^2$$

$$A = \frac{\pi d^2}{4} = 18.42$$

$$d^2 = \frac{18.42 \times 4}{\pi} = 23.44$$

$\therefore$ $$d = 4.84 \text{ m}$$

$\therefore$ Diameter of well to give 15 litres/sec discharge under 4 m depression head = 4.84 metres.

17.7. ADVANTAGES AND DISADVANTAGES OF OPEN WELL IRRIGATION

Advantages

1. Whenever necessity is felt, in any tract locally, well may be sunk to start open well irrigation. Much consideration need not be given to any of the other factors which are given proper weightage while introducing canal irrigation.
2. There is no need of constructing many and expensive hydraulic structures. The cost of well irrigation project is therefore much less.
3. When the water is withdrawn from the subsoil formation by means of wells the waters table obviously lowers and waterlogging of the land is prevented.
4. The water is used more economically as cultivator has to put in labour for lifting water.
5. The water can be used at any time depending upon the choice of cultivator and water needs of crops.
6. As the water is assured for whole of the year, provided groundwater conditions are favourable, two to three corps can be grown on the same field in one year.
7. Maintenance cost of a well is less. Also as the well is situated usually in the middle of the field water losses in transit are less.
8. By constructing number of wells in any tract intensive irrigation of some valuable crops can be done.
9. In well irrigation duty realized is higher.
10. Well water irrigates the uncommanded patches of the culturable land in the canal irrigated tract. Thus it assists the canal irrigation.
11. Well water is cooler in summer season and warmer in winter season. This water when applied to the crops, tries to neutralize the bad effects of the hot or cold season.

Disadvantages

1. To make the water available for irrgation purposes it is necessary to lift it from underground. For lifting the water power is reqired. Thus the well irrigation is much dependent on the availability of power or trouble free working of the machinery which is very rare thing.

2. Sometimes cost of well water is so high that the returns obtained from it are not justifiable.
3. Availability of water from the wells depends on groundwater storage. The discharge from well is low and area commanded is less.
4. The water in well is static and therefore it is free of supsended silt. Water carrying silt together with some useful suspended salts is very beneficial to the crops. Well water thus lacks in this respect.

QUESTIONS FOR REVISION

1. How will you differentiate between shallow, deep and artesian wells?
2. What factors are taken into account while selecting the site for a well?
3. Define the following:
   (*a*) Depression head
   (*b*) Cricle of influence
   (*c*) Specific yield of a well
   (*d*) Mota
4. (*a*) How the yield of an open well is determined?
   (*b*) Derive and expression for specific yield of a well.
5. Explain the following methods of lifting water:
   (*i*) Mote or Churus
   (*ii*) Persian wheel
6. What are the advantages and disadvantages of open well irrigation?

❑❑❑

18

Tube Well Irrigation

18.1. INTRODUCTION

It has been already mentioned that tube well irrigation comes under lift irrigation. Tube well is a small diameter hole drilled in the subsoil formation. The cross-sectional area of this type of well is small and unless some mechanical power is used for lifting water, rate of water withdrawal will be low. The velocity of water withdrawal should be more to maintain high discharge. Obviously the limit of critical velocity is exceeded. Under such conditions the soil particles are dislodged. To prevent clogging of the bore hole a metal pipe with perforations is driven in the hole. It prevents the soil particles from entering the hole. To increase the efficiency of the tube well, in the space between perforated metal tube and soil some straining device is always provided. The perforations in the metal pipe are necessary only for the depth of water bearing subsoil formations. Other portion of the pipe is generally kept plain. Even though the perforations prevent average soil particles from entering the hole, finer particles enter the hole. To keep the hole clean to avoid settling of particles it is essential to keep the velocity higher (say not less than 1 metre/sec). The tube well may be taken down to 50 metres in the water bearing stratum. When mechanical power is used to lift water average discharge of tube well is 0.04 m^3/sec. The factors which affect the location of tube well are more or less, the same as those discussed for open wells. Tube well irrigation is very suitable where the subsoil formation is suitable for storing water. Water bearing layer may be of unlimited extent (unconfined) or it may be alternate to some type of impervious layers or may be confined between two impervious layers.

In Haryana, Uttar Pradesh, Punjab and Bihar States tube well irrigation is extensively in practice. The obvious reason is that most of the area is made up of alluvium which holds the water to a great extent and there are massive groundwater reservoirs.

18.2. TYPES OF TUBE WELLS

Depending upon the mode of water contribution to the bore hole from the subsoil formation following main types of tube wells may be recognized:

- Strainer type tube wells.
- Cavity type tube wells.
- Slotted type tube wells.

18.2.1. Strainer Type Tube Wells

It has been already mentioned above that in tube wells the metal pipe driven in ground is perforated to allow only clear water to enter the hole. It is obvious that if no other means is adopted the performations in the metal tube will have to be made very fine. It is very costly process. As an alternative wire net may be wrapped on the cylindrical frame of small diameter but it is liable to break as it is very delicate. So the best and most commonly adopted practice is to provide a pipe with fairly big perforations and surrounding that is a wire net or a strainer with smaller openings. The wire net with finer openings excludes the objectionable soil particles from entering the tube well. Total area of the openings in the metal tube and in the strainer is kept the same. This is because if the area of opening is same the velocity of inflow will be the same. Moreover there is some annular space left between the perforated metal tube and the strainer. If the space is not left the strainer may rest directly over the tube and consequently the open area of perforations will be reduced, Fig. 18.1 shows the section of a strainer type tube well. Normally the mesh size of the wire net or strainer is kept equal to D_{60} to D_{70} of the surrounding soil. This type of well derives water from one aquifer of unlimited extent or from a confined aquifer or from number of aquifers.

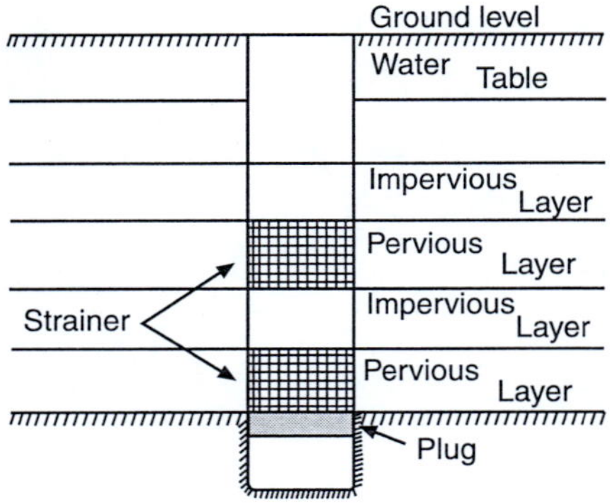

Fig. 18.1. Strainer type tubewell

The perforated pipe extends only for the aquifer portions of the formation while for other portion the pipe is kept plain. At the bottom a tube well is plugged. The plug is kept a little above the bottom. This procedure prevents failure of the plug due to the weight of the well. The strainer type tube well is normally suitable and therefore very widely used. When a term tube well is used it refers to strainer well unless otherwise stated.

18.2.2. Cavity Type Tube Well

In this type water contribution to bore hole takes place through the bottom layer only. From Fig. 18.2, it is clear that in principle it is similar to the category of the deep wells under open wells.

It derives water from the pervious layer underlying the hard impervious layer. The tube well is taken down till it penetrates the impervious or mota layer and reaches the water bearing layer.

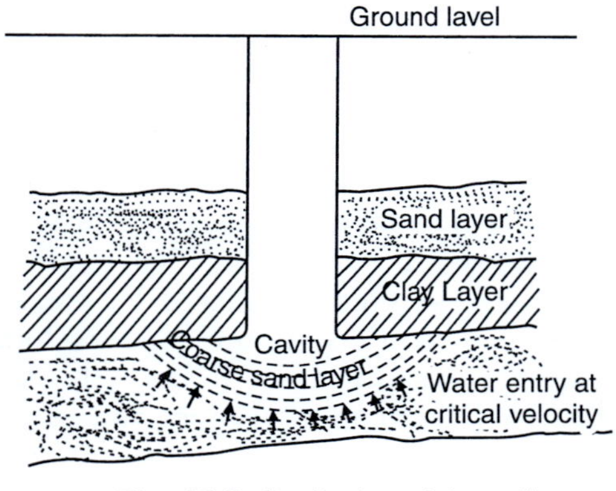

Fig. 18.2. Cavity type tubewell

In the initial stages when the water is pumped out fine sand comes in the tube well with the water and consequently a hollow or cavity is formed at the bottom. The bottom of the cavity for some thickness is thus made free of finer particles. After the cavity formation only clear water enters the tube well. Since the rate of pumping is more, the velocity of water entering the coarse sand layer is critical but when it comes in the hollow the velocity is reduced. Finally the water enters the tube well at the bottom with a velocity lower than the critical velocity. The main difference between a strainer and cavity tube well is that in the former the inflow is radial whereas in the later it is spherical.

18.2.3. Slotted Type Tube well

Sometimes the nature of subsoil formation is not anticipated correctly. Obviously bore hole driven for constructing strainer well will be a failure. If a mota formation is present cavity well may be resorted to. But if neither of the conditions are existing the slotted tube well can be rightly constructed. There should be of course an aquifer present at the bottom. In the bore hole (say 36 cm diameter) a 15 cm diameter education pipe is lowered till it reaches the bottom. The bottom of the education pipe is slotted as shown in Fig. 18.3.

The size of the slots may be 25 mm × 3 mm with 12 mm spacing. As the slots are quite wide, to avoid sand entry in the pipe a filter of shingle is provided at bottom, surrounding the slotted pipe portion.

Finally before withdrawing the 36 cm diameter pipe casing the shingle is poured in the annular space between the education pipe and the casing pipe. The development of this well is done gradually with the compressed air. Thus the slotted tube well, unlike strainer well receives inflow only at bottom through shingle shrouding.

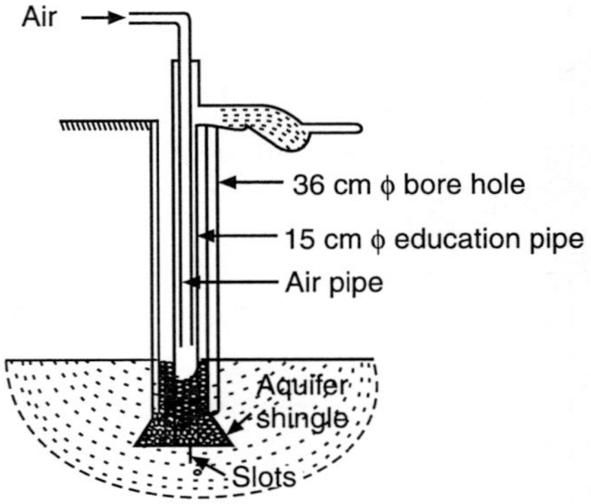

Fig. 18.3. Slotted type tubewell

18.3. DRILLING METHODS FOR TUBE WELLS (CONSTRUCTION OF TUBE WELLS)

The operation consists of sinking a hole in the crest of an earth. The diameter of the hole varies according to requirement. The hole is generally in vertically downward direction, however, certain requirements may demand that the hole should be oblique or horizontal. Though in principle drilling is extremely simple it requires skill and care in actual practice.

Most commonly adopted methods of boring are:

- percussion boring method, and
- rotary boring method,
  - (*i*) hydraulic rotary boring method.
  - (*ii*) reverse rotary boring method.

18.3.1. Percussion Boring Method

This method is suitable for the soft and fissured rock formations. In purely soil formations this method gives very high working rate. This method consists in breaking and pulverizing the subsoil material by series of repeated blows with a cutter made of hard metal. The pulverized material mixes with water and it is then removed. Sometimes this method is also called cable tool method. Boring can be done manually or mechanically. Bores as big as 30 cm in diameter and over 200 to 300 m deep can be very conveniently drilled which meet the normal requirement.

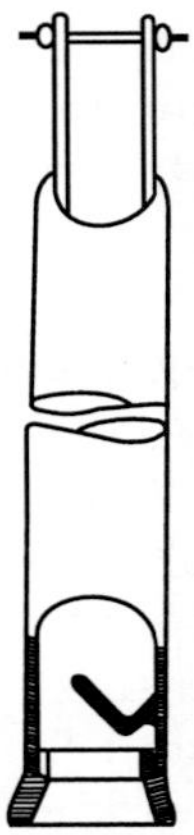
Fig. 18.4. Plunger with a flap valve

The blows are given by means of a plunger. The plunger consists of a hollow metal tube. A cutter is fixed to the bottom end of the plunger by riveting or welding. A ball valve made of steel is also provided at the bottom of the plunger. The valve is such that it allows a slurry of pulverized soil material in water to enter the plunger. Once the slurry enters the plunger the valve closes and slurry is prevented from coming out. Thus the valve has only one way action. Sometimes flap valve may also be provided in the plunger. The purpose served is exactly the same as that of ball valve. Figure 18.4 shows a flap valve. Thus plunger performs the function of a sludger and a bailer also.

The plunger can be lowered and raised in two ways:

- by rope system, and
- by rod system.

In rope system the upper end of the plunger is attached to a rope. The rope runs over a pulley. The plunger is lifted up and released suddenly to give a blow. Rod system is similar in principle to the rope system. Only difference is rod replaces the rope and consequently the rod operating machinery is also changed. The disadvantage of rod system is, time is wasted in increasing or decreasing the rod length. The rod length can be increased or decreased by screwing or unscrewing small rod lengths. The actual boring procedure by manual percussion method is as described below.

A pit is dug at the site where the tube well is to be sunk. The casing pipe with a cutter shoe is inserted in the pit. A platform is clamped to the casing pipe. The platform is loaded by means of locally available material filled in jute bags.

Over the casing pipe a tripod is erected and a pulley is fixed to it centrally. A rope runs over the pulley. One end of the rope is attached to the plunger.

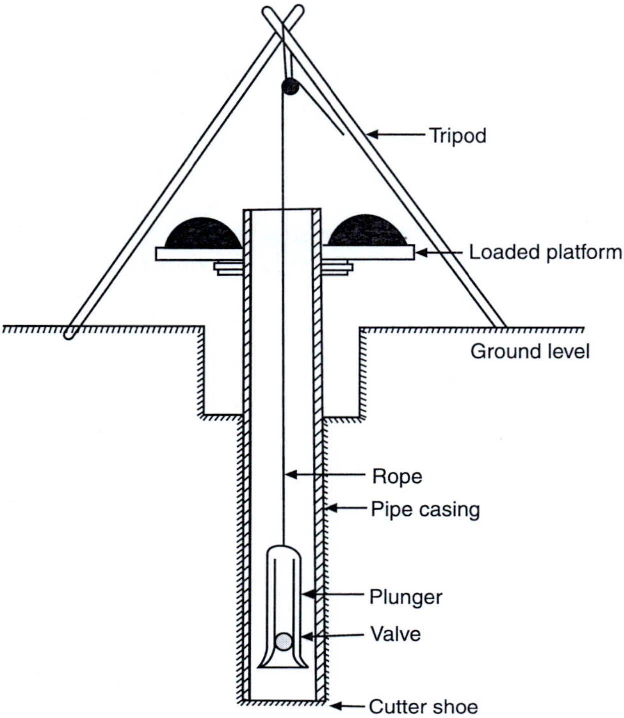

Fig. 18.5. Percussion boring (manual method)

The diameter of the plunger is slightly less than the casing pipe (say by 6 cm). The arrangement is clear from Fig. 18.5.

Before starting actual boring some water is poured in the hole. As the plunger gives blows during each blow slurry gets into the plunger. Blows are given repeatedly till the plunger is full of slurry. The plunger is then taken out and the slurry is removed by turning plunger upside down. The plunger is again lowered and the process repeated. Thus the casing pipe goes on sinking. When the length of casing pipe goes below the ground level sufficiently, additional pipe may be attached at the top of the first pipe. In order to increase working rate machine percussion, is used.

The material coming out is inspected carefully and a record is prepared, it is called logging. From the well log position of aquifers can be computed correctly. When the tube well casing pipe reaches a predetermined depth the platform is removed and a pipe with strainers at the determined levels of aquifers is lowered in the hole. After lowering it up to required depth the strainer pipe is clamped. It gives support and prevents it form falling into the bottom of the hole. Then shrouding* is started. In the beginning about 60 cm length of shrouding is done. Then the pipe casing is slowly removed by 30 cm. Then again 30 cm of shrouding is done and again pipe casing is lifted by about 30 cm. Thus the shrouding and pipe withdrawal is done slowly, successively and in small lifts of about 30 cm till the whole pipe casing is withdrawn. The amount of material required for shrouding per 30 cm length

*Shrouding is a coarse material filled in the annular space between the strainer and the casing pipe. It acts as a filter and prevents average soil particles from entering into the bore hole. Such a filter is also required in sandy and unconsolidated aquifers.

can be accurately calculated before hand. It will depend naturally on the thickness of the gravel pack. Normally thickness of the gravel pack varies between 7.5 cm to 25 cm. The thickness of gravel pack should be such that it would not allow even finest particles to move.

Machine percussion or cable tool method: The machine used for tube well drilling is called a drilling rig. The rig for cable tool method is a truck mounted assembly from consideration of mobility and consists of a mast, a multiline hoist, a walking beam and an engine. Figure 18.6 shows the assembly. The string of tools includes drill bit, drill stem, drilling jars which serve as connecting links and rope socket for connecting drill line. Figure 18.7 shows components of drilling tool. Total weight of the tool varies from 100 to 2000 kg. because different types of bits are required for different rock formations. Length of drill bit varies from 1 to 3 m whereas drill stem is 2 to 10 m long.

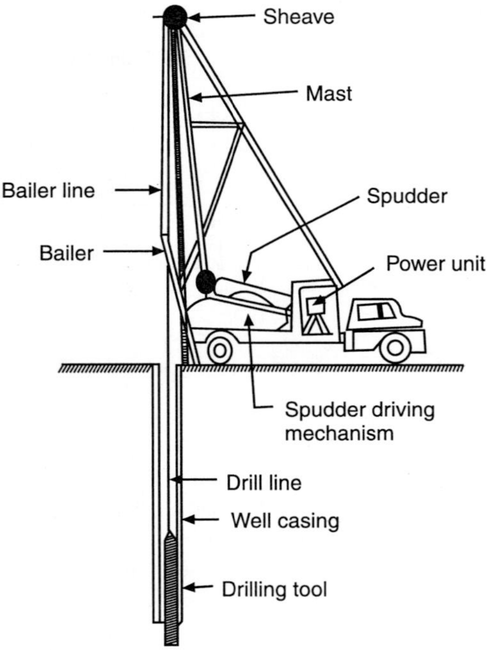

Fig. 18.6. Truck mounted cable tool machinery

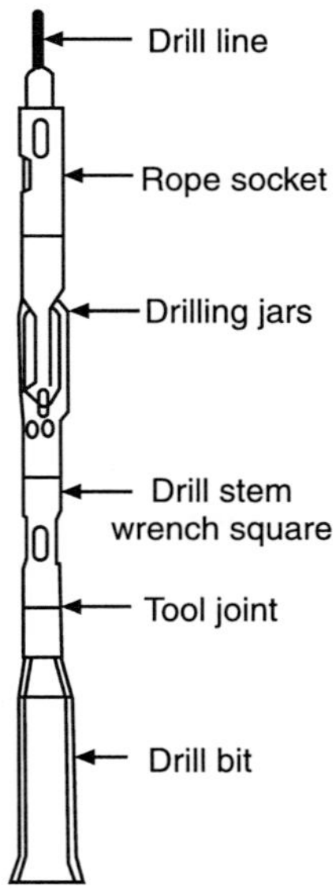

Fig. 18.7. Components of drilling tool

The operation in principle is similar to manual method. The cable tool bit acts as a crusher for drilling in consolidated rocks. The drilling is performed by repeated blows of the drilling tool which makes about 40 to 60 strokes in one minute. The drill line is rotated so that a round hole is drilled. As in the manual method water is added to the bore to form the slurry if the same is not present in the subsoil formation. After the bore is drilled by 1.25 to 1.5 m the drilling tool is removed and the slurry is taken out of the hole by means of sand pump or bailer. The bailer has one way valve which permits slurry to enter the bailer but does not allow to escape. After the bailer is filled it is raised and emptied at surface. The length of the bailer also ranges from 3 to 12 m.

In unconsolidated formations the well casing is inserted and simultaneously sunk to full depth to avoid caving in of the material. The rate of drilling depends on the type of sub-soil formation met with, diameter of the well and the depth of hole mainly. In solid crystalline rock formations the drilling rate could be as low as 2 to 3 m per day. Drilling rate in loose flowing fine sand formations is equally low because it fills the hole as soon as material is bailed out. To check the sand inflow the hole can be kept filled with water. Drilling in unconsolidated formation with boulders is quite difficult because the boulders not only deflect the hole but they are hard to drill and prevent sinking of well casing. In soft formations like sand stone or sandy clay, the drilling rate could be as high as 20 to 30 m per day. As already explained the bore is carefully logged to ascertain the position of various types of formations met with.

18.4. ROTARY BORING METHOD

18.4.1. Hydraulic Rotary Boring Method

This method is generally called rotary boring method. This method can be successfully used for rock as well as unconsolidated formation. The method is therefore used for water well as well as oil well drilling. Oil wells are generally very deep and this method is well suited for them because unlike other methods the drilling rate is not dependent on the depth of the hole. In this method drilling is done by means of rotating bits attached to lower end of hallow steel pipe. The steel pipe is attached to a square section of a steel rod at the top which is called kelly. Kelly fits in a rotating table at the surface. The rotating table is rotated by power. The powdered rock and cuttings are removed by continuous circulation of drilling fluid.

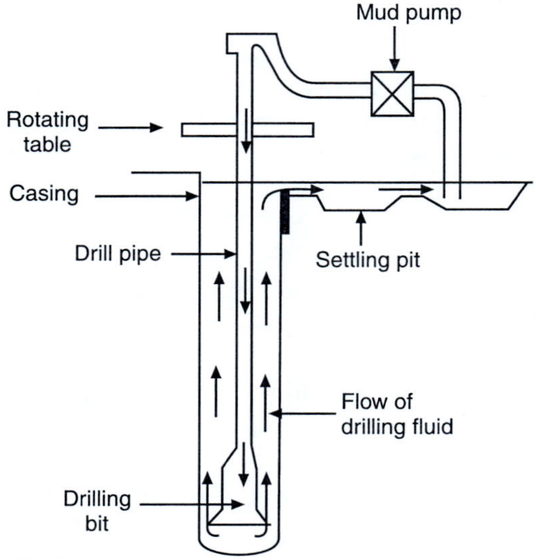

Fig. 18.8. Schematic diagram of rotary method

The hydraulic rotary drilling rig consists of a derrick or mast, a rotating table, a pump for injecting drilling mud, a hoist and an engine. The drill pipes are seamless steel tubings generally available in 6 m lengths. The external diameter of the pipes ranges from 6 to 12 cm. Generally adequate size of pipe is used because well drilling requires large quantity of drilling fluid in circulation. The drilling bits attached to the lower end of the drill pipe is provided with short nozzles to direct the jets of drilling fluid down the faces of the blades of drilling bit. Figure 18.8 shows schematic diagram of rotary boring method.

The rotating table in which the kelly fits in closely turns the drill. The drilling bit cuts through the formation and as the hole deepens the drill rod slides down. At this stage the kelly is unfastened and pulled up to attach new length of pipe.

The drilling fluid or bentonite* (clay) slurry is pumped down through the drill pipe and out through the nozzles in the bit. The mud then rises to the surface through the annular space between drill pipe and the bore and removes the rock fragments and cuttings with it. The drilling fluid performs following functions:

(*i*) It supports the walls of the bore and prevents caving.

(*ii*) It removes cuttings from the bore hole.

***Drilling fluid or bentonite slurry:** It may be anything from muddy water to specially prepared slurry of special purpose materials. Main constituents of drilling mud are water, suspended non-colloidal solid particles like sand and silt, colloidal solids like (clay) bentonite and sometimes dissolved chemicals. Bentonite is natural earth material in the form of hydro-silicates of aluminium with a high percentage of montmorillonite. Bentonite has property of absorbing large amount of water and becoming imprevious on swelling which is made use of in hydraulic rotary drilling. For water well drilling simple available type of drilling muds are used because high artesian pressure or extremely unstable formations are not met with.

(*iii*) It checks the inflow of groundwater into the well while drilling is in progress.

(*iv*) It cools the bit and lubricates the drill stem.

(*v*) It prevents the cuttings from settling down on the bottom of the bore.

(*vi*) It softens the underground formation and speeds up drilling.

Since there is likelihood that the drilling mud may seal low pressure water bearing formations it is necessary to control quantity of mud in the water.

Once the drilling fluid comes to the surface, it is taken in a settling pit where rock fragments settle down. The cleaned fluid is recirculated through the hole.

As boring progresses detailed logging of the subsurface formations met with is done. After the well is drilled to required depth a well pipe with strainer and blind lengths of determined sizes is lowered. Since the bore walls are coated with colloidal mixture of bentonite it becomes necessary to wash the walls. It is called back-washing. For back-washing drill pipe with a collar of the size of the well pipe attached above the drill bit is again inserted. The pump forces the water containing calgon (sodium hexa-metaphosphate) down the drill pipe. The water rushes through the strainers and the calgon disperses the clay colloids deposited on the bore walls. To increase efficiency of back-wash the drill pipe is moved up and down to create surging action.

The drilling rate by this method depends on type of subsurface formation penetrated and type of rig equipment used. Contrary to the cable tool method the rate of drilling by hydraulic rotary method does not depend upon the depth of the hole.

The rate of drilling in consolidated rock formations may vary from 10 to 15 m per day whereas in unconsolidated formations it may reach 100 to 150 m per day. The main advantages of hydraulic rotary method are the following:

(*i*) The drilling can go on continuously.

(*ii*) The drilling rates are quite high.

(*iii*) Casing pipe is not required. The mud forms a clay lining on the bore wall and it prevents caving.

(*iv*) When bore proves unsuccessful it can be abandoned right away as removal of casing pipe etc., is not involved.

18.4.2. Reverse Rotary Boring Method

This method is called reverse rotary method because the flow of drilling fluid is reversed as compared to the (hydraulic) rotary boring method. The drilling rig used for reverse rotary method is similar to that used for rotary boring. There are however, two variations. The first is that the drill pipe is of larger diameter (say 15 cm) and second is that large capacity special pump with open blade rotors is used. The pump allows large gravel to be discharged. The large diameter drill pipe enables as big 12 cm diameter size stones to be lifted up to the surface. As a drilling fluid only water is generally used. It moves into the bore hole through the annular space between the drill pipe and walls of the bore. The water picks up the cuttings and the mixture is sucked upward by the pump through the drill pipe. Due to force of suction the rising fluid has large velocity and it lifts with it big particles. Schematic diagram of the method is shown in Fig. 18.9.

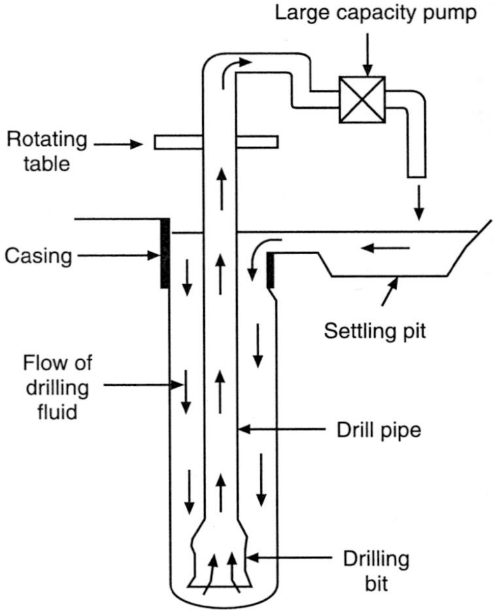

Fig. 18.9. Schematic diagram of reverse rotary method

At the surface the mixture is discharged into a settling pit. The water picks up fine particles from the subsurface formations and it is not necessary to add bentonite or any other mud to the water. The level of the drilling fluid inside the annular space should be kept upto the ground surface to prevent caving in of the hole.

By this method it is possible to drill wells of big diameter say upto 150 cm. This is the cheapest method of drilling large diameter wells in soft unconsolidated formations made of sand, silt or soft clay.

Selection of drilling method: A choice of particular drilling method depends upon:

- purpose of the well;
- quantity of discharge required;
- depth of water table:
- type of subsurface formation; and
- type of equipment available.

Shallow wells of small diameter may be bored in unconsolidated formation by an auger operated manually.

Small capacity wells in unconsolidated formations may be drilled by self jetting well point method or water jet boring method.

All deep tube wells are constructed by drilling. Rotary methods are generally better for geological investigations whereas cable tool or percussion method is superior for study of water quality.

18.5. DEVELOPMENT OF TUBE WELLS OR TUBE WELL EDUCATION

18.5.1. Definition and Need

A tube well is not completely ready for use just after construction. The next important step is to develop the drilled tube well. The tube well can function successfully only after proper development. It is the process by which the finer particles from around the screen are removed to increase the permeability of the formation through which water moves towards the well. Development serves following functions:

- It *clears* the water bearing formation clogged by the mud in the drilling operation.
- It causes the gravel pack and surrounding formation to settle and to get compacted against the screen, thus it makes the tube well structure *stable.*
- It serves to breakdown the bridging of sand grains across the screen openings and in the surrounding gravel pack and aquifer formation and makes the well *efficient.*
- It helps in obtaining *sand free water* by stabilising the sand formation around the screen.
- It helps in reducing head losses near the screen.
- It brings the well to its maximum capacity that is *maximum yield is available at minimum drawdown.*
- It gives a measure of available water supply and helps in determining the required characteristics of a pump and power unit to be installed.
- It increases useful life of the screen or strainer.

18.5.2. Methods of Development

When the water bearing strata in which well is drilled is made up of sand and gravel or alluvium, development it accomplished by removing finer particles from the area surrounding the tube well. The following are the methods commonly adopted for development of a tube well:

- Development by pumping
- Development by compressed air

- Development by surging
- Development by back-washing
- Development by high velocity jetting
- Development by using chemicals.

18.5.2.1. Development by Pumping

It is the simplest and most common method of removing fine particles. In this method water is ultimately pumped from the well at a rate equal to or higher than the design discharge. So it is a case of overpumping. A variable speed pump of large capacity is used. Water is withdrawn at a very slow rate in the beginning. Then the rate of withdrawal is increased in steps. In between the steps the rate of withdrawal is kept constant until no further sand particles are removed. Pumping should be continued until the maximum discharge is reached and no further sand particles are withdrawn. The water withdrawal is then stopped and the water level is allowed to rise to its normal position. The procedure is again repeated until no further sand particles are removed.

In the initial stages of pumping if the speed is maintained high fine particles get sucked in with great force and may clog the perforations in the pipe or the filter media. It may cause failure of tube well.

The limitations of this method are the following:

- It induces velocity in radial direction only and therefore the fine particles are removed in one direction only. It is partly overcome by intermittent pumping which agitates water in the well.
- It removes fine particles from limited area surrounding the screen.
- It requires pumps of larger capacity than conventionally available.

Owing to above limitations this method is not found very effective for developing large capacity wells. For small wells, however, it is quite suitable and commonly used.

18.5.2.2. Development By Compressed Air

The main components of this assembly are a air pipe (air line) of smaller diameter and a drop pipe of bigger diameter. The drop pipe is also called discharge pipe because the assembly is similar that of air lift pump assembly. An air compressor is directly connected to an air tank which in turn is connected to the air pipe through a quick opening valve. In this method an assembly of air pipe of smaller diameter and drop or discharge pipe surrounding the air pipe is introduced into the well till it reaches nearly bottom of the first strainer pipe. The air pipe is so adjusted that its bottom end is about 30 cm above the lower *end of discharge pipe*. It is called pumping position of the air pipe.

The air is then compressed into the well to start pumping. The pumping is continued till the pumped water is free from sand. At this point air entry is cutoff by closing the valve. The tank is brought to full pressure by keeping the compressor on. In the mean time the air pipe is so lowered that now it emerges below the lower end of discharge pipe by 30 cm. It can be called back-washing position of the air pipe. Now the valve is quickly opened to allow sudden rushing entry of compressed air into the well. Because of heavy rush of air a surge of water is created. It forces the well water into the aquifer through the screen. The surge agitates the aquifer and dislodges the fine sand particles. The air pipe is again raised inside the discharge pipe, *i.e.*, to a pumping position and as the pumping starts the direction of flow is reversed and now water enters the well through the soil. The entering water brings with it dislodged fine sand particles. The process of alternate airlift pumping and surging is continued till the aquifer is fully developed and sand flow is stopped.

In this manner full aquifer is developed taking 1 to 2 m length of the screen at a time.

18.5.2.3. Development by Surging

A surge is formed by the reciprocating movement of a plunger in the well. The water moves alternately into the soil and comes out in the well during downward and backward stroke respectively. The speed of the plunger is slowly increased. The plunger is operated in the casing pipe provided above the screened portion of the well. The repeated application of surging force draws the fine particles into the well, leaving coarser particles intact in the aquifer. Sometimes to increase efficiency of development a dispersing agent like calgon (sodium hexa-metaphosphate) is added to well water. Alternate surging and bailing is continued to draw sand from aquifer and to remove that water from the well respectively till no sand is drawn into the well.

18.5.2.4. Development by Back-washing

As the name suggests it is a process in which the water is made to flow into the aquifer formation from the well through the screen. The back-washing causes agitation of the formation and breaks down the bridging of sand particles. Back-washing thus helps in effective removal of fine particles. Various methods can be used for creating back-wash or causing reverse flow. Main methods are the following:

(*a*) **Intermittent pumping method:** When pumping is started and stopped intermittently it produces rapid changes in the pressure head in the well. When the pumping is stopped suddenly the water column in the well falls down causing reverse flow. The process is repeated till development is done which can be ascertained by sand pumping.

(*c*) **Back-washing with bailer:** In this method water is fed into the well as fast as possible and then it is bailed out with sand pump or bailer. The movement of water agitates the formation around the well. Faster rates of feeding and bailing ensure effective agitation and vigorous suction of fine material respectively. Bailed water can be reused after allowing the sand to settle down in a settling tank.

(c) **Back-washing with air pressure:** The principle adopted is similar to one adopted for compressed air development method with some modification. In this method in addition to air pipe and discharge pipe assembly there is one more small air pipe. The air pipe and discharge pipe assembly serves the purpose of air lift pump whereas the small air pipe compresses the air into the sealed well to create reverse flow. In this method the well is sealed at top after inserting the assembly of air and discharge pipes and another small air pipe. The compressed air is fed to both the pipes through a three-way valve.

The air lift pump assembly functions as usual to pump out the water. When clear water comes out of the well pumping is stopped by cutting air supply. When the water level regains static level air supply is diverted through smaller single air pipe. Compressed air admitted into the well causes reverse flow through the screen into the formation. The back-wash agitates and dislodges the sand particles in the formation. When water level in the well goes below and air starts escaping through discharge pipe the air valve is turned to activate air lift pump assembly. The pumping starts and sand and water is pumped out. The process is repeated till the well is developed thoroughly. Sometimes dispersing agent is used to accelerate the process as already explained. It may be seen that although back-washing is essential it alone cannot effectively develop the well unless it is combined with surging, bailing or pumping.

18.5.2.5. Development by High Velocity Jetting

It is one of the most effective methods of development. In this method high velocity jets released through the jetting tool pass through the screen and the formation behind the screen gets agitated. It loosens the fine particles which can be removed from the well by pumping or bailing the well water. Figure 18.10 makes the procedure clear.

This method has following advantages:

- The energy is concentrated over a small area and is therefore more effective.
- Each portion of the screen can be covered selectively and completely.
- The process is simple and does not require elaborate arrangement or special equipment.
- There is little chance of over development by this method.

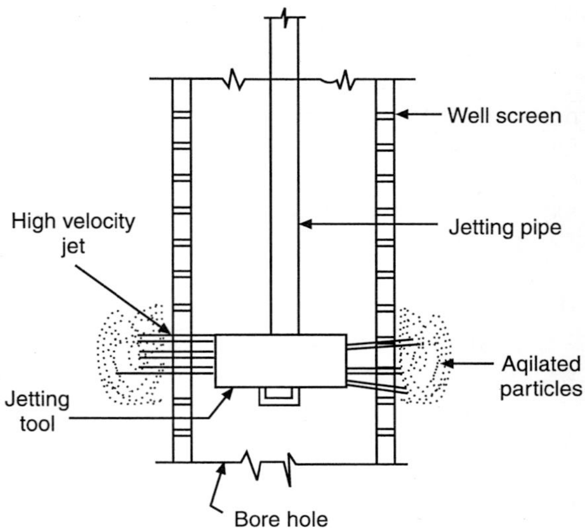

Fig. 18.10. High velocity jetting method

18.5.2.6. Development by Using Chemicals

Dispersing agents are many times added to the water used for back-washing or jetting. The dispersing agents counteract the property of clay to stick to sand particles. The common dispersing agents which are quite effective are various polyphosphates like Tetrasodium pyrophosphate. Sodium tripolyphosphate, sodium hexametaphosphate (calgon) and sodium deptaphosphate. Once the dispersing agent neutralises colloidal property of clay it can be easily removed by surging and back-washing.

Lately two chemicals, namely, hydrochloric acid and solid carbon dioxide (also called dry ice) are used for development of wells. In the well to be developed hydrochloric acid is added and the well is sealed at top. A back-wash is created by forcing compressed air in the well. As a result hydrochloric acid mixed water enters formation around the screen. The top seal is now removed and dry ice or solid carbon dioxide is dropped into the well. The sublimation takes place and in the process carbon dioxide is released. As a result high pressure is built up in the well. When the pressure is released muddy water is forced up and is thrown out of the well in the form of a jet. Thus the development to well is accomplished.

This method is also used for removing incrustation and corrosion of the well screens.

When a well is drilled in a sedimentary rock formation development is done by dissolving the cementing material partly. Thus bigger cavities and openings are formed around the well.

When a tube well penetrates rocky formation the development is done by fracturing the surrounding rock formation to create additional openings. Then more water enters the well through the fractured rock formation.

18.6. TYPES OF STRAINERS

To exclude average soil particles from entering the bore hole strainers are used. Following are some common types of the strainers:

Cook strainer: It consists of a solid drawn brass tube. The wedge shaped slots are cut on the surface of the tube. The slots are horizontal as shown in Fig. 18.11.

The slots are cut from inside of the tube. Thus the openings are wide inside of the tube. The width of slot varies from 15/100 to 40/100 of a mm depending upon the coarseness of sand. This strainer is quite good but it is very expensive as the slot cutting is a difficult process.

Tej strainer: Only difference between the Tej and Cook strainer is in former the brass sheet is first slotted and then bent to form a tube. The tube is then brazed. As there are joints this strainer is weaker. It is cheap and hence widely used.

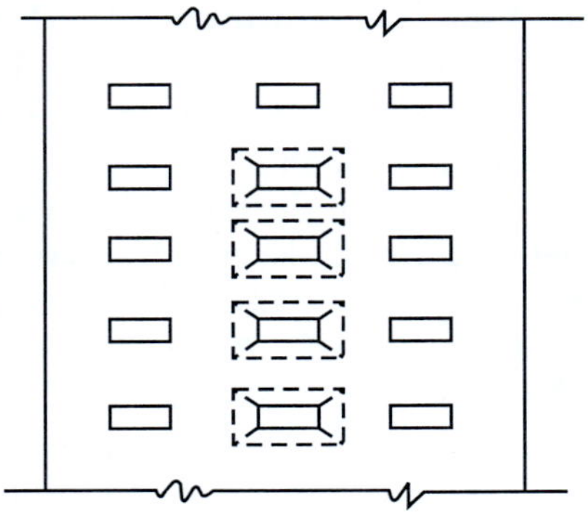

Fig. 18.11. Cook strainer

Brownlie strainer: It is made of a convoluted tube of steel. There are perforations in the tube. The wire mesh of copper surrounds the steel tube. Figure 18.13 shows this type of strainer.

The convolution may be done in various ways as shown in Fig. 18.12 (*a* and *b*). This type is designed in such a way that there is some space left between the perforated tube and the surrounding wire-mesh.

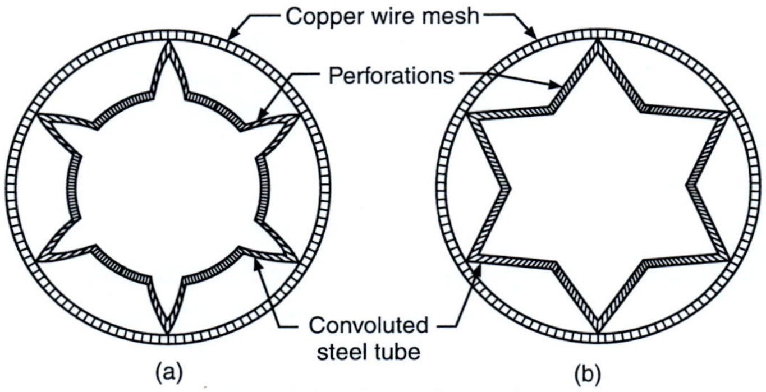

Fig. 18.12. Brownlie strainer

Ashford strainer: It consists of a perforated tube with a wire mesh surrounding it as shown in Fig. 18.13. To keep the space between the mesh and the tube a wire is wound around the tube. The wire-mesh is soldered to this wire. The wire mesh can be protected by a wire net above all. This type is delicate because there are many loose parts.

Legget strainer: In this type cutters are provided in the tube. By operating the cutters from the ground surface perforations in the tube can be cleaned.

Phoenix strainer: It is made of mild steel tube. To avoid corrosion of the tube it is plated with cadmium. The slits are made from inside.

Layne and Bowler strainer: It consists of a perforated steel or wrought iron tube. A wedge shaped steel wire is wound over this tube at suitable pitch. Though this type of strainer is robust it is costly.

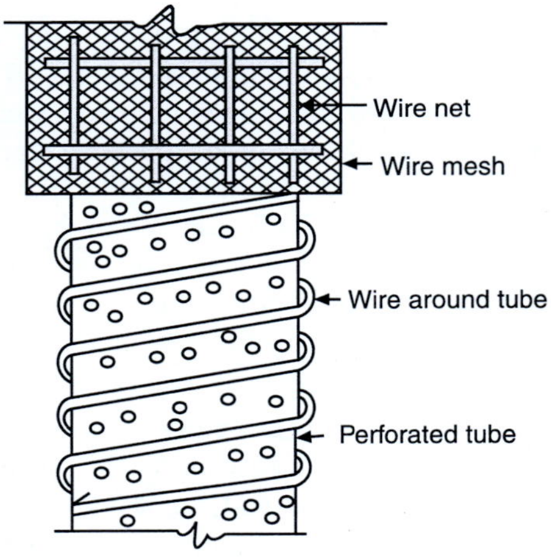

Fig. 18.13. Ashford strainer

18.7. DEFINITIONS OF TERMS USED IN TUBE WELL TECHNOLOGY

Before proceeding with the subject matter it is necessary to understand the meaning of various terms used in tube well technology. While in previous chapters definitions of some common terms have already been given even at the risk of repetition it is felt useful to student to reiterate them here.

Porosity: It is a measure of the voids present in a rock or soil mass. It is expressed as a ratio of volume of voids and total volume of the mass. The extent of voids is dependent on the size, shape, gradation and manner of arrangement of particles forming the soil or rock mass.

$$\text{Porosity } (n) = \frac{\text{Volume of voids}}{\text{Volume of soil mass}} \times 100$$

In general a medium having more than 20% porosity is considered porus and porosity less than 5% is considered small. Average porosity for various sedimentary materials is given below (Table 18.1).

It may be seen that porosity also indicates measure of water-bearing capacity of any medium. The relationship between permeability and porosity is, however, not straight forward. The size of effective opening between the grains is more important from permeability considerations. For example, sand formations with large rounded or angular grains may have smaller porosity than clays but they are more pervious and, therefore, good aquifers than the clay formations.

Table 18.1.

Material	*Porosity (%)*	*Specific Yield (%)*
Soils	50 – 60	
Clay	45 – 55	1 – 10
Mixed sand		
(*a*) Medium to coarse	35 – 40 }	10 – 30
(*b*) Fine to medium	30 – 35 }	
Gravel	30 – 40	15 – 30
Gravel and sand	20 – 25	15 – 25
Sand stone	10 – 20	5 – 15
Shale and lime stone	1 – 10	1/2 – 5

Grain size classification of aquifer material: The aquifer material may be classified into various categories according to the following ranges of the particle (Table 18.2).

Effective diameter: It is an index of the measure of the fineness of an aquifer and is used for designing various intake components of a tube well. For permeability d_{10} (90% grains retained) or d_{17} (83%, retained) is generally taken as the effective size.

Average grain size: Similarly for the design of the gravel pack or shrouding d_{50} (50% grains retained on the seive) has been designated as average grain size.

Uniformity coefficient: It is a ratio of d_{60} and d_{10} of any given soil sample. That is:

$$Cu = \frac{d_{60}\text{(40\% grains retained on seive)}}{d_{10}\text{ (90\% grains retained on seive)}}$$

Table 18.2.

Classification	*Particle size (mm)*
Fine gravel	above 2.0
Coarse sand	0.5 to 2.0
Medium sand	0.25 to 0.5
Fine sand	0.05 to 0.25
Silt	0.002 to 0.05
Clay	below 0.002

For poorly graded sands $Cu < 4$

Whereas for well graded material $Cu > 4$.

Note: In fact particle size parameters (mentioned above) can be obtained conveniently by passing the sample through a set of IS seives from no. 75 onwards and weighing the material retained on each seive. Then the cumulative weight passing through each seive is plotted on a semi-logarithmic graph paper. On the graph ordinate represents percent weight in ordinary scale and abscissa represents size of seive opening on logarithmic scale. The resulting smooth curve gives particle size distribution.

Safe yield: The amount of water that can be withdrawn from an aquifer without producing adverse effect is called safe yield from the aquifer. Obviously the withdrawn amount of water gets replenished from precipitation through recharge areas.

Overdraft: The amount of water withdrawn in excess of safe yield is called overdraft. Withdrawals in excess of replenishable amount has to come from ground water reservoir. Naturally overdraft will result in permanent lowering of water table which is also called mining of ground water. Excessive pumping causes overdrafts and will have following deterimental effects:

- decrease in yield due to lowering of water table;
- interference with other ground water structures may create shortage else where;
- overpumping may lead to salt water intrusion if the well is in vicinity of sea; and
- deep pumping may sometimes produce water of inferior quality.

Specific yield and specific retention: It has been already mentioned that porosity indicates capacity of the formation to hold water when fully saturated. It is also already mentioned that all water is not capable of free drainage. Part of the water in the pores drains out whereas rest is retained in pore spaces by molecular and capillary forces. Therefore the water that an aquifer can offer for withdrawal is the one free to flow by gravity.

Specific yield: Specific yield of the soil or rock mass is a ratio of the volume of water that can be obtained from an aquifer and total volume of the mass.

$$\text{Specific yield } (S_y) = 100 \times \frac{W_y}{y} \text{ or } \frac{\text{Volume of water drained}}{\text{Volume of soil mass}}$$

Specific yield is also therefore called effective porosity. Representative specific yield values for various sedimentary materials is also given in Table 18.1 along side porosity for comparison.

Specific retention: If a unit volume of dry porous material is saturated and then permitted to drain by gravity, the volume of water released is less than that required for saturation. The volume of water retained in the material is held by capillary action and molecular forces against the pull of gravity. The specific retention can be defined as the ratio of the volume of water retained by saturated soil mass after drainage and the total volume of the soil or rock mass.

$$\text{Specific retention } S_r = 100 \times \frac{W_r}{y} \text{ or } \frac{\text{Volume of retained water}}{\text{Volume of soil mass}}$$

From the definitions of porosity, specific yield and specific retention it can be stated that

$$n = S_y + S_r \quad \text{or}$$

$$\text{porosity} = \text{Specific yield} + \text{Specific retention}$$

like porosity, specific yield also depends upon grain size, shape, gradation, distribution of pores, manner of arrangement of particles etc.

Specific capacity: It is a term which provides measure of productivity of the well. It is defined as the ratio of the pumping rate under steady flow condition (in other words yield of tube well) and the drawdown in a well. Thus it is the yield of the well per metre drawdown.

$$\therefore \qquad S_c = \frac{Q}{h}$$

where S_c is specific capacity;

Q is pumping rate at steady flow condition or yield of well; and

h is drawdown in the well below static water level.

It may be noted that maximum yield from a well is obtained at maximum drawdown. Maximum drawdown occurs when water level in the well is lowered to the bottom of the well. It is, however, seen that the ratio of pumping rate (Q) and corresponding drawdown (h) viz. discharge per metre drawdown records substantial reduction as discharge reaches maximum value. For optimum well characteristics, in fact the product of well yield and specific capacity should be maximum. It is seen that this occurs at about 67 per cent of the maximum drawdown. Using this fact a design practise of providing a screen approximately for the lower one-third length of homogeneous unconfined aquifer has been adopted.

Storativity or Storage Coefficient: It is also called Storativity. Storage coefficient indicates the water yielding capacity of an aquifer. It is defined as the volume of water released from or absorbed into storage by an aquifer per unit surface area of aquifer per metre decline or rise of head respectively.

From the definitions of specific yield and storage coefficient it can be said that for unconfined aquifer (water table condition) the storage coefficient equals specific yield provided gravity drainage is complete.

For unconfined aquifers, the storage coefficient depends on the volume of water thrust out of the pores due to elastic compression of aquifer resulting from change in hydrostatic pressure caused by pumping.

Table 18.3.

Soil Type	***Coefficient of permeability (mm/hr)***
Clay	less than 0.05
Silty sand	4 to 60
Fine sand	40 to 1600
Clean gravel	30,000 and more

Coefficient of permeability: The permeability is also called hydraulic conductivity. It is defined as the velocity of flow of water through a porous medium under a unit hydraulic gradient. It indicates the ease with which water can flow through soil mass. Mathematically,

$$K = \frac{Q/A}{(h_1 - h_2)/L}$$

Obviously coefficient of permeability 'K' has the dimensions of velocity. It is expressed in length dimension per unit time. Table 18.3 gives an idea of the typical ranges of permeability values for common type of formations.

Coefficient of Transmissibility or Transmissivity: It is generally designated by 'T'. As the term hydraulic conductivity or permeability fails to describe adequately the flow characteristics of an aquifer, CV. Theis introduced the term *transmissivity* T = Km, which is equal to the average permeability times the saturated thickness of the aquifer to clarify this deficiency. Transmissivity has dimensions of L^2/t.

The coefficient of transmissibility or transmissivity of an aquifer is the rate of flow through the entire thickness of a saturated aquifer of unit width under unit hydraulic gradient.

Therefore, $\qquad T = m.K$

where m is saturated thickness of aquifer, and K is coefficient of permeability.

18.8. YIELD OF TUBE WELL USING DUPIT'S THEORY

Definitions of unconfined and confined aquifers are already given above. To make it very simple, unconfined aquifer is that water bearing layer which is not interposed by any impervious layer. The water bearing stratum is unlimited in extent as far as depth of the tube well is concerned. The pressure at the top surface of this aquifer is atmospheric. That is the top of aquifer represents water table.

Conversely confined aquifer is that water bearing layer which lies between two impervious layers. One at top and other at bottom. The pressure at the top of the confined aquifer is not atmospheric but it is positive hydrostatic pressure.

It may be appreciated here that in case of unconfined flow the top flow boundary is a free surface and it is unknown. But in case of confined flow all boundaries are known initially. Approximate analytical solution to this problem was given by Dupit and the theory is, therefore, known after his name.

Theoretically the discharge estimation for the strainer type tube well (It is most common in occurrence) is done after making following simplifying assumptions:

- Water table is horizontal and uniform everywhere in a vertical section.
- The inclination of the free surface is equal to hydraulic gradient and it does not change with depth.
- The bottom of the strainer reaches a horizontal impervious layer.
- The well penetrates homogeneous isotropic aquifer.
- There is no flow in the aquifer other than that towards the well. That is the flow is radial. It is also assumed that the flow is in horizontal direction.
- Total discharge remains constant across the surface of a series of concentric cylinders.
- The flow through soil is assumed to be steady and laminar. Therefore Darcy's law is applicable.

Most of the assumptions are not justified and hence the formula derived gives approximate discharge. But as no other mathematical basis could be given still this formula is used for finding out the discharge. The results obtained are reasonably accurate.

18.9. CALCULATION OF DISCHARGE IN UNCONFINED AQUIFER BY DUPIT'S FORMULA FOR STEADY RADIAL FLOW

Discharge through **unconfined aquifer** can be calculated from the formula

$$Q = \frac{\pi.\, K(H_1^2 - H_2^2)}{2.303 \log_{10} R/r} \qquad \text{(refer Fig. 18.14)}$$

where

R is radius of circle of influence in metres. (Also called radius of zero drawdown)

r is radius of tube well in metres.

H_1 is height of static water-table above impervious layer in metres.

H_2 is height of water level in the tube well in metres after depression.

Q is total discharge of the tube well.

K is coefficient of permeability of the soil.

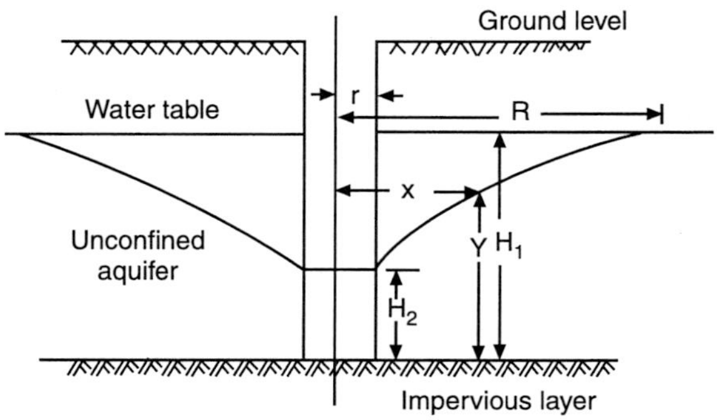

Fig. 18.14. Tube well in unconfined aquifer

18.10. CALCULATION OF DISCHARGE IN CONFINED AQUIFER BY DUPIT'S FORMULA FOR STEADY RADIAL FLOW

Discharge through **confined aquifer** can be calculated from the formula

$$Q = \frac{2\pi. K.m.(H_1 - H_2)}{2.303 \log_{10} R/r}$$

where

m is thickness of confined aquifer. It is assumed that m is constant (that is confined aquifer is of uniform thickness). Other notations are same as in the case of unconfined aquifer. The well penetrates aquifer fully.

In the above equation $K.m.$ in called coeff. of transmissibility (T).

$$\therefore \quad (H_1 - H_2) = \frac{2.303}{2\pi.T}. \log_{10} \frac{R}{r}$$

The above equation is called equilibrium equation. It can be used to determine variation of piezometric head radially outward from the well.

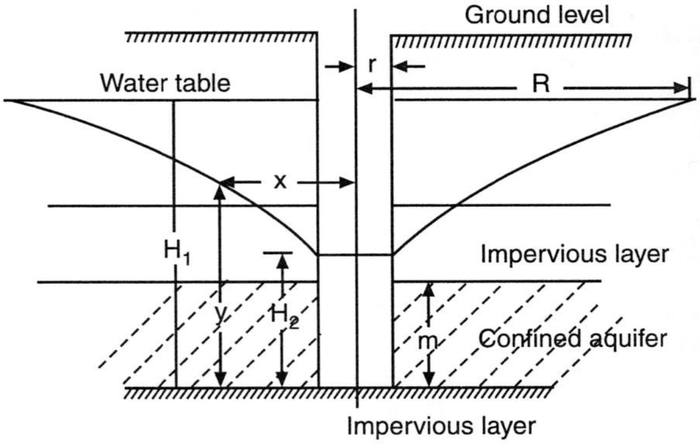

Fig. 18.15. Tubewell in confined aquifer

In the upper limit $(H_1 - H_2) = s$ which represents drawdown in the well.

Similarly, drawdown in the aquifer at any distance r_x from the well is given by

$$s_x = (H_1 - H_x)$$

$$= \frac{2.303 Q}{2\pi T}. \log_{10} \frac{R}{r_x}$$

Further, knowing the values of head at any two points located at distance r_a and r_b from the well (say in two observation wells) during pumping test the value of K or T can be obtained by using same equation viz.

$$K = \frac{2.303\, Q}{2\pi.\, m(H_a - H_b)}. \log_{10} \frac{r_a}{r_b}$$

or

$$T = \frac{2.303\, Q}{2\pi\, (H_a - H_b)}. \log_{10} \frac{r_a}{r_b}$$

This method of finding out permeability of the aquifer is called Thiem's method.

Although selection of radius of influence in practice is arbitrary and approximate it does not affect discharge values much over a wide range. Generally radius of influence values fall between 150 to 300 m.

Sichardt had given following relationship to calculate value of R.

$$R = 3000\ s.\ \sqrt{K}$$

where K is permeability is m/sec and s is drawdown in metres.

It is already mentioned that due to interference of an adjacent well the discharge of the well reduces. For two identical wells in confined aquifer at a discharge w apart the discharge of each well is given by expression.

$$Q_1 = Q_2 = \frac{2\pi\, T(H_1 - H_2)}{2.303 \log_{10} \dfrac{R^2}{r.w}}$$

It may be noted that when $R = w$ there will be no effect due to interference of adjacent well.

18.11. CALCULATION OF DISCHARGE IN CONFINED AQUIFER USING THEIM'S FORMULA FOR STEADY RADIAL FLOW

Discharge through confined aquifer can be calculated from the formula

$$K = \frac{2.303\, Q \log \dfrac{r}{r_w}}{2\pi\, m\,(h - H_2)}$$

Consider that the recharge to the aquifer within the influence zone of the pumped well equals the rate of discharge of the well so that the drawdown remains stabilised and therefore steady state exists.

Also $h - H_2 = \dfrac{2.303\, Q}{2\pi\, T} \log_{10} \dfrac{r}{r_w}$

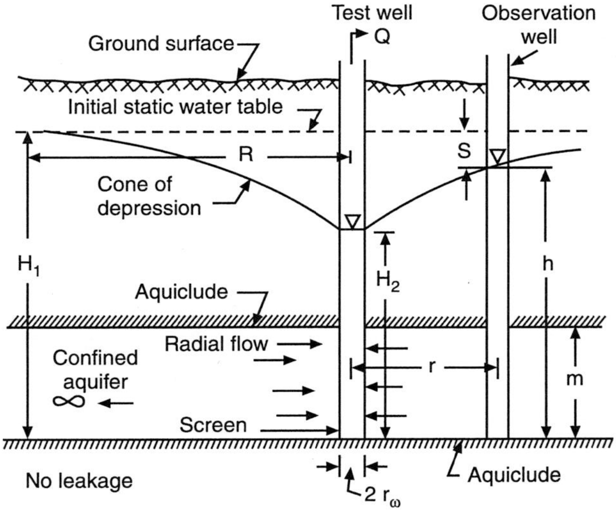

Fig. 18.16. Radial flow to a well in confined aquifer

where K = coefficient of permeability

m = thickness of aquifer

r_w = radius of the well

T = transmissibility of aquifer = $K.\ m$

The above equation is called **equilibrium or Thiem's equation** and is used to determine piezometric head at any point at a radial distance r from the centre of the well.

Carrying the logic further if the piezometric heads in two observation wells say h_1 and h_2 at two points r_1 and r_2 distance radially away respectively from the centre of the pumped well are measured during the pumping test, coefficient of permeability 'K' can be easily calculated. The formula can be written as follows: ($r_2 > r_1$)

$$K = \frac{2.303\, Q}{2\pi m(h_2 - h_1)} . \log_{10} \frac{r_2}{r_1} \qquad ...(4)$$

The above method is popularly called Thiem's method.

18.12. CALCULATION OF DISCHARGE IN UNCONFINED AQUIFER USING THEIM'S FORMULA FOR STEADY FLOW

Referring Fig. 18.17 and considering the steady state condition as mentioned in earlier article, the discharge at any distance r towards the well is given by applying Darcy's formula in conjunction with simplifying assumptions made by Dupit as follows:

$$Q = K\, A\, i = 2\pi r K h \frac{dh}{dr}$$

Integrating the equation (1) between limits $h = H_2$ at $r = r_w$ h = head at any distance r

$$h^2 - H_2^2 = \frac{2.303Q}{\pi.K}\log_{10}\frac{r}{r_w} \qquad ...(a)$$

Equation (*a*) can be used to determine the distribution of head radially outward from the well. If the values of head h_1 and h_2 at any two observation wells at a distance r_1 and r_2 respectively ($r_2 > r_1$) from the test well are measured, coefficient of permeability K can be measured by substituting values in equation (*a*) above. Then

taking limit when $h = H_1$ at $r = R$, (the radius of influence) equation (*a*) will become

$$H_1^2 - H_2^2 = \frac{2.303Q}{\pi.K}\log_{10}\frac{R}{r_w} \qquad ...(b)$$

It may be mentioned that in equation (*a*) as well as (*b*) H_2 is head at well and therefore is equal to the depth of water in the well.

Sichardt's formula may be used to calculate radius of influence R. It is expressed below to recapatulate:

$$R = 3000s\sqrt{K}$$

where R is radius of influence in metres

s is drawdown at well in metres

K is coeff. of permeability in m/sec.

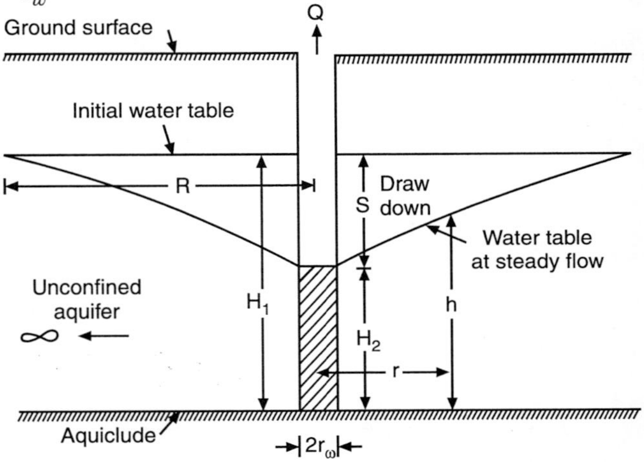

Fig. 18.17. Radial flow to a well in unconfined aquifier

Problem 18.1. A tube well is 0.46 m in diameter. The unconfined aquifer is of 18 m depth. After drawdown depth of water is 12 m in the well. Permeability of soil is 24.50 m/day. Radius of circle of influence is 275 metres. Calculate discharge of the tube well.

Solution: Formula to be used is

$$Q = \frac{\pi.K(H_1^2 - H_2^2)}{2.303\log_{10} R/r}$$

K = 24.50 m/day;

R = 275 m; r = 0.23 m.

H_1 = 18 m; H_2 = 12 m.

$$\therefore \quad Q = \frac{\pi \times 24.5(18^2 - 12^2)}{2.303\log_{10}\frac{275}{0.23}}$$

$$= \frac{\pi \times 24.5 \times 180}{2.303 \times \log_{10} 1200}$$

$$= \frac{\pi \times 24.5 \times 180}{2.303 \times 3.079}$$

= 2,000 m^3/day.

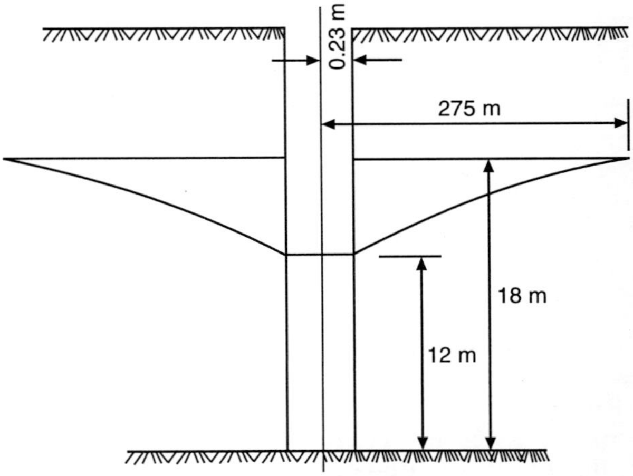

Fig. 18.18. Problem 1

∴ Discharge of the tube well is 2000 m^3/day or 23.15 litres/sec.

Problem 18.2. **A tube well is driven fully in a confined aquifer of 24 m thickness. The aquifer is met 25 m below ground level. The static water table is 15 m below ground level. The discharge of tube well is found to be 6000 m^3/day when depression head is 12.25 m. Permeability is 24.50 m/day. Find the diameter of the tube well.**

Solution:

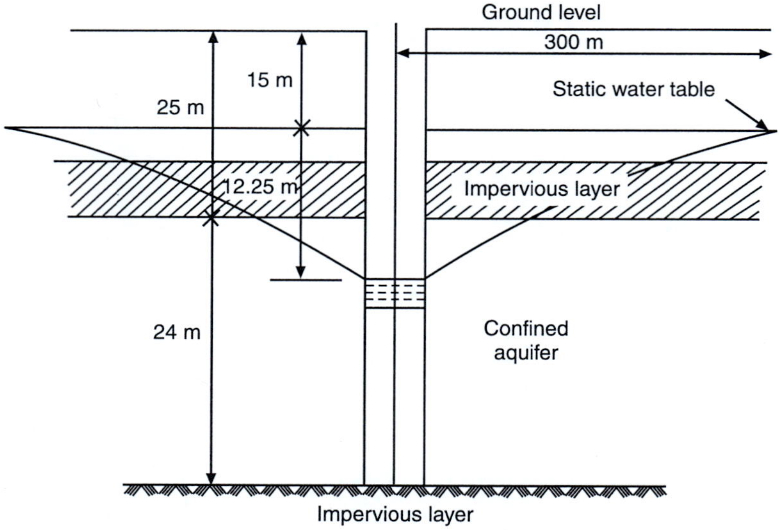

Fig. 18.19. Problem 2

Formula to be used is, $Q = \dfrac{2\pi.K.m.(H_1 - H_2)}{2.303 \log_{10} \dfrac{R}{r}}$

K = 24.50 m/day; m = 24 m; $(H_1 - H_2)$ = 12.25 m;
R = 300 m; Q = 6000 m^3/day;

$$\therefore \qquad 6000 = \frac{2\pi \times 24.5 \times 25 \times 12.25}{2.303 \log_{10} \dfrac{300}{r}}$$

$$\text{or} \qquad \log_{10} \frac{300}{r} = \frac{2\pi \times 24.5 \times 25 \times 12.25}{2.303 \times 6000} = \frac{2.4\pi}{2.303}$$

$$\log_{10} 300 - \log_{10} r = \frac{2.4\pi}{2.303}$$

$$\text{or} \qquad 2.47712 - \log_{10} r = 3.275$$

$$\log_{10} r = -0.72788 \quad \text{or} \quad \bar{1}.20212$$

$$r = 0.16 \text{ m}$$

Ans. Diameter of well is 0.32 metres.

Problem 18.3. **Two identifical tube wells tap a confined aquifer of 10 m thickness and are spaced at 80 m distance from each other. The radius of influence may be taken as 240 m whereas the radius of each well is 12 cm. The coefficient of permeability of the aquifer was found out to be 60 m/day.**

(*a*) Calculate the discharge of each well under the 4 m depression head when working simultaneously;

(b) What will be the discharge under similar drawndown when only one tube well is working;

(c) What difference do you expect in the discharge of one well by keeping another idle.

Solution: (a) When both the tube wells are functioning discharge formula to be used is

$$Q_1 = Q_2 = \frac{2\pi K \cdot m(H_1 - H_2)}{2.303 \log_{10} \dfrac{R^2}{r \cdot w}}$$

where $K = 60$ m/day; $m = 10$; $(H_1 - H_2) = 4$ m

$R = 240$ m; $r = 0.12$ m and $w = 80$ m

$$\therefore \quad Q_1 = Q_2 = \frac{2\pi \times 60 \times 10 \times 4}{2.303 \log_{10} \dfrac{240 \times 240}{0.12 \times 80}} = \frac{4800\pi}{2.303 \log_{10} 6000}$$

$$= \frac{4800\pi}{2.303 \times 3.778} = 1733 \text{ m}^3\text{/day.}$$

(b) When only one tube well is working discharge formula to be used is

$$Q = \frac{2\pi K \cdot m(H_1 - H_2)}{2.303 \log_{10} \dfrac{R}{r}} = \frac{2\pi \times 60 \times 10 \times 4}{2.303 \log_{10} \dfrac{240}{0.12}}$$

$$= \frac{4800\pi}{2.303 \log_{10} 2000} = \frac{4800\pi}{2.303 \times 3.301} = 1983 \text{ m}^3\text{/day.}$$

(c) When only one tube well is working it gives more yield since interference from other well is not existing. The increase in discharge for the well will be 1983 – 1733 = 250 m^3/day.

Problem 18.4. In an unconfined aquifer a fully penetrating tube well is sunk. The depth of static water table above the impervious layer is 17 metres. Discharge of the tube well is 0.08 m^3/sec. Three observation wells were simultaneously read. Their distance from centre of pumping well is 35, 70 and 150 metres respectively. Observed drawdown after 16 hours was 0.6 m, 0.4 m and 0.3 metres respectively. Using observed data calculate the coefficient of permeability and transmissivity of the aquifer.

Solution: It is a steady state case pertaining to unconfined aquifer. Hence we can use the formula:

$$K = \frac{Q \log_e \dfrac{r_2}{r_1}}{\pi[h_2^2 - h_1^2]} \qquad \text{... (A)}$$

From observed data, $Q = 0.08$ m^3/sec and $H = 17$ metres

Also, $s_1 = 0.6$ m at $r_1 = 35$ m

$s_2 = 0.4$ m at $r_2 = 70$ m ; and

$s_3 = 0.3$ m at $r_3 = 150$ m

We know that, $h_1 = H - s_1 = 17 - 0.6 = 16.4$ m

$h_2 = H - s_2 = 17 - 0.4 = 16.6$ m

$h_3 = H - s_3 = 17 - 0.3 = 16.7$ m

Using values h_1 and h_2 and r_1 and r_2 and putting them in equation (A) above

We get $$K = \frac{0.08 \log_e \frac{70}{35}}{\pi[(16.6)^2 - (16.4)^2]}$$

Coefficient of permeability, $K = 0.0035$ m/sec.

Now $T = K\,m$

or $T = K\,H$ (in this problem)

$\therefore$ $T = 0.0035 \times 17$

Transmissivity, $T = 0.061$ m^2/sec.

18.13. WATER LIFTING ARRANGEMENT FROM A TUBE WELL

In tube wells water is normally available at greater depths. Secondly diameter of a tube well is much smaller as compared to an open well. Obviously water can be lifted from tube wells only with the help of pumps. There are many types of pumps available.

For deep open wells and tube wells suction lift is normally more than 8 metres. Also lift further increases due to fluctuations in water table and depression head. However, it is many times possible to reduce the suction lift to permissible limit of 7.6 m by installing the pump in the pump sump below ground level. A single stage centrifugal pump with horizontal spindle is commonly used in such cases. The water lifting arrangement is shown in Fig. 18.20.

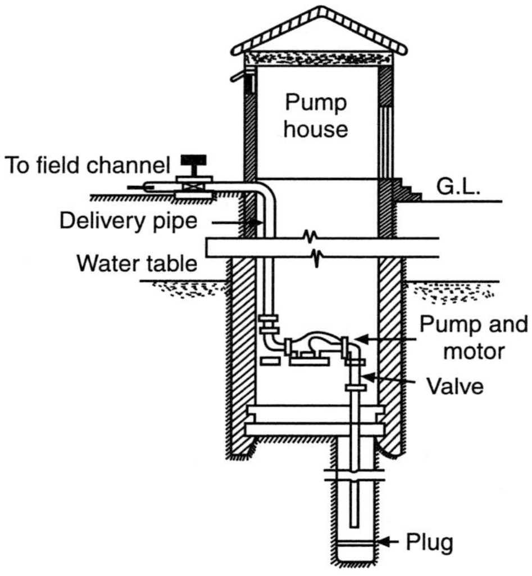

Fig. 18.20. Water lifting arrangement

This arrangement has following advantages:

- It is very easy to install.
- It is relatively trouble free arrangement.
- Cost of installation is moderate.
- It is capable of attaining good efficiency.

It has the main disadvantage that the suction lift is limited. In tube wells when the lift increases appreciably this arrangement becomes useless. As a result increasingly other types of pumps which can operate under high heads are coming into use. Choice of a particular type of pump depends on the following factors:

- Diameter of the tube well.
- Depth of the well and depth of water table below ground surface.
- Variations in water table.
- Expected drawdown in the well.
- Quality of water.
- Duration of pumping per day.
- Pumping capacity.
- Requirement of power for working of pumps.
- Capital and maintenance cost of the pump.

Apart from single stage and multistage* centrifugal pumps other common types of pumps used to lift water from the tube wells are:

*Multistage pumps comprise several impellers on one shaft in series arranged to generate greater head than can be efficiently obtained from a single impeller.

- Vertical turbine pump;
- Submersible motor pump;
- Jet pump; and
- Air lift pump.

18.13.1. Vertical Turbine Pump

This type of pump is used to pump water from deep tube wells because it is a multistage diffuser type pump. A vertical turbine pump consists of a vertical spindle electric motor at ground level. From it water flow pipe and a drive shaft within the pipe reaches the lowest water level. The pump is attached to the lower end of the water flow pipe and remains submerged under water. The turbine pumps are available in varying sizes ranging from 10 to 35 cm in diameter. The pump length varies from 20 to 60 cm. Number of stages depends upon the head to be developed. In high head pumps the number of stages may reach 20 stages or more. This type of pump is capable of discharging 2000 l/sec and can develop heads upto 500 m.

18.13.2. Submersible Motor Pump

This is a multistage centrifugal type pump. The impellers and diffusers are arranged in series to discharge water against sufficient head. The entire assembly comprising the pump and electric motor is kept submerged under water in the well. The delivery of water is under pressure and there is no suction condition. The pump is placed above the motor with water inlet in-between. The electric motor is connected to the control system on the ground by means of special submarine type electrical cables.

This pump can be installed in a tube well of as small diameter as 10 cm. The pump can be installed at any depth subject to horsepower and pipe size. Unlike turbine pump it does not need long drive shaft. Also sump well or pump house etc., is also not required.

18.13.3. Jet Pump

It is nothing but a single stage centrifugal pump aided by a jet or an ejector nozzle. The pump is located at surface with foot valve reaching lowest water level in the well. The ejector nozzle enters suction pipe just above the foot valve. Ejector is fed by another pressure pipe which receives part of the pumped water. The water flows down under pressure and enters suction pipe through the nozzle and helps to improve the flow into the pump.

The pump can be installed in a tube well having 10 to 20 cm diameter. Jet pumps are used to obtain small discharges upto 3.3 litres/sec. They are suitable for lifts between 6 to 50 metres.

18.13.4. Air Lift Pumps

Due to low efficiency this method is rarely used. It is useful for removing objectionable gas, highly corrosive or abrasive water which damages parts of the pump submerged in the well. It is also used for testing or development of wells.

18.14. IDEAL TUBE WELL

An ideal tube well is one which satisfies following requisites:

- Tube well should be located in the area centrally or at such a point that it gives good command. It is very essential requirement because the discharge of tube well is low. It goes upto say 0.04 m^3/sec.

- Water derived from the tube well should be suitable for the crops grown in the area.
- The tube well should be driven in a good aquifer having sufficient storage.
- The tube well should have ample power supply for working of pumps.
- The tube well should have sufficient storage to fulfil the inadequacy of the canal irrigation. The tube well should help in improving the ground water conditions.
- There should be some annular space left between the perforated tube and the straining device.
- The area of openings in the tube well and in the straining device should be equal.
- The velocity of discharge should not be less. If rate of pumping is kept low silting may occur.
- The working of the tube well should be trouble free.

18.15. WELL TROUBLES

1. If the rate of water withdrawal from the well is increased to such an extent that it is more than the rate of water entry into the well then the yield of well will be reduced to a great extent. Naturally it is sort of a well failure. To avoid this situation the rate of pumping or water withdrawal should be reduced at and after the equilibrium stage. Equilibrium stage is one where rate of water entry equals the rate of water withdrawal.
2. The strainer pipe may collapse or cave in. Then the well will be blocked either partially or completely. In such a situation the strainer pipe will have to be changed.
3. The strainer pipe may corrode. Due to corrosion pipe will start leaking and the water from the tube well may escape into the ground. Harmful water from upper portion of the subsoil may also find way in the tube well. The pipe will have to be replaced under such circumstances. If the tube well diameter is big enough then a strainer pipe of smaller diameter may be introduced in the existing well.
4. Sometimes wire mesh may corrode or some sort of crust may develop on the mesh. Only method of prevention of corrosion is to use wires plated with suitable material. Incrustation can be reduced by acid treatment.
5. Some times polluted water may enter the tube well from the surface or through the bottom. This may cause contamination of tube well water. To avoid this trouble the well is to be sealed.

18.16. COMPARISON OF WELL IRRIGATION AND CANAL IRRIGATION

Following points are in favour of well irrigation:

1. The well irrigation system may be introduced at any time when cultivator feels it necessary.
2. Rainfall even if scanty does not affect tube well discharge immediately.
3. Water is more economically used. To give examples, firstly, when rainfall starts, irrigation may be stopped without delay. Secondly, as water charges are based on the actual volume of water supplied to the cultivators, they use it cautiously.
4. The patches of land which are not commanded by canal, can be very easily brought under irrigation by sinking wells provided good underground water conditions prevail at the site.
5. As the source is situated in the middle of the area to be irrigated, the water losses in transit and handling are less.

6. Duty of well water is higher than canal water.
7. It improves the underground water-table conditions. Waterlagging is reduced to quite a good extent by well irrigation.
8. Unlike canal water well water is warm in winter and cool in summer. This condition is more suitable for crop growth.

Following points are not in favour of well irrigation:

1. As the water is to be lifted up before taking it in the water courses the working expenses are high.
2. As machinery is used for lifting water it is always liable to give troubles.
3. The water is free from silt. The silt if present in irrigation water helps crop growth. The well water lacks in nutritional property due to absence of silt load.
4. Life of a tube well is shorter than flow irrigation system.

QUESTIONS FOR REVISION

1. Explain with sketches the method of construction of tube well. Under what conditions tube well irrigation is possible ? Draw section of a tube well.
2. Describe various types of strainer with their relative advantages and disadvantages.
3. Compare and contrast well irrigation with canal irrigation.
4. Write shorts notes on:
   (*i*) Cavity type tube well
   (*ii*) Slotted tube well
   (*iii*) Tube well education
   (*iv*) Shrouding
   (*v*) Well troubles

19 Design of Diversion Headworks and Other Regulatory Structures

19.1. WEIR DESIGN DATA

General layout and description of canal headworks and also other canal regulatory structures have been already described. It is needless to mention that a weir or a barrage is the most important component of a diversion headworks. Therefore it is necessary to understand the principles on which a weir is designed. Following data is essentially required to design a weir :

(*i*) **Design flood discharge :** Maximum flood discharge data of a river recorded in the past is collected. Using this information maximum probable flood discharge is anticipated. It may be taken as a design flood discharge.

(*ii*) **Stage-discharge curve :** The water levels attained during the floods and other seasons are collected. A graph, showing the stage of the river versus the discharge is plotted.

(*iii*) **Silt data :** Along with discharge measurement the amount of silt carried by the river water is also measured. The analysis of the silt is done to know its properties.

(*iv*) **FSL** and full supply discharge of the off-taking canal.

(*v*) **Cross-section of the river:** It is very essential to get full information about the cross section of the river at the weir site and also upstream and downstream of the site. Various river cross-sections are plotted covering upstream and downstream portions.

(*vi*) **Amount of afflux:** When the weir obstructs the flow of water in the river the water heads up on the upstream side of the weir. This rise is maximum during floods. The amount by which the level rises above the normal level is known as afflux. This rise extends on the upstream for a sufficient length. The amount of afflux determines the height and the section of the guide banks and protection bunds used for training the river. Afflux can be calculated using Molesworth formula. By providing high afflux it is possible to reduce the length of actual waterway required. But there is also a limit upto which waterway can be reduced.

(*vii*) **Pond Level:** The weir stores water on its upstream side in front of a head regulator to create a still pond. The canal gets water from this pond through the head regulator. Hence, it is essential to keep the level of the pond some 0.75 to 1 m above the FSL of the canal to provide sufficient working head. Of course it is essential to keep some margin for future silting up of the pond. The pond level determines the crest level and the height of the shutters to be provided.

(*viii*) Waterway: It is the width provided at the site for the river water to flow. In other words, it is the length of the weir. Approximate waterway to be provided between the abutments may be calculated from Lacey's regime perimeter formula:

$$P_w = 4.825\,Q^{\frac{1}{2}}$$

P_w is the wetted perimeter, but in this case it denotes the length of the weir between the abutments in metres.

Q is the design flood discharge in m^3/sec.

To avoid the risk of overtopping of the training works the weir length is generally increased.

When possible the weir is constructed stretching full width of the river. The ratio of actual width of the waterway provided and the P_w calculated from Lacey formula is called *looseness factor*. A looseness factor vary from 1.1 to 1.6 depending upon maximum flood discharge observed in the river and the discharge intensity per metre width of the weir.

(*ix*) Crest level: Determination of the crest level of the weir is closely related to afflux, waterway or intensity of discharge and the pond level. It is determined by the permissible maximum value of afflux attained during the maximum flood. Crest level, waterway, pond level and afflux values are generally determined in coordination with each other as is permissible from the considerations already mentioned. For example, having decided waterway L, discharge intensity q, amount of afflux A, crest level can be worked out as below. (Refer Fig. 19.1)

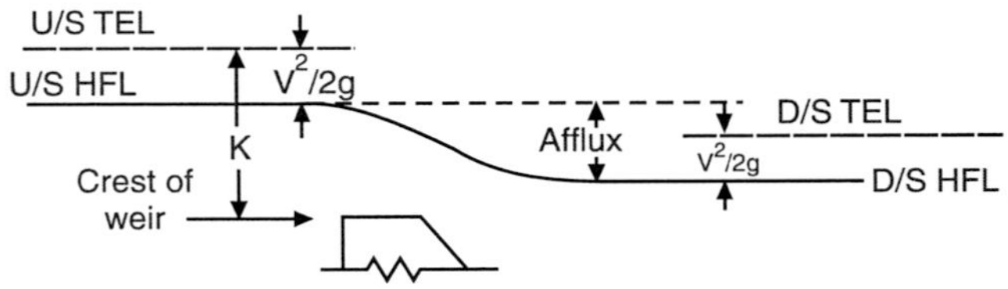

Fig. 19.1. Determination of crest level

Under the flood conditions when the river is expected to attain regime conditions:

Velocity of approach $V = \dfrac{q}{R_1}$

Where R_1 is scour depth

Also, velocity head $= \dfrac{V^2}{2g}$

Discharge over the crest of the river is given by the formula

$$q = 1.7\,(K)^{3/2}$$

where K is height of total energy line (TEL) above the crest. It is equal to static head + velocity head over the crest.

Therefore once the elevation of TEL above crest of weir is known crest is given by (TEL – K).

TEL above crest can be worked out as follows:

Level of TEL on the D/S of the weir = HFL (pre weir) + $V^2/2g$

$$\left.\begin{array}{l}\text{Level of TEL on the U/S}\\ \text{of the weir or over the crest}\end{array}\right\} = \text{D/S TEL + Afflux}$$

(*x*) **Scour Depth:** Scour depth in the river during high floods can be calculated by using Lacey's scour depth formula:

$$R_1 = 1.35\left(\frac{q^2}{f}\right)^{\frac{1}{2}}$$

Using this value R_1, velocity of approach can be calculated from relation:

$$V = \frac{q}{R_1}$$

19.2. DESIGN OF WEIR ON SOLID FOUNDATION

When the weir is to be constructed on a good foundation, following design criteria is found to be quite sufficient:

Dimensions of the body wall and the crest of weir:

- It is designed as retaining wall to store the water against it. Obviously main two forces are water pressure and self-weight of the weir. Thus the design will be similar in principle to that of a solid gravity dam.
- The resultant of the water pressure (*P*) and self-weight (*W*) should remain within the middle-third of the weir base.
- There should be sufficient width at the top for working of the shutters.

According to Bligh dimensions of body wall are

$$B = \frac{H+d}{\sqrt{\rho}}$$

and $$a = 0.552 \times \left(\sqrt{H} + \sqrt{d}\right)$$

From Fig. 19.2 it is clear that *B* is the base width, *a* is top width and *H* is the height of weir above its floor level.

'*d*' is the height of the flood water above the weir crest or height of the shutter, whichever is more.

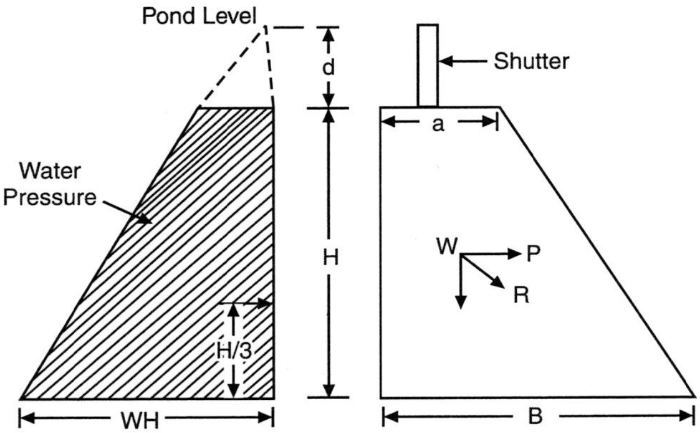

Fig. 19.2. Dimensions of a body wall of the weir

Upstream curtain wall is generally kept 2 to 2.5 m deep whereas downstream curtain wall should be deeper say 2.5 to 3 metres.

Upstream apron length according to Bligh is $2.208\sqrt{CH}$.

where C is a creep coefficient and H is the head of water against the weir. The thickness is generally 0.3 m over 0.3 to 0.5 m foundation concrete. The only requirement of upstream apron is, it should be watertight.

Downstream apron length according to Bligh is $2.201\,C\sqrt{H/13}$. The thickness of the downstream apron should be sufficient to withstand uplift pressure below it. It may be calculated from the formula

$$t = 4/3\,.\,\frac{H-h}{\rho-1}$$

where t is the thickness of apron in metres. $(H - h)$ is the residual uplift head at the point under consideration. Obviously H is the total head causing seepage and h is the head lost by creeping water in its path upto the point under consideration.

ρ is specific gravity of the material used in the apron floor.

General practice of providing dimensions of body wall under-sluices, fish ladder are already given earlier. Most of the big weirs are, however, constructed on permeable foundations and required special considerations.

19.3. CAUSES OF FAILURE OF WEIRS

The criteria of considering only water pressure and self-weight in the design calculations was found to be not sufficient for the weirs to be founded on the permeable foundations. Many weirs designed and constructed on the pervious foundation on the basis of retaining wall design criteria failed or damaged in course of time. The problem was analysed by many research workers. The main causes of failure were found out to be the following:

1. **Piping or undermining:** Undermining or piping takes place due to excessive percolation of water below the foundation. The water percolates through the foundation under pressure and when it emerges out at the downstream end it may retain sufficient force to dislodge and to lift the soil particles. Thus progressively the foundation becomes weaker and the structure above finally subsides in the hollow formed in the foundation.
2. **Failure due to uplift:** When the percolation water exerts excessive upward pressure and when the apron is not sufficiently strong, it may fail by rupture of its part. Uplift pressure is not important consideration for the upstream apron. The obvious reason is that the downward weight of stored water on the upstream apron is sufficient to neutralize the effect of the uplift presure. Naturally the thickness of the upstream apron is minimum. At the same time it should be watertight.

19.4. BLIGH'S CREEP THEORY FOR DESIGN OF WEIR ON PREMEABLE FOUNDATION

Bligh assumed that the water which percolates into the foundation creeps through the joint between the profile of the base of weir and the subsoil. Of course water also percolates into the subsoil. He then stated that this percolating water loses its head enroute. The seeping water finally comes out at the downstream end. According to Bligh water travels along vertical, horizontal or inclined path without making any distinction.

The total length covered by the percolating water till it emerges out at the downstream end is called a creep length. It is clear from the knowledge of hydraulics that the head of water lost in the path of percolation is the differene of water levels on the upstream and the downstream ends. Also, an imaginary line which joins the water levels on the upstream and the downstream end is called a hydraulic gradient line. Figure 19.3 (a, b) gives the full explanation of Bligh's theory.

In Fig. 19.3 (a) arrows show the path followed by the creeping water.

$B = L$ = total creep length and h/L is the head lost in creeping.

Loss of head per unit creep length will be h/L and it is hydraulic gradient.

To increase the path of percolation vertical cutoffs or sheet piles can be provided, Fig. 19.3 (b).

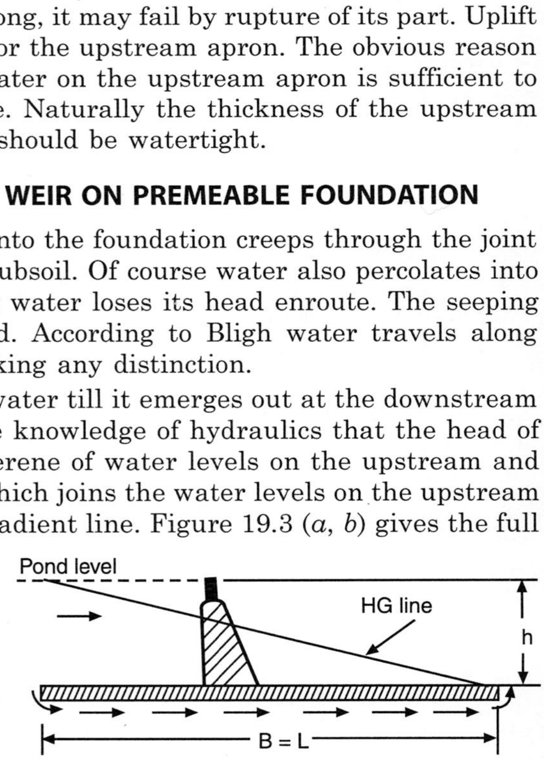

Fig. 19.3. (a) Bligh's creep theory

Bligh took vertical and horizontal path of percolation in the same sense. So now

Total creep length,

$$L = B + 2d_1 + 2d_2 + 2d_3$$

$$\therefore \quad HG = h/L$$

$$\text{or} \quad = \frac{h}{B + 2d_1 + 2d_2 + 2d_3}$$

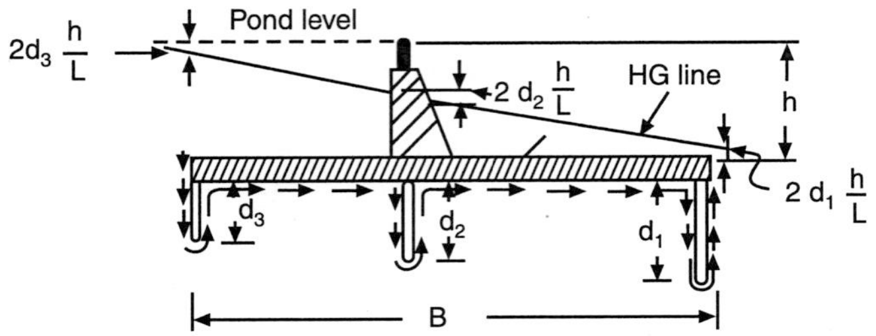

Fig. 19.3. (b) Bligh's creep theory

When the water follows a vertical path the loss takes place in a vertical plane at same section. This loss is proportional to the length of the vertical path. For example, for cutoff d_1, loss will be $\frac{h}{L} \times 2d_1$ and it takes place in its plane. Loss of head at other cutoffs may be calculated in the same way.

The reasons for failures are already mentioned. Bligh gave the criteria for the safety of a weir against piping and uplift separately and is as follows:

The structre is safe against piping when the percolating water retains negligible upward pressure when it emerges out at the downstream end of the weir. Obviously the path of percolation should be sufficiently long to provide safe hydraulic gradient. It depends on the soil type.

This condition is provided by equation

$$L = CH$$

where L is creep length or path of percolation;
C is Bligh's creep co-efficient for soil; and
H is head of water against the weir.

Table 19.1 gives values of C for various soil types.

Table 19.1.

Soil Type	*Value of C*
Silt and sandy clay	18
Micaceous sand (North Indian river)	15
Coarse grained sand	12
Sand mixed with boulder, gravel and shingle	5 to 9
Loamy soil	5 to 9
Gravel	5

Obviously for no piping

$$\frac{H}{L} \le \frac{1}{C} \quad \text{or}$$

$$\text{Hydraulic gradient} \le \frac{1}{C}$$

To make the apron floor safe against the uplift pressure Bligh gives following criteria:

From Fig. 19.4 it is clear that the uplift pressure at any point is represented by the ordinate between the bottom of apron floor and the hydraulic gradient line.

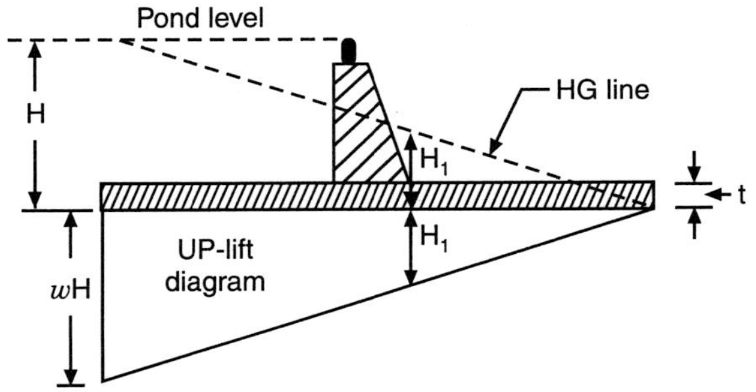

Fig. 19.4. Bligh's creep theory

Hence, uplift pressure $= wH_1$

where w is the density of water and H_1 is the ordinate between HG line and bottom of the apron floor.

Downward force exerted by the material in the apron is given by

$$t \cdot w \cdot \rho$$

where t is the thickness of apron floor and ρ is the specific gravity of material used in the apron.

The stable condition is attained when

$$wH_1 = t \cdot w \cdot \rho$$

or $$t = \frac{H_1}{\rho} \qquad ...(1)$$

It is clear from Fig. 19.4 that H_1 can only be known when t is known. Hence to determine 't' following algebrical manipulation may be done. From equation (1)

$$H_1 = t \cdot \rho$$

$$(H_1 - t) = t \cdot \rho - t = (\rho - 1) \cdot t$$

or $$\frac{(H_1 - t)}{\rho - 1} = t \qquad ...(2)$$

Where $(H_1 - t)$ is the ordinate between dotted HG line and top of the apron. It can be known easily and hence depth of apron can be calculated from equation (2). Now adding factor of safety of 4/3 to equation (2), the expression finally becomes

$$t = \frac{4}{3}\left(\frac{(H_1 - t)}{\rho - 1}\right) \qquad ...(3)$$

For economy provide greater apron length on the upstream side which requires minimum practical thickness. Of course on the downstream side some minimum length of apron is required to protect the bed of the channel.

19.5. LIMITATIONS OF BLIGH'S THEORY

Bligh's theory has sevaral limitations. They are:

- In his theory Bligh made no distinction between horizontal and vertical creep lengths.
- The idea of exit gradient has not been considered.
- The effect of varying lengths of sheet piles not considered.
- No distinction is made between inner or outer faces of the sheet piles.
- Loss of head is considered proportional to the creep length which in actual is not so.
- The uplift pressure distribution is not linear as assumed but in fact it follows a sine curve.
- Necessity of providing end sheet pile not appreciated.

19.6. THEORY OF FLOWNETS OR THEORY OF SEEPAGE FLOW THROUGH SOILS

Weirs designed and constructed on the basis of Bligh's theory also failed due to undermining of the subsoil. As a result it was thought essential to study the problem of weirs on the permeable foundations more elaborately. Theory of flownets provides a remarkable solution to the problem.

Briefly the theory is as follows:

It is already explained that the flow through soils is governed mostly by Darcy's law. It states that

$V = K \cdot \dfrac{H}{L}$; where V is the velocity of flow through the soil.

This can also be expressed as

$V = K \cdot \dfrac{\partial H}{\partial L}$; where ∂H is the head loss in the subsoil over the elementary length ∂L. If we consider three dimensional flow, *i.e.,* the flow is not only in one direction but in all the three directions the easiest way of studying the problem is to resolve the velocity into three components. Let u, v, w denote the velocities parallel to X, Y, Z axis respectively.

Now the Darcy law can be written as

$$u = K\,\frac{\partial \text{H}}{\partial X};\; v = K\,\frac{\partial \text{H}}{\partial Y};\; w = K\frac{\partial \text{H}}{\partial Z} \qquad ...(1)$$

Flow of water through the soil is incompressible and obeys continuity equation. That is

$$\frac{\partial u}{\partial X} + \frac{\partial v}{\partial Y} + \frac{\partial w}{\partial Z} = 0$$

Substituting values of u, v and w from (1)

$$\frac{\partial^2 K \cdot H}{\partial X^2} + \frac{\partial^2 K \cdot H}{\partial Y^2} + \frac{\partial^2 K \cdot H}{\partial Z^2} = 0 \qquad ...(2)$$

Equation (2) is nothing but Laplace's equation

Thus, Laplace's equation also holds good for flow of water through soils. For simplicity we can take the two dimensional flow in a vertical plane. Then the Laplace's equation is

$$\frac{\partial^2 K \cdot H}{\partial X^2} + \frac{\partial^2 K \cdot H}{\partial Z^2} = 0$$

If we represent $K \cdot H$ by ϕ then

$$\frac{\partial^2 \phi}{\partial X^2} + \frac{\partial^2 \phi}{\partial Z^2} = 0$$

where ϕ is called flow potential.

The equation represents two sets of curves. They intersect each other orthogonally. One set of curves are called stream lines. They indicate the path followed by the seepage water. Other set of curves are called equipotential lines. They are the lines which join points of equal potential. The flownet can be constructed conveniently for most of the hydraulic structures graphically by trial and error method. The method consists of following steps:

(*a*) draw a cross-section through a pervious stratum and a hydraulic structure;

(*b*) make first trial to construct a flownet;

(*c*) make second trial adjustment of constructed flownet.

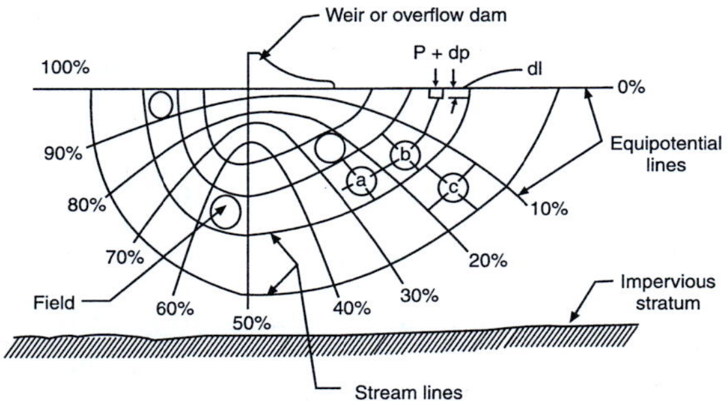

Fig. 19.5. Flownet for a weir on previous soil

If required more trials may be taken to draw the flownet finally.

The procedure can be vividly understood with the help of Fig. 19.5. Upstream ground surface represents one equipotential line since all the points of the surface are under the same head. Similarly downstream ground surface represents another equipotential line since all the points are under same head. Let the head of water stored by the dam is H. Then upstream ground surface represents equipotential line with 100% head. Total head is lost by the time it reaches downstream end. Naturally downstream ground surface represents equipotential line with zero head. The base of the dam and the side of the cutoff pile represents first stream line or flow line. It is the first joint through which water seeps as correctly pointed out in Bligh's theory.

In case there exists an impervious stratum in the foundation it obviously represents last stream line. Thus, simply by drawing a cross-section of the hydraulic structure the shape of the extreme stream lines and equipotential lines is ascertained.

Now all the intermediate streamlines and equipotential lines can be drawn by trial and error method graphically with the help of following properties of the curves:

- The shape of the successive flow lines represent gradual transition from one to the other.
- The flow lines and the equipotential lines must intersect each other at right angles.
- The flow lines must start and end at right angles to the ground surface upstream and downstream respectively.
- If an impervious stratum does not exist the flow line gradually adopts semi-elliptical shape.
- The equipotential lines must start and end at right angles to the first and last flow lines respectively.
- Each square obtained by intersection of flow lines and equipotential lines is called a field.
- If the curves are drawn properly a circle can be drawn in each field which touches all the four sides of the field.

The flownet may be constructed with innumerable curves. However, for practical purposes the flownet should consist of a limited number of curves as shown in Fig. 19.5. Each field is perfectly an elementary square.

The amount of seepage can be calculated making use of the flownet. Referring Fig. 19.5:

Let us consider three elementary fields with average dimensions a, b and c.

Let head lost in field a is ΔH_1 and head lost in field b is ΔH_2.

Discharge passing through the same flow channel is always the same. Let it be Δq_1 per unit length of the dam.

Then according to Darcy's law

$$\Delta q_1 = K \cdot \frac{\Delta H_1}{a} \times a = K \cdot \frac{\Delta H_2}{b} \times b$$

or $$\Delta H_1 = \Delta H_2 \quad \text{...(1)}$$

Now considering the field b and c between the two equipotential lines.

Let the discharge through the flow channel of field b be Δq_1 and that through the flow channel of field c be Δq_2.

Using Darcy's law

$$\Delta q_1 = K \cdot \frac{\Delta H_2}{b} \cdot b = K \cdot \Delta H_2$$

and $$\Delta q_2 = K \cdot \frac{\Delta H_2}{c} \cdot c = K \cdot \Delta H_2$$

$\therefore$ $$\Delta q_1 = \Delta q_2 \qquad \ldots(2)$$

From the derivations (1) and (2) two inferences can be drawn for a flownet with elementary square fields.

(*i*) The potential drop interval between seuccessive equipotential lines is the same. Thus if the total seepage head is H and if there are 'N_p' potential drops then the potential drop interval is constant and is equal to $\frac{H}{N_p} = \Delta H$.

(*ii*) The seepage discharge through all the flow channels is the same.

$\therefore$ Total seepage discharge $q = \Sigma K \cdot \frac{H}{N_p}$

or $$= K \cdot H \frac{N_f}{N_p}$$

where N_f is total number of flow channels.

Now total seepage discharge below the dam $(Q) = K \cdot H \frac{N_f}{N_p} \cdot L$

where L is the length of the dam.

19.7. CRITERIA FOR SAFETY AGAINST PIPING

It is already mentioned that the piping starts from the downstream end. Naturally if we consider the case at d/s end and find it stable there will be no piping at all. To test the stability let us take an elementary cylinder at the downstream end. Its volume is $dA \cdot dl$. The intensity of pressure on two faces being p and $p + dp$.

Total force in the direction of flow is $= p \cdot dA - (dp + p)\, dA = -\, dp \cdot dA$

$\therefore$ Force per unit volume $= \frac{-dp \cdot dA}{dA \cdot dl} = \frac{-dp}{dl}$

The negative sign indicates that the pressure decreases in the direction of flow.

At the downstream end the flow line is vertical and hence the force is also acting vertically upwards. It tries to lift the soil particles. The stabilizing force is the self-weight of the soil particles. Assuming that the soil particles are submerged under water, the weight of soil per unit volume is given by the formula

$$\gamma = w\,(1 - n)\,(G - 1)$$

where w is density of water

n is percentage of pore spaces

G is sp. gravity of soil particles

The soil particles are stable

When $$\gamma > -\frac{dp}{dl}$$

When $\gamma = -\frac{dp}{dl}$ the condition is called critical. It can be rewritten as

$$w(1 - n)(G - 1) = -\frac{\partial p}{\partial l}$$

dividing both sides by w

$$(1 - n)(G - 1) = -\frac{dH}{\partial l} \quad \text{since } p = wH$$

$\frac{\partial H}{\partial l}$ is nothing but hydraulic gradient.

When it is equal to $(1 - n)(G - 1)$ it is called critical exit gradient. Thus criteria for safety against piping is that the value of the exit gradient should be less than $(1 - n)(G - 1)$.

For alluvial soil considering $n = 40\%$ and $G = 2.65$ and $w = 1$

$$\frac{\partial H}{\partial l} \text{ critical} = 1\,(1 - 0.4)\,(2.65 - 1)$$

$$\cong \text{unity}$$

Permissible hydraulic gradient at exit, however, cannot be allowed to reach unity value because scour, wave action, sudden increase or reduction of head, high spring level etc., in practice increase exit gradient. The exit gradient increases rapidly as the cavity (depth exposed) increases. It is, therefore, necessary to provide sufficient factor of safety. Following factor of safety to critical values of exit gradient are recommended:

Fine sand	.....	6 to 7
Coarse sand	.....	5 to 6
Shingle	.....	4 to 5

19.8. DR. A.N. KHOSLA'S THEORY

Weirs designed and constructed on Bligh's theory also failed due to undermining of subsoil. Dr. Khosla investigated the problem and drawn following conclusions:

1. The outer faces of the end sheet piles were much more effective than the inner ones and the horizontal length of floor.
2. The intermediate sheet piles if smaller in length than outer ones were ineffective except for local redistribution of pressures.
3. Undermining of the floors started from the tail end. If the hydraulic gradient (HG) at exit was more than the critical gradient* for the particular subsoil the soil particles would move with the flow of water thus causing progressive degradation of the subsoil, resulting in cavities and ultimate failure.
4. It was absolutely essential to have a reasonably deep vertical cutoff at the downstream end to prevent undermining.

Thus, it was recognized that there is an urgent necessity for research on the problem. Research was done on the prototypes by inserting pressure pipes at suitable locations. A continuous and detailed record of the observations of pressures in pipes was maintained.

As a result following facts were established:

(*a*) The flow of water through the subsoil is in stream lines and therefore susceptible of mathematical treatment.

***Critical exit gradient:** The hydraulic gradient at the downstream end where water emerges out is called exit gradient. When the vertical pressure due to particular exit gradient is such that it equals submerged weight of soil particles the exit gradient is called a critical exit gradient. For safety exit gradient should always be less than this critical value. The critical gradient varies from 1/4 to 1/7 for different formations.

(*b*) The ratio (ϕ) of uplift pressure (P) at any point along the base of a particular weir founded on permeable soil to the total head (H) is constant. It is independent of:

- Head
- Nature of subsoil as long as it is homogeneous
- Upstream and downstream water level
- Temperature, provided it is uniform throughout the subsoil.

However, ratio (ϕ) varies with:

- Silt deposit or scour upstream or downstream of impervious floor.
- Temperature which varies from point-to-point in the subsoil and in different seasons of the year

(*c*) Law of loss under the floor was nearly a straight line law and for the sheet piles something like logarithmic.

Dr. Khosla broke the complicated weir profile into number of simple and common profiles. He then studied each profile independently for finding out uplift pressures.

He thus made it clear that the loss of head does not take place uniformly in proportion to the length of creep. It actually depends on the profile of the base of the weir. Secondly he also established that the safety against undermining is not obtained by flat hydraulic gradient but the exit gradient should be kept below critical value.

19.9. DESIGN PRINCIPLES FOR WEIRS ON PERMEABLE FOUNDATION

It is already mentioned that the two main factors to be considered in design of weirs on permeable foundation are **uplift pressure** and **exit gradient**. It is also shown that the sheet piles alter the pressure distribution under the floor.

(*i*) To counteract uplift pressure it is necessary to provide suitable thickness of the floor at different points. The criteria for determining thickness is already given in article 19.4.

(*ii*) The exit gradient can be kept within permissible limit by providing suitable depth of the sheet pile at downstream end of the floor.

(*iii*) Since the scour holes can occur on the upstream or the downstream side of the weir sheet piles are essentially required at u/s as well as d/s end of the floor to prevent failure of the weir by slipping of the subsoil into the scour holes by simple earth pressures. In other words u/s and d/s sheet piles prevent failure of weir due to undermining of the foundation.

(*iv*) The pressures under the downstream floor increase as the depth of the d/s sheet pile increases. The u/s pile line has little effect in reducing these pressures as the spacing between the two is generally more.

(*v*) The intermediate sheet pile neither prevents undermining nor it alters pressure distribution substantially to reduce uplift pressure under d/s floor. The main purpose of providing intermediate sheet pile is to create second line of defence in case of failure of either u/s or d/s sheet piles.

(*vi*) Generally the sheet piles are taken down upto the depth of normal flood scour. The depth of scour (R) is given by Lacey formula. It is measured below maximum flood level.

(*vii*) The depth of d/s sheet pile is determined in such a way that in addition to its depth reaching normal flood scour depth for that section of the work it gives safe exit gradient in conjunction with a suitable length of the floor.

(*viii*) Commonly adopted practice in respect of intermediate pile is to provide a single and deep intermediate sheet pile under the crest of the weir. It is, however, not obligatory. More than one intermediate sheet piles can also be provided say one u/s of the crest and the other d/s of the crest under the toes of the glacis.

(*ix*) The level of the d/s floor is so fixed that the hydraulic jump even under worst conditions occur on the sloping glacis itself. The depth of sheet pile will obviously be the depth below the top of the d/s floor to the bottom of the normal scour hole given by Lacey formula.

(*x*) The slopes between 1 in 3 to 1 in 5 for the glacis are considered to be suitable for achieving economy as well as maximum dissipation of energy.

(*xi*) The disturbance of a standing wave dies out at a distance of about 5 times height of the jump. Therefore the length of the downstream horizontal impervious floor should be at least 5 times the height of the hydraulic jump.

(*xii*) The thickness of the floor calculated to counteract uplift pressures is adequate only when the entire thickness of the impervious floor is considered as one solid mass. If the floor is laid in layers and if they separate full pressure can be transmitted through cracks. It is obvious that the floor will not be safe in such a situation. Hence, it is essential that the floor is laid as one solid mass in a single layer. Alternatively steel reinforcement can be used to tie the layers of the floor to make them inseparable.

(*xiii*) The abutments and the wing walls, as usual for any hydraulic structure, are designed considering pressure due to dry soil, pressure due to saturated soil and hydraulic pressure acting in conjunction with saturated soil pressure.

(*xiv*) The foundations of abutments and flank walls between any pair of pile lines should go down to the level of the bottom of those pile lines. Similarly, the foundations of the u/s and d/s return walls of the flank should go down to the levels of the bottom of corresponding pile line. Also the return walls should be taken back into the bank equal to twice the depth of the corresponding sheet pile below the floor level.

(*xv*) At the upstream end of the impervious floor further protection by concrete blocks over loose stone for a length equal to the depth of scour below the bed of the river is generally given. It is generally 1.2 m thick and concrete blocks over 60 cm deep graded stones.

(*xvi*) On the downstream, after the impervious floor an inverted filter varying in length from 1.5 to 2 times the depth of scour below the river bed is provided. The inverted filter is generally 60 cm thick and is to be protected by 1 m to 1.2 m deep concrete blocks.

(*xvii*) The upstream and downstream bed protection is flexible and it tries to adjust to slight subsidence. The protection is, however, meant to be immovable and not intended to cover the scoured face in the same way as is done by launching aprons. This loose protection, therefore, needs constant maintenance.

(*xviii*) Upstream and downstream of the weir/barrage provision of guide banks are essential to direct the flow through the waterway. The arrangement of guide bunds and launching aprons is already dealt earlier.

19.10. DESIGN PROCEDURE FOR WEIRS ON PERMEABLE FOUNDATION

The design procedure of the weirs on the permeable foundations can be summarised as below. It is already described how to decide afflux, pond level, crest level, discharge per foot run, length of waterway etc., from the observed data on maximum flood levels etc. On the basis of the above mentioned principles barrage parameters can be calculated. They are:

- Total length of impervious floor;
- Depth and location of sheet piles;
- Exit gradient at the end of impervious floor;

- Depth of d/s end sheet pile in relation to exit gradient and flood scour;
- Level and length of horizontal part of the d/s impervious floor;
- Thickness of the d/s impervious floor;
- Arrangement of flanks;
- Length and thickness of u/s block protection;
- Length and thickness of d/s inverted filter; and
- Arrangement of guide bunds.

19.11. DESIGN PRINCIPLES FOR HEAD REGULATORS

The head regulator is constructed at the head of a main canal which takes off from the river pond created by the barrage or the weir. It is, therefore, subjected to similar pressures as those existing in the river upstream of the barrage. The main design principles are enumerated below.

(*i*) The crest level of the head regulator should be fixed higher than the crest of the undersluices of the barrage with or without the silt excluder by 1.25 to 2 m to avoid silt entry into the canal.

(*ii*) The crest level and the waterway required by the head regulator are interrelated because the discharge to be passed down into the canal and the pond level of the barrage are already fixed.

(*iii*) Generally for the head regulator broad crest with sloping downstream glacis are provided. So the waterway required can be calculated by corresponding hydraulic formula. The discharge formula applicable will be

$$Q = 1.7(L\text{-}knH)H^{3/2}$$

where Q = design discharge of the canal in cumec

L = length of waterway in m

H = head causing flow in m

k = a constant which varies from 0.01 to 0.03. It depends on the shape of pier nose or cut water.

n = number of end contractions.

(*iv*) The regulation of discharge is done by providing gates. The gates can be provided singly or in two tiers, one rising and other falling. The gates are fitted in the grooves made in the pier body walls.

(*v*) A downstream sloping glacis is provided upto such level that the hydraulic jump occurs on the slope itself under different discharge conditions in similar manner as for sloping glacis of a barrage or a weir.

(*vi*) The horizontal impervious floor or the cistern beyond the d/s sloping glacis is provided at least for a length, 5 times the maximum height of the jump, *i.e.*, $5\ (D_2 - D_1)$.

(*vii*) The canal may remain closed during highest flood. This constitutes worst condition. For such a condition the stability and safety of glacis and floor has to be checked against uplift pressures.

(*viii*) A reinforced concrete mat may be adopted to resist uplift pressure to reduce excessive floor thickness otherwise required. Under such conditions if found necessary the piers may be extended beyond the sloping glacis on to the cistern floor to stabilise the mat against bending.

(*ix*) As uplift consideration necessitates sheet piles or concrete cut-offs may be provided below the regulator floor and the glacis and floor checked for safety against uplift and piping. It is necessary to determine exit gradient and to see that it is within safe limits.

(*x*) Since the water level on the upstream rises above the pond level upto high flood level a suitable RCC breast wall of sufficient height and strength stretching full waterway of the regulator is required to prevent spilling of flood flows.

(*xi*) A bridge over the head regulator should be provided to house working platform for operation of gates.

(*xii*) As provided for the weir RCC block protection (talus) over stone pitching and inverted filter should be provided beyond the upstream and downstream ends respectively of the regulator impervious floor to prevent scouring. Lenth of d/s talus may vary from 4 to 5 times FSD of the canal.

(*xiii*) The regulator piers should be designed adopting usual design criteria and should be stable against maximum overturning moment caused by flood flows.

(*xiv*) The length of the waterway together with width of piers and abutments may be at variance to the bed width of the canal taking off. It is neccessary to provide suitable wing walls to smoothly attain full canal section on the downstream.

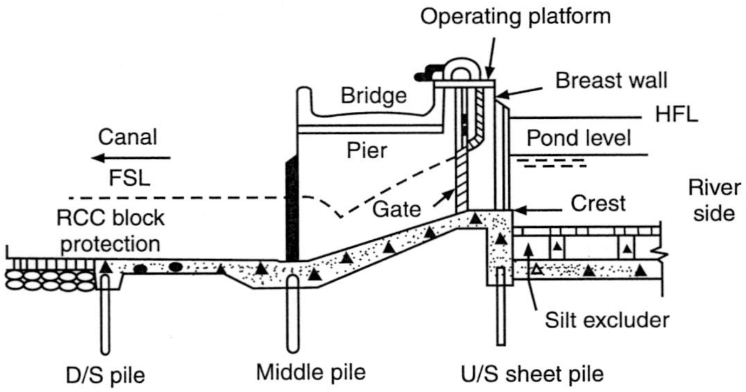

Fig. 19.6. Typical section of a head regulator

Figure 19.6 shows a typical section of a head regulator.

19.12. DESIGN PRINCIPLES FOR CROSS REGULATORS AND DISTRIBUTARY HEAD REGULATORS

Unlike the head regulator, cross regulator and distributary head regulator are constructed across the canals which carry predetermined discharges on the upstream as well as dowstream side. Also the site conditions are moderate and the structures are not subject to unprecedented forces. The design criteria for these structures are, therefore, not so rigorous. General principles of design are mentioned below.

(*i*) Crest level of the cross regulator should generally be kept in level with the upstream bed level of the canal. Some times from economic considerations the canal section is flumed at the regulator site to reduce width of the concrete floor. In such cases the crest may be raised little above the bed level. A sloping glacis and a cistern in such a situation is necessary on the downstream side of the crest.

(*ii*) Crest level of a distributary head regulator is generally kept 0.3 to 1.0 m higher than the bed level of the parent channel (or crest of cross regulator). A sloping glacis with 1 in 2 slope is provided on the downstream to connect the downstream floor.

(*iii*) The waterway required in case of cross regulators and distributary head regulators is obtained from a hydraulic formula used for determining the discharge over a submerged or a drowned weir because water level on either side of the crest is above the crest level. Since the velocity of approach is generally small, neglecting the head due to velocity of approach the formula applicable will be

$$Q = \frac{2}{3} Cd_1 \cdot L \cdot \sqrt{2g}\, (H_1 - H_2)^{3/2} + Cd_2 \cdot L\, H_2 \sqrt{2g\, (H_1 - H_2)}$$

or $$= \frac{2}{3} Cd_1 \cdot L \cdot \sqrt{2g}\, h^{3/2} + Cd_2 \cdot L \cdot H_2 \sqrt{2gh}$$

In the above equation first part gives the discharge through the free portion and the second part gives discharge through submerged portion which is similar to discharge through submerged orifice.

In the above equation

Q = discharge over crest in cumec

L = length of waterway in m

H_1 = depth of water over crest in the u/s in m

H_2 = depth of water over crest in the d/s in m

h = difference in water level ($H_1 - H_2$) causing the flow in m (It is also = u/s FSL – d/s FSL)

Cd_1 = 0.577 and is a constant, and

Cd_2 = 0.80 and is another constant

(*iv*) Regulation of discharge is done by providing gates. Depending upon the height of shutters and convenience of working they may be provided in one length or in two tiers.

(*v*) A downstream sloping glacis where required can be given 2:1 (Horizontal: Vertical) slope. In this case also glacis is provided upto such level that under extreme condition the hydraulic jump occurs before the toe of slope. This is the level at which d/s horizontal impervious floor is laid.

This level is given by the relation: downstream floor level = d/s TEL — E_{f2}

where TEL = total energy line on downstream

$\cong$ d/s FSL $\left(\text{Since } \frac{V^2}{2g} \text{ is quite small}\right)$

and E_{f2} = energy of flow after the jump

(the hydraulic jump calculations are dealt in detail in the following article)

Where the above relation gives value of downstream floor level more than the bed level of the canal, the d/s floor level may be taken to be equal to the bed level there.

(*vi*) The impervious horizontal floor or cistern length beyond d/s sloping glacis may be taken to be 5 times the height of the jump *i.e.*, 5 (D_2–D_1). It however, should not be less than 60% of the total impervious floor length.

(*vii*) Since uplift pressures and piping are important considerations in these structures also suitable cut-offs on the upstream and downstream should be provided.

As a rule of thumb following depths for cut-off may be sufficient:

$$\text{u/s cut-off depth} = \left(\frac{\text{depth of flow on u/s}}{3} + 0.6\right)\text{m}$$

$$\text{d/s cut-off depth} = \left(\frac{\text{depth of flow on } d/s}{2} + 0.6\right)\text{m}$$

(*viii*) Knowing the depth of downstream cut-off total length of impervious floor can be obtained from the knowledge of the safe hydraulic gradient and the head causing seepage as follows:

$$G_E = \frac{H}{d} \frac{1}{\pi \sqrt{\lambda}}$$

Knowing values of safe hydraulic gradient G_E, maximum static head causing seepage flow H, and depth of d/s cut-off d value of $\frac{1}{\pi\sqrt{\lambda}}$ can be calculated. Using Khosla's curve for hydraulic gradient value of $\alpha\left(=\frac{b}{d}\right)$ for known value of $\frac{1}{\pi\sqrt{\lambda}}$ can be read.

Thus, total length of impervious floor can be calculated.

(*ix*) The thickness of impervious floor should be sufficient to withstand uplift pressures. From considerations of construction facility a minimum thickness of the floor ranging from 0.3 to 0.5 m is, however, considered essential.

(*x*) When the regulator is closed the water pressure on the u/s face acts on the piers. The piers should be strong enough to withstand this pressure.

(*xi*) On the piers from the considerations of economy as well as working facility a bridge should be provided to allow working space for operating gates and to allow passage of traffic.

(*xii*) Beyond the impervious floor on the u/s and d/s, c.c. block protection over stone pitching and inverted filter should be provided to prevent scouring of bed. It should be maintained by frequent repairs to its shape. The length of u/s and d/s protection may be taken equal to the depth of corresponding cut-offs which are supposed to reach bottom of the scour hole. A thickness of 1 m is considered sufficient in most cases for this flexible protection. Figure 19.7 shows a typical section of a distributary head regulator.

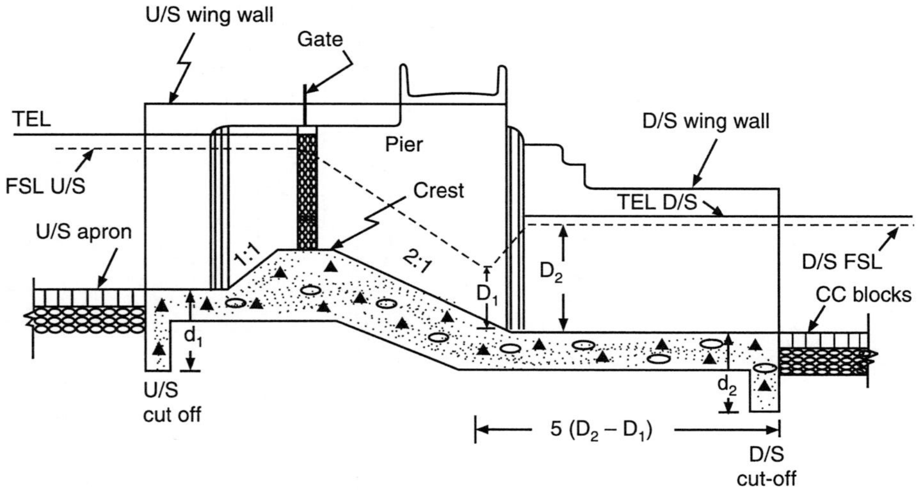

Fig. 19.7. Typical section of a distributary head regulator

19.13. PRINCIPLE OF HYDRAULIC JUMP AND ITS USE IN DESIGN OF IMPERVIOUS FLOOR

In weirs, regulators and other hydraulic structures over or through which the flow passes down, energy dissipation is important consideration. It calls for suitable design of downstream works like sloping glacis, horizontal floor or cistern and other energy dissipators. Design of these works involves determination of the elevation of the horizontal floor, and length of impervious floor or cistern. These dimensions can be worked out from the knowledge of the hydraulic jump elements like pre and post jump energy of flow, depth of flow, and critical water depth for the given intensity of discharge and energy to be dissipated or head loss in hydraulic jump.

Under appropriate conditions when a shallow stream moving with high or hypercritical velocity meets a slowly moving flow of sufficient depth an abrupt rise in the water surface takes places. This abrupt rise is called hydraulic jump. In other words hydraulic jump in an open channel is an abrupt transition from the water depth $D_1 < D_c$ to $D_2 > D_c$. The jump elements can be calculated knowing H_L and q from the following formula. Refer Fig. 19.8.

$$H_L = \frac{(D_2 - D_1)^3}{4\, D_1\, D_2}$$

$$D_2 = \frac{D_1}{2} + \sqrt{\frac{2q^2}{gD_1} + \frac{D_1^2}{4}}$$

$$D_C = \left(\frac{q^2}{g}\right)^{1/3}$$

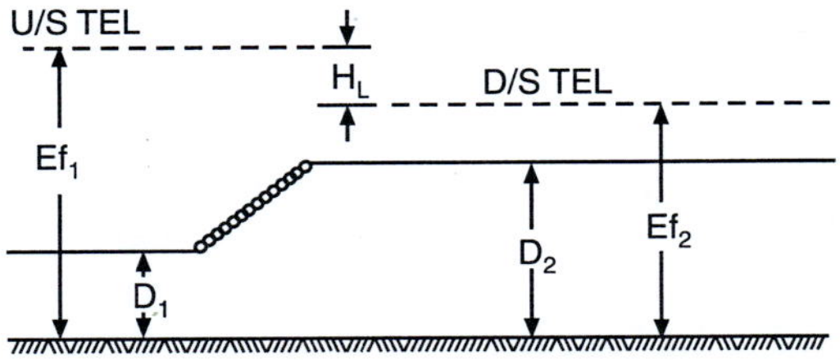

Fig. 19.8. Elements of hydraulic jump

where D_1 = pre-jump depth

D_2 = post-jump depth (conjugate depth)

Ef_1 = total energy of flow at pre-jump section

Ef_2 = total energy of flow at post-jump section

H_L = Head loss in hydraulic jump, or = energy to be dissipated

= $Ef_1 - Ef_2 - hf$

(h_f is usually neglected)

q = intensity of discharge

g = acceleration due to gravity

D_c = critical water depth

With the known values of q and H_L it is rather cumbersome to find out D_1, D_2, Ef_1, Ef_2 from the above equations. The help of curves can be taken to facilitate calculations. Blench has prepared curves to give Ef_2 for different values of H_L and q, it is given in Fig. 19.9.

To find out values of D_1 and D_2 IS 4997 gives curves in terms of dimensionless parameters like $\frac{H_L}{D_c}$; $\frac{D_2}{D_1}$ and $\frac{D_1}{D_c}$. Thus once D_c is calculated from the formula D_1 can be

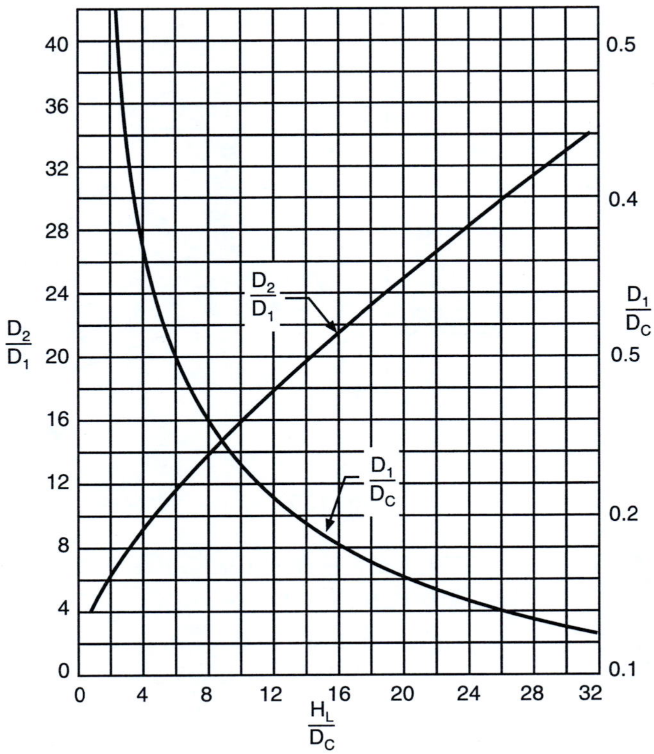

Fig. 19.10. Curve for conjugate depths (Horizontal Apron)

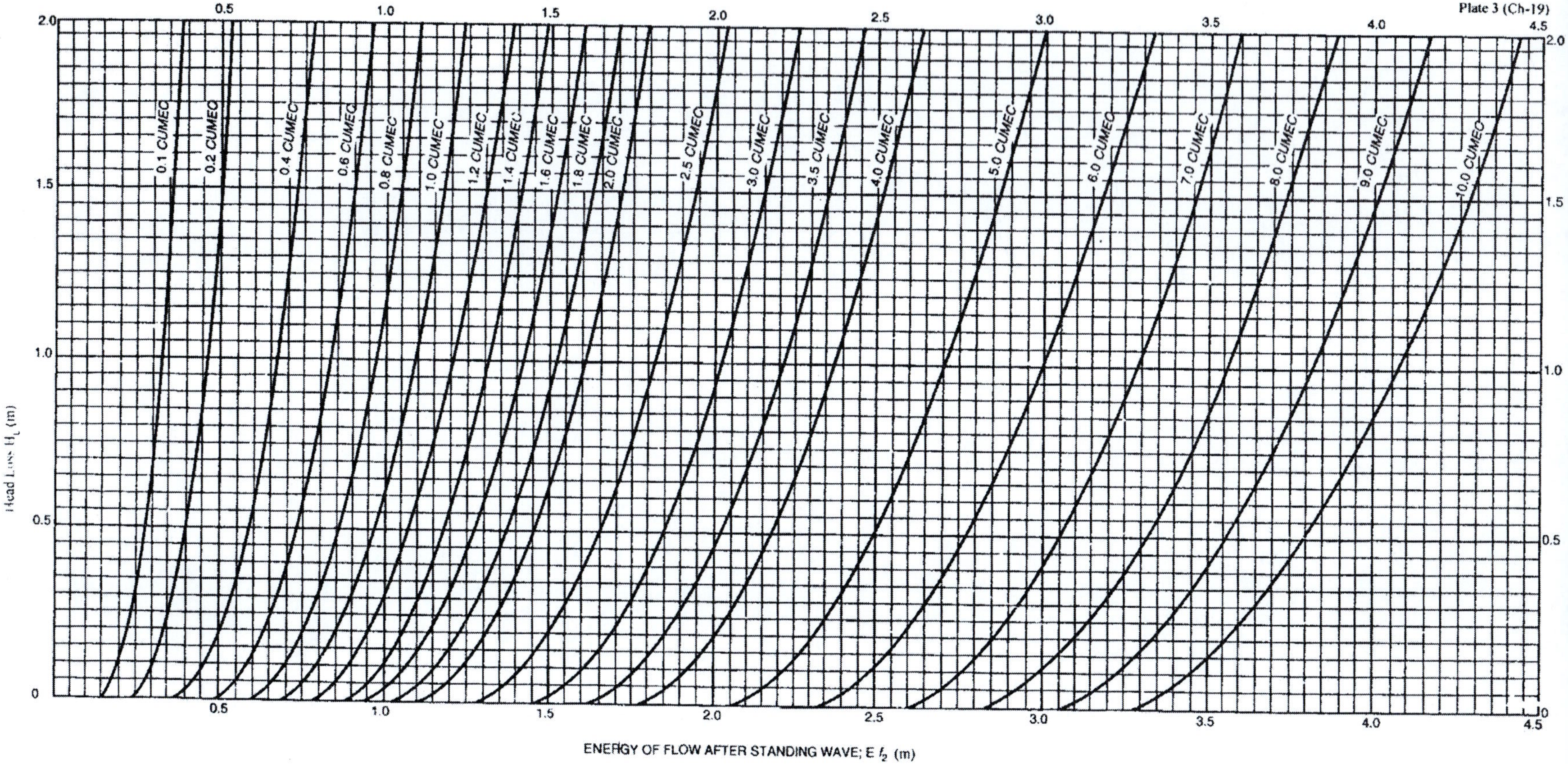

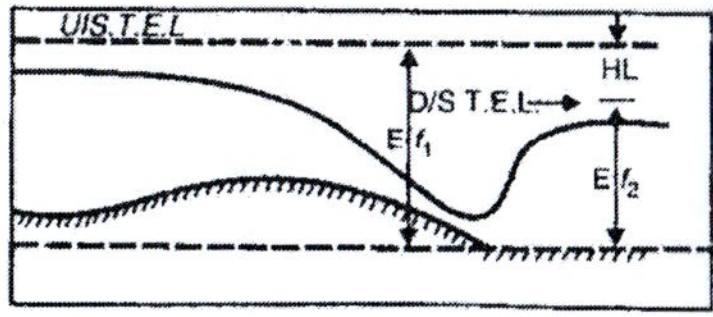

Fig. 19.9 Blench's curve

read from the $\frac{D_1}{D_c}$ curve given in IS 4997. Using this value of D_1, D_2 can also be calculated from another $\frac{D_2}{D_1}$ curve. The curves are given in Fig. 19.10.

The disadvantage of using this curve is that any error commited in finding D_1 by interpolation will be reflected in the value of D_2 and therefore, in all further calculations. To avoid carry over of such interpolation error two CWC engineers C. Chinnaswamy and E. Sundaraiya have prepared two separate curves on the same principle but providing relation between head loss factor $\left(\frac{H_L}{D_c}\right)$ and $\frac{D_2}{D_c}$ and $\frac{D_c}{D_1}$ respectively. These curves can be adopted to find out the values of D_1 and D_2 with benefit and are given in Fig. 19.11.

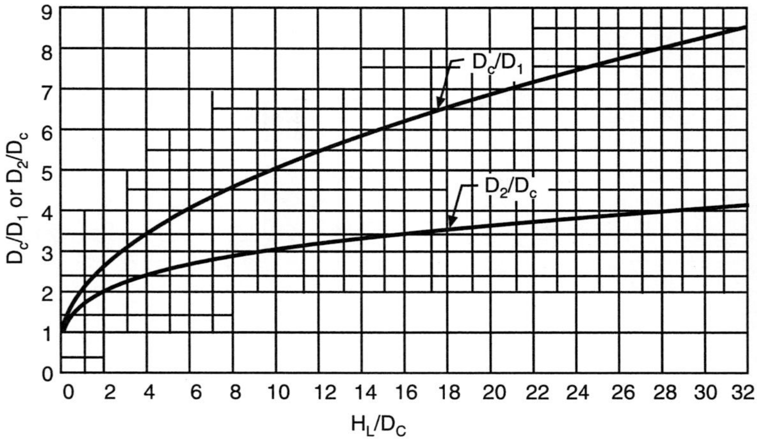

Fig. 19.11. Head loss factor versus depth factor

It may be clarified here that the hydraulic jump does not remain stable on smooth horizontal floor and tends to move downwards. A situation may come when the pre-jump hypercritical depth may prevail on the downstream protection works and may damage it. To avoid such a situation sloping glacis is provided and it is carried upto such a level, in other words the level of the horizontal floor is so fixed, that a stable hydraulic jump is formed on the glacis and is contained within the impervious pucca horizontal floor.

The level or elevation of the horizontal floor may be calculated by either deducting d/s specific energy (Ef_2) from d/s total energy line (TEL) or deducting D_2 from the d/s water level. It ensures the formation of hydraulic jump on the glacis.

In order to ensure turbulence free flow on the d/s the length of the horizontal impervious floor should be equal to the length of the jump. The length of jump may be taken as 5 times the difference of conjugate depths, *i.e.*,

Length of jump L_j = Length of horizontal impervious floor = $5\,(D_2 - D_1)$

The length of the stilling basin can be reduced by providing appurtenances like dentated sill, chute blocks, basin block in the middle of the basin, etc.

19.14. TYPES OF FALLS

Various types of falls commonly constructed on the various canal systems are the following:

19.14.1. Notch Fall

It consists of a body wall constructed across a canal. On the body wall there are notches in between the piers. The notches may be trapezoidal or rectangular in shape. The sill of the notches is in level with the upstream canal bed above the fall. Thus, depth discharge relationship of normal canal section is maintained at the fall also. Hence, the fall can be used for measuring the discharge of a canal. As the sill of the notch is at bed level there is no silting. Some energy dissipation device is provided below the fall (Fig. 19.12).

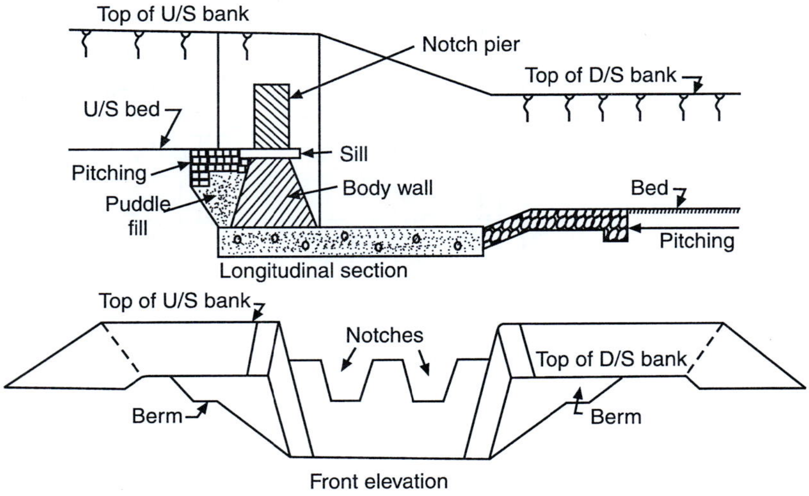

Fig. 19.12. Trapezoidal notch fall

19.14.2. Sarda Type Fall

It is a fall with a raised crest. The body wall is constructed like a weir (Fig. 19.13). Below the fall suitable device is provided for dissipating excess energy of falling water. This type of falls were constructed on the Sarda canal in Uttar Pradesh and hence the name. As the crest of the fall is raised silting of the upstream canal is possible.

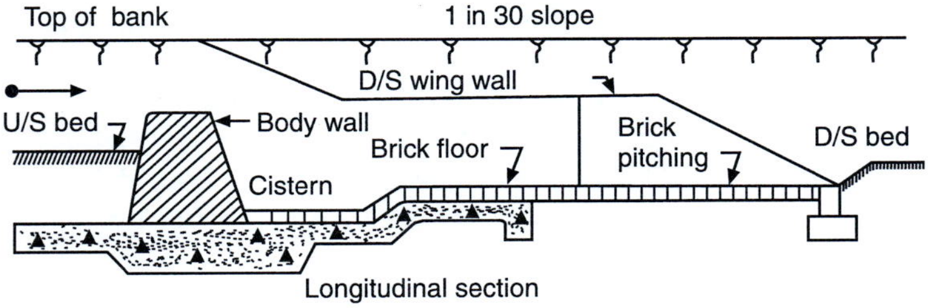

Fig. 19.13. Sarda fall

19.14.3. Glacis Fall

In this type on the d/s side after the crest a sloping glacis is provided. The fall may be flumed or unflumed. In the flumed type length of the body wall of a fall is less than the normal canal width. The section is restricted at the site of the fall. The narrowing of the section is done gradually. Two main falls come under this category. They are the following:

- Montague type fall.
- Inglis type fall.

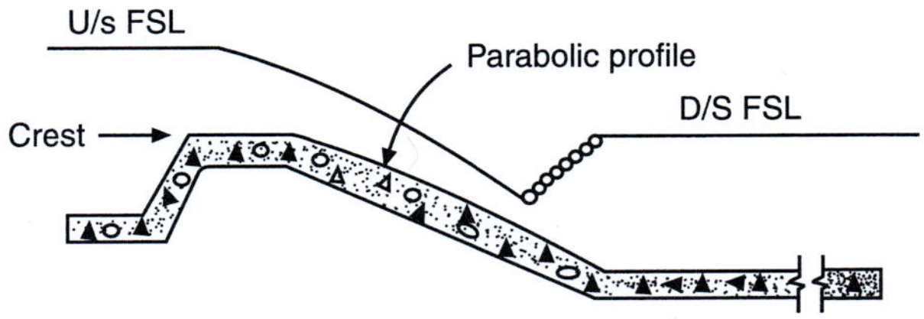

Fig. 19.14. Montague type fall

Both are very similar in features. In both the falls hydraulic jump occurs on the downstream sloping face. It destroys the energy. The d/s sloping face is not straight but has parabolic glacis in Montague type fall (Figs. 19.14).

19.14.4. Fall Regulators

They are designed as a fall-cum-regulator. Generally cross regulator is very well combined with a fall. They are constructed in such a way that the regulating gates can be arranged to suit the water level upstream of the fall.

19.14.5. CDO Type Fall

It is a flumed vertical drop fall. It is widely constructed in Punjab.

19.15. COMPONENTS OF A FALL STRUCTURE

Irrespective of a type of fall, a typical fall structure can be said to consist of the following main parts as shown in Fig. 19.15. They are:

- U/s approach
- Throat/crest/body wall
- D/s glacis/cistern
- D/s expansion
- Energy dissipators.

It is clear that every fall does not necessarily require all components mentioned above because provision of a particular component depends upon the type of fall adopted, available site conditions and design criteria. However, Fig. 19.15 gives good understanding about the way various parts of the fall are provided. Referring Fig. 19.15 meanings of alphabetical dimensions which are not very commonly used are as follows:

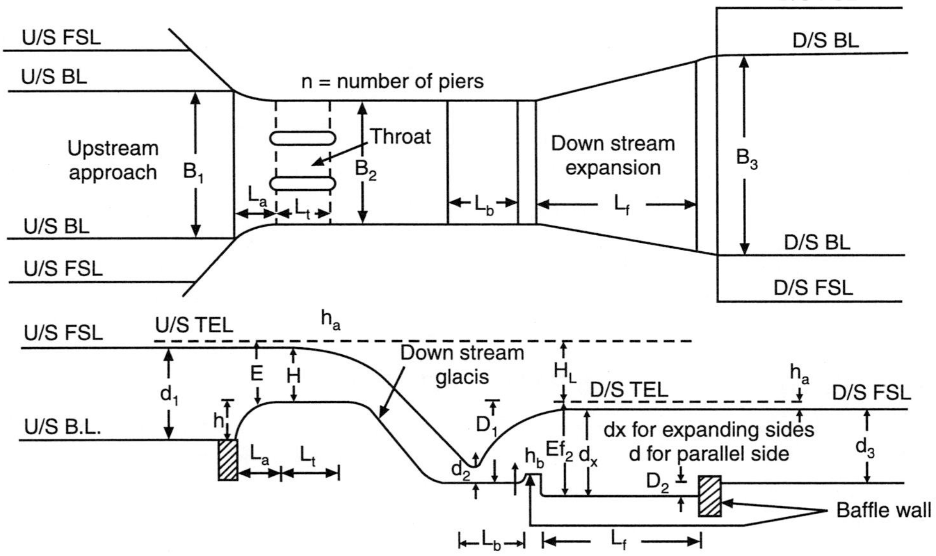

Fig. 19.15. Components of a fall structure

Bt = Clear throat width

d_2 = Hypercritical depth at formation of hydraulic jump

dx and d = Subcritical depth in the canal d/s after formation of hydraulic jump

E = Depth of crest below u/s TEL = $(H + ha)$

H = Depth of crest below u/s FSL

D_1 = Difference of crest level and d/s floor level.
D_2 = Depth of cistern below d/s bed level
Ef_2 = Energy of flow after formation of hydraulic jump
h = Height of crest above u/s bed level.
h_b = Height of baffle wall
H_L = Loss of head = u/s TEL – d/s TEL
≈ u/s FSL – d/s FSL
L_a = Horizontal length of u/s glacis
L_t = Length of crest (along axis of the canal)
L_b = Length of baffle platform
L_f = Length of cistern/horizontal impervious floor
q = Discharge intensity over crest. $= \dfrac{Q}{B_2}$

19.16. SELECTION OF TYPE OF FALL

The main considerations in selection of a type of fall are: (*i*) the height of drop, and (*ii*) the discharge passing over the fall. In other words the amount of energy to be dissipated dictate the type. The type which dissipates energy most satisfactorily should be selected.

When full energy dissipation does not take place on the masonry structure of the fall, the issuing jet still possesses higher velocities than what the soil can withstand. In these circumstances provision of baffle becomes obligatory.

Where the bed material is hard enough to withstand scouring action of the strong current, the type of fall that envisages dissipation of just surplus energy could be adopted.

For unflumed conditions the baffle design is well suited under clear over fall conditions especially where the soil is easily erodible.

For flumed conditions vertical type is not suitable because effective dissipation is difficult and harmful scouring is always expected. In such cases the glacis fall either with straight sloping floor or with baffle wall may prove useful.

In drowned fall conditions fall structure with a baffle may not be necessary and straight glacis with 3 to 4 rows of friction blocks and a deflector may be adopted for a flumed condition and a vertical fall for unflumed condition may be adopted. For canals with discharges below 8 cumec the choice may be based on cost consideration alone.

19.17. METHODS OF ENERGY DISSIPATION BELOW FALLS

It is already mentioned that the water which spills and flows over the crest of a fall acquires high velocity. The reason for this abnormal increase in the velocity is that whole potential energy is transformed into kinetic energy. This high velocity flow has a tendency to erode the bed and to penetrate the open cracks in the canal bed. This erosion then endangers the fall structure. Hence, it is very essential to protect the bed of the canal below the fall. The canal bed may be protected by adopting any one or combination of more than one of the devices explained below. The devices may be classified as those creating cistern, roughening and hydraulic jump.

19.17.1. Provision of a Water-cushion

When a water-cushion is provided below the fall it serves two purposes.

- Firstly it reduces the intensity of impact of the falling sheet of water.
- Secondly it dissipates the energy of flow.

Water cushioning can be successfully achieved by providing a still water pocket or pond below the fall. To create a pond or a pocket of still water a cistern may be designed. It is nothing but a depression in the bed of a canal immediately below the fall. Actually suitable length and depth of cistern are not amenable to theoretical treatment but it is matter of wide experience in the field and model studies. However, following formulae give good basis for cistern design:

Referring Fig. 19.16, $E = \frac{V^2}{2g} + D_1$

According to Bahadurabad Irrigation Research Institute formula (in metric units):

$$L_C = 5 \cdot \sqrt{E \,.\, H_L} \text{ and}$$

$$x = \frac{1}{4} (E \cdot H_L)^{2/3}$$

According to Etchevery formulae

$$L_C = 3 \sqrt{E \cdot H_L} \text{ and}$$

$$x = \frac{1}{6} L_C$$

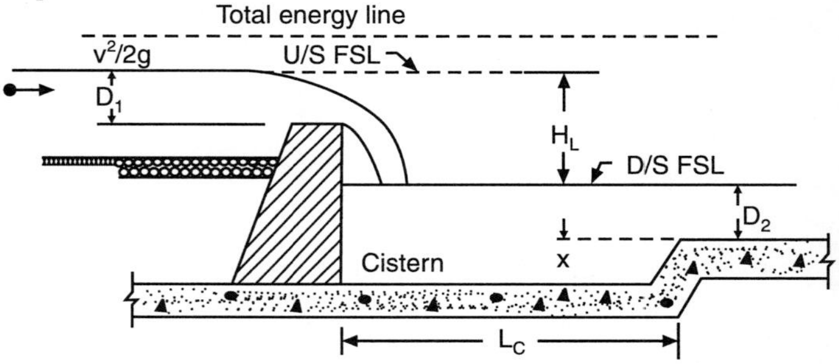

Fig. 19.16. Water cushion below a fall

According to Montague formulae (generally used in Punjab)

$$L_C = 4 \cdot Ef_2 \text{ to } 6 \cdot Ef_2\text{; and}$$

$$x = \frac{1}{2} Ef_2$$

where Ef_2 is the total energy of flow downstream of a fall.

19.17.2. Baffle Wall

It is an obstruction constructed across the canal downstream of the fall. It is in the form of a wall of low height. It heads up the water just upstream of it. Thus it tries to create a water cushion on the upstream. Many times when the flow conditions are favourable hydraulic jump may occur. Leliavsky has given formula to calculate the dimensions of a stilling pool and end sill (Fig. 19.17)

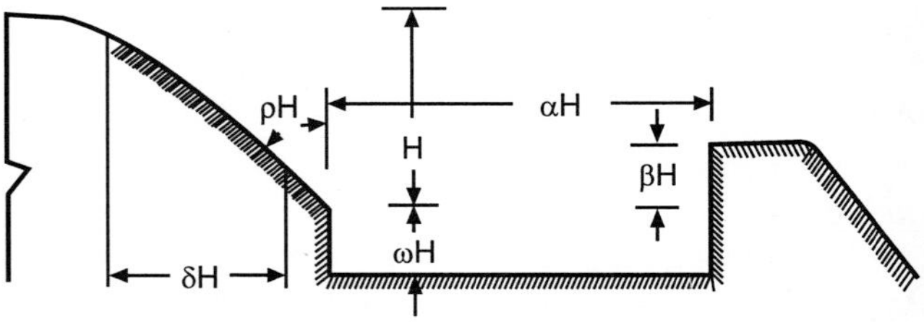

Fig. 19.17. Design of stilling pool

$$\beta = \sqrt{\frac{q}{5H^{1.5}}}$$

where q = Discharge in m³/sec per metre width of a fall.

H = Drop between sill level of fall crest and basin bed level in m.

β = Height of end sill expressed in terms of H.

Leliavsky recommends following proportions for designing a stilling pool:

$$\omega = 0.084; \quad \beta > 0.05$$

$$\alpha = 0.50 \text{ to } 0.80; \quad \rho = 0.15; \text{ and } \quad \delta = 0.75$$

19.17.3. Biff Wall

It is an end wall of the cistern. It is a vertical wall with a horizontal projection extending in the cistern (Fig. 19.18).

Due to the projection, the flow of water returns back in the cistern. It creates an obstruction to the fast moving water down the fall. As a result the energy of flow is dissipated.

19.17.4. Deflector

It is a short wall constructed at the end of a downstream apron (Fig. 19.19).

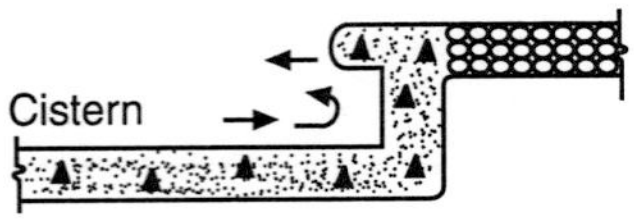

Fig. 19.18. Biff wall

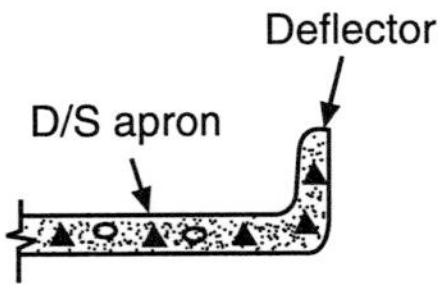

Fig. 19.19. Deflector

This end wall deflects the high velocity flow of water. Due to deflection the velocity of flow in the direction of motion is reduced.

Roughening devices create frictional resistance to flow and reduce the velocity. Some devices are mentioned below.

19.17.5. Staggered Blocks

They are nothing but rectangular blocks or cubes generally made of concrete. They are arranged in a staggered manner on the downstream horizontal apron (Fig. 19.20). They deflect the high velocity flow in a lateral direction. It provides an obstruction to high velocity downstream flow and the energy of flow is dissipated effectively. They are very commonly used below the falls to dissipate the energy in combination with a cistern.

19.17.6. Ribbed Pitching or Cellular Pitching

Its construction is based on the principle that the rough wetted perimeter lowers the velocity of flow appreciably due to increased frictional resistance. To roughen the wetted perimeter pitching may be provided with one brick on edge and the next brick on end. This type of pitching is provided on the downstream side of the fall (Fig. 19.21). This device is found out to be cheap and it dissipates the energy effectively.

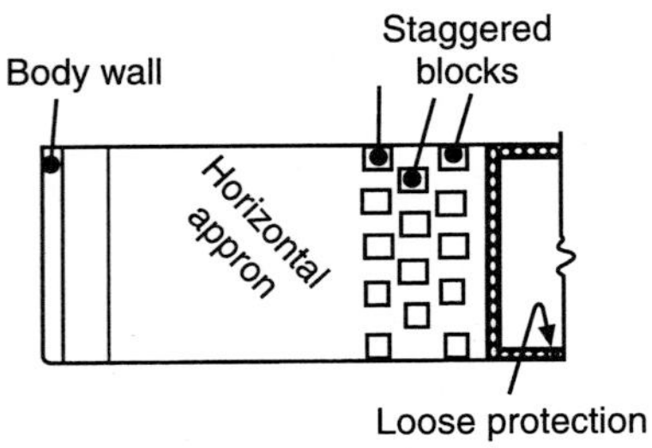

Fig. 19.20. Plan of staggered blocks

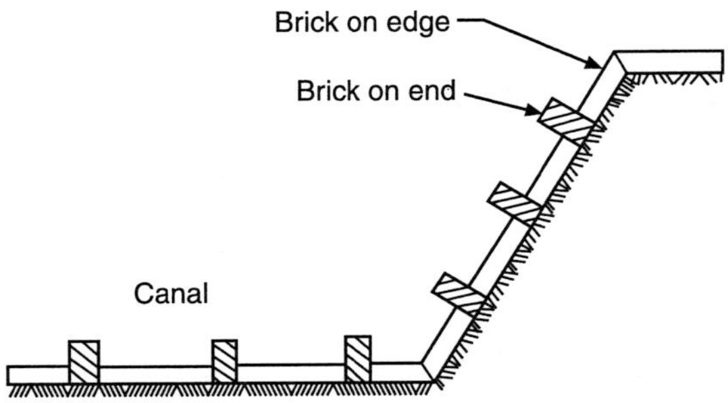

Fig. 19.21. Ribbed pitching

19.17.7. Hydraulic Jump on Sloping Glacis

Hydraulic jump or standing wave is considered to be the most effective means of dissipating energy and reducing hypercritical velocity to normal velocity in the canal downstream of a fall. To ensure formation of the hydraulic jump it is essential that the d/s depth of water flowing with subcritical velocity should bear the following relation to hypercritical depth of flow at the toe of the glacis (Refer Fig. 19.15).

$$d_x = \frac{d_2}{2} + \sqrt{\frac{2q^2}{gd_2} + \frac{d_2^2}{4}}$$

Neglecting the frictional resistance of the glacis and using values of q and H_L and Blench curves given in Fig. 19.9 energy of flow below the standing wave (Ef_2) can be calculated.

1. Dimensions of a cistern for straight glacis: The cistern level may then be obtained by subtracting 1.25 d_x from d/s FSL or 1.25 Ef_2 from the d/s TEL.

In case the natural surface level is lower than the cistern level determined as above the natural surface should be adopted as cistern level.

It has been found that energy is not completely dissipated in the hydraulic jump and therefore it is necessary to provide sufficient length of cistern to avoid damage to the bed and banks of a canal. In case of glacis fall without baffle platform a cistern length equal to 5 Ef_2 is considered to be enough for good earthen bed and 6 Ef_2 for erodible and sandy soils.

The cistern should be joined to the designed bed d/s with a up slope of 1 in 5.

2. Dimensions of a cistern for glacis with baffle wall at end: Referring Fig. 19.15 the dimensions of a baffle platform and a baffle wall may be determined from the following relations:

Height of baffle wall, $h_b = d_c - d_2$

where d_c (critical depth) $= \left(\frac{q}{g}\right)^{1/3}$

d_2 can be calculated using Fig. 19.11 with known values of H_L and D_C.

Thickness of baffle wall $= \frac{2}{3}\ h_b$

Length of baffle platform $= 5.25\ h_b$.

The baffle platform should join the toe of glacis with a radius equal to depth of water above crest and to the baffle wall with radius $R = \frac{2}{3}\ h_b$.

The cistern length $= 5\ d_x$

where d_x is conjugate or sequent depth after hydraulic jump.

The cistern should be depressed below d/s bed level by 0.1 (d/s FS depth) subject to minimum of 15 cm for distributaries and minors and 30 cm for main canals and branch canals.

19.18. DESIGN PRINCIPLES FOR SARDA TYPE FALL

This type of falls are constructed on Sarda canal in Uttar Pradesh. It is a fall with raised crest and with vertical impact. The soils in Sarda command comprised sandy stratum overlain by sandy-clay on which depth of cutting was to be kept minimum. This made it obligatory to provide number of falls with small drops. In Sarda type falls (q) discharge intensity varied from 1.6 to 3.5 cumec/m and drop varied from 0.6 to 2.5 m.

Crest dimensions: This type of fall is not flumed.

For canal discharge 15 cumec and more

Crest length of fall = Bed width of the canal.

For distributaries and minors

Crest length of fall = Bed width + Depth of flow.

Body wall: When the discharge of a canal is less than 14 m^3/sec the section of body wall is kept rectangular (Fig. 19.22 (*a*)).

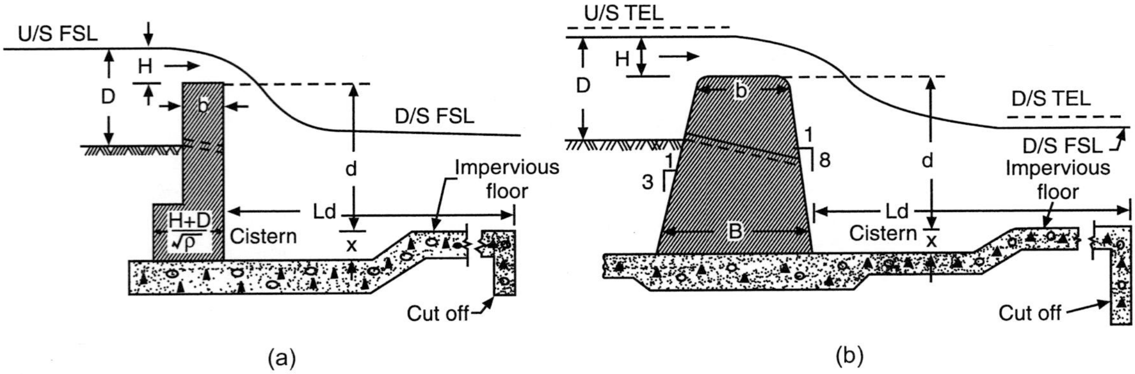

Fig. 19.22. Sarda type fall

When the discharge of a canal is more than 14 m^3/sec the section of the body wall is kept trapezoidal with upstream batter 1 : 3 and downstream batter 1 : 8.

For rectangular body wall:

$$\text{Top width } `b' = 0.552\sqrt{d}$$

$$\text{Base width } `B' = \frac{H+d}{\sqrt{\rho}}$$

For trapezoidal body wall Top width $b = 0.522\sqrt{(H+d)}$

The edges are rounded with a radius of 0.3 m.

Base width B is determined by the batter given to u/s and d/s sides.

Here H is depth of water above the crest of the fall in metres. (It includes velocity of approach also).

d is the height of the crest above the downstream bed level in metres.

Discharge over crest: The discharge formula used in this type of fall under free fall condition is:

$$Q = CLH^{3/2}\left\{\frac{H}{b}\right\}^{1/6}$$

where L is length of crest in m and Q is discharge in cumec.

Value of C for trapezoidal crest is 2 and for rectangular crest 1.85.

For submerged flow conditions (above 33% submergence) neglecting velocity of approach the discharge is given by the following formula:

$$Q = \frac{2}{3}\ Cd \cdot L \cdot \sqrt{2g}\ (H_L)^{3/2} + Cd \cdot L \cdot h_2\ \sqrt{2g \cdot H_L}$$

where $Cd = 0.65$

H_L = drop in water surface

and h_2 = depth of d/s water level over top of crest.

Crest level: The height of crest above the upstream bed level is fixed in such a way that the depth of flow u/s of the fall is not affected.

From the discharge formula mentioned above since Q is known value of H can be calculated.

$\therefore \quad R \cdot L$ of crest $= F \cdot S \cdot L$ on the u/s $- H$.

The stability of body wall should be tested by usual procedure when the drops exceeding 1.5 m are to be designed. In the body wall drain holes may be provided at the u/s bed level to dry out the canal during closures for maintenance, etc.

Cistern dimensions: Dimensions of the cistern may be fixed from the Bahadurabad Research Institute formula given in article 19.17, *i.e.*,

$$L_C = 5\sqrt{E \cdot H_L} \text{ and}$$

$$X = \frac{1}{4}(E \cdot H_L)^{2/3}$$

Total length of impervious floor: As for any hydraulic structure total length of the impervious floor should be designed on the basis of Bligh's theory for small structures and Khosla's theory for other works.

Maximum seepage head is experienced when on the u/s water is upto the crest level of the fall and there is no flow on the d/s side. Referring Fig. 19.22 maximum seepage head is given by '*d*'.

Length of d/s impervious floor: The maximum length of the d/s impervious floor is given by the following relation (Refer Fig. 19.22).

$$L_d = 2D + 2.4 + H_L \text{ in metres.}$$

The balance of impervious floor may be provided under the body wall and on the u/s.

Thickness of the floor: The d/s floor should be made thick enough to resist uplift pressures. However, minimum thickness of 0.3 to 0.6 m (depending upon the size of the drop) of concrete under 35 cm of brick masonry may be provided on the d/s. On the u/s brick masonry is not necessary. The brick on the edge laid on the d/s impervious concrete floor provide additional strength and affords easy repairs to the floor.

Cut-off: A sufficient depth of cut-off below the floor should be provided at the d/s end of the floor for providing safety against steep exit gradient. The depth of cut-off may range from 1 to 1.5 m. Sometimes deeper cut-offs may be necessary to reduce horizontal floor length to satisfy Khosla's principle of exit gradient. For falls having 1 m and above head on the crest should be provided more cut-offs. Cut-off at u/s end of floor is also provided which may be smaller in depth.

Other protective works: Provision of other accessories like upstream wings, staggered blocks on the cistern floor, downstream wings, bed and side pitching is generally done on the basis of thumb rules. For big structures, however, actual design calculations may be done. For general arrangement see Fig. 19.13.

Upstream wing walls: For small falls upto 14 cumec the upstream wings may be splayed at 1: 1. For higher discharges u/s wing walls are kept segmental with a radius equal to 6 H and continued thereafter tangentially merging into the banks. The wings may be embedded into the bank for about 1 m.

Downstream wing walls: For the length of the cistern the d/s wing walls are kept vertical from the crest. Thereafter they are wasped or flared to a slope of 1 : 1. An average splay of 1 in 3 for attaining the required slope is given to the top of the wings. The wings may be taken deep into the banks.

Staggered blocks: Staggered block of height dc should be provided at a distance of 1.0 dc to 1.5 dc from the d/s toe of the crest for clear falls.

In case of submerged falls the blocks may be provided at the end of the cistern.

A row of staggered cubical blocks of height equal to 0.1 to 0.13 of depth of water should invariably be provided at the end of the d/s impervious floor.

Bed and side pitching: The d/s bed pitching with bricks 20 cm thick over 10 cm ballast is provided horizontally for a length of 6 m. Thereafter for lengths upto 5 to 15 m for falls varying from 0.75 to 1.5 m may be provided with down slope of 1 in 10.

The side pitching with bricks on edge with 1 : 1 slope is provided after the return-wing on the downstream. A toe wall should be provided between the bed pitching and the side pitching to provide a firm support to the latter.

19.19. DESIGN PRINCIPLES FOR STRAIGHT GLACIS FALL

Crest dimensions: Clear width of crest 'B_t':

It is already mentioned that vertical falls should be full width falls, *i.e.*, the width of the crest should be same as bed width of the canal because increased intensity of discharge due to fluming creates scour on the downstream.

Unlike vertical falls the glacis falls can be flumed when combined with bridge so as to economize in the cost.

It is quite rational to select such (q) discharge per metre run of crest width which with the height of drop (H_L) available gives value of total energy on the d/s (Ef_2) equal to *F.S.* depth of the canal. (It can be read from Blench curves). It does not require deep cistern on d/s and avoids construction difficulty particularly when subsoil water level is high. The throat width may be rounded off to next half metre. The fluming thus calculated may not, however, exceed limits given below subject to the condition that overall width of fall crest is not more than bed width of the canal on the downstream.

Drop (H_L)	Permissible fluming ratio
upto 1 m	66%
1 to 3 m	75%
above 3 m	85%

Refer Fig. 19.23:

$$\text{Crest level} = \text{u/s } TEL - E$$

In case of full width falls and sometimes in flumed falls if crest level works out unreasonably high fluming may be done or increased if already flumed so that crest is not higher than 0.4 D_1 above the u/s bed as otherwise it will increase afflux at low supplies and may cause alternate silting/scouring.

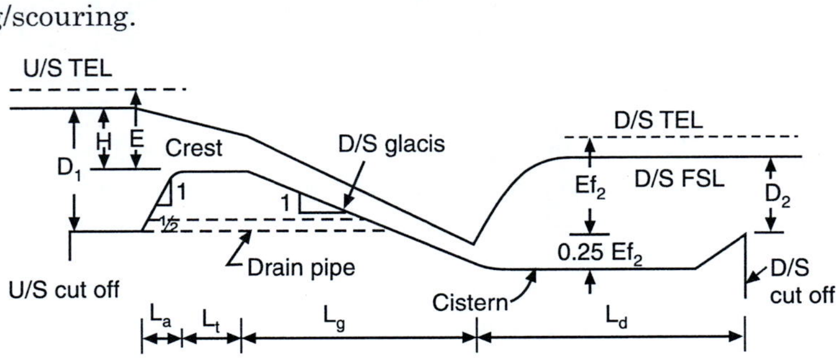

Fig. 19.23. Straight glacis fall

The value of E is given by discharge formula

$$Q = 1.84\ B_t \times E^{3/2}$$

where B_t is clear width of crest. Therefore if n piers are provided in between

$$\text{effective } B_t = (B_t - 0.2\ n\ H)$$

and E is depth of crest below u/s TEL.

$$\text{Length of crest } (L_t) = \frac{2}{3}\ E.$$

The crest is joined to u/s and d/s canal bed with sloping glacis.

The u/s glacis (for non-meter falls) is given a slope of $\frac{1}{2}$: 1. The u/s crest end is kept curved with a radius of $\frac{E}{2}$,

The d/s glacis is given 2 : 1 slope and it joins the cistern d/s with curve having radius equal to E.

Cistern dimensions: As already explained in article 19.17.7 earlier (Ref. Fig. 19.23)

$$R.L. \text{ of cistern} = \text{d/s TEL} -1.25\ Ef_2 = \text{d/s FSL} -1.25\ D_2$$

$$\text{Length of cistern} = 5\ Ef_2 \text{ for good earthen bed}$$

or $$L_d = 6\ Ef_2 \text{ for erodible sandy soils.}$$

The cistern should be joined to the designed d/s bed with a up slope of 1 in 5 (1 : 5)

This arrangement enables formation of hydraulic jump on the sloping glacis.

Provisions of Cut-offs

The cut-offs should be invariably provided at the upstream end of upstream glacis and at the downstream end of the downstream cistern. The width of each curtain wall may be kept 0.4 m. The depth may be as follows:

$$\text{Depth of u/s cut-off} = \frac{D_1}{3}$$

$$\text{Depth of d/s cut-off} = \frac{D_2}{2}$$

The minimum depth should however be 0.5 m.

Total length of impervious floor: The total length of floor should be such that with the depth of curtain walls as fixed earlier give permissible exit gradient. Khosla's curve for exit gradient may be used for this purpose.

The length of floor between u/s and d/s cut-offs so determined if appears excessive the downstream cut-off may be further deepened suitably to achieve adequate floor length.

It may be noted that the total impervious floor length comprises:

- length of cistern;
- horizontal length of d/s glacis;
- crest length along the axis of the canal; and
- horizontal length of u/s glacis.

In case some little length is still remains to be provided as per earlier calculations it may be provided on the u/s side of u/s glacis.

Thickness of the floor: Minimum thickness on the u/s may be from 0.3 to 0.6 m. Floor thickness in the glacis and cistern should be sufficient to withstand uplift pressure safely.

U/s approach and u/s protection

(*i*) If the fall combines with it functions of a discharge meter as well, the side and bed approaches to the crest have necessarily to be gradual and smooth so as to avoid eddies and impact losses and to reduce concentration of flow.

In non-meter falls, however, the side walls may be splayed at an angle of 45° from the upstream edge of the crest. The walls are carried straight into the canal berm for a length of at least 1 m.

(*ii*) As already explained the bed approach may be by means of u/s glacis having $\frac{1}{2}$: 1 slope and joining tangentially the u/s end of crest with a radius equal to $\frac{E}{2}$.

(*iii*) Protection of bed and sides by stone or dry brick pitching may be done for a length of (D_1 + 0.5) m. The bed pitching may be laid at a slope of 1 in 10.

D/s expansion and d/s protection

(*i*) On the downstream parallel and vertical walls are provided upto the toe of the glacis.

(*ii*) The expansion afterwards should be gradual so that the expanding flow adheres to the sides and scour due to formation of back-rollers on sides is prevented. A rectangular hyperbolic expansion given by Mitra's equation for hyperbolic expansion is generally adopted.

If this expansion works out too long, side splay of about 1 in 5 may be adopted. For small falls to effect economy expansion with side splay of 1 in 3 considered sufficient.

(*iii*) Side walls in expansion may be flared out from vertical to 1 : 1 if the earth fill behind is not problematic like black cotton soil. In such cases the side walls may be designed as vertical gravity walls.

(*iv*) Side protection consisting of 20 cm thick dry brick pitching for a length of 3 D_2 should be provided. It should rest on a toe wall 1½ brick thick and of depth equal to $D_2/2$ subject to minimum of 0.5 m depth.

(*v*) A deflector wall of height $D_2/10$ above d/s bed may be provided at the downstream end of the cistern. The minimum height should be 15 cm. Thickness of the deflector wall may be kept 0.4 m.

(*vi*) With provision of deflector wall at the end of the floor d/s bed pitching beyond the floor is not necessary.

Friction blocks as energy dissipators: Friction blocks are found to be most effective. In case of flumed straight glacis falls (without baffle) four rows of friction blocks may be provided. As already mentioned they are staggered in plan. The u/s edge of the first row of the friction block is located at a distance of 5 times the height of the blocks (5 · h) from the toe of the glacis. The dimensions of the blocks may be as follows:

Let, height of the blocks = h

$$h = \frac{D_1}{8}$$

$$\text{Length of the block} = 3\ h$$

$$\text{Width of the block} = \frac{2}{3}\ h$$

$$\text{Distance between rows} = \frac{2}{3}\ h$$

When glacis is provided with baffle only two rows of friction blocks is sufficient upto 2 m fall. The u/s edge of the first row may be located at 1/3 length of d/s expansion from the end of the cristern floor.

19.20. DESIGN PRINCIPLES FOR CROSS DRAINAGE WORKS

The main objective of design is to determine the dimensions and arrangement of structures proposed to be constructed. In cross drainage works, unlike other hydraulic structures, suitable structures have to be provided for the canal as well as the natural drain to carry one water conveyance channel over, below or through the other appropriately. Obviously in a design procedure for a cross drainage work determination of hydraulic parameters of the natural drainage as well as the canal crossing the former constitute important consideration. The arrangement of structures and their structural strength are also of equal importance. The important steps involved in the design of a cross drainage work could be summarized as follows:

(A) Determination of drainage parameters

- Design flood discharge and high flood level (HFL)
- Waterway required for the drain.
- Velocity of flow through the barrel.
- Height of opening or clearances.
- Number of spans or openings.

(B) Determination of canal parameters

- Size of canal waterway or its contraction.
- Design of transitions.
- Bank connections.

(C) Determination of hydraulic forces

- Head loss through barrels.
- Uplift pressure on the roof of the barrel or trough.
- Uplift pressure on the floor of the barrel.

After determining the above parameters structural design can be done.

19.21. DESIGN PRINCIPLES FOR AQUEDUCT

(*i*) **Estimation of design (maximum) flood discharge of a drain:** The drain to be crossed may be small or like a river. In all cases correct assessment of maximum flood or peak flow of a drain should be obtained before hand.

(*ii*) **Waterway requirement for a drain:** Lacey's regime perimeter equation gives good basis for calculating the drainage waterway. The equation is

$$P_w = 4.825\ Q^{1/2}$$

where, P_w is the waterway to be provided for the drain at the site in metres.

Q is flood discharge of the drain in m³/sec. As the piers reduce the actual waterway available, the length between the abutments (P_w) may be increased by 20 per cent.

When the waterway is fixed from Lacey's regime perimeter equation, the regime condition in the drain upstream and downstream of the structure is not disturbed appreciably.

To confine the drainage water to the desired waterway guide banks may be constructed.

(iii) **Velocity of flow through barrel:** The velocity of flow through the barrel may range from 1.8 m/sec to 3 m/sec. The reason for selecting this range is that the lower velocities may cause silting in the barrels. Whereas when the velocity is higher than 3 m/sec the bed load may cause abrasion of the barrel floor and subsequently it may be damaged.

(iv) **Height of opening:** Once the waterway discharge and velocity are fixed the depth of flow may be obtained easily.

There should be sufficient headway or clearance left between the HFL and the bottom of the canal bed. A clearance of 1 m or half the height of the culvert, whichever is less would be sufficient.

Hence, Height of opening = Depth of flow + Clearance or headway.

(v) **Number of spans:** After determining the total length of an aqueduct between the abutments number of spans to be provided may be fixed on the basis of the following two considerations:

- Structural strength required, and
- Economical consideration.

For example, when arches are used the number of spans to be provided may be more. When the cost of construction in the foundation is rather high, small number of spans should be adopted and then RCC beams may be used.

(vi) **Canal waterway:** Generally fluming ratio* is taken to be 1/2. This ratio is adopted in such a way that the velocity of flow in the trough does not go above critical velocity limit. Generally velocity of flow should not be more than 3 m/sec. This precaution is taken to avoid the possibility of formation of a hydraulic jump. The obvious reason is that when hydraulic jump forms it absorbs energy. In this process valuable head is lost and large stresses are produced in the structure.

(vii) **Length of contraction or approach transition:** Once the width at throat is fixed length of contraction can be determined after knowing the convergence ratio. The convergence ratio is generally taken as 2 : 1 (horizontal : lateral), *i.e.*, not steeper than 30°.

(viii) **Length of expansion or departure transition:** Length of expansion on the downstream side of the aqueduct may be fixed after knowing the expansion ratio. The expansion ratio is generally taken as 3 : 1 (horizontal : lateral), *i.e.*, not steeper than 22.5°.

To maintain streamlined flow and also to reduce loss of head the transitions are generally made up of curved and flared wing walls. The design of transition can be worked out by making use of any of the following three methods:

- Hind's method;
- Mitra's hyperbolic transition method;
- Chaturvedi's semi-cubical parabolic transition method.

It may be noted that whereas Hind's method may be used when the water depth in the normal section and the flumed trough also varies, the remaining two methods can be used only when water depth remains constant in the normal canal section as well as the trough section.

***Fluming ratio:** It is a ratio of the width of restricted canal over the drain and that of a normal canal width.

***(ix)* Bank connections:** It is already mentioned that an aqueduct requires four sets of wing wall, (two for the canal and two for the drain (Fig. 19.24).

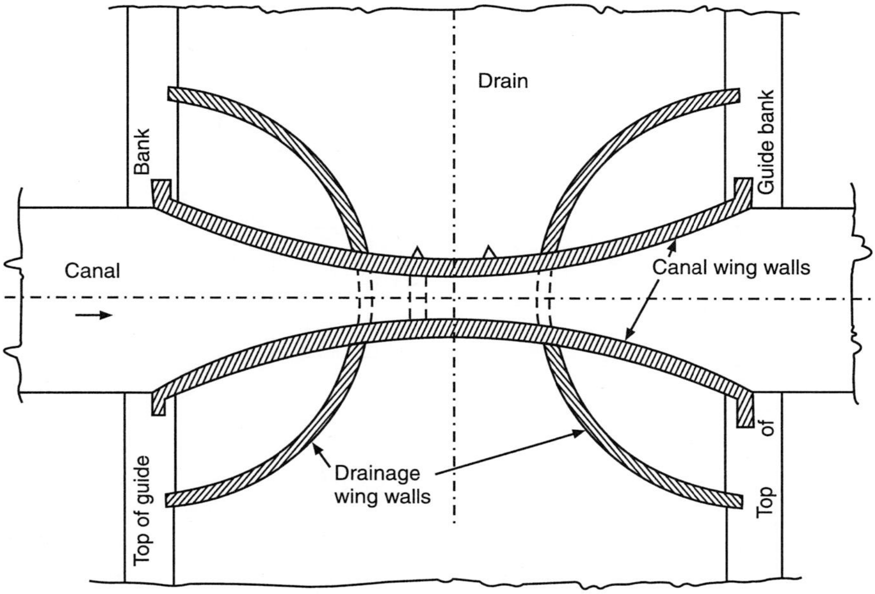

Fig. 19.24. Arrangement of wing walls

Canal wing walls on the upstream and downstream side of the aqueduct protect and retain the earth in the canal banks. The foundation of the canal wing walls should not be left in the embanked earth. The wing walls should be based on the sound foundation in the natural ground. In the transitions the side slopes of the natural section (generally 11/2 : 1) are warped to conform with shape (generally vertical) of the trough over the drain.

Drainage wing walls are provided on the upstream and downstream of the barrel to protect and to retain the natural sides of the drain. As the bed of the drain gets scoured during floods the drainage wing walls should be taken deep into the foundation below maximum scour depth. The wing walls should be taken back sufficiently into the top of the guide banks. The wing walls should be designed to permit smooth entry and exit of the flow in the drain.

19.21.1. Hind's Method for Design of Transition

This method is based on the premise that there is minimum loss of head, flow is streamlined and normal flow conditions in the canal are restored before the canal discharges pass down on to the earthen section immediately after curved and flared transitions. In Fig. 19.25 the contraction or approach transition, the throat portion and the expansion or departure transition are shown. It may be seen that sections 1-1, 2-2, 3-3 and 4-4 indicate start of contraction, end of contraction, start of expansion and end of expansion respectively. Thus the contraction or approach transition lies between sections 1 and 2, throat between sections 2 and 3 and expansion or departure transition between sections 3 and 4. Upto section 1 and beyond section 4 the canal flows under its normal conditions and therefore the canal parameters at these two points are equal and already known. So also conditions of flow and canal parameters are same between sections 2 to 3 which represents throat or trough portion. The design procedure can be outlined as follows:

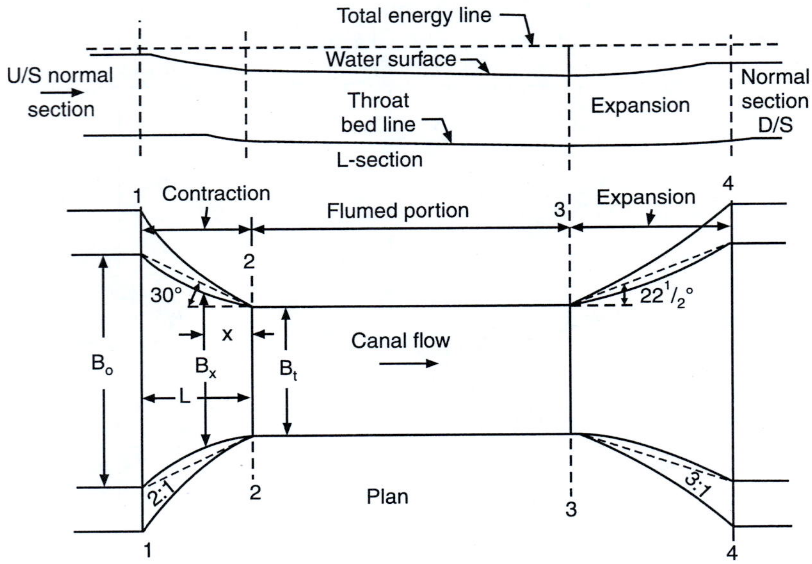

Fig. 19.25. Transition of canal water-way

Let D and V with appropriate subscripts refer to depths and velocities at four sections. Also since canal levels and dimensions are already known at section 4-4:

Step 1: TEL at section 4-4 = Water surface elevation + $\dfrac{V_4^2}{2g}$

where water surface elevation at sec. 4-4 = Bed level + D_4

(Remember TEL is abbreviation of total energy line)

Step 2: TEL at sec. 3-3 = (TEL at sec 4-4) + (energy loss between sec. 3 and 4)

Energy loss between sections 3-3 and 4-4 takes place due to expansion of streamlines and also due to friction. Neglecting loss due to friction which is small and taking loss due to expansion to be

$$0.3\left(\frac{V_3^2 - V_4^2}{2g}\right)$$

$$\text{TEL at sec. 3-3} = (\text{TEL at sec. 4-4}) + 0.3\left(\frac{V_3^2 - V_4^2}{2g}\right)$$

$\therefore$ Water surface level at sec. 3-3 = (TEL at sec. 3-3) – $\left(\dfrac{V_3^2}{2g}\right)$ and

bed level at sec 3-3 = (Water surface level at sec. 3-3) – (D_3).

Step 3: Similarly, TEL at section 2-2 = (TEL at sec. 3-3) + (Head loss between sections 2 and 3)

Between sections 2-2 and 3-3 canal flows through trough section of same dimension and therefore flow is uniform. Loss of head in the trough is only due to friction losses. The loss of head can be calculated using Manning's equation.

$$Q = A \cdot \frac{1}{N} R^{2/3} \cdot S^{1/2}$$

Water surface level at sec. 2-2 = (TEL at sec. 2-2) – $\left(\frac{V_2^2}{2g}\right)$ and

bed level at sec. 2-2 = (Water surface level at sec. 2-2) – (D_2)

Since depth and velocity are same in the trough from sections 2 to 3 it may be noted that the total energy line, water surface line and bed line are parallel to each other.

Step 4: On the similar lines

TEL at sec. 1-1 = (TEL at sec. 2-2) + (Head loss between secs. 1 and 2)

Between sections 2-2 and 1-1 energy loss takes place due to contraction of streamlines and friction. Neglecting friction loss, loss due to contraction may be taken to be

$$0.2\left(\frac{V_2^2 - V_1^2}{2g}\right)$$

$$\therefore \quad \text{TEL at sec. 1-1} = (\text{TEL at sec. 2-2}) + 0.2\left(\frac{V_2^2 - V_1^2}{2g}\right)$$

Water surface level at sec. 1-1 = (TEL at sec. 1-1) – $\left(\frac{V_1^2}{2g}\right)$ and

bed level at sec. 1-1 = (Water surface level at sec. 1-1)–(D_1)

Step 5: As mentioned in the first four steps the bed level, water surface level and level of total energy line can be determined at the four sections. Now the TE line, water surface line and the bed line can be drawn as follows:

(*a*) Now the total energy line can be drawn by joining these points at four sections by a straight line.

(*b*) The bed line may also be drawn as straight lines between adjacent sections if the fall or rise of bed level is small. The corners should be rounded off. In case the drop in the bed line is appreciable the bed lines should be joined with a smooth tangential reverse curve.

(*c*) It is now clear that between any two consecutive sections the drop in water surface level may result due to: (*i*) drop in TE line between the two sections; (*ii*) increased velocity head in contraction; and (*iii*) decreased velocity head in expansion.

This drop in water surface gets negotiated by two parabolic curves. As shown in Figs. 19.26 and 19.27 for contraction (approach transition) and expansion (departure transition) this is achieved by convex upwards curve followed by concave upwards curve in the former transition and concave upwards curve followed by a convex upwards curve in the latter transition.

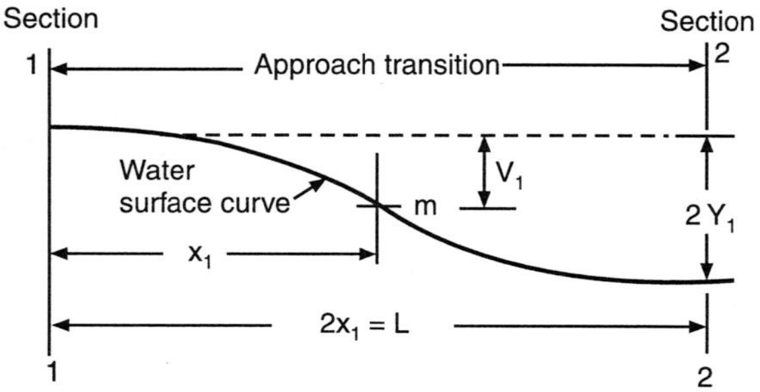

Fig. 19.26. Water surface curve in approach transition

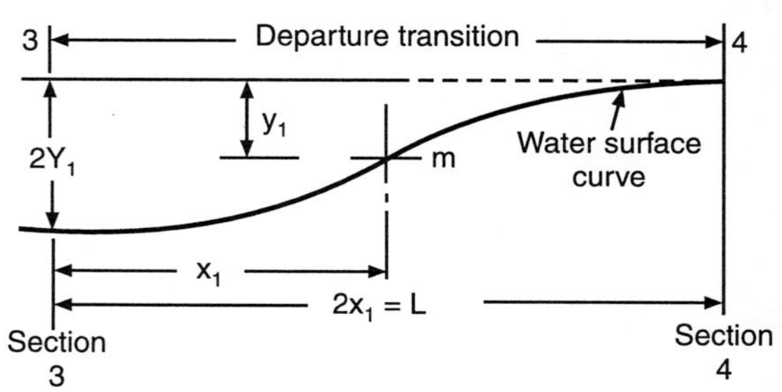

Fig. 19.27. Water surface curve in departure transition

It may be seen from Figs. 19.26 and 19.27

L = Length of transition (contraction or departure) = $2x_1$ and

$2y_1$ = Total drop or rise in water surface.

The point m is the midpoint of transition length and is situated to divide the total drop as well as length equally.

Taking water surface at section point as the origin equation of the parabola is given by

$$y = cx^2$$

Substituting the known values of y_1 and x_1

$$c = \frac{y_1}{x_1^2}$$

With this value of c parabolic water surface curves can be plotted starting from section points which represent origin.

The equation to be used for plotting is now reduced to

$$y = \left(\frac{y_1}{x_1^2}\right) \cdot x^2$$

Thus, water surface profile can be plotted.

Step 6: Velocity and area of flow at various points can be obtained as follows:

(*i*) Velocity head at any point is given by the difference between T.E.L. and water surface.

$\therefore$ Velocity head h_v = TEL – WS line

Also $$h_v = \frac{v^2}{2g}$$

So velocity (V) at each point = $\sqrt{2g \cdot h_v}$

(*ii*) The area of flow at any point can now be obtained by simple formula

$$A = \frac{Q}{V}$$

With known values of A and D other dimensions of the trapezoidal channel can be calculated using the formula

$$A = BD + SD^2$$

where B is bed width and $S : 1$, *i.e.*, (H: V) is the side slope.

In case of flared wing walls, the side slopes are gradually brought to vertical from an initial slope. The value of side slope at any intermediate section in the transition length can be interpolated in proportion to the length of transition achieved upto that point.

19.21.2. Mitra's Hyperbolic Transition Method

This method is based on the principle that

- alongwith discharge the depth of flow in the canal is also constant; and
- rate of change of velocity per unit length of transition is constant throughout the length of transition.

From Fig. 19.25 it may be seen that:

B_o = normal bed width of canal;

B_t = bed width in the throat or trough;

B_x = bed width at any distance x from the extremity of the trough;

and L = total length of transition.

Also let V_o V_t and V_x be the velocities at the corresponding sections.

From second principle

$$\frac{V_t - V_x}{x} = \frac{V_t - V_o}{L} \quad \text{... (1)}$$

From first principle

$$Q = V_t \times B_t \times D$$
$$= V_x \times B_x \times D = V_o \times B_o \times D$$

Since Q and D are constant

Let $$\frac{Q}{D} = K$$

$\therefore$ $$K = V_t \times B_t = V_x \times B_x = V_o \times B_o$$

or $$V_t = \frac{K}{B_t}$$

$$V_x = \frac{K}{B_x} \text{ and}$$

$$V_o = \frac{K}{B_o}$$

Substituting these values in equation (1) above

$$\frac{\left(\frac{K}{B_t} - \frac{K}{B_x}\right)}{x} = \frac{\left(\frac{K}{B_t} - \frac{K}{B_o}\right)}{L}$$

or $$B_x = \frac{B_o B_t L}{L B_o - (B_o - B_t)x}$$

Thus B_x gives bed width at any distance x from the throat.

19.21.3. Chaturvedi's Semi-cubical Parabolic Transition Method

It states that (Refer Fig. 19.25 for notations)

$$x = \frac{L B_o^{3/2}}{B_o^{3/2} - B_t^{3/2}} \left\{1 - \left(\frac{B_t}{B_x}\right)^{3/2}\right\}$$

By assigning suitable values to B_x corresponding values of x can be obtained from the above equation.

19.22. DESIGN PRINCIPLES FOR SIPHON AQUEDUCT

It is clear that the siphon aqueducts are basically different from ordinary aqueducts. As such above mentioned criteria for aqueduct design is not sufficient in the design of siphon aqueducts. In addition to the above considerations the following criteria should be adopted while designing the siphon aqueducts:

(*i*) Discharge through siphon barrel: The head which causes flow (it also represents head loss in barrel) through the inverted siphon barrel may be obtained from Unwin's formula:

$$h = \left(1 + f_1 + f_2 \frac{L}{R}\right) \frac{V^2}{2g} - \frac{V_a^2}{2g}$$

where h is the head causing flow, it is also the loss of head in the barrel in m.

L is the length of barrel in m.
R is hydraulic mean radius of barrel in m.
V is velocity of flow through barrel in m/sec.
V_a is velocity of approach in m/sec, it is generally neglected.
f_1 is a coefficient for loss of head at entry and generally taken as 0.505.
f_2 is a coefficient which accounts for friction in the barrel.

$$f_2 = a\left(1 + .305\frac{b}{R}\right)$$

where a and b are constants.
Following Table 19.2 gives values of a and b for different surfaces:

Table 19.2.

Type of surface		a	b
Smooth iron pipes	— — —	0.00497	0.084
Incrusted iron pipes	— — —	0.00996	0.054
Smooth cement plaster	— — —	0.00316	0.10
Ashlar brick work	— — —	0.00401	0.23
Rubble masonry or stone pitching	— — —	0.00507	0.83

When the velocity of approach is small, the term representing the loss of head in friction in the barrel $\left(i.e., f_2 \cdot \frac{L}{R} \cdot \frac{V^2}{2g}\right)$ may be neglected.

As already mentioned velocity of flow through the barrel in generally limited to 2 to 3 m/sec.

Thus, since all the values are known head loss in barrel or head causing flow can be calculated. This value when added to the High Flood Level (HFL) on the d/s of the aqueduct gives u/s HFL.

Adding free board to the u/s HFL we can obtain the top of river protection works like guide bunds and marginal bunds.

***(ii)* Uplift pressure on the roof of the barrel:** As the barrel runs full during floods there exists positive pressure in the barrel. Due to the positive pressure in the barrel the roof is subjected to the uplift pressure. The uplift pressure diagram for the roof may be drawn knowing the pressure head on the u/s and d/s side of the barrel. Pressure head on the d/s side of the barrel is equal to the height of the water level above the bottom of the roof. Pressure head on the u/s side may be obtained by adding the loss of head in the barrel to the pressure head on d/s side. The loss of head may be obtained from Unwin's formula. Figure 19.28 shows the profile of hydraulic gradient line that may exist. It may be seen that maximum uplift pressure occurs at the u/s end of the barrel roof.

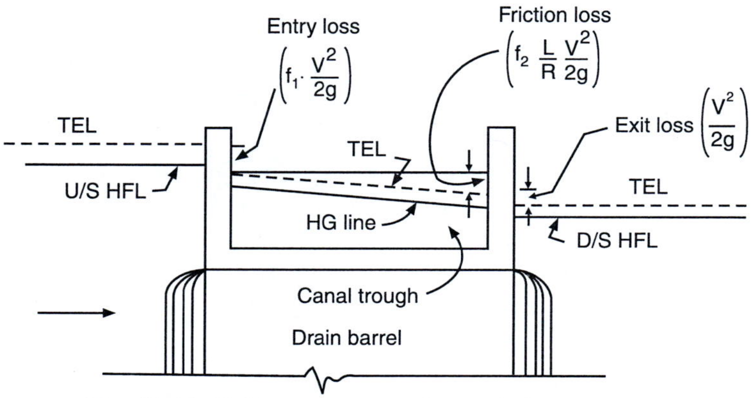

Fig. 19.28. Uplift pressure on the roof of the barrel

While designing the trough it is necessary to consider two extreme conditions, namely:

- The barrel runs full during maximum flood and there is no water in the canal trough. This condition gives maximum uplift pressure acting on the trough.
- The canal trough is carrying full discharge but the barrel does not run full and hence there is no uplift on the roof of the barrel.

In order to limit the thickness of the trough it is advisable to provide reinforced concrete roof with reinforcement at the bottom to take load of the canal trough and reinforcement at top to resist uplift pressure by bending.

***(iii)* Uplift pressure on the floor of the barrel:** Unlike other hydraulic structures aqueducts are subjected to two different types of uplift pressures from two different sources. They are the following:

(*a*) Static uplift pressure due to rise in water table: The water table many a times rises up to bed level of the drain. Particularly in case of siphon aqueduct whose floor bed is depressed below the bed of the drainage, static uplift pressure acts on the floor bed. The uplift pressure is equal to the difference of the bed level of the drain and that of the floor level of the barrel.

(*b*) Uplift pressure due to seepage of canal water to the drain: Since there exists a difference of level between canal water level and the drainage water level seepage flow takes place where conditions are favourable. This seepage head is maximum when canal runs with full capacity and there is no flow in the drain below.

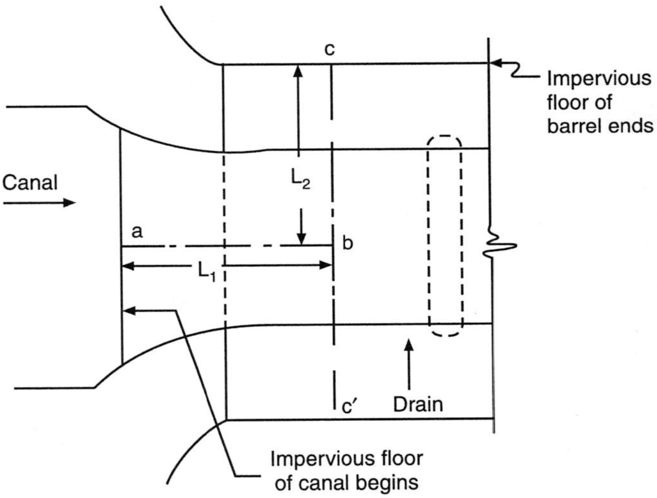

Fig. 19.29. Uplift pressure due to canal seepage

As shown in Fig. 19.29 the seepage flow in this case is not simple but flow pattern is three dimensional every where. The seepage flow starts from either sides of impervious canal trough bed and reappears on either sides of the impervious barrel floor in the drain. Since no approximation to two dimensional flow is possible Khosla's theory cannot be applied strictly. Solution by intricate "method of relaxations" is possible but it is too laborius. For design purposes Bligh's creep theory principle as explained below may be applied. For major works, however, it is essential to check the results of preliminary design so obtained by model studies.

Referring to Fig. 19.29.

Taking the case of the first barrel where seepage will be maximum, total creep length = (creep length *a-b*) + (creep length *b-c*)

$$\therefore \qquad L = L_1 + L_2$$

Total seepage head = canal FSL – d/s bed level of drain = H_s

$$\therefore \text{ Residual seepage head at } b = \frac{H_s}{L} \times L_2$$

Total residual seepage head at b may be considered to design the thickness of the entire floor of all the barrels.

The floor thickness of the barrel is in fact designed considering total uplift pressure created by static uplift condition and canal seepage flow mentioned above.

In order to limit the thickness of the floor RCC construction may be adopted as then part of the pressure is resisted by weight of the floor and remaining by bending strength of the floor. In such arrangement the pressure gets transferred to the piers and is resisted by the entire weight of the superstructure.

When it is seen that uplift pressure is very high it can be reduced by providing suitable safe guards. They are:

(*a*) Increase the length of impervious floor of the canal bed so that creep length is increased;

(*b*) Provide drainage holes or relief holes in the floor of the barrel in conjunction with inverted filter below the floor. To avoid choking of the relief holes and filter below the drain sill the relief holes should be provided with flap valves.

19.23. DESIGN PRINCIPLES FOR SILT CONTROL STRUCTURES

Silt control structures like silt excluders and silt ejectors are built in the canal system where the silting is likely to be excessive. Silting may occur to such as extent as to diminish the capacity of the canal largely. It is necessary, however, to appreciate that the canals on alluvial soils are designed considering the regime condition of the canal that may get established in due course as a result of silt grade and silt charge transported by the canal water. Obviously the aim of the silt control structure is really not to exclude or extract all the silt transported by the canal or the river but to allow only that amount of silt to pass down as is suited for the canal regime. The excluder or extractor has therefore to be designed with only that much efficiency as to regulate the silt entry consistent with the canal regime requirement. Owing of the fact that amount of silt exclusion has to be arbitrary a theoretical design of silt control structures is not always possible. It has therefore to be done on the basis of past experience of such structures. General guidelines are mentioned below. It is necessary to assess the behaviour of the structure on the model.

19.23.1. Design Criteria for Silt Excluders

(*i*) The approach to the excluder should be straight in which the silt can settle into lower layers of flow. Artificial curves to the river flow should be avoided because it may cause turbulence and disturb bed concentration. The slope of approach channel bed should be flattest enough to carry the heaviest grade of silt likely to approach the work.

(*ii*) If the flow is concentrated near head regulator better efficiency can be achieved in silt exclusion. The excluder tunnels may cover appropriate number of scouring sluice bays of the barrage.

(*iii*) Usually 4 to 6 tunnels may be provided for excluders. The number of tunnels is, however, governed by conditions of approach, length of canal regulator, discharge of the canal and the discharge to be escaped through the tunnels.

(*iv*) Sizes of the tunnels may be fixed to escape 25 to 30% of the canal discharge at not less than 3 m/sec velocity of flow. Height of the tunnel is fixed considering depth of flow available in the pocket. A common size of tunnel could be 2 m wide and 3 m high if other conditions are fulfilled.

19.23.2. Design Criteria for Silt Ejectors or Silt Extractors

(*i*) Approach to the ejector should be straight and the bed should be flattest enough to carry the heaviest grade of silt to the tunnel entrance.

(*ii*) Approach channel should be lined for a length of 3 to 4 times depth of flow in the channel upstream of the tunnel entrance.

(*iii*) Where possible the bed of the channel may be depressed by about 30 cm gradually.

(*iv*) The tunnel entrance should be so designed that there is no disturbance at entry point.

(*v*) To prevent silting in the tunnels velocity of flow in the tunnel can be increased by subdividing the tunnel into convenient compartments.

(*vi*) To avoid sudden increase in velocity roof slab of the tunnels may be constructed as a cantilever slab extending on U/S side beyond the edge of the piers.

(*vii*) Velocity of flow in the tunnel to evacuate the silt laden flow has to be higher and should be around say 3 m/sec.

(*viii*) Escape discharge of the ejector may range between 15 to 25% of canal discharge. This is, however, purely arbitrary value and has to be related to experience on the existing structures in the similar situations.

(*ix*) The size of each tunnel is fixed considering depth of flow, discharge of the canal and the discharge to be escaped through the tunnels. The height of tunnel at entrance may be kept between 1/4 to 1/5 of the depth of water in the canal.

QUESTIONS FOR REVISION

1. (*a*) What are the causes of failure of weirs on permeable foundation ?
   (*b*) What are the main points of Bligh's creep theory ?
   (*c*) What are its shortcomings ?
2. (*a*) How Dr. Khosla's theory is an improvement over Bligh's creep theory ?
   (*b*) What are the criteria for safety against uplift pressure in case of weirs on permeable foundation ?
3. Explain breifly principle of hydraulic jump and its use in design of impervious floor.
4. Explain briefly principles for design of Sarda type fall.
5. (*a*) Differentiate between aqueduct and siphon aqueduct.
   (*b*) Differentiate between silt excluder and silt ejector.

❐❐❐

20

Hydropower Stations

20.1. INTRODUCTION

To meet irrigation demand water has to be spread over land which the crops utilise consumptively and the water gets lost. But in generation of hydropower only benefit is taken of potential energy of water. Once that potential is used water is available for other uses. Thus irrigation and power generation objectives can be combined gainfully.

Hydropower has been occupying an important place in power supply industry ever since its inception in the country. This has been mainly due to the fact that hydropower forms the cheapest source of power inspite of the technological advances that have taken place in the field of thermal and nuclear power. India is blessed with numerous sites where hydropower can be generated economically. The surface water resource can be harnessed gainfully for power generation provided that continuous and abundant water supply with reasonable potential energy or head of water or drop is available at a particular site. The water with sufficient potential energy is taken over the blades of a turbine which provides motive power for the generator directly coupled on a common shaft. The water then passes down and is disposed off suitably. Thus the basic objective of any hydropower project is to provide motive power to the generators which generate electricity. A complete hydroelectric installation includes hydraulic structures for interception, collection and storage of water, for its diversion to the turbines and for its disposal after the use. In addition, it includes power house where turbines together with control apparatus and electricity generators with switch gear, transformers etc., are installed.

20.2. LAYOUT OF A HYDROPOWER INSTALLATION

The nature, size and arrangement of various structures which constitute hydro-power installation depend upon topographical, geological and hydrological factors. In principle, however, there is similarity in the types of structures required for generation of electricity from the water power. A hydro power consists of the following works which form the nucleus of every installation:

1. A structure to intercept and to regulate the free flow of water. It may be in the form of a dam or a weir or a barrage.
2. Head race or intake conduit together with gates, trash racks, surge tank, etc., to control the entry of water from the reservoir on to the turbines. The intake arrangement may be in the form of a tunnel, or open channel or penstock or combination thereof.

3. A power station where motive power obtained from water flow is utilised to generate electrical energy. It houses:
   (*a*) turbines together with controlling devices, and
   (*b*) generators and other regulatory apparatus like switch gear, transformers etc. Transmission system is generally considered out of the purview of a hydro-power plant.
4. Tail race or exit conduit to lead the water back into the river or artificial channel for its subsequent use for irrigation or other purposes.

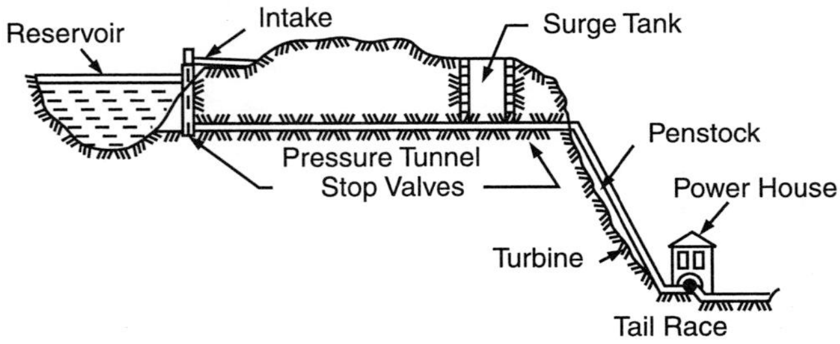

Fig. 20.1. High head installation with pressure tunnel

General arrangement of various types of work have been shown in Figs. 20.1, 20.2, 20.3 and 20.4. It may be seen that in fig. 20.1, water is taken from a reservoir, built in the upper reaches of a river valley in a remote place, by means of a pressure tunnel to the power house located down below at a suitable place to utilise the high head available at the site. The main conduit may run for several kilometres before it ends near the power house.

Alternatively when the situation permits an open head race channel can be built with a flat gradient to bring the water down from the remote reservoir. The channel ends in a forebay or a head pond. From the forebay the water is taken down to the power house by penstock. After use the water is conveyed back to the river by a tail race channel.

In a medium head installation all the head is created by an intercepting dam. Figure 20.2 shows the hydro power installation at a dam site with medium head. Here the main conduit is short and consists of penstock taken through the dam and the power is developed locally at the dam site.

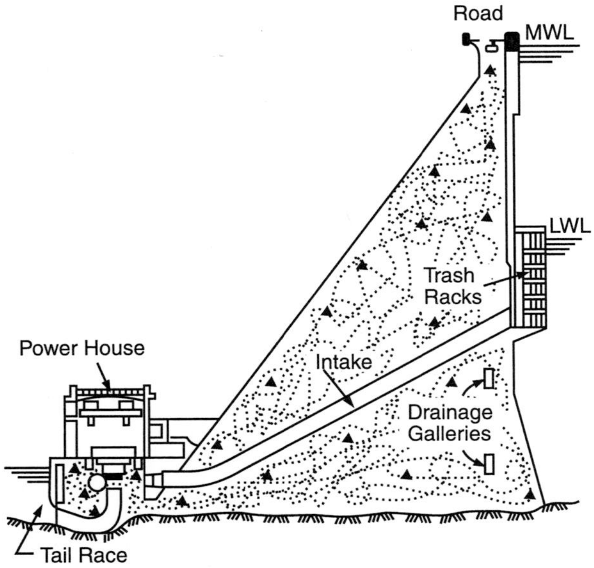

Fig. 20.2. Medium-head installation at a dam site

In canals and lower reaches of the rivers where large discharge at very low head is available low-head hydro power plants are constructed. Fig. 20.3 shows a conventional low-head installation.

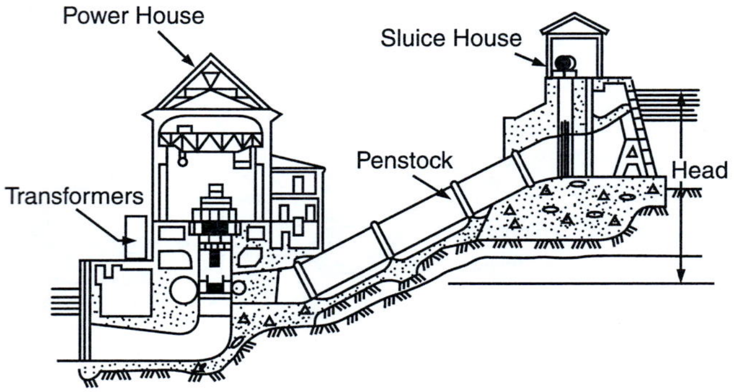

Fig. 20.3. Low-head installation on a canal

Recently a tubular type of turbine has been put in use on big canals in our country. It utilises large discharges at very low-heads of the order of 6 metres. Fig. 20.4 shows low-head installation with tubular turbine on an irrigation canal.

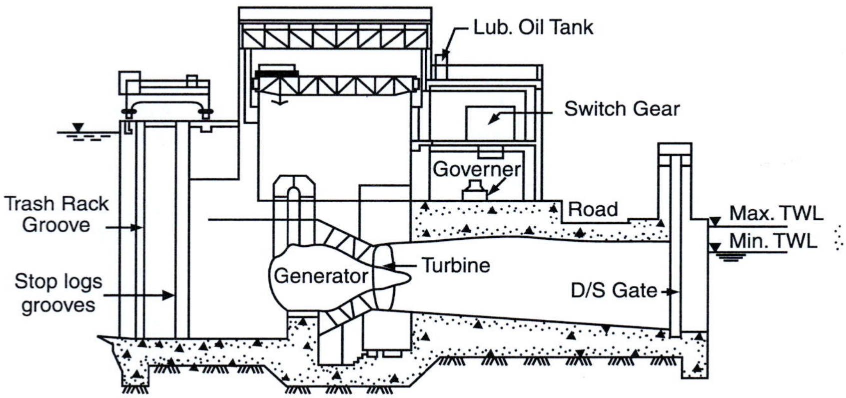

Fig. 20.4. Low-head installation using tubular turbine

Thus main components of hydro power plant could be listed as below:

- Fore bay
- Intake arrangement
- Head race or intake conduit (it includes penstock)
- Surge tank
- Turbines and generators
- Power house
- Tail race (it includes draft tube)
- Transformers and transmission lines etc.

20.3. FOREBAY AND INTAKE STRUCTURES

As the name suggests forebay is an enlarged body of water in front of intake. The reservoir acts as forebay when penstock take water directly from it. When canal leads water to the turbines the section of the canal in front of turbines is enlarged to create forebay. The forebay temporarily stores water for supplying the same to the turbines.

The water cannot be allowed to pass as it comes in the reservoir or the canal. At intake gates are provided with hoist to control the entry of water. In front of the gates trash racks are provided to prevent debris, trees, etc., from entering into the penstock. Rakes are also provided to clean the trash racks at intervals.

20.4. HEAD RACE OR INTAKE CONDUITS

They carry water to the turbines from the reservoir. The choice of open channel or a pressure conduit (Penstock) depends upon site conditions. The pressure conduit may be in the form of a flared intake passage in the body of the dam or it may be a long conduit of steel or concrete or sometimes a tunnel extending for few kilometres between the reservoir and the power house. The pressure conduit does not follow the ground contours and any gradient is given to suit the site conditions. The velocity of water in the power conduit is also higher than in the open channel. Upto about 60 metres head the velocity may range between 2.5 to 3.0 m/sec. For higher heads the velocity may be still higher. Sometimes it is convenient or economical to adopt open channel partly or wholly as the main conduit. The head race canal may lead water to the turbines or to the penstocks and is usually adopted in low-head installations where head losses are relatively important. The advantage of an open channel is that it could be used for irrigation or navigation purposes.

20.5. SURGE TANK

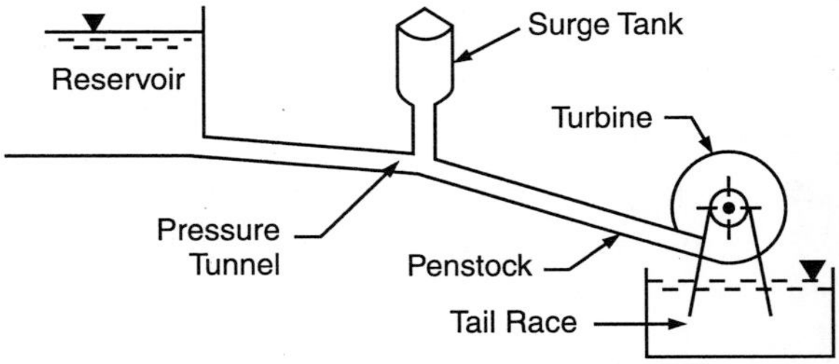

Fig. 20.5. Surge tank in hydro power plant

A surge tank is a storage reservoir fitted at some opening made on a long pipe line or penstock to receive the rejected flow when the pipe line is suddenly closed by a valve fitted at its steep end, see Fig. 20.5. A surge tank, therefore, relieves the pipe line of excessive pressure produced due to its closing, thus eliminating the positive water hammer effect. It is done by admitting in the surge tank a large mass of water which otherwise would have flown out of the pipe line, but returns to the tank due to closure of pipe end. It also serves the purpose of supplying suddenly an additional flow whenever required by the hydraulic prime movers at any instant. The surge tank is mostly employed in a water power plant or in a large pumping plant to control the pressure variations resulting from rapid changes in the flow.

In the case of water power plant, when there is sudden reduction of load on the turbine, it becomes necessary for the governor to close the turbine gates for adjusting the flow of water in order to keep the speed of the turbine constant. However, the water is already on its way to the turbine. When the turbine gates are closed, the moving water has to go back. A surge tank would then act as a receptacle to store the rejected water and thus avoids water hammer. On the other hand when there is an immediate demand on the turbine for more power, the governor re-opens the gates in proportion to the increased load, thus, making it necessary to supply more water. For a long pipe it takes a considerable time before the entire mass of water can be accelerated. The surge tank which is generally located near the turbine will meet the suddenly increased demand of water till such time the velocity in the upper portion of the line acquires a new value.

Similarly for a large pumping plant with a long delivery pipe, a surge tank can also be employed to control the pressure variations on the delivery side, which result due to sudden shut down or starting up of a pump. When the pump is started, most of the intial flow from the pump enters the surge tank thus reducing the water hammer effect in the delivery pipe. On the other hand when the pump is shut down suddenly, the surge tank provides extra space to accommodate water which would come back, thus relieving water hammer pressure.

Functions of surge tank: The surge tank thus serves the following purposes:

- Control of pressure variations resulting from rapid changes in pipe line flow, thus eliminating water hammer effect.
- Regulation of flow in power and pumping plants by providing necessary accelerating or retarding head.

Location of surge tank. Theoretically a surge tank should be located as close to a power or pumping plant as possible. The ideal place in case of power plant is at the turbine inlet, but it is seldom possible in case of medium and high head installations because it will have to be made very high. In order to reduce its height, it is generally located at a junction of pressure tunnel and a penstock (see Fig. 20.5) or on the moutain side.

20.6. TURBINES AND GENERATORS

Turbine converts hydraulic energy into mechanical energy. The mechanical energy developed by a turbine is used in running an electric generator. It is directly coupled to the shaft of the turbine. The generator develops electric power. A turbine consists of a wheel called *runner*. The runner is provided with specially designed blades or buckets. The water possessing large hydraulic energy strikes the blades and the runner rotates.

Water turbines may be classified under two types, namely:

- Impulse or velocity turbines, and
- Reaction or pressure turbines.

20.6.1. Impulse Turbine

In the impulse turbine, all the available potential energy or head is converted into kinetic energy or velocity head by passing the water through a contracting nozzle or by guide vanes before it strikes the buckets. The wheel revolves free in air and water is in contact with only a part of wheel at a time. The pressure of water all along is atmospheric. In order to prevent splashing and to guide the water discharged from the buckets to the tail race, a casing is provided. An impulse turbine is essentially a low-speed wheel and is used for relatively high heads. Pelton wheel, Turgo impulse wheel and Girard turbine, are some types of impulse turbine. In the Pelton wheel water strikes the runner tangentially.

20.6.2. Reaction Turbine

In a reaction turbine, only part of the available potential energy is converted into velocity head, at the entrance to the runner. The balance portion remains as a pressure head. The pressure at the inlet of the turbine is much higher than the pressure at the outlet. It varies throughout the passage of water through the turbine. Mostly the power is developed by the difference in pressure acting on front and back of runner blades. Only small part of power comes from the dynamic action of velocity. Since the water is under pressure, the entire flow from head race to tail race takes place in a closed system.

Francis and Kaplan turbines are two important types of reaction turbines. In Francis turbine there is inward radial flow of water. In modern Francis turbine the flow enters radially inward but leaves in parallel direction to shaft at centre. It is called mixed flow.

In Girard, propeller and Kaplan turbines the flow is axial or parallel to the axis of the turbine shaft.

Selection of a suitable type of turbine depends primarily upon the available head and the quantity of wate required. The turbines may be classified as follows with reference to type of power plant:

- Low head turbine (less than 30 m);
- Medium head turbine (30 to 160 m);
- High head turbine (upto and over 1000 m);

Low head turbines are Propeller turbine and Kaplan turbine. These turbine use large quantity of water. **Medium head turbines** are modern Francis turbines. Impulse turbines are **high head turbines**. These turbines require relatively less quantity of water.

20.7. POWER HOUSE

The purpose of the power house is to support and house the hydraulic and electrical equipment. The power house is readily divided into two parts as follows:

- The substructure to support the equipment and to provide the necessary water-ways.
- The superstructure or building to house and protect the equipment.

Substructure: The substructure may form an integral part of the dam and intake structure. In other cases the substructure may be remote from the dam, the dam intake and power house being entirely separate structures. The substructure is built exclusively of concrete and is enforced with steel where necessary.

Super-structure: The generating room, the main portion of the power house, contains the main units and their accessories, and usually there is a power or hand operated overhead crane which spans the width of the power house. The switch board and operating stand are usually near the middle of the station, either at floor level or, for better visibility, on the second floor or at a level above the main floor. Usually an auxiliary bay or section of the power house will be required upstream from the main units for the switches, bus connections, and outgoing lines. If transformers are located inside the station, these will also be in the auxiliary bay, commonly at floor level and shut off the main floor by steel doors or shutters.

A travelling crane is an important part of the power house equipment. In fixing the elevation of the crane rail above the floor, it is essential that sufficient headroom is provided for lifting and carrying along any of the various machine parts.

20.8. TAIL RACE AND DRAFT TUBE

The channel into which the turbine discharges in case of impulse wheel and through draft tube in case of reaction turbine is called a tail race.

The suction pipe or draft tube is nothing but an airtight tube fitted to all reaction turbines on the outlet side. It extends from the discharge end of the turbine runner to about 0.5 metres below the surface of the tail water level. The straight draft tube is generally given a flare of 4 to 6 degrees to gradually reduce the velocity of water. The suction action of the water in this tube has same effect on the runner as an equivalent head so that the turbine develops the same power as if it were placed at the surface of the tail water.

The tail race of the impulse wheel is commonly an approximately rectangular passage, running from a point under the wheel to a point outside the power house foundations where it enters the exit channel or the river. Because of the small discharge of the impulse wheel, as well as higher allowable velocity, the tail race passage is much smaller than that of the reaction turbine.

In case of the reaction turbine the width of the tail race channel under the power house depends upon the unit spacing and thickness of piers and walls between the unit bays. The depth of the tail race channel depends upon the velocity which is generally taken to be about 1 metre per second.

Where the power house is close to the river, the tail race may be the river itself. In other case a tail race channel of some length may be provided to join the turbine pit with the river.

20.9. DEFINITIONS OF GENERAL TERMS

1. Base Load and Firm Demand: The demand for power varies from time to time depending upon the utilisation of electricity required by the consumers at any instant. If a curve is plotted to show variation of this load against time, it will be found that there are some well defined peaks or periods during which demand is high and also period of low and average demands. When the generator is capable of supplying energy according to a pre-arranged continuous programme covering not only the periods during which the generator works or remains idle but also when it will be under maintenance and over-haul, the load taken up by the generator is called base load and the demand is described as firm demand.

2. Peak Load: It is defined as that part of the load which is greater than 4/3 times the average lead.

3. Load Factor: It is the ratio of the average load to the peak load over any given period.

$$\text{So, Load factor} = \frac{\text{Average load}}{\text{Peak load}}$$

The load factor can be calculated on daily, weekly, monthly or annual basis.

Sometimes load factor is also calculated using either installed capacity instead of peak demand or the average number of kilo-watt-hours (kWh) generated instead of those consumed.

4. Gross and Net Head: Output and capacity of any hydropower installation depends upon flow and net head. The **Gross head** at any instant is the difference between the water levels in the forebay and in the tail race. The **Net head** at any instant is obtained by deducting the conduit losses from the gross head. The losses in conduit vary with the flow through it. Obviously net head varies simultaneously with the load and the reservoir level.

5. Specific Speed: The specific speed 'N_S' of a turbine is defined as the r.p.m. at which the runner would run if it were so reduced in size, without changing the design, that it would develop one horse power under one metre head. Runners having same specific speed have similar characteristics of performance. As a result specific speed becomes convenient tool to indicate type of a runner.

6. Storage and Pondage: The firm output of a hydropower station during any period depends upon the availability of definite volume of water at definite rate of flow. The firm output is assured by creating a storage reservoir to retain the surplus run-off for use as and when there is demand. The reservoirs can be operated in different ways as mentioned below :

Storage: In the wet season when the run-off is very much above the demand the excess water is stored for use during dry season. Sometimes the storage capacity may be big enough to maintain supply of water for hydropower station in the event of a dry year following a normal year.

Pondage: It is nothing but a short term storage. For example, over a short period such as a day or a week the river flow may be high enough to meet average demand. Obviously the flow will be in excess during hours of low demand but will be inadequate during the hours of peak demand. In such a situation if the excess flow during low demand hours is stored it can meet the demand during peak periods. The short term storage done for this adjustment is called pondage. Depending upon the period under consideration the pondage could be daily pondage or weekly pondage.

QUESTIONS FOR REVISION

1. What are important components of a typical Hydropower plant ? Draw a sketch of a medium head hydro-power plant at toe of the dam.
2. What is a surge tank ? Where it is located ? Describe its functions.
3. Differentiate between the following:
   (*a*) Impulse turbine and reaction turbine. (*b*) Storage and pondage.
   (*c*) Base load and peak load.

Index

Index

D...

E...

R...

S...

T...

U...

V...

W...

Y...

❏❏❏

Win Prizes

ATTENTION READERS

Win Prizes

We request for your frank assessment regarding various aspects of the book as given below :

WATER RESOURCES ENGINEERING

By: S.R. Sahasrabudhe

Please Fill up the Sheet in Neat and Clean Writing

(i) What topics of your syllabus, important for your examination point of view, are not covered in this book ?

....

....

....

....

(ii) What are the chapters wherein the treatment of the subject matter is not systematic or organised or updated ?

....

....

....

....

(iii) Have you come across misprints/mistakes/factual inaccuracies in the book? Please specify the chapters and the page numbers.

....

....

....

....

(iv) What other books on the same subject have you found/heard better than the present book? Please specify in terms of price and quality

....

....

....

....

(v) Further suggestions and comments for the improvement of this book.

....

....

....

....

(PS: If need be please attach a separate sheet to write your views)

OTHER DETAILS

(i) Who recommended you this book ? (Please tick (✓) in the box near the option relevant to you.

☐ Teacher ☐ Friends ☐ Bookseller

....

....

....

....

(ii) Name of the recommending teacher, his designation and address :

....

....

....

....

(iii) Name and address of the bookseller you purchased the book from :

....

....

....

....

(iv) Name and address of your institution (Please mention the University or Board as the case may be)

....

....

....

....

(v) Your name and complete postal address :

....

....

....

....

(vi) Course you are enrolled in :

....

....

....

....

The best assessment will be awarded each month: The award will be in the form of our publications as decided by the Editorial Board.

Please mail the filled up coupon at your earliest to :

S.K. KATARIA & SONS®

4885/109, Prakash Mahal, Dr. Subhash Bhargav Lane,

Opposite Delhi Medical Association, Daryaganj, New Delhi–110002 (INDIA)

Phone: +91-11-43551243, Mobile: +91-9871775858, 7678425209

e-mail: katariabook@gmail.com, katariabooks@yahoo.com

Website: www.skkatariaandsons.com